An Introduction to

EMBRYOLOGY

B. I. Balinsky, Dr.Biol.Sc.

PROFESSOR OF ZOOLOGY

UNIVERSITY OF THE WITWATERSRAND

JOHANNESBURG, SOUTH AFRICA

An Introduction to

EMBRYOLOGY

W. B. Saunders Company

PHILADELPHIA   1960   LONDON

Preface

The teaching of embryology has long been an established feature at universities throughout the world, both for students in biology and students in medical sciences. Although overshadowed during a large part of the twentieth century by the rapid development of genetics and cytology, embryology has also made rapid advances, especially as an experimental science—as experimental or physiological embryology. It is realized now that embryology is a branch of biology which has a most immediate bearing on the problem of life. Life cannot be fully accounted for without an understanding of its dynamic nature, which expresses itself in the incessant production of new organisms in the process of ontogenetic development.

In the midst of the rapid changes of outlook that the experimental method has brought with it, it has been difficult to coordinate the older data of purely descriptive embryology with the new discoveries. This has hampered the teaching of embryology and is to this day reflected in the subdivision of most textbooks of embryology into two groups. Books of the first type deal with the classic "descriptive" embryology and are written mainly for the use of medical students. Short chapters on experimental embryology are appended to some of them, but these chapters are extremely brief and not organically connected with the description of the morphology of the developing embryos. The second type of book deals with experimental embryology or "physiological" embryology. These books are written for advanced students and the basic facts of development are more or less taken for granted, so that a student cannot profitably proceed to the study of such a book without previously making himself familiar with "descriptive" embryology from one of the books of the first type.

In the course of many years' teaching of embryology to university students I have endeavored to present embryology as a single science in which the descriptive morphological approach and the experimental physiological approach are integrated and both contribute to the understanding of the onto-genetic development of organisms. This integrated approach to development is now incorporated in the present book. Data of a more purely physiological and biochemical nature are adduced inasmuch as it is practicable to treat them in a book that does not presuppose an advanced knowledge of bio-chemistry in the student.

The subject of embryology is interpreted in my book in a broad sense, as the science dealing with ontogenetic development of animals, and includes therefore such topics as postembryonic development, regeneration, meta-morphosis and asexual reproduction, which are seldom handled in students' textbooks at any length. Lastly I believe that embryology cannot be presented adequately without establishing some connection with genetics, inas-much as processes of development are under the control of genes. The connection between inheritance and development is therefore also indicated in the text.

With such a wide scope, my book can only be "an introduction to embry-ology." The whole field could not be covered in the same detail as is custom-arily given in textbooks dealing with only one aspect of the science of embryology. The student having studied this book, however, will be prepared to understand and appreciate special information in any section of the science which would be of interest to him in his further studies.

The first draft of this book was written in 1952, and duplicated copies of the manuscript were used by my students during subsequent years. This gave me an opportunity to convince myself of the usefulness of the book and also to eliminate some defects in the original text. For the present printed edition the book has been completely revised and brought up to date. An extensive study of special literature up to the end of 1958 has been carried out for this purpose (as can be seen from the list of references). Later publi-cations could not be included in the text.

In illustrating the book I have drawn on my own experience in embryo-logical work wherever practicable, but of course most of the illustrations have been reproduced from other sources.

In preparing the book for print I have been assisted by a number of per-sons to whom I should like to express my gratitude on this occasion. In the first place I wish to thank all the authors and publishers who have kindly agreed to the reproduction of figures used to illustrate this book, as well as colleagues in many countries who by sending me reprints of their publications have facilitated the arduous task of keeping track of current embryological literature.

Of my immediate collaborators and friends I am most profoundly indebted to Dr. Margaret Kalk of the University of the Witwatersrand for reading the whole text of the book and for many valuable suggestions and helpful criticism. I am indebted to Dr. H. B. S. Cooke of the same University for his expert advice on the preparation of illustrations for the book. I am very grateful for the invaluable assistance of Mrs. E. J. Pienaar, who has typed the manuscript, has assisted me in preparing the index, and has been of great help on diverse occasions during the work on the manuscript and on the proofs. I should like to thank Miss R. J. Devis, Mrs. E. du Plessis and Mr. M. J. de Kock for their help in preparing the illustrations for the book.

Last but not least I should like to express my gratitude to the staff of the W. B. Saunders Company, whose friendly encouragement has done much to bring this book to its present form.

Johannesburg   B. I. BALINSKY

Contents

ix

PART IV. ORGANOGENESIS

PART V. DIFFERENTIATION AND GROWTH

**PART VI.   MORPHOGENETIC PROCESSES IN THE
LATER PART OF ONTOGENESIS**

PART I

The Science of Embryology

CHAPTER **1**

The Scope of Embryology

and Its Development

as a Science

1–1 ONTOGENETIC DEVELOPMENT AS THE SUBJECT
MATTER OF EMBRYOLOGY

The aim of this book is to make the student familiar with the basic facts
and problems of the science of **embryology.** The name "embryology" is
somewhat misleading. Literally it means: the study of **embryos.** The term
"embryo" denotes the juvenile stage of an animal while it is contained in
the egg (within the egg membranes) or in the maternal body. A young
animal, once it has hatched from the egg or has been born, ceases to be an
embryo and would escape from the sphere pertaining to the science of em-

bryology, if we were to keep strictly to the exact meaning of the word. Although birth or hatching from the egg is a very important occasion in the life of the animal, it must be admitted that the processes going on in the animal's body may not be profoundly different before and after the hatching from an egg especially in some lower animals. It would be artificial to limit the studies of the juvenile forms of animal life to the period before the animal is hatched from the egg or born. It is customary, therefore, to study the life history of an animal as a whole, and accordingly to interpret the contents of the science of embryology as **the study of the development of animals.**

Now the word "development" has to be qualified in turn. In the sphere of biology with which we are concerned here, the term "development" is used with two different meanings. It is used to denote the processes that are involved in the transformation of the fertilized egg, or some other rudiment derived from a parent organism, into a new adult individual. The term development may, however, also be applied legitimately to the gradual historical transformation of the forms of life, starting with simple forms which might have been the first to appear, and leading to the contemporary diversity of organic life on our planet. Development of the first type may be distinguished as individual development or **ontogenetic development.** Development of the second type is the historical development of species or **phylogenetic development.** Phylogenetic development is often referred to as evolutionary development or simply **evolution.** Accordingly we will define embryology as **the study of the ontogenetic development of organisms.** In this book we will be dealing only with the ontogenetic development of multicellular animals, the **Metazoa.**

In multicellular animals the typical and most widespread form of ontogenetic development is the type occurring in sexual reproduction. In sexual reproduction new individuals are produced by special **generative cells** or **gametes.** These cells differ essentially from other cells of the animal, in that they go through the process of maturation or **meiosis,** as a result of which they lose half of their chromosomes and become **haploid,** whereas all other cells of the parent individual, the **somatic cells,** are, as a rule, **diploid.** Once a cell has gone through the process of meiosis, it can no longer function as an integral part of the parent body but is sooner or later extruded to serve for the formation of a new individual. In multicellular animals there exist two types of sex cells: the female cells or **ova,** and the male cells or **spermatozoa.** As a rule the two cells of the opposite sexes must unite in the process of fertilization before development can start. When the two gametes (the ovum and the spermatozoon) unite, they fuse into a single cell, the **zygote,** which again has a diploid number of chromosomes. The zygote, or fertilized ovum, then proceeds to develop into a new adult animal.

Side by side with sexual reproduction there exists in many species of animals a different mode of producing new generations—asexual reproduction. In asexual reproduction the offspring are not derived from generative cells (gametes) but from parts of the parent's body, consisting of somatic cells. The size of the part which is set aside as the rudiment of the new individual may be large or small, but in the Metazoa it always consists of more than one cell. The development of an animal by way of asexual reproduction obviously belongs in the same category as the development from an egg and should be treated as a special form of ontogenetic development. It will be dealt with in Chapter 20. To distinguish between the two forms of ontogenetic development, the term **embryogenesis** may be used to denote the development from the egg, and the term **blastogenesis** may be used for the development of new individuals by means of asexual reproduction.

1–2   THE PHASES OF ONTOGENETIC DEVELOPMENT

From what has already been said it is clear that the processes leading to the development of a new individual really start before the fertilization of the egg, because the ripening of the egg and the formation of the spermatozoon, which constitute the phase of **gametogenesis,** create the conditions from which the subsequent embryogenesis takes its start. In both oogenesis and spermatogenesis meiosis, by discarding half of the chromosomes, singles out the set of genes which are to operate in the development of a particular individual. The cytoplasmic differentiations of the spermatozoon enable it to reach the egg by active movement and to fertilize it. On the other hand, the egg cell accumulates in its cytoplasm substances which are used up during development—either directly, by becoming transformed into the various structures of which the embryo consists, or indirectly, as sources of energy for development. The elaboration in the egg cell of cytoplasmic substances to be used by the embryo, and their placing in correct positions, are essential parts of what occurs during the first phase of development.

The second phase of development is **fertilization.** Fertilization involves a number of rather independent biological and physiological processes. First of all, the spermatozoa must be brought into proximity with the eggs if fertilization is to occur. This involves adaptations on the part of the parents, ensuring that they meet during the breeding season, discharge their sex cells simultaneously in cases of external fertilization, or copulate in cases of internal fertilization. Next, the spermatozoa must find the egg and penetrate into the egg cytoplasm. This entails a very finely adjusted mechanism of physicochemical reactions which serve to direct the spermatozoa to the egg. Next the egg is **activated** by a spermatozoon and starts developing. A further rearrangement of the organ-forming substances in the egg is among the first changes that take place in the egg after fertilization. (See section 3–5.)

The third phase of development is the period of **cleavage.** The fertilized egg is still a single cell, since the nucleus and cytoplasm of the spermatozoon fuse with the nucleus and cytoplasm of the egg. If a complex and multicellular organism is to develop from a single cell, the egg, the latter must give rise to a large number of cells. This is achieved by a number of mitotic cell divisions following each other in quick succession. During this period the size of the egg does not change, the **cleavage cells** or **blastomeres** becoming smaller and smaller with each division. No far-reaching changes can be discovered in the substance of the developing embryo during the period of cleavage, as if the preoccupation with the increase of cell numbers excludes the possibility of any other activity. The whole process of cleavage is dominated by the cytoplasmic organoids of the cells: the centrosomes and achromatic figures. The nuclei multiply but do not interfere with the processes going on in the cytoplasm. The result of cleavage is sometimes a compact heap of cells, but usually the cells are arranged in a hollow spherical body, a **blastula,** with a layer of cells, the **blastoderm,** surrounding a cavity, the **blastocoele.**

There follows the fourth phase of development, that of **gastrulation.** During this phase the single layer of cells, the blastoderm, gives rise to two or more layers of cells known as the **germinal layers.** The germinal layers are complex rudiments from which are derived the various organs of the animal's body. In the higher animals the body consists of several layers of tissues and organs, such as the skin, the subcutaneous connective tissue, the layer of muscles, the wall of the gut, and so on. All these tissues and organs may be traced back to three layers of cells—the aforementioned germinal layers. Of these the external one, the **ectoderm,** always gives rise to the skin epidermis and the nervous system. The layer next to the first, the **mesoderm,** is the source of the muscles, of the blood vascular system, of the lining of the secondary body cavity (the **coelom,** in animals in which such a cavity is present), and of the sex organs. In many animals, in particular in the vertebrates, the excretory system and most of the internal skeleton are also derived from the mesoderm. The third and innermost germinal layer, the **endoderm,** forms the alimentary canal and the digestive glands. In the Coelenterata the mesoderm is missing as a separate germinal layer, and only ectoderm and endoderm are present.

The germinal layers are produced by the disappearance of a part of the blastoderm from the surface and its enclosure by the remainder of the blastoderm. The part that remains on the surface becomes ectoderm; the part disappearing into the interior becomes endoderm and mesoderm. The disappearance of endoderm and mesoderm from the surface sometimes takes the form of a folding-in of part of the blastoderm, so that the simple spherical body becomes converted into a double walled cup, as if one side of the wall of an elastic hollow ball had been pushed in by an external force. This infolding or inpushing of endoderm and mesoderm is known as **invagination,** and the resulting embryo is known as a **gastrula**—whence the term **gastrulation.** The way in which the endoderm and mesoderm become separated from each other in the interior of the gastrula varies a great deal in different animals, and cannot be described in this general review. (See Chapter 5.) If the gastrula is formed by invagination, the cavity of the double walled cup is called the **archenteron,** and the opening leading from this cavity to the exterior is called the **blastopore.** In animals in which the gastrula is formed in a different way—not by invagination—the cavity (archenteron) and the opening of the cavity to the exterior (blastopore) may still appear later on.

The archenteron, or part of it, eventually gives rise to the cavity of the alimentary system. The fate of the blastopore differs in the three main groups of Metazoa. In **Coelenterata** it becomes the oral opening. In **Protostomia** (including Annelida, Mollusca, Arthropoda and allied groups) it becomes subdivided into two openings, one of which becomes the mouth and the other the anus. In **Deuterostomia** (including Echinodermata and Chordata) the blastopore becomes the anal opening, the mouth being formed later on as an independent perforation of the body wall. The whole of the lining of the alimentary canal does not always consist of endoderm: in all groups of animals the ectoderm may be invaginated secondarily at the oral or at both oral and anal openings to become a part of the alimentary canal. The parts of the alimentary canal lined by ectoderm are known as the stomodeum (adjoining the mouth) and proctodeum (adjoining the anus).

With the formation of the three germinal layers, the process of subdivision of the embryo into parts with specific destinies commences. This subdivision is continued in the next (fifth) phase of development, the phase of **organogenesis** (organ formation). The continuous masses of cells of the three germinal layers become split up into smaller groups of cells, each of which is destined to produce a certain organ or part of the animal. Every organ begins its development as a group of cells segregated from the other cells of the embryo. This group of cells we will call the **rudiment** of the respective organ. The rudiments into which the germinal layers become subdivided are called **primary organ rudiments.** Some of these are very complex, containing cells destined to produce a whole system of organs, such as the whole of the nervous system, or the alimentary canal. These complex primary organ rudiments later become subdivided into **secondary organ rudiments**—the rudiments of the subordinated and simpler organs and parts. The formation of the primary organ rudiments follows so closely on the processes of gastrulation that the two processes can hardly be considered apart. Dynamically they are linked into one whole, and will be described in conjunction in the following pages. The chapters on organogenesis will then be concerned with the later development of the primary organ rudiments and with the formation of the secondary organ rudiments. With the appearance of primary and secondary organ rudiments the embryo begins to show some similarity to the adult animal, or to the larva if the development includes a larval stage.

The sixth phase of development is the period of **growth** and **histological differentiation.** After the organ rudiments are formed they begin to grow and greatly increase their volume. In this way the animal gradually achieves the size of its parents. Sooner or later the cells in each rudiment become histologically differentiated, i.e., they acquire the structure and physicochemical properties which enable them to perform their physiological functions. When the cells in all the organs, or at least in the vitally important organs have become capable of performing their physiological functions, the young animal can embark upon an independent existence—an existence in which it has to procure food from the surrounding environment.

In rather rare cases (in the nematodes for instance) the young animal emerging from the egg is a miniature copy of the adult animal and differs from the latter only in size and in the degree of differentiation of the sex organs. In this case the subsequent development consists only of growth and the maturation of the gonads. It is more usual, however, for the animals emerging from the egg to differ from the adult to a greater extent: not only the gonads, but other organs may not be fully differentiated, or they may even be absent altogether and have to develop later. Sometimes the animal emerging from the egg possesses special organs which are absent in the adult but which are necessary for the special mode of existence of the young animal. In this case the young animal is called a **larva.** The larva may lead a different mode of life from the adult, and therein lies one of the advantages of having a larval stage in development. The larva undergoes a process of **metamorphosis** when it is transformed into an animal similar to the adult. The metamorphosis involves more or less drastic changes in the organization of the larva, depending on the degree of difference between the larva and the adult. During metamorphosis new organs may develop, so that morphogenetic processes become active again after a more or less prolonged period of larval life.

A secondary activation of morphogenetic processes may be produced in a different way. Many animals possess a considerable plasticity and may be able to repair injuries sustained from the environment or caused experimentally. Lost parts may be **regenerated,** and this means that the developmental processes may sometimes be repeated in an adult or adolescent organism.

The asexual reproduction of animals involves the development of new parts and organs in animals that had already achieved the adult stage.

All morphogenetic processes occurring in the later life of the animal, after the larval stage, or even when the adult stage has been achieved, will be dealt with as constituting a seventh and last phase of development.

1–3   HISTORICAL REVIEW OF THE MAIN TRENDS OF THOUGHT IN EMBRYOLOGY*

Descriptive and Comparative Embryology.   Although the correct understanding of ontogenetic development could be achieved only after the establishment of the cell theory, fragmentary information on the development of animals has been obtained since very ancient times. Aristotle had described the development of the chick in the egg as early as 340 B.C. Many observations on development of various animals, especially of insects and vertebrates, were made in the seventeenth and eighteenth centuries. However, the data of embryology were first presented in a coherent form by

* Further references: Nordenskiöld, 1929; Needham, Part II, 1931; Singer, 1931; Hall, 1951; Gabriel and Fogel, 1955; Needham, 1959.

Karl Ernst von Baer (1828). In his book "Die Entwicklungsgeschichte der Tiere, Beobachtung und Reflexion" Baer not only summed up the existing data and supplemented them by his original investigations but also made some important generalizations. The most important of these is known as **"Baer's law."** The law can be formulated thus: "More general features that are common to all the members of a group of animals are, in the embryo, developed earlier than the more special features which distinguish the various members of the group." Thus the features that characterize all vertebrate animals (brain and spinal cord, axial skeleton in the form of a notochord, segmented muscles, aortic arches) are developed earlier than the features distinguishing the various classes of vertebrates (limbs in quadrupeds, hair in mammals, feathers in birds, etc.). The characters distinguishing the families, genera and species come last in the development of the individual. The early embryo thus has a structure common to all members of a large group of the animal kingdom and may be said to represent the basic plan of organization of that particular group. The groups having a common basic plan of organization are the **phyla** of the animal kingdom.

Baer's law was formulated at a time when the theory of evolution was not recognized by the majority of biologists. It has been found, however, that the law can be reinterpreted in the light of the evolutionary theory. In its new form the law is known as the **biogenetic law** of Müller-Haeckel. Müller propounded the law in its new form and supported it by extensive observations on the development of crustaceans (1864). Haeckel (in 1868) gave it the name of biogenetic law and contributed most to its wide application in biology.

According to Baer's law the common features of large groups of animals develop earliest during ontogeny. In the light of the evolutionary theory, however, these features are the ones that are inherited from the common ancestor of the animal group in question; therefore, they have an ancient origin. The features that distinguish the various animals from one another are those that the animals have acquired later in the course of their evolution. Baer's law states that these features in ontogeny develop at later stages. Briefly, the features of ancient origin develop early in ontogeny; features of newer origin develop late. Hence the ontogenetic development presents the various features of the animal's organization in the same sequence as they evolved during the phylogenetic development. **Ontogeny** is a recapitulation of **phylogeny.** The repetition is obviously not a complete one, and the biogenetic law states that: "Ontogeny is a shortened and modified recapitulation of phylogeny." The shortening of the process is evident not only from the fact that what had once taken thousands of millions of years (phylogeny) is now performed in a matter of days and weeks (ontogeny), but also from the fact that many stages passed through in the original phylogenetic development may be omitted in ontogeny. The modifications arise mainly because the embryo at any given time is a living system which has to be in harmony with its surroundings if it is to stay alive. The embryo must be adapted to

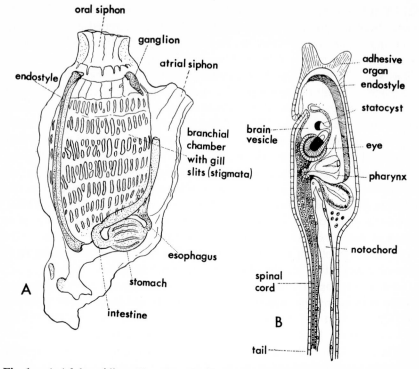

Fig. 1. *A,* Adult ascidian, *Ciona intestinalis. B,* Larva of ascidian, lateral view. (After Kowalevsky, from Korschelt, 1936.)

its surroundings, and these adaptations often necessitate the modification of inherited features of organization. A good example of such adaptations is the placenta in mammals. The placenta is a structure developed by the embryo to establish a connection with the uterine wall of the mother and thus provide for the nutrition of the embryo. This structure, though developed rather early in the life of the embryo, cannot possibly bear any relationship to any organ of ancestors of mammals. It is obviously an adaptation to the special conditions in which a mammalian embryo develops.

Even if the repetition of features of their ancestors in the ontogenetic development of contemporary animals is not complete, still even the fragmentary repetition of certain ancestral characters may be very useful in elucidating the relationships of animals. As an example of this we may consider the formation of gill pouches in the ontogenetic development of all vertebrates. In the aquatic vertebrates, such as Cyclostomata and fishes, the gill clefts serve as respiratory organs. In the adult state of terrestrial vertebrates, the gill pouches have disappeared completely or have been modified out of all recognition, and the function of respiration has been taken over by other organs—the lungs. Nevertheless, the gill pouches appear in the embryo (Fig. 155). In amphibians whose larvae are aquatic, the gill pouches at least temporarily serve for respiration. In reptiles, birds and mammals the gill pouches of the embryo do not serve for respiration at all. Their formation can only be explained as an indication that the terrestrial vertebrates have been derived from aquatic forms with functional gills. The paleontological evidence fully confirms this conclusion.

The systematic position of some animals cannot be recognized, owing to profound modification acquired as a result of adaptation to very special conditions. Here the knowledge of the development sometimes throws unexpected light on true relationships. The adult ascidian is a sessile animal with no organs of locomotion and a nervous system of a very primitive nature. The adult animal had been classed as a near relative of molluscs until Kowalevsky (1866) discovered the larvae of the ascidians. The larvae of ascidians possess a well developed dorsal brain and spinal cord, a definite notochord, and lateral bands of muscles (in short organs that are typical for the vertebrates). The ascidians are therefore considered as belonging to the same phylum as the vertebrates, the phylum Chordata (Fig. 1).

In the adult parasitic animal *Sacculina* the organization of the animal is very much simplified in connection with the easy life that the parasite enjoys; it is reduced practically to a shapeless sack producing eggs and a system of branched rhizoids, by means of which the parasite is attached to its host, the crab, and sucks the host's body fluids on which it feeds. It would be impossible to place *Sacculina* in any group of the animal kingdom if its development were not known (Fig. 2). However, the larva of *Sacculina* is a typical arthropod, bearing a close similarity to the larvae of the lower crustaceans, the Entomostraca (Delage, 1884). A rather similar larva is also found in the barnacles (*Cirripedia*), which, though possessing jointed legs like other arthropods, have lost the segmentation of the body in the adult state. The

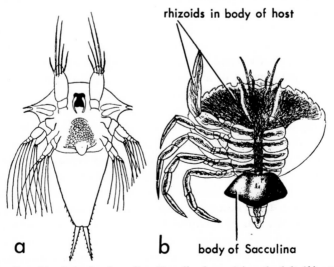

rhizoids in body of host

a   b   body of Sacculina

Fig. 2.   Parasitic cirripede, *Sacculina*. Nauplius larva (*a*) and adult (*b*) attached to a crab. (After Delage, from Parker and Haswell, Text-book of Zoology, Macmillan, London, and from Korschelt, 1936.)

attachment of the starfish larva, the brachiolaria, to the substrate while it is metamorphosing into the definitive form is an indication that the free-living echinoderms have been derived from sessile forms. This conclusion is again borne out by the evidence of paleontology.

Following the principles of Baer's law and the biogenetic law, embryologists have systematically investigated the development of animals belonging to all the major groups of the animal kingdom. As a result of very extensive and painstaking investigations, a magnificent edifice of **comparative embryology** has been built up.

Explaining Development: Theories of Preformation and Epigenesis. Neither the description of morphological transformations occurring in the embryo, nor the comparison of embryos and larvae among themselves and with the adult animals exhausts all the problems presented by the ontogenetic development of animals. The fundamental problem presented by the existence of cyclical ontogenetic changes is the question: Why does ontogeny occur at all? What are the forces which produce the changes? How is it that, starting from a simple spherical cell, the process always ends in producing a highly complex and specific structure which, though varying in detail, reproduces with astonishing perseverance the same or almost the same adult form?

Attempts at solving this basic problem have been made ever since the human mind recognized the existence of development. For a long time the explanations proposed were purely speculative.

Aristotle attempted to give a solution to the problem of ontogeny along the general lines of his philosophical teaching, distinguishing between the **substance** and the **form** of things. The form appears in this concept as the creative principle. Aristotle further supposed that the substance for the development of a child is given by the mother (in the form of nutrition), but that the creative principle is supplied by the father. He thus accounted also for the necessity of fertilization. Although this treatment of the phenomena of development is completely contradictory to what we now know of the material basis of development (the part played by the ovum and the spermatozoon), still the concept of a creative principle has turned up repeatedly in the teachings of embryologists up to the twentieth century.

In the seventeenth and eighteenth centuries, when all biological sciences developed rapidly, together with the physicochemical sciences, there existed a widespread theory explaining the ontogenetic development of animals. This was the **theory of preformation.** The theory of preformation claimed that if we see that something develops from the egg, then this something must actually have been there all the time but in an invisible form. It is common knowledge that in a bud of a tree the leaves, and sometimes also the flowers with all their parts, can be discovered long before the bud starts growing and spreading and thus exposing to view all that before was hidden inside, covered by the superficial scales of the bud. Furthermore, it was known that in a chrysalis of a butterfly the parts of the butterfly's body—the legs, the wings, etc.—can be discovered if the cuticular coat of the chrysalis is carefully removed a few days before the butterfly emerges from the chrysalis.

Something of this sort was supposed to exist in the egg. All the parts of the future embryo were imagined to be already in the egg, but they were transparent, folded together and very small, so that they could not be seen. When the embryo began to develop, these parts started to grow, unfold and stretch themselves, become denser and therefore more readily visible. The embryo, and therefore indirectly also the future animal, was **preformed** in the egg. Hence the theory is called the theory of preformation.

When spermatozoa were discovered in the seminal fluid, the relative significance of the ova and spermatozoa had to be accounted for. It is obvious that a preformed embryo cannot be present in both the egg and the spermatozoon. The preformationists were split therefore into two rival schools: the ovists and the animalculists. The latter name comes from the word **animalcule,** as the spermatozoa were then called. The ovists asserted that the embryo was preformed in the egg. The spermatozoa then seemed superfluous, and in fact they were declared parasites living in the spermatic fluid. On the other hand the animalculists declared that the embryo was preformed in the spermatozoon, and that the egg served only to supply nutrition for the developing embryo. A lively discussion arose, which eventually ended in favor of the ovists. The final victory of the ovists was due to the discovery of the parthenogenetic development in some insects, e.g., the aphids (Bonnet, 1745). If the egg could develop without fertilization it was clear that the embryo could not be preformed in the spermatozoon.

The theory of preformation, although very popular in its time, did not satisfy all biologists, and opposing views, denying the existence of a preformed embryo in the egg, were proposed. The most important contribution in this field was the theory of **epigenesis,** proposed by Caspar Friedrich Wolff (1759). In favor of his theory Wolff adduced his own observations on the formation of the chick embryo. In the earliest stages of the development of the chick he could not find any parts of the future embryo. What is more, he found that the egg was by no means devoid of any visible structure; there was a structure present, but it was different from that of the later embryo. Wolff found that the substance of which the embryo is built up is granular. Presumably the granules must have been the cells or their nuclei. These granules were later arranged into the layers which we now call the *germinal layers*. Wolff saw that by the formation of local thickenings in some parts of these layers, by thinnning out in others, by the formation of folds and pockets, the layers are transformed into the body of the embryo. He concluded, therefore, that in the early egg there does not exist a preformed embryo, but only the material of which the embryo is built. This material did not represent an embryo any more than a heap of bricks represents the house that will be built of these bricks. In both cases there had to be an architect who would use the material for a purpose that he had in mind. In the case of the developing embryo the architect was represented by a vital force, perhaps not essentially different from the "creative principle" postulated by Aristotle.

Experimental Embryology.   Wolff's observations, however, could not be

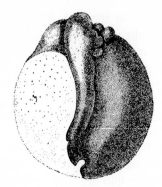

Fig. 3. Half embryo produced by W. Roux by killing one blastomere of a frog's egg with a hot needle. (After Roux, from Morgan, 1927.)

considered as final in deciding between the alternative theories of preformation and epigenesis. In spite of what he actually observed it was still conceivable that organs and parts of the body of the future embryo were represented in the egg by discrete particles, qualitatively different among themselves. The granules which he saw might have been different in their properties. Even if the transformation of such qualitatively different parts into the organs of the embryo should have been more complicated than was envisaged by the crude preformistic theory, the principle of preformation might well have held true in spite of the apparent homogeneity of the material of which the embryo was supposedly made. Observation alone could not go any further towards the solution of this problem, and further progress could be achieved only with the aid of experiment.

One of the experiments which is relevant to the above problem is the separation of the two cells into which the egg is divided at the beginning of development. If the theory of preformation is correct, we should expect that one of the two first blastomeres, containing one half of the egg material, should develop into an embryo lacking one half of its organs and parts. If on the other hand the substances contained in the egg are but the building material used for the construction of the embryo, then it is conceivable that a half of the material might be sufficient for making a complete embryo even if it may have to be on a diminished scale, just as the bricks prepared for the construction of a big house may be used for building two houses of a smaller size.

The first embryologist to see this way of solving the problem was Wilhelm Roux (1850–1924). Accordingly he proceeded to test one of the first two cleavage cells in the common frog for its ability to develop. To achieve his end Roux destroyed one of the two cleavage cells with a red hot needle (1888). The embryo that was derived from the surviving cleavage cell was found to develop, at least at first, as if it were still forming a half of a complete embryo. In other words, the developing embryo was defective, as it should have been according to the theory of preformation (Fig. 3). It was found later, however, that the technique used by Roux was too crude. The damaged cleavage cell was not removed, and it was the presence of this

Fig. 4.   Separation of the first two blastomeres in the sea urchin, resulting in the development of two whole embryos. (After Driesch, from Gabriel and Fogel, 1955.)

damaged cleavage cell, as was later found out, that caused the defects in the surviving embryo. If the two cleavage cells of the egg were separated completely, two whole and, except for their size, normal embryos could develop, one from each of the two cleavage cells. This result was first found by H. Driesch (1891), working on sea urchin eggs (Fig. 4), and later by Endres (1895) and Spemann (1901, 1903), working with eggs of newts. Eventually the experiment was repeated by Schmidt (1933) on the frog, the same animal that served for the experiments of Roux. Schmidt found that if the two cleavage cells were completely separated, each could develop into a whole embryo (Fig. 5).

The first experiments on the developing embryo were followed by many others, and soon a new science was born: **experimental embryology** (Roux 1905).

Experimental embryology, in contrast to comparative embryology or descriptive embryology, uses experiment as a method of investigation. However, the use of the experimental method in itself does not create a science or a branch of science. New branches of science are created by novel viewpoints and novel problems set before science. It was the problem of what ontogenetic development actually is, what the driving forces behind it are, that necessitated the application of experiment after the methods of speculation and of pure observation were found to be impotent in solving the problem.

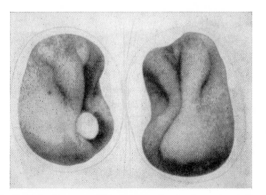

Fig. 5.   Separation of the first two blastomeres in the frog, resulting in the development of two whole embryos. (From Schmidt, 1933.)

Naturally, experiment can be used for purposes other than those of discovering the nature of the driving forces in ontogenetic development. Even in the case of embryonic development there may be many ways in which experiment can be applied. The embryo in every stage of its development is a living organism and, as such, performs all the basic functions of living matter. It is, for instance, subject to the processes of metabolism, and these processes may be studied by experimental methods without the investigator's being concerned with answering the question of why a simple egg is converted into a complex new individual. The supply of nutrition to the embryo, whether through the medium of the egg or directly from the maternal organism, as in mammalian development, is another such problem that may be, and is, studied by physiological methods. Some of these researches may be concerned with the problem of ontogenetic development. Others may be purely physiological in the ordinary sense of the word, the fact that the embryo is a developing system being irrelevant to the ends aimed at in the investigation. Recently, the more physiological ways of approach to the developing embryo have been treated as a special branch of embryology: **chemical embryology** (Needham, J., 1931); (Brachet, J., 1950b) or **physiological embryology** (Lehmann, 1945).

In this "introduction to embryology" all three branches of embryology, i.e., the descriptive and comparative embryology, the experimental embryology, and the physiological embryology, will be dealt with together as far as possible.

PART **II**

The Egg, Its Origin
and Beginning
of Development

CHAPTER **2**

Gametogenesis

Gametogenesis is the first phase in the sexual reproduction of animals. The essential process during this phase is the transformation of certain cells in the parents into specialized cells: the eggs, or ova, in the female and the spermatozoa in the male (or in the female and male organs in hermaphrodite animals).

In both sexes the initial cells giving rise to the gametes are very similar and, as a rule, not essentially different from other cells of the body except that these cells are not involved in any of the differentiations serving to support the life of the parent individual. In both sexes the first step in the production of gametes is a more or less rapid proliferation of cells by ordinary mitosis. The proliferating cells in the testes are known as the **spermatogonia;** the proliferating cells in the ovaries are called **oogonia.** Once proliferation ceases, the cells are called **spermatocytes** in the male and **oocytes** in the female. They now enter into a stage of growth and later into a stage of maturation. Although the stage of proliferation is not essentially different in

19

the male and female, the processes of growth and maturation in the two sexes differ to a very great extent.

2–1   SPERMATOGENESIS

The increase in size of a spermatocyte during the stage of growth is very limited, and it soon enters into the stage of maturation and as a result of two meiotic divisions gives rise to four cells—the spermatids. The spermatids, though having a haploid set of chromosomes, are still not capable of functioning as male gametes. They have to undergo a process of differentiation to become the spermatozoa.

In the course of this the organization of the cell changes in a most radical way: the nucleus of the spermatid shrinks by losing water from the nuclear sap, and the chromosomes become closely packed in a small volume and form the main part of the **head** of the spermatozoon. The cytoplasmic parts

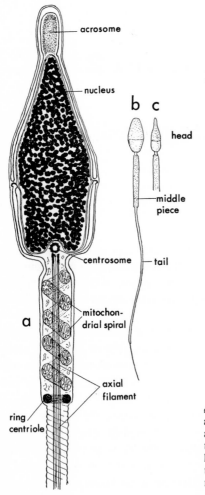

Fig. 6. A mammalian spermatozoid. *a*, Semidiagrammatic drawing (redrawn after an electron micrograph by Burgos and Fawsett: J. Biophys. & Biochem. Cytol., 1955). *b, c,* The same, as seen in the light microscope; the head is seen from the flattened side in *b* and from the narrow side in *c.*

give rise to the **acrosome,** which forms the most anterior structure on the head of the spermatozoon, and to the **middle piece** and the **tail.** The acrosome is developed at the expense of the laminar parts of the Golgi bodies.

The middle piece of the spermatozoon contains the centrosome, lying next to the nucleus, and the proximal part of the axial filament, which is continued in the tail. The mitochondria of the spermatid become concentrated around the axial filament in the middle piece of the spermatozoon and join to form a spiral body surrounding the axial filament between the centrosome and the beginning of the tail. The middle piece of the spermatozoon also contains a small amount of undifferentiated cytoplasm.

The tail of the spermatozoon consists mainly of the axial filament, covered by a very thin layer of cytoplasm which, however, does not reach the tip of the tail. The axial filament is now known to consist of several longitudinal fibers surrounded by another fiber coiled around the others in a tight spiral (Fig. 6). A considerable part of the cytoplasm of the spermatid is not used in the differentiation of the spermatozoon but becomes constricted off from the other parts and is discarded.

As we will see later, most of the specially differentiated parts of the spermatozoon are only necessary so that it can reach the egg and penetrate into the egg cytoplasm. There is no evidence that any part of the spermatozoon, except for the nucleus and the centrosome, plays any role in the development of the egg once it is fertilized.

The spermatozoa of different animals vary greatly in structure: the head may be more or less elongated; the acrosomes may sometimes have very complicated shapes; the tail may have a cytoplasmic undulating membrane; or the tail may be absent altogether. It is not necessary for us, however, to consider all these different forms of spermatozoa.

For further information on the differentiation and fine structure of spermatozoa see Bishop and Austin (1957).

2–2 OOGENESIS. GROWTH OF THE OOCYTE

The period of growth in the female gametes is a very prolonged one, and the increase in size is very considerable. In frogs a young oocyte may be about 50 μ in diameter, and the fully developed egg in many species is between 1000 μ and 2000 μ in diameter. If *Rana pipiens,* in which the diameter of the mature egg is about 1500 μ, is taken as an example, the increase in size of the oocyte is by a factor of 27,000. This growth takes place over a period of three years. The young oocytes start growing after the tadpoles metamorphose into young froglets. One-year-old and two-year-old frogs do not yet have mature eggs, but by the third year the eggs are ready and the frogs may spawn for the first time. Every year a new batch of oocytes is produced as the result of oogonial divisions, but these do not mature until three years later, so that oocytes of three generations may be contained in the ovary at the same time. The growth of the oocytes is fairly slow during the first two seasons, but becomes much more rapid in the summer of the third

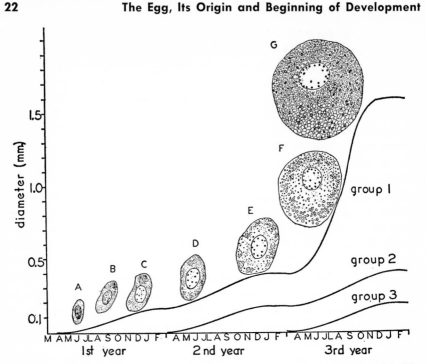

Fig. 7.   Growth of the frog oocytes during the first three years of the female's life. The curves show the increase in diameter of three generations of oocytes; the drawings represent the changes in size and structure of oocytes of the first generation. (Modified from Grant, 1953.)

year of the frog's life, so that by the autumn the eggs reach their maximum dimensions (Fig. 7) (Grant, 1953).

In other animals the growth of oocytes may proceed at a much higher rate and take a shorter time for completion. Oocytes reach their full size in 16-day-old mice. In mammals, of course, the mature eggs are much smaller than in amphibians. The oocyte of a mouse grows from a size of about 20 μ in diameter into the ripe ovum of about 70 μ in diameter, an increase by a factor of only 43 as against an increase of 27,000 in the frog.

It has been held for a long time that the proliferation of the oogonia in mammals is restricted to the intra-uterine period of life, and that all the eggs produced by a mammalian female throughout her reproductive life are derived from oocytes already present at birth. Many investigators believe, however, that new oocytes may be produced from the germinal epithelium of the ovary at later periods of life. In fact, the opinion has been put forward that all the oocytes which are not transformed into mature ova at any one estrus cycle degenerate, and a new lot of oocytes develop from one estrus cycle to another. (For a review of this controversy see Pincus, 1936.)

In many groups of animals, notably in the chordates, the oocytes are surrounded during the whole time of their growth and maturation by special cells of the ovary, the **follicle cells.** In mammals the follicle cells are derived from the germinal epithelium of the ovaries, and initially the young oocyte is surrounded by one layer of follicle cells, forming a simple cuboidal epithelium around the oocyte. Later, the number of follicle cells increases greatly, the cells becoming arranged in several rows. A membrane, the **zona pellucida,** develops between the follicle cells and the surface of the oocyte. As the egg approaches maturity an eccentric cavity appears in the mass of the follicle cells. This cavity is filled with fluid secreted presumably by the cells of the follicle. The follicle at this stage is known as a **Graafian follicle** (Fig. 8). The oocyte is surrounded by follicle cells not only in mammals but in other vertebrates as well, though due to the larger size of the egg the follicle cells are not so conspicuous. It is believed that the follicle cells actively assist the growth of the oocyte by secreting substances which are taken up by the oocyte. In support of this view is the fact that protoplasmic outgrowths of follicle cells penetrate through the zona pellucida and reach the surface of the oocyte. In the last stages of egg maturation these processes may become withdrawn (Sadov, 1956).

In insects the nutrition of the growing oocyte is mediated by special **nurse cells** or **nutritive cells** (Fig. 9). These may be next to the oocyte, or they may be located at the proximal ends of the ovarian tubules, connected to the oocytes by long cytoplasmic processes. The follicular or nurse cells may be completely used up during the growth of the oocyte (in some annelids) (Fig. 10), or they may be completely engulfed in the cytoplasm of the oocyte (in the snail *Helix*).

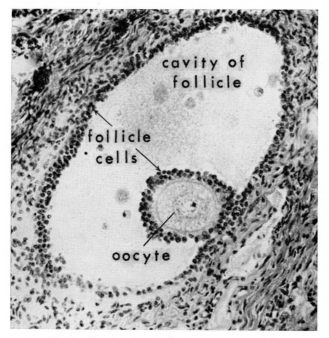

Fig 8.   Graafian follicle in the ovary of a bitch.

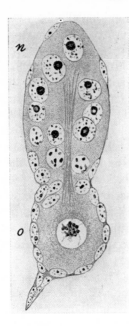

Fig. 9. Ovary of *Aphis*, with oocyte (below) connected to nurse cells (above) by nutritive root. (After de Baer, from Wilson, E. B., 1925.)

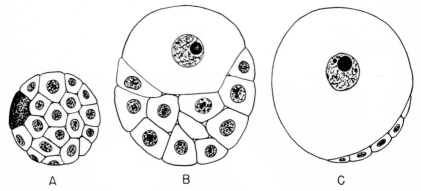

A   B   C

Fig. 10. Three stages in the development of the oocyte of the leech *Piscicola*, which uses up the follicle (nurse) cells as it grows. (From Jörgensen, in Willier, Weiss and Hamburger, 1955.)

The young oocyte, just emerged from the oogonial mitotic divisions, does not possess special cytoplasmic zones any more than does any undifferentiated cell of the animal's body. Neither is it able to develop into a new individual. The special organization of the egg cell, which makes it capable of developing into a new individual, is elaborated gradually and usually over a considerable period of time. The changes that the future cell undergoes during this period can in part be directly observed in microscopic preparations, but some of the changes are still completely unknown to us.

2–3   CHANGES IN THE NUCLEUS OF THE OOCYTE

Simultaneously with the growth of the oocyte its nucleus enters into the prophase of the meiotic divisions; the homologous chromosomes pair together. But the subsequent stages of meiosis are postponed until the end of the period of growth. Instead, the nucleus of the oocyte increases in size, though not nearly to the same extent as the cytoplasm. The increase in size is due mainly to the production of very large amounts of nuclear sap, so that the nuclei of advanced oocytes appear to be bloated with fluid and are often referred to as **germinal vesicles.** The chromosomes at the same time may increase in length, but the amount of deoxyribonucleic acid in the chromosomes does not increase in proportion to the enlargement of the nucleus. In oocytes of animals having large eggs the chromosomes acquire a very characteristic appearance: thin threads or loops are thrown out transverse to the main axis of the chromosomes, making the chromosomes look like lamp brushes: **lamp brush chromosomes.**

The formation of these lateral outgrowths of the chromosomes may result in a greater surface of contact between the chromosomes and the surrounding materials, and may have something to do with the enormous turnover of substances involved in the growth of the oocytes. There is, however, no direct proof that any substances which are deposited in the cytoplasm of the growing oocytes are originally produced by the lamp brush chromosomes or in their immediate vicinity.

The nucleoli in the germinal vesicle seem to be very actively involved in the metabolism of the growing and maturing oocyte. The nucleolus of the oocyte increases greatly in size and becomes very conspicuous against the background of the vesicular nucleus. In many animals, in amphibians in

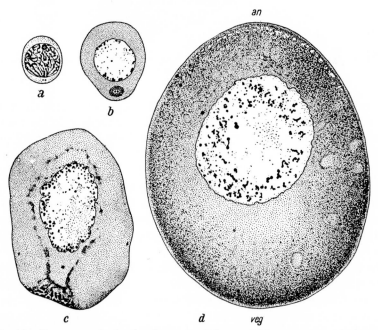

Fig. 11.   Four stages in the development of the amphibian oocyte. *a, Ambystoma,* young oocyte with chromosomes in the "bouquet stage." *b, c, d, Rana,* beginning of deposition of yolk. (After Witschi, from Kühn, 1955.)

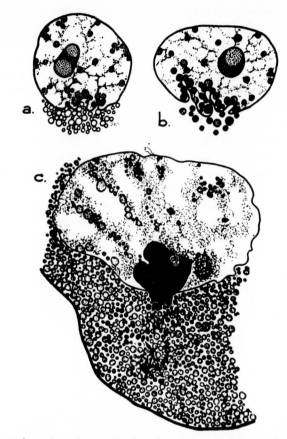

Fig. 12. Extrusion of nuclear substances into the cytoplasm during oogenesis of *Limnaea stagnalis*. (From Bretschneider and Raven, 1951.)

particular, the nucleolus becomes fragmented and gives rise to many smaller nucleoli. Most of these become localized on the periphery of the nucleus, immediately underneath the nuclear membrane (Fig. 11). There are many reports, both in the older and more recent literature (see, for instance, Bretschneider and Raven, 1951; Logachev, 1956), that nucleolar material may penetrate through gaps in the nuclear membrane and enter the cytoplasm (Fig. 12). As the nucleoli consist mainly of ribonucleic acid, the penetration of nucleolar material into the cytoplasm may account, at least in part, for the increase in cytoplasmic ribonucleic acid in oocytes. (See section 2–4.)

As the oocyte nucleus grows, a remarkable change occurs in its ability to be stained by agents currently used for the demonstration of the presence of deoxyribonucleic acid. The chromosomes of the young oocyte can be clearly stained with methyl green or the Feulgen reagent, but at later stages the staining becomes paler and paler until it may be lacking altogether. When the oocyte reaches maturity and approaches the reduction divisions, the chromosomes reappear in a very much contracted form and may again be stained by the Feulgen reagent or by methyl green.

This seems to be very disconcerting, since the deoxyribonucleic acid is presumed to be the essential component of the chromosomal genes—the carriers of heredity. Accordingly, the inability of oocyte nuclei to be stained with Feulgen reagent has been interpreted by some biologists (Makarov, 1951) as a proof that there is no continuity of nuclear material during oogenesis, and that the genes, if they consist of deoxyribonucleic acid, must disappear during the growth of the oocyte and be formed anew at a later stage. This would be a very serious contradiction of the chromosomal theory of heredity in organisms. There is, however, sufficient evidence that deoxyribonucleic acid is present in the nucleus at all stages, and that the failure of the Feulgen reagent to demonstrate its presence is due to the extreme dispersal of the chromosomal material in the oocyte nucleus. It is known that the Feulgen reaction can be positive only if the amount of the deoxyribonucleic acid in the preparation is not below a certain minimum. Concentrations of deoxyribonucleic acid under 0.67 mg. per ml. cannot be detected by the Feulgen method in a 10 μ section (Pollister, after Sze, 1953). Actually, in some animals such as the newt, in which the chromatin particles in oocyte chromosomes do not become too small, they can be stained throughout the period of growth of the oocyte. In other animals the chromosomal material of the oocyte may be concentrated, by high speed centrifugation, at the centrifugal pole of the oocyte nucleus, and then it gives a clear positive reaction with the Feulgen reagent (Brachet, J., 1950b, p. 63).

After the oocyte completes its growth, it is ready for the reduction divisions. The chromosomes of the oocyte at this stage have become greatly contracted and concentrated toward the center of the germinal vesicle. The nuclear membrane breaks up and the chromosomes are carried to the periphery of the oocyte. Here the achromatic spindle is formed, and half of the chromosomes, with a small quantity of cytoplasm, are extruded from the egg, forming the **first polar body.** There follows, either immediately or after a more or less lengthy interval, the second meiotic division, leading to the extrusion of the **second polar body.** Half of the remaining chromosomes are removed from the egg, leaving it with only a haploid set of chromosomes.

2–4 ACCUMULATION OF FOOD RESERVES IN THE CYTOPLASM OF THE OOCYTE

During the growth of the oocyte not only does the amount of cytoplasm increase in quantity, but it changes in quality by the elaboration and regular distribution of various cell inclusions and specially modified parts of the cytoplasm which, as we will see, are essential for the development of the embryo. Of these the food supplies stored in the cytoplasm of the egg should be mentioned first.

The most usual form of food storage in the egg consists of granules of **yolk.** Yolk is not a definite chemical substance, but rather a morphological

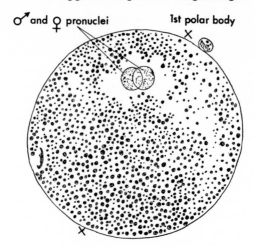

Fig. 13. Oligolecithal egg of *Amphioxus* shortly after fertilization. The animal and vegetal poles are indicated by crosses. (After Kerr, 1919.)

term; the chemical substance may not be the same in all cases. In *Amphioxus,* amphibians, fishes and in many invertebrates the chief constituent of the yolk is a phosphoprotein called **vitellin.**

Eggs with a small amount of yolk are called **oligolecithal.** In *Amphioxus* and in invertebrates with a small amount of yolk (for instance, in echinoderms), the yolk granules are very fine and fairly equally distributed in the cytoplasm of the egg (Fig. 13).

In amphibian eggs the yolk is found in the form of rather large granules, usually described as the **yolk platelets.** The yolk platelets have an oval shape and are flattened in one plane. The cytoplasm is densely packed with them. The amphibian egg is much larger, so that not only the relative but also the absolute amount of yolk is far in excess of that found in the eggs of *Amphioxus* or the echinoderms. The distribution of yolk in the amphibian egg is distinctly unequal: the yolk platelets are densest in the lower part of

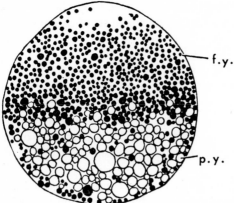
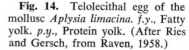

Fig. 14. Telolecithal egg of the mollusc *Aplysia limacina.* f.y., Fatty yolk. p.y., Protein yolk. (After Ries and Gersch, from Raven, 1958.)

the egg, and there is relatively more cytoplasm in the upper part of the egg. Eggs of this type are known as **telolecithal** (Fig. 14).

In addition to the yolk platelets the amphibian egg contains stored supplies in the form of lipoid and glycogen. Lipoid is distributed throughout the cytoplasm of the egg in the form of organized inclusions, the **lipochondria** (Holtfreter, 1946), which consist of an internal core of lipoid surrounded by a thin protein coat. The lipochondria are much smaller than the yolk platelets and have a less regular shape (Fig. 15). Glycogen is present in the egg cytoplasm in the form of small granules. Fat may also be stored in the egg for the nourishment of the developing embryo, but its quantity is relatively small in the lower vertebrates.

In a mature amphibian egg the yolk constitutes roughly 45 per cent of the dry weight, lipoids 25 per cent, and glycogen 8.1 per cent. Only about 20 per cent of the dry weight of the mature egg is active cytoplasm (Barth and Barth, 1954).

Cyclostomes, elasmobranchs, ganoids and the lungfishes have eggs with a distribution of food reserves much the same as in amphibians, though the amount of yolk may be greater, especially in some selachians. In the higher teleosts, on the other hand, the yolk is completely segregated from the cytoplasm. The yolk constitutes the greater part of the egg, and the cytoplasm is restricted to a thin surface layer covering the yolk and thickened on the

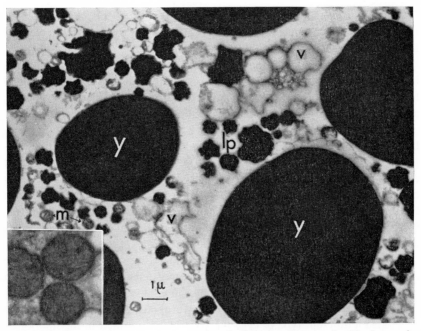

Fig. 15. Electron micrograph of cytoplasmic inclusions in an unfertilized egg of a frog *Xenopus laevis*. *y*, Yolk; *lp*, lipochondria; *m*, mitochondria; *v*, vacuoles. Inset, A group of mitochondria under higher magnification.

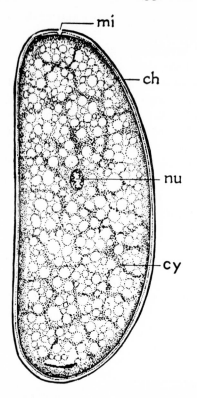

mi

ch

nu

cy

Fig. 16.   Insect egg, diagrammatic. *nu,* Nucleus; *cy,* cytoplasm; *ch,* chorion; *mi,* micropyle. (From Johannsen and Butt, 1941.)

uppermost side of the egg in the shape of a cytoplasmic cap. The nucleus of the egg in this case lies inside this cytoplasmic cap.

In the bony fishes the fat may be present in the form of large fat droplets inside the mass of yolk. The number and size of the fat droplets is typical for different families of fishes. Sometimes only one large droplet is present; at other times there is a large number of smaller droplets.

Fat plays an important role in the eggs of reptiles and birds. The yolk of a bird's egg consists of about one-third protein, in the form of vitellin, and two-thirds a fatty substance, known as **lecithin.** Most of this is neutral fat, but there is also a considerable amount of lipoid present. The yolk of a bird's or a reptile's egg, as in the case of the bony fishes, lies in a compact mass in the interior of the egg, and the cytoplasm is restricted to a thin layer on the surface, with a thickened cap of cytoplasm on the upper side. As in bony fishes this cytoplasmic cap also contains the nucleus of the egg cell.

Some of the invertebrates have also developed eggs in which the relative amount of yolk is high and more or less segregated from the cytoplasm. In cephalopods and some gastropods among the molluscs the eggs are telolecithal, much as in the lower vertebrates. Arthropods, especially insects, have developed a different type of egg: the yolk is concentrated in the interior of the egg and the cytoplasm is distributed as a thin coat on the external surface; however, there is also an island of cytoplasm in the center of the egg. This island, surrounded on all sides by yolk, contains the nucleus of the egg cell (Fig. 16). Eggs of this type are called **centrolecithal.**

The arrangement of various substances and cellular constituents in the advanced oocyte and, later, in the egg shows a polarity, that is, an unequal distribution in respect to what may be called the two opposite poles of the egg and in respect to the main axis of the egg—the line connecting the two poles. The nucleus of the egg is approximated to one pole of the egg, which is termed the **animal pole.** The opposite pole is termed the **vegetal pole,** because the accumulation of yolk at that pole serves for the nutrition of the developing embryo. When the cell undergoes meiosis the nucleus of the egg cell approaches the animal pole of the egg, and the polar bodies are always discharged at the animal pole. This process serves to distinguish the animal pole in oligolecithal eggs if the concentration of the yolk at the vegetal pole is not very distinct, as is often the case, and also in centrolecithal eggs.

Eggs in which the yolk is completely segregated from the cytoplasm, as in the bony fishes, reptiles and birds, have a thickened cytoplasmic cap denoting the animal pole of the egg.

The specific gravity of the yolk is higher than that of the cytoplasm. As a result, the vegetal pole of the egg rotates downward when the egg is suspended free in water. The animal pole comes to lie on the top.

The growth of the oocyte, accompanied by the accumulation of food reserves, is a complicated synthetic process, and a special structural and chemical mechanism is elaborated in the oocyte for this purpose. The synthetic activities leading to the accumulation of food supplies in the oocyte appear to be initiated in a special area of the cytoplasm which has been named, somewhat inappropriately, the **yolk nucleus.** This consists of the centrosome and the centrioles, lying next to the nucleus on one side, and of a surrounding mantle of Golgi bodies and mitochondria. The granules of lipoids are the first to be formed in the immediate vicinity of the yolk nucleus, and glycogen granules appear shortly after that.

The yolk nucleus does not remain in the original compact form for very long. Soon it breaks up, and parts of the nucleus, consisting of Golgi bodies and mitochondria, spread outward and eventually become distributed on the periphery of the oocyte. Yolk platelets start forming at this stage. The oocyte by this time, in the case of the frog, has increased in diameter from 50 μ to about 400 μ. As a result of dispersion of the yolk nucleus, the yolk platelets appear first just underneath the cortex of the cytoplasm of the oocyte. As more and more yolk platelets are produced, they fill the cytoplasm from outside inward. In the last stages of the growth of the oocyte the yolk platelets are formed around the nucleus, so that the cytoplasm becomes filled up with yolk. The yolk platelets are not all of the same size. At the animal pole of the oocyte, where the polar bodies are to be given off later, the yolk platelets remain relatively small, and are not packed quite so densely. At the vegetal pole they reach a larger size (about 1.5 μ in length) and they come to lie very close to each other, leaving little cytoplasm in between. The inte-

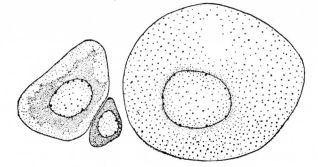

Fig. 17. Distribution of ribonucleic acid (shown by stippling) in three stages of development of amphibian oocytes. (From Brachet: Chemical Embryology, copyright 1950 by Interscience Publishers, Inc., New York.)

rior of the oocytes, the area surrounding the nucleus, is fairly closely packed with yolk platelets which are not quite so large as at the vegetal pole (see Wittek, 1952). Lipochondria, fat droplets and yolk granules are somehow connected in their origin with the yolk nucleus and its parts, when the latter are distributed to the periphery of the oocyte. However, the nature of this connection is not yet clearly known.

It has been suggested that the fat droplets are generated in connection with the Golgi bodies, and that yolk platelets have direct relation to mitochondria, or that mitochondria are actually transformed into yolk platelets. Some recent work has made it seem probable that yolk granules or at least some types of yolk granules may be produced by Golgi bodies in the same way that the Golgi bodies are responsible for secretory activity in some glandular cells. It was found that in the oocytes of *Limnaea* the Golgi apparatus breaks up into numerous small Golgi bodies, and that yolk granules can be first traced as particles appearing inside the Golgi bodies, surrounded by the osmiophilic substance of the bodies. As the yolk granules grow, the layer of osmiophilic substance stretches and is thinned out until it disappears altogether (Bretschneider and Raven, 1951). Under the electron microscope the osmiophilic substance of the Golgi bodies appears as a system of double membranes and vesicles surrounding a central mass of more homogeneous cytoplasm. It is presumably in this central mass of cytoplasm that the yolk granules are laid down. In some electron micrographs of sea urchin oocytes one can see large yolk granules surrounded by numerous layers of double membranes which may be recognized as the osmiophilic parts of the Golgi bodies (Afzelius, 1956). Further information on the origin of yolk granules may be expected when modern methods are applied to oocytes of a greater variety of animals.

It is now generally accepted that nucleic acids contained in the nucleus and cytoplasm of cells play a very important part in the synthesis of the most essential constituents of living matter. It is not astonishing, therefore, that large amounts of nucleic acids, mainly in the form of ribonucleic acid, but possibly with some amount of deoxyribonucleic acid as well (Sze, 1953), are present in the cytoplasm of oocytes. The cytoplasm of young amphibian oocytes stains red with pyronin, indicating the presence of large amounts of ribonucleic acid.

As the yolk platelets begin to accumulate at the periphery of the oocyte, the area of cytoplasm rich in ribonucleic acid becomes restricted to the deeper parts of the cytoplasm, and still later to the area around the nucleus (Fig. 17). Eventually the ribonucleic acid in the cytoplasm is reduced to very small quantities. In eggs having smaller amounts of yolk, such as the echinoderm or mammalian eggs, however, quite a considerable amount of ribonucleic acid remains even in fully mature eggs.

The intense synthetic activity in the oocyte requires energy, and this is derived through the increased oxidations going on in the ooplasm during this period. It has been found that young frog oocytes, before the beginning of yolk formation, absorb 0.69 cu. mm. of oxygen per cubic millimeter of oocyte. At the stage when the yolk platelets begin to form, the oxygen consumption rises to 1.5 cu. mm. per cubic millimeter. In older oocytes nearing maturity the oxygen consumption per cubic millimeter falls to 1.2 cu. mm. This is still a high value, considering that a large part of the volume of the oocyte is filled with inert yolk; thus the remaining active cytoplasm continues to respire at a high rate (Mestcherskaia, 1935, cited after Brachet, 1950).

Parallel with increase in respiration goes increased enzymatic activity, as indicated by measurements of the activity of the enzyme dipeptidase. Dipeptidase is the enzyme which breaks the peptide linkage between two amino acids in a protein molecule (e.g., alanine and glycine) and may be involved in the reorganization of protein molecules. It has been found that dipeptidase activity per unit of volume of protoplasm is low in very young frog oocytes,

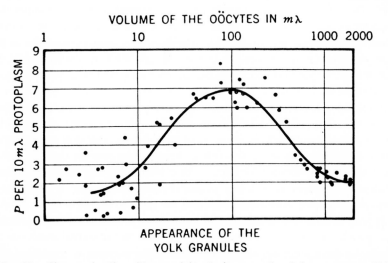

Fig. 18.   Changes in dipeptidase activity during growth of frog oocytes. (After Duspiva, from Brachet: Chemical Embryology, copyright 1950 by Interscience Publishers, Inc., New York.)

but that it increases with the growth of the oocyte. It reaches a maximum at the stage when yolk appears in the subcortical layer of the cytoplasm, and then it decreases, probably as the result of a relative decrease of the active cytoplasm, which is gradually replaced by inert yolk (Fig. 18).

2–5   ORGANIZATION OF THE EGG CYTOPLASM

Yolk and the other substances obviously destined to serve as food for the developing embryo are not the only ones that are laid down in the oocyte during its growth. A characteristic feature of the mature eggs in many animals is the presence of granules of pigment, which are elaborated during the growth of the oocyte.

In the oocytes of the ascidian *Cynthia partita,* besides the yolk, which accumulates at the vegetal pole, there appears in the later stages a yellow pigment in the form of granules distributed all over the surface of the oocyte in a thin cortical layer of cytoplasm. In the sea urchin *Paracentrotus lividus* a similar distribution of pigment granules is observed; only the pigment is red. We will see later that the yellow pigment of *Cynthia partita* and the red pigment of *Paracentrotus lividus,* although uniformly distributed over the surface of the maturing egg, will later be concentrated and come to lie in specific parts of the embryo—in the muscles and mesenchyme of the former, and in the walls of the gut of the latter. Although it does not necessarily follow that the pigment is in any way a precursor of muscular or intestinal differentiation, it may well be that the pigment is an indicator of some specialization in the cytoplasm of the oocyte, which eventually leads to specific types of differentiation of parts of the embryo.

In the eggs of most amphibians there is present a greater or lesser amount of dark brown or black pigment. Depending on the amount of pigment, the egg may appear to be light fawn, through various shades of brown, to pitch black. However, the young oocytes have no pigment. Pigment granules start being formed somewhat later than the yolk platelets, in oocytes which have grown to about one half of the final diameter. The greatest number of pigment granules becomes located in the cortical layer of cytoplasm of the mature oocyte, but quite considerable amounts of pigment are distributed in the interior.

A remarkable feature in the distribution of pigment in the amphibian egg is that it is not uniform. There is much more pigment in the animal hemisphere of the oocyte than in the vegetal hemisphere. The difference may be very marked, so that while the animal hemisphere may be dark brown or black, the vegetal hemisphere appears clear white, although in reality a small number of pigment granules is practically always present in the vegetal hemisphere as well. The transition from the dark to the light areas is fairly sharp, but there is always a zone of intermediate, gradually fading, pigmentation, which can be referred to conveniently as the **marginal zone.** In the interior of the mature egg the distribution of the pigment is correlated with the distribution of the yolk.

In cross section of a ripe amphibian egg (Fig. 19) it may be seen that the vegetal half of the egg is filled by a densely packed mass of yolk containing very little pigment. This mass of yolk is slightly concave on the top. The center of the egg is taken up by a roughly lens-shaped mass of cytoplasm with middle-sized yolk platelets and a moderate amount of pigment. This zone also contains the nucleus in the immature oocyte. On the outer edges of this interior mass of protoplasm lies a ring-shaped area containing large amounts of pigment. This area coincides with the marginal zone on the surface of the egg but lies deeper. The ring is thicker on one side of the egg, where it also reaches nearer to the surface. This side of the egg corresponds, as we will see later, to the future dorsal side of the embryo, and thus, in conjunction with the differences along the main axis of the egg, indicates a plane of bilateral symmetry. Lastly, on top of the interior mass of protoplasm lies, like an inverted saucer, a layer of cytoplasm of the animal hemisphere (the "animal cap"), which is rich in pigment, especially in the cortical layer, and relatively poor in yolk.

The pigment granules in themselves may perhaps not be very important for the development of the embryo; in fact, there are some species of amphibians which do not have any pigment in their eggs (the large European crested newt *Triturus cristatus* and some frogs making foam nests, like the African *Chiromantis xerampelina*) and yet develop in the same way as related species having pigmented eggs. However, the uneven distribution of the pigment may be considered as an indicator of qualitatively different areas in the cytoplasm of the egg. This can be corroborated by some further observations. We have seen that the cytoplasm of young frog oocytes contains large amounts of ribonucleic acid. It has been observed that in oocytes 400 to 500 μ in diameter the ribonucleic acid is especially concentrated in the surface layer of the animal hemisphere of the oocyte. With the development of yolk platelets in the subcortical layer of cytoplasm, the ribonucleic acid concentration disappears, but it seems very probable that the localization of pigment in the animal hemisphere reflects some peculiarities in the cytoplasm—peculiarities which are related to the distribution of the ribo-

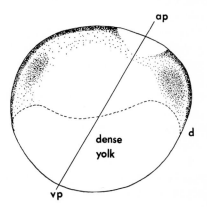

Fig. 19. Distribution of yolk and pigment in the ripe egg of a frog. Median section. *ap*, Animal pole; *vp*, vegetal pole; *d*, dorsal side of the egg. (After Lehmann, 1945, and Wittek, 1952.)

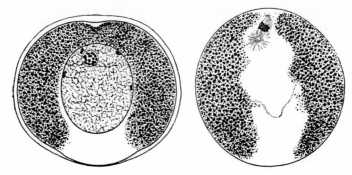

Fig. 20.   Oocytes of *Dentalium* just before the beginning of meiotic divisions (left) and in the stage of first meiotic division (right). (From Wilson, 1904.)

nucleic acid in the preceding stage (Wittek, 1952; Kemp, 1953, p. 493).

In the elephant tusk mollusc *Dentalium,* very distinct cytoplasmic areas can be seen in the egg even before it leaves the ovary. The fully grown oocyte contains a pigment varying in different individuals from olive green to brick red. However, the pigment does not encroach on two areas at the opposite sides of the egg, which are therefore colorless. One of the two colorless areas lies on the side with which the egg is attached to the wall of the ovary. While the oocyte is in the ovary, this part is somewhat drawn out, so that the oocyte is more or less pear-shaped. The second pigment-free area is on the free surface of the oocyte, and it is in the center of this area that the polar bodies are given off, thus marking this side as the animal pole.

The egg is released from the ovary with the germinal vesicle still intact, and it proceeds to the meiotic divisions only afterwards. After release from the ovary, the egg partially rounds up but remains slightly flattened from animal to vegetal pole. In a vertical section through the egg (Fig. 20) the light area at the vegetal pole (the pole that was attached to the wall of the ovary) is seen to be made up of clear protoplasm which contains no yolk granules. This protoplasm reaches inward to the nucleus and surrounds it, and at the outer edges it is continuous with a very thin cortical layer covering the whole of the egg surface. There is also a small patch of clear cytoplasm at the opposite pole; this is the place where the polar bodies will be given off. The rest of the cytoplasm is filled by rather densely packed yolk granules. The yolk-free and unpigmented cytoplasm at the vegetal pole may be termed the **vegetal polar plasm.** The pigment-free **animal polar plasm** is only partly free from yolk. The significance of the two polar plasms for the development of the embryo will be considered in a later section (4–6).

It has been stated already that the oocytes of higher mammals do not increase to a very great extent during the period of growth. This is because the embryo receives its nutrition from the maternal body, and there is in consequence no need to accumulate large amounts of food supplies in the egg. Much importance can be attached to the development of those features which bear a relation to the differentiation of parts of the embryo from the egg. Already in very early mammalian oocytes there exists a polar organization: the nucleus of the oocyte lies nearer to one pole of the egg, the animal pole. A "yolk nucleus" is present in the younger oocytes as an area of basophilic cytoplasm containing ribonucleic acid and a concentration of mitochondria. This basophilic area lies next to the nucleus, nearer to the vegetal pole but somewhat to one side of the main axis. In middle-sized oocytes the basophilic area next to the nucleus becomes less conspicuous, and the mitochondria are now found distributed next to the cortex of the egg. They become especially numerous on one side of the egg—opposite from where the yolk nucleus was to be seen. The mitochondria are accompanied by large amounts of basophilic granules containing ribonucleic acid, and the ground substance of the cytoplasm is here very dense and rich in proteins. The zone in question covers about half the surface of the egg, on one side of the main axis, reaching from the animal to the vegetal pole. The opposite side of the egg surface appears to be richer in water, and contains numerous vacuoles, the latter being especially distinct in some animals such as the mole.

This distribution of cytoplasmic constituents shows that the mammalian oocyte not only has a polarity, but actually possesses a bilateral organization (Fig. 21). The side of the egg rich in ribonucleic acid corresponds to the dorsal side of the embryo, and the side with vacuolated cytoplasm corresponds to the ventral side (Jones-Seaton, 1950; Dalcq, 1954). The position of the animal pole is reaffirmed at the time of the meiotic divisions, as shown in Fig. 21.

Of all the developments in the cytoplasm of the oocytes perhaps the most important are those which concern the surface layer or the **cortex** of the oocytes. In many animals, and possibly in all, the cortex contains a layer of special structures, the **cortical granules.** These granules break up during fertilization and supply the material for the development of the fertiliza-

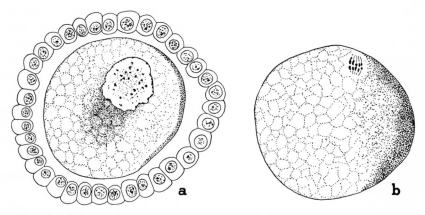

Fig. 21.   Oocytes of the rat before the first meiotic division (*a*) and during the first meiotic division (*b*). Ribonucleic acid is shown by the close stippling. (From Jones-Seaton, 1950.)

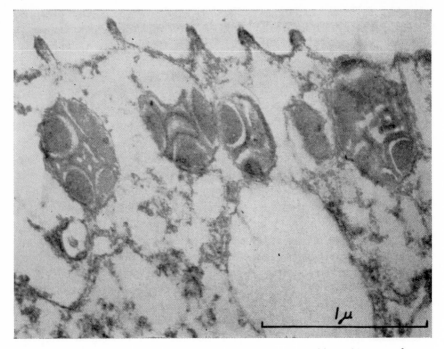

Fig. 22. Cortical layer in the unfertilized egg of the sea urchin *Echinus esculentus,*
with cortical granules and microvilli on the surface. (From Afzelius, 1956.)

tion membrane. (See section 3–5.) In sea urchin eggs the granules in
question are originally formed in the deeper layers of the oocyte cytoplasm. As the oocyte reaches maturity, the cortical granules move outward
and come to lie in the superficial layer of cytoplasm which is approximately
1 μ thick. During their migration the cortical granules increase in size and
undergo structural changes. With the aid of the electron microscope it can
be seen that the immature granules have outer membranes, and the interior
of each consists of a system of concentric or possibly spirally arranged
denser lamellae separated by less dense interstices. In the mature granules,
parts of the interior become filled by a more homogeneous substance, and
the lamellae become less regular (Fig. 22). The mature granules are about
0.8 μ in diameter and are quite closely packed in the cortical layer.

In the frog oocyte the cortical granules arise in the surface layer of cytoplasm of the oocyte at about the same time when the yolk granules are
formed in the somewhat deeper layers. The granules differ from those of
the sea urchin egg in that they do not have an internal lamellar structure
but consist of a homogeneous substance surrounded by a thin membrane.
The mature granules are about 2 μ in diameter (Fig. 23).

Electron microscopic studies of oocytes, besides showing the details of the
structure of the cortical granules, have revealed also that the surface layer
of the cytoplasm of the oocytes is drawn out into numerous fingerlike proc-

esses, the **microvilli.** These are especially numerous and dense in frog
oocytes, and in the ovary they interdigitate with similar processes on the
inner surface of the follicle cells surrounding the oocyte. A similar inter-
digitation of microvilli of the outer surface of the oocyte with those of the
inner surface of the follicle cells has been discovered in mammals (Sotelo
and Porter, 1958). Microvilli can also be seen in sea urchin eggs, but they
are less numerous and not so close to one another. With a light microscope
individual microvilli cannot be seen, and the zone of microvilli appears as a
radially striated layer, which has long been known as the **zona radiata.** The
presence of the microvilli greatly increases the surface area of the oocyte.
In the frog oocyte such an increase in surface area has been estimated to
be by a factor of about 35. It would appear that the increase in area
facilitates metabolic turnover between the oocyte and its environment. In
the frog and in the mammal the intimate association of the microvilli of the
oocyte with those on the surface of the follicle cells is in agreement with the
assumption that the follicle cells pass nutritive substances to the oocyte
(see p. 23).

In the last stage of maturation the nuclear membrane of the oocyte breaks
down and the chromosomes move to the surface at the animal pole to take
part in the maturation divisions. At the same time the nuclear sap merges
with the cytoplasm of the oocyte. It has been observed in some animals,

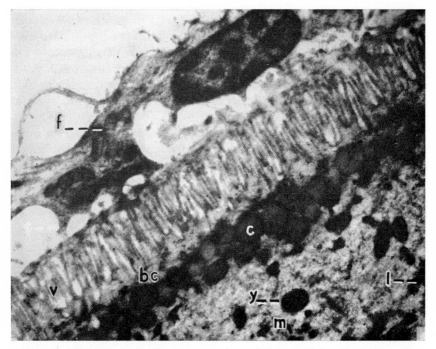

Fig. 23.   The surface of a frog oocyte and an adjoining follicle cell (*f*). The microvilli of the follicle cell and the oocyte interdigitate (*v*). *bc,* Basal cortical cytoplasm; *c,*
cortical granules; *y,* yolk; *l,* lipochondria; *m,* mitochondria. (From Kemp, 1956.)

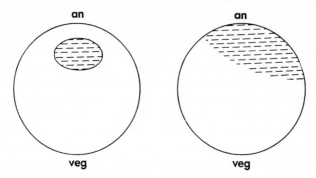

Fig. 24. Distribution of sulfhydryl groups in the frog egg before rupture of the germinal vesicle (left) and after maturation (right). (After Brachet, from Kühn, 1955.)

that the nuclear sap does not mix completely with the cytoplasm, but that it forms a more or less separate mass, thus increasing the diversity of the cytoplasmic areas of the eggs. In the ascidian *Cynthia partita* the nuclear sap released from the germinal vesicle floats up toward the animal pole of the egg and gives rise to a large yolk-free zone just under the cortex of the egg at the animal pole. In amphibians the nuclear sap differs from the cytoplasm of the oocyte in that its proteins contain large amounts of SH– groups (due to the presence of the amino acid, cysteine). The reaction with ammonium or sodium nitroprusside, which causes a red color to appear in the presence of SH– groups, is restricted in the mature oocyte to the germinal vesicle. Once the nuclear membrane breaks down, the red staining with nitroprusside reagent is observed in the cytoplasm at the animal pole, obviously due to the admixture of the nuclear sap released from the germinal vesicle. After fertilization the cytoplasm containing the SH– spreads out along one side of the egg which will become the dorsal side, thus accentuating the bilateral symmetry of the egg (Fig. 24) (Brachet, 1950).

The fully grown egg is a relatively large cell, always larger than the average somatic cell in the animal. When large quantities of foodstuffs are stored in the egg cell, it may attain giant proportions. The yolk of a hen's egg, which represents the egg cell in this case, has an average weight of 55 gm. and is, of course, by no means the largest of its kind. Very large eggs are also found in reptiles and in some sharks. However, the quantity of active cytoplasm in such large eggs is comparatively rather small, and the nucleus, although larger than in the somatic cells, never increases in proportion to the bulk of the whole egg.

As a general rule mature eggs are spherical in shape, though elongated eggs are not infrequently found, especially among insects. Among vertebrates oval-shaped eggs are found in the hagfish *Myxine* and in the ganoid fishes. The elongated shape of the bird's egg, on the other hand, is not due to an elongated shape of the egg cell itself. The egg cell, which is the yellow of the egg, is in this case spherical.

Before concluding this chapter on the organization of the egg and the proc-

esses which provide for the arrangement of the various cytoplasmic substances in the egg, we must return once more to the polarity of the egg and
consider its origin. The polarity of the egg is discernible from the position
of parts in the egg cell: the nucleus, yolk and other cytoplasmic substances.

Many efforts have been made to elucidate what factors are responsible for
the unequal distribution of these parts. It has been claimed that the polarity
of the egg may be imposed on it by the direction of flow of the nutrient
substances during the growth of the oocyte. It has been mentioned that in
the molluscs and echinoderms the vegetal pole of the egg develops from
that end of the oocyte by which it is attached to the wall of the ovary. The
nutrient substances, at the expense of which the oocyte grows, presumably
enter the ovary from outside, from the body cavity. It would stand to reason, then, that greater amounts of yolk might be deposited in the part of
the cell nearest to the proximal surface of the ovary, thus causing this part
to become the vegetal pole. This explanation would not hold, however, for
oocytes which are surrounded by follicle cells from all sides, as are the
oocytes of amphibians or mammals. It has been suggested that the course
of the nearest blood vessel supplying parts of the ovary with nourishment
might cause the parts of the oocyte nearest to the vessel to develop into the
vegetal pole. But according to the views of Child (1941), the animal pole,
as the more active one, should develop from that part of the oocyte which
has a better oxygen supply, and on this principle the part of the oocyte
nearest to a blood vessel should become the animal pole.

Actually, there does not seem to be a very clear connection between the
position of the animal and vegetal pole of the egg and the course of the
blood vessels. In view of the differences in the structure of ovaries in different animals, it would seem rather hopeless to try to find a common factor
in the environment of the growing oocyte which could be held responsible
for the origin of polarity of the egg.

Polarity is, however, a phenomenon which is found not only in egg cells
but in other cells as well. In epithelial cells there is a distinct difference
between the proximal end of the cell (the end resting on the underlying
basement membrane) and the distal end (which forms the free surface of
the epithelium). In nerve cells the polarity of the cell takes the form of the
opposite differentiations of axon and dendrites. In mesenchyme cells the
polarity shows itself in the direction of movement of the cells, in their
possessing an anterior and a posterior end when in movement (Holtfreter,
1947a, b). We should expect that the young oocyte also possesses a polarity,
whatever the exact physicochemical nature of this polarity should prove
to be. This intrinsic polarity may well serve as a basis for the distribution
of cellular constituents in the ripening oocyte.

We have seen that the building up of food supplies (lipochondria and
yolk) in the egg starts in connection with the yolk nucleus, and that the
yolk nucleus is essentially an accumulation of Golgi bodies and mitochondria around the centrosome of the oocyte. It is therefore the position of

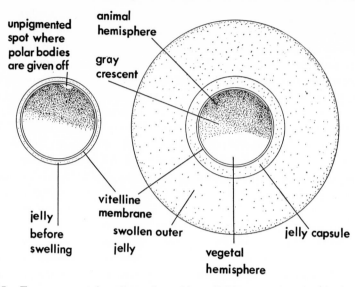

Fig. 25. Frog egg as taken from the oviduct (left), and after fertilization, with swollen jelly-membrane (right).

the centrosome in respect to the nucleus that may, in the last instance, be responsible for the polarity of the oocyte and, subsequently, the egg. The line drawn through the centrosome and the nucleus determines an axis of polarity which may well be the future main axis of the egg. It is worth noting that this polarity finds an expression not only in the arrangement of cytoplasmic inclusions, but in the intimate structure of the nucleus itself. In the early leptotene stage of the meiotic prophase, the chromosomes are arranged in a definite way, converging to that side of the nucleus which is nearest to the centrosome. This stage is known as the "bouquet stage" (Fig. 11, p. 25). Obviously, this orientation conforms to the primary polarity of the oocyte.

What has been said so far refers to the polarity along the main axis of the egg (along the axis going from the animal to the vegetal pole). A similar sort of polarity may be responsible for the difference between the side which is to develop into the dorsal side of the embryo, and that which is to develop into the ventral side of the embryo. However, even less is known about the origin of this polarity than about the origin of polarity along the main axis.

2–6   THE EGG MEMBRANES

The eggs of all animals, with the exception of some coelenterates, are enclosed by membranes which may differ in their origin and properties.

The innermost membrane is the **vitelline membrane.** Except for some coelenterates, this membrane is always present. The vitelline membrane is produced by the cytoplasm of the egg itself, and in the unfertilized eggs it adheres closely to the surface of the egg. The vitelline membrane is always very thin and transparent. It is built of a fibrous protein which has been identified as keratin. After fertilization the vitelline membrane becomes separated from the cytoplasm of the egg, and a fluid is secreted by the egg cell into the space between itself and the membrane. The space inside the vitelline membrane, which after fertilization becomes the fertilization membrane (see section 3–2), is termed the perivitelline space, and the fluid filling it is called the perivitelline fluid.

In many invertebrates (ctenophores, many worms, and echinoderms) the vitelline membrane is the only membrane present. Among the chordates this is also true of the Enteropneusta and of the Acrania. However, in many invertebrates and in all the higher vertebrates the vitelline membrane is supplemented by one or more additional membranes deposited on the outer surface of the egg. The insect egg is covered by a thick membrane, the **chorion** (Fig. 16), which consists of chitin and is secreted by the follicle cells surrounding the egg cell in the ovary. The amphibian egg is surrounded by a layer of jelly (Fig. 25), which protects the egg and sometimes serves to make the eggs adhere to one another and to submerged objects such as water plants. This jelly is secreted as the eggs pass through the oviducts. When the amphibian egg is deposited in water, the jelly absorbs water and swells. In the oviparous sharks and rays the egg is surrounded in the oviducts (in the special parts called the **shell glands**) by a hard shell of a complicated shape. The shell is drawn out into long twisted horns which serve to entangle the eggs among the seaweed. The eggs of mammals possess, besides the vitelline membrane, a thick glassy membrane which is secreted by the cells of the egg follicle and is called the **zona pellucida** (Fig. 26). When the mammalian egg escapes from the ovary it is also surrounded by a

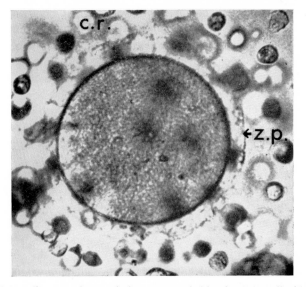

Fig. 26. Mammalian egg after ovulation, surrounded by the zona pellucida (*z.p.*) and the corona radiata (*c.r.*). (After Hamilton, 1943.)

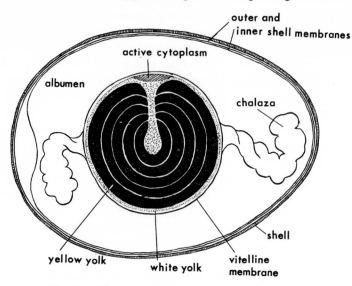

Fig. 27.   Diagram of hen's egg. (After Lillie, 1919b.)

layer of follicle cells, which is peeled off later as the egg descends the oviduct.

However the most complicated egg membranes are found in the eggs of birds, where no less than five membranes can be distinguished (Fig. 27). The innermost egg membrane is the vitelline membrane—a very thin membrane covering the surface of the yellow of the egg (which is the true egg cell). The next membrane is the white of the egg. Eighty-five per cent of the egg white is water; the rest is a mixture of several proteins, mostly albumens, which make up 94 per cent of the dry weight. Next to the egg white come two layers of **shell membranes.** These consist of fibers of keratin matted together. Over most of the surface of the egg the shell membranes are in contact with each other, but at the blunt end of the egg they are separated: the inner membrane adheres to the egg white, and the outer membrane adheres to the shell. In between there is a space filled with air. The outermost membrane is the **shell.** This consists chiefly of calcium carbonate ($CaCO_3$), about 5 gm. in a hen's egg. The shell is pierced by a great number of fine pores which are filled by an organic (protein) substance related to collagen. In an average hen's egg the pores have a diameter of 0.04 to 0.05 mm., and the total number is estimated at about 7000. The membranes of a bird's egg are secreted one after another as the egg proceeds down the oviduct. The whole process takes slightly longer than 24 hours. After the egg has been released from the ovary, it quickly passes into the oviduct and descends through the oviduct for about three hours, during which most of the egg white is secreted and envelops the egg cell. The lowest portion of the oviduct is widened and is termed the uterus. Here the egg remains for 20 to 24 hours, while the remainder of the egg white is secreted and eventually the shell membranes and the shell itself.

The membranes of a bird's egg serve not only for the protection of the egg cell; the egg white serves also as an additional source of nourishment and is gradually used up in the course of the development of the embryo.

2–7   THE DEVELOPING EGG AND THE ENVIRONMENT

In every stage of its development the embryo is a living organism, and like every living organism it requires foodstuffs for its vital processes. It assimilates its food and metabolizes organic substances to gain energy for its maintenance and for performing the processes together constituting its development. Furthermore, the new individual has to increase in size to a variable, but usually very considerable, degree until it approximates the structure of its parents. The growth process again involves the consumption and processing and partial combustion of a vast quantity of foodstuffs. Since the egg is a single cell and does not have any of the organs which an adult uses to procure and utilize its food supply, the supply and utilization of food for the developing embryo have to be organized along lines that are quite special and specific for the process of embryonic development.

The primary source of nutrition in the eggs of most animals is the reserve material stored in the egg cell during its development in the ovary (see section 2–4). This material is metabolized; it is partly broken down as a source of energy and partly transformed into the substances of which the various organs of the new individual are built up in the course of development. In addition to materials contained in the egg, various substances may be taken up from the environment. The nature and the amount of extraneous materials used by the embryo depend very largely on the environ-

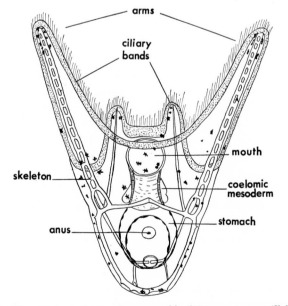

Fig. 28.   Pluteus—larva of a sea urchin (*Tripneustes gratilla*).

ment in which the embryo develops. A vast number of animals rather early attain a stage in which the young individual can start feeding and thus become self-supporting. This is often greatly facilitated by the egg's developing in the first instance into a **larva,** instead of directly into an adult.

The sea urchin egg may be taken as an example of a small egg with relatively little yolk. It develops in sea water and after a very short time (35 to 40 hours) produces a larva, called a **pluteus,** which has an alimentary canal of three parts: a foregut, opening to the exterior by a mouth, a dilated stomach and a hindgut with anus. The pluteus swims freely in the water with the aid of a ciliary band drawn out into loops along the edges of arms supported by a complicated calcareous skeleton. It feeds on minute planktonic organisms, mainly algae, which are injected into the gut by ciliary action of the stomodeum (Fig. 28). Table 1 gives an idea of the turnover of substances during the first 40 hours of development—up to the formation of the pluteus.

Table 1. Chemical Composition of Sea Urchin Egg and Embryo (in % of Wet Weight) (after Ephrussi and Rapkine, from Lehmann, 1945)

Substance	Unfertilized Egg	Blastula, 12 Hours	Pluteus
Water	77.3	77.3	78.8
Dry substances	22.7	22.7	21.2
Protein nitrogen	10.7	10.2	9.7
Total carbohydrates	5.43	5.46	3.4
Glycogen	5.13	4.8	traces.
Fat	4.82	4.43	3.69
Ash	0.34	2.07	3.56

Some points may be made from this table: The developing embryo takes up some water from the environment and also quite a considerable quantity of mineral substances which are mainly utilized in building up the calcareous skeleton. The embryo loses a large part of its carbohydrates (including almost all of the glycogen) and some fat. There is only a very slight decrease of nitrogen; according to some more recent investigations (Gustafson and Hjelte, 1951), the total nitrogen of the sea urchin egg does not change during the period up to the pluteus stage.

Water is absorbed as a general rule by embryos developing in an aquatic medium (see Table 3). The amount absorbed may be quite considerable, as can be seen in the case of an egg-laying dogfish *Scyllium canicula*. Additional ash (salts) is also taken up from surrounding sea water by marine invertebrates (coelenterates, molluscs, crustaceans and echinoderms). Among fishes the developing embryos of the elasmobranchs absorb salts, but the embryos of some teleost fishes, such as the Salmonidae, do not absorb salts from the surrounding medium although they take in water. The same is true of the developing embryos of the Amphibia. The animals whose eggs develop in fresh water cannot depend on the surrounding medium for the intake of salts because fresh water does not contain the necessary salts (especially Na and K ions) in sufficient quantity.

The turnover of substances in the developing egg of a frog will serve as an example of an animal developing in fresh water (see Table 2).

Table 2.   Chemical Composition of the Egg and Embryo of a Frog; Weights in Micrograms (after Barth and Barth, 1954, from various sources)

Substance	Egg	Tadpole soon after Hatching (Stage 20)
Total nitrogen	162	159
Extractable nitrogen (= nitrogen of active protoplasm)	40	42.8
Non-protein nitrogen (in nucleic acids ?)	4	5.0
Total carbohydrates	104	58
Glycogen	73.2	31

Table 2 shows that there is very little increase in the amount of active cytoplasm during the early development of the embryo, somewhat more in the amount of nucleic acid nitrogen (about 25 per cent), and that simultaneously there is a very considerable decrease of the total carbohydrates, due almost exclusively to the loss of glycogen, which is thus the main source of energy for the maintenance and the development of the embryo. The increase of water is considerable. It is not included in Table 2, but according to some available data it accounts for a 75 per cent increase of the total weight of the embryo during the corresponding stages. There is also some uptake of salts, particularly calcium, from the environment.

An entirely new situation is faced by animals that have abandoned the aquatic medium and become completely terrestrial. Some of these have attempted a compromise by returning to the water for egg laying. This is true of most of the amphibians. Other amphibians (many terrestrial frogs, some salamanders, and the Gymnophiona) lay their eggs on land but in damp places—in burrows underground, etc.—where the eggs can absorb the minimal quantities of water that are necessary for their development. Even among the reptiles there are some whose eggs take up water from the environment. This is the case in turtles. For instance, the turtle *Malaclemys centrata,* whose egg weighs 10.58 gm., absorbs 3.07 gm. of water during its development. This is made possible because the eggs are laid in damp sand.

In the eggs of other reptiles and of birds there is no longer a possibility for the absorption of water from the exterior. The egg membranes have become watertight. On the other hand, the loss of water from the egg by evaporation is reduced to a minimum. The only substance that is taken from without is the oxygen necessary for the oxidative processes in the egg. Otherwise, the egg has become a closed system, developing at the expense of the substances stored inside the egg itself. Such an egg, which has become self-sufficient (except for oxygen intake), is called a **cleidoic** egg (Needham, J., 1931). (Cleidoic means boxlike.)

Independently of the vertebrates, the terrestrial arthropods have also evolved cleidoic eggs. Some of the stages of this evolution can still be traced among contemporary insects. For instance, in the eggs of grasshoppers (*Melanoplus* and *Locusta*) there is an intake of water during development. In *Melanoplus* the water content increases from 2.5 mg. to 4.7 mg. In the eggs of other insects, however, no water can be taken in, and throughout the development of the eggs there is only a certain amount of water loss by evaporation. Such eggs are therefore as much cleidoic as the egg of a bird.

There is a fairly obvious advantage in the young individual's being more advanced in development and growth when it emerges from the egg. This advancement may be achieved by the amassing of greater and greater supplies of foodstuffs in the egg. Increase of food supplies in the egg may be correlated with the elimination of a larval stage—as in cephalopods, among the molluscs, and in elasmobranchs, reptiles and birds, in the vertebrate phylum.

The eggs of birds present an example of a very abundant supply of the egg with food material for the nourishment of the embryo. The chicken which hatches from the egg is already essentially a bird. The principal difference, besides the small size, lies in the development of the feathers. The body of a newly hatched chicken is covered with down, which is a simplified form of a feather. As soon as the down is replaced by typical feathers, the chicken acquires all the typical features of a bird.

A very special though widespread method of increasing the chances of survival of the offspring consists in retaining the eggs in the mother's body and letting them develop there to a greater or lesser degree. The eggs are usually retained in the oviducts (which are then usually referred to as **uteri**), but in some cyprinodont fishes (*Poecilia* and *Girardinus*) the eggs begin their development while still in the ovaries (see Needham, J., 1942). Sometimes, as in some salamanders, the young hatch from the eggs inside the mother's body. Occasionally, as in the adders, the eggs are laid intact, but the young begin to hatch as soon as the eggs are laid. This phenomenon is known as **ovoviviparity.**

In typical cases of ovoviviparity the embryo is nourished by the food stored in the egg. A next step in evolution may be made when the embryo absorbs some substances present in the fluids filling the oviducts. The degree to which such oviducal fluids are used for the nourishment of the embryo is very variable. In different species of the sharks and rays, all possible gradations may be found between such forms in which the embryo depends chiefly on the food supplied in the egg, and such forms in which the embryo depends chiefly on the food supplied by the mother. The proportions of these two sources of nourishment may be best estimated by comparing the weight of organic substances in the egg and in the newly born offspring. Table 3 presents a comparison of the turnover of materials in the eggs of an oviparous and an ovoviviparous fish (as well as a viviparous species).

Table 3.   Chemical Composition of the Eggs and of Later Stages of Development in Three Species of Dogfish (After Ranzi from Needham, J., 1942)

Species		Composition of Egg, and of Young at Hatching or Birth (Weight in Grams)	
		Egg	Young
Scyllium canicula	Total weight	1.3	2.7
(oviparous)	Organic substances	0.61	0.48
	Water	0.68	2.15
Mustelus vulgaris	Total weight	3.9	60.6
(ovoviviparous)	Organic substances	1.9	8.9
	Water	1.9	49.8
Mustelus laevis	Total weight	5.5	189.0
(viviparous)	Organic substances	2.8	32.0
	Water	2.6	152.0

It will be seen that in the development of an egg-laying fish the turnover of substances is very similar to that in the embryo of the sea urchin or frog: there is a decrease in organic substances, due to their combustion, and an increase of water content which makes the young at hatching exceed the weight of the egg. In the ovoviviparous fish, however, the uptake of organic substances from the maternal body not only compensates for their loss through combustion but brings about a total increase of the amount of organic material during development. There is also a very considerable intake of water, so that the over-all increase in weight of the embryo is many times that observed in a species in which the embryo depends on the egg reserves as the sole source of its organic materials. Ovoviviparity in several groups of animals seems to have been a transitional phase in the development of true **viviparity.**

True viviparity is shown when the embryo establishes a direct connection with the maternal body, so that the nutrition can pass from the mother to the embryo without the intermediate state of being dissolved in the uterine fluid. The connection is established through a special organ, the **placenta,** which is an outgrowth of the embryo. Viviparity and placentae have been developed in several groups of the animal kingdom independently of each other. Placentae are found in the protracheates (*Peripatus*), in the tunicates (*Salpa*), in several elasmobranchs and in the placental mammals. The mode of origin and the structure of the placenta is different in each case mentioned. The placenta of the mammals will be described later, in section 7–3.

The supply of the nutrition to the embryo through a placenta appears to be a very much more efficient method than the absorption of nourishment from the uterine fluids. This fact may be illustrated by the third example in Table 3, that of the dogfish *Mustelus laevis,* in which a placenta is developed when the yolk sac of the embryo becomes connected to the walls of the uterus (see Chapter 7). In this particular case the increase in organic materials during intra-uterine life is more than tenfold as compared with the roughly fourfold increase of organic substances in *Mustelus vulgaris,* taken as an example of the ovoviviparous fish.

In mammals, which have developed the placenta to a degree of perfection not found in other animals, the increase in weight of the embryo is very much greater, especially as the eggs of mammals are very small as compared with the eggs of viviparous fishes.

CHAPTER **3**

Fertilization[*]

Fertilization is the fusion of two **gametes** (sex cells), a male and a female one, followed by the joining together of the nuclei of the two gametes. The fusion of the gametes in the Metazoa activates the egg so that it starts to develop, and the joining together of the nuclei of the egg and the spermatozoon results in the endowment of all the cells of the developing new organism with carriers of hereditary properties derived from the maternal and paternal organisms.

It should be pointed out here that the place of fertilization in the life cycle of various organisms is not always the same: although in the Metazoa and Metaphyta the fertilization starts the egg on its way of development into an embryo, in many protozoans the zygote enters into a dormant state, which may be interrupted later, usually by changes in the environmental conditions. Similarly, the fusion of the nuclei of the two gametes is deferred

* Further references: Loeb, 1913; Lillie, 1919; Rothschild, 1951 and 1956; Tyler and others, 1957.

51

in some organisms (especially fungi), and occurs many cell generations later. There are thus two essentially independent aspects to fertilization: one is the **activation** of the egg, the other is the intermingling of the paternal and maternal hereditary characters in the offspring. The latter aspect of fertilization is also known as **amphimixis** and really falls within the sphere of the science of genetics.

We shall now consider the consecutive steps which lead to the fertilization of the egg and the changes in the egg occurring during and after fertilization, laying the main stress on those aspects of fertilization which contribute to the transformation of the egg into a new organism.

3–1   APPROACH OF THE SPERMATOZOON TO THE EGG

After both the eggs and the spermatozoa are discharged into the surrounding (aquatic) medium, or after the spermatozoa have been introduced into the genital ducts of the female, in the case of internal fertilization, the first step is the encounter of the spermatozoon and the egg. This encounter is brought about by the swimming movements of the spermatozoa. It would have been tempting to suggest that the spermatozoa are attracted by the eggs, or that their movement in space is somehow directed toward the eggs. In spite of numerous investigations, no such attraction or directing of the path of the spermatozoa of animals has been proved. The movements of the spermatozoa are entirely at random, and the spermatozoa collide with the eggs as a matter of pure chance. That this chance encounter occurs regularly in nature is partly a result of the enormous number of spermatozoa produced by the male gonads, and partly the result of the egg's being a relatively very large target, so that it can be hit fairly easily.

A very fine chemical mechanism now comes into play. In the presence of ripe eggs, or even in the water in which ripe eggs of the same species have been lying for some time ("egg water"), the spermatozoa become "sticky" and adhere to the surface of the egg or its membrane and even to each other. The mutual adhesion of the spermatozoa results in their clumping or **agglutination.** The agglutination is more easily observed with sperm of some animals (especially sea urchins) than with sperm of others (for instance starfish), and it also depends to a large extent on environmental conditions. But the sticking of sperm to eggs of the same species is found in most animals, and possibly it is always present. The change in the properties of the spermatozoa in egg water is due to a substance which dissolves into the surrounding water either from the egg itself, or from the egg membranes.

This substance has been studied in detail by F. R. Lillie (1919a), who called it **fertilizin.** Fertilizin is probably always present in the surface layer of the egg and may also be contained in the egg membranes. The layer of jelly surrounding the ripe eggs, outside the vitelline membrane, of sea urchins and other echinoderms consists wholly of fertilizin. (It is not definitely known whether in this case the jelly is produced by the egg itself

or by the follicle cells surrounding the egg in the ovary.) Chemically, fertilizin is a glycoprotein or mucopolysaccharide. As a protein it contains a
number of amino acids, and as a polysaccharide it includes molecules of
one or more monosaccharides. The monosaccharides (glucose, fucose,
fructose or galactose) are esterified by sulfuric acid, as shown in the
formula (Runnström, 1952):

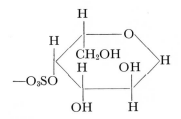

   Both the amino acids and the monosaccharides vary from one species
to another, so that it is more correct to speak of **fertilizins** rather than of
one fertilizin found in different animals. The molecules of the fertilizins
are quite large—the molecular weight is about 300,000—and each molecule
may have more than one "active group" by which it may become attached
to the surface of the spermatozoon (see Metz, 1957, and Rothschild, 1956).
The surface layer of the cytoplasm of the spermatozoon contains another
substance (or substances) known as the **antifertilizin** (or antifertilizins).
The antifertilizins can be extracted from the spermatozoa by heating, freezing and thawing or acidifying the water. Their properties show that they
are acid proteins with a fairly small molecule—molecular weight about
10,000. The remarkable peculiarity of the fertilizins and antifertilizins is
that they combine in a specific way; that is, the egg fertilizin of any species
reacts best with the sperm antifertilizin of the same species. Reactions with
other species, although possible, are very much weaker, and even so they
occur only when two species are fairly nearly related to one another. The
reaction between fertilizin and antifertilizin is thus very similar to reactions
which take place between the substances in the serum of immunized animals
(antibodies) and the foreign substances which caused the immunization
(antigens) (Tyler, 1948). In both cases the bond between the two comple-
mentary substances depends on the spatial arrangement of the atoms on
certain parts of the molecules in such a way that the shape of the surface of one molecule fits closely onto the surface of the other. The
two bodies correspond to one another as a template to the model. Another
often-used analogy is the correspondence of the key to the lock. When the
latter analogy is used, it may be pertinent to remember that sometimes a
lock may be opened by a wrong key if it is sufficiently similar to the right
one. This finds its counterpart in the cross reactions of fertilizins and antifertilizins of sufficiently nearly related species.
   The reaction between the fertilizin and antifertilizin molecules accounts
for both the agglutination of the spermatozoa by egg water and the adhesion

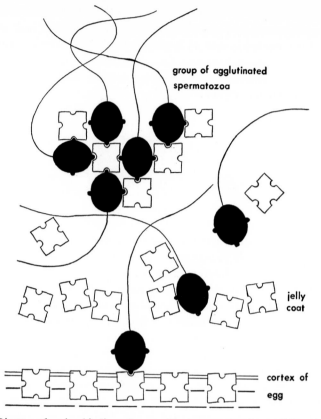

Fig. 29. Diagram showing binding of sea urchin spermatozoa by particles of fertilizin.
(Modified after Rothschild, 1951.)

of the spermatozoa to the egg surface or the egg membranes. The reason
for the agglutination of the spermatozoa is that the fertilizin particles have
more than one reactive spot on their surface so that one fertilizin particle
may become attached to two or more spermatozoa, thus binding them
together (Fig. 29). An adhesion of a spermatozoon to the surface of the
egg or to the egg membrane is the result of the establishment of a bond
between the antifertilizin molecule or molecules on the surface of the
spermatozoon, and the fertilizin molecules embedded in the surface layer
of the cytoplasm or the egg membrane. It will be recalled that in the sea
urchin the ripe egg is surrounded with a jelly membrane consisting of
fertilizin molecules!

We have seen (section 2–6) that the surface of the ripe egg is very
seldom naked (as in coelenterates), but usually surrounded by membranes,
or follicle cells or both (as in mammals). The spermatozoon has to penetrate through these before it can reach the egg.

The mechanism of penetration is chemical. The spermatozoon produces
substances of an enzymatic nature, known under the general name of sperm

lysins, which dissolve the egg membranes locally and make the path clear for the spermatozoon to enter the perivitelline space and reach the egg. In mammals the ovulated ripe egg is surrounded by the follicle cells which in the ovary were nearest to the egg, and at the time of fertilization they surround the egg as the **corona radiata** (see pp. 43–44). The spermatozoon penetrates through the corona with the aid of an enzyme, **hyaluronidase,** which is produced by it and which dissolves the mucopolysaccharide **hyaluronic acid,** which is the substance cementing the follicle cells together. In this way the spermatozoid penetrates between the cells of the corona radiata. It also makes a narrow canal through the zona pellucida, which lies inside the corona radiata, by dissolving the substance of the zona, and so reaches the surface of the egg. Extracts containing large amounts of hyaluronidase may remove all the follicle cells from the surface of the egg, but this does not occur during normal fertilization; the corona persists for some time after the penetration of the spermatozoon into the egg. It is dispersed later, before the implantation of the egg, but this is not due to the action of the spermatozoa.

The sperm lysins are produced presumably by the acrosome of the spermatozoon. This is in accord with the origin of the acrosome; we have seen that the acrosome is developed from the Golgi bodies of the spermatid. There is clear evidence that in a normal animal cell at least some of the secretory activity is performed by the Golgi bodies. The production of sperm lysins is thus one of the aspects of the secretory activity of these bodies.

The sperm lysins certainly differ from one animal group to another. In some cases the dissolution of the egg membranes may be brought about by simpler means. Thus it is believed that the jelly coat of echinoderm eggs

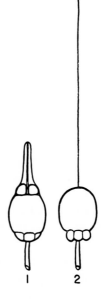

Fig. 30. Change in the structure of a mollusc (*Mytilus edulis*) spermatozoon caused by "egg water." Left, untreated spermatozoon; right, "activated" spermatozoon with acrosomal filament. (From Colwin and Colwin, 1957.)

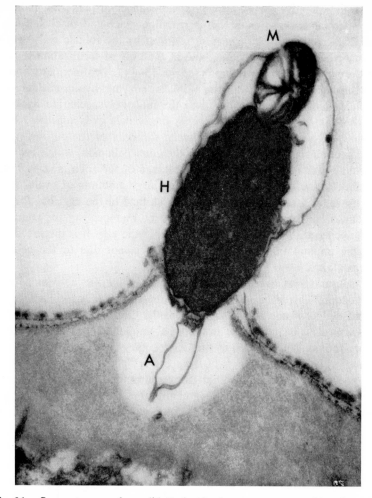

Fig. 31. Spermatozoon of annelid *Hydroides hexagonus,* approaching the egg and penetrating the vitelline membrane. *A,* Acrosomal filament; *H,* head of spermatozoon; *M,* middle piece of spermatozoon. (From Colwin, Colwin and Philpott, 1957.)

may be dissolved as a result of the acidification of sea water by the carbon dioxide developed by the spermatozoa in the course of their respiration.

In the case of eggs with very thick and resistant membranes, such as the chorion of fishes or insects, the sperm cannot reach the egg at any point but must penetrate through a special canal (the micropyle), or canals, left in the egg membrane (see section 2–2).

The agglutination of the spermatozoa is not the only change caused in them by the egg water or the proximity of ripe eggs. Under the influence of substances diffusing from the eggs (whether the substance in question is fertilizin, or some other compound, has not yet been determined) the structure of the spermatozoon becomes changed, as can best be seen by using the electron microscope. The main change concerns the acrosome, the peripheral part of which collapses and becomes extruded and partially dissolved in water (possibly releasing the lysins which were referred to earlier). The central part of the acrosome, which appears rather transparent in electron micrographs, elongates and becomes transformed into a long (1 to 75 μ) thin filament, or possibly a tube, which has considerable rigidity and protrudes forward from the head of the spermatozoon (Fig. 30). This structure has been called the **acrosomal filament** (Dan and Wada, 1955). The spermatozoa which have developed acrosomal filaments are considered to have been activated and are then ready to penetrate into the egg. As the spermatozoon approaches the egg, the acrosomal filament is pushed through the jelly (where present) and through the vitelline membrane, the pathway for it being cleared by the action of the lysins, as explained above (Fig. 31). Eventually, the tip of the acrosomal filament touches the surface membrane of the egg cytoplasm and triggers off the next phase of the process of fertilization.

3–2   THE REACTION OF THE EGG

Immediately after the acrosomal filament of the spermatozoon touches the surface of the egg, the cytoplasm of the egg bulges forward at the point of

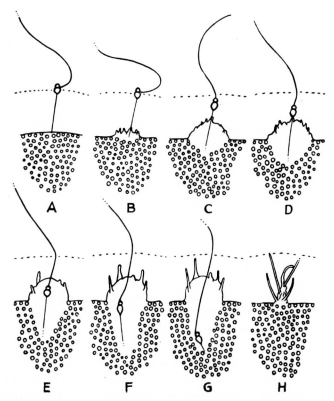

Fig. 32.   Sperm entry into the egg of the sea cucumber *Holothuria atra*, with formation of fertilization cone by the cytoplasm of the egg. (From Colwin and Colwin, 1957.)

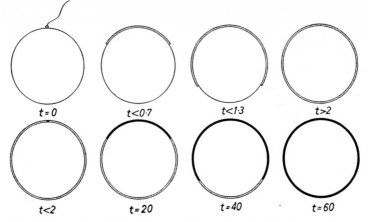

Fig. 33. Propagation of the primary and secondary fertilization reaction over the surface of a sea urchin egg; $t =$ time in seconds after entry of sperm. (From Rothschild, 1956.)

contact, producing a process of hyaline cytoplasm, the **fertilization cone.** The fertilization cone may be in the form of a more or less simple conical protrusion (Fig. 32), or it may consist of several irregular pseudopodia-like processes, or in some cases it may take the form of a cytoplasmic cylinder stretching forward along the acrosomal filament (Colwin and Colwin, 1957). Whatever its shape, the fertilization cone gradually engulfs the spermatozoon and then begins to retract, carrying the spermatozoon inward, surrounded by the hyaline cytoplasm of the fertilization cone.

Even before the spermatozoon thus penetrates into the interior of the egg, the egg shows signs that it has become profoundly changed, that it has been **activated.** By this statement we mean that the egg is now capable of starting on its way to develop into an embryo. The first observable changes in the egg, however, bear only an indirect relation to the formation of the embryo. They are: (1) a change in the surface of the egg cytoplasm which prevents the entry of further spermatozoa, and (2) the elevation of a **fertilization membrane.**

Both changes start at the point of penetration of the spermatozoid and gradually spread over the whole surface of the egg (Fig. 33). The first is the more rapid, and in sea urchins it has been found that it spreads over the whole surface of the egg in under two seconds. After this the receptivity of the egg to penetration by a spermatozoon is reduced to about 1/20 of what it had been before contact with the first spermatozoon (Rothschild, 1956). Nevertheless, after this first phase of the reaction the egg is not totally resistant to the entry of further spermatozoa. The complete "sealing off" of the egg surface to spermatozoa is achieved during the second phase of the reaction, which is much slower and reaches the whole of the surface of the egg about one minute after the contact with the first spermatozoon. Although the first reaction does not seem to have a clearly established morphological counterpart, the second or slower reaction can be traced by numerous changes in the cytoplasm of the egg, and especially in the cortical granules mentioned in section 2–5. These start swelling rapidly and then "explode," ejecting their contents onto the surface of the egg cytoplasm into the narrow space between the cell membrane and the vitelline membrane. Part of the contents of the cortical granules adheres to the inner surface of the vitelline membrane and fuses with it, transforming the vitelline membrane into a very much stronger **fertilization membrane.** In animals whose eggs possess a chorion instead of a vitelline membrane, the chorion is also reinforced by the substance of the cortical granules and becomes harder, or, as is sometimes said, becomes "tanned." In the sea urchins a part of the material of the cortical granules which does not enter into the formation of the fertilization membrane remains near the surface of the egg cytoplasm and envelops this surface by a thin **hyaline layer.**

The fertilization membrane becomes lifted from the surface of the egg, and a fluid-filled space appears between them—the **perivitelline space.** The fluid in the perivitelline space probably accumulates because of the osmotic action of some hydrophilic substances released into the perivitelline space and presumably also derived from the extruded cortical granules. The water may be absorbed from the surrounding medium or from the egg cytoplasm itself. The latter event would lead to a shrinkage of the egg cytoplasm—a fairly widespread phenomenon observed in particular in the fertilized eggs of mammals.

As a result of the swift propagation of the first surface reaction of the egg, followed by the second, slower cortical reaction and the elevation of the fertilization membrane, only one spermatozoon penetrates into the egg in the greater part of the subdivisions of the animal kingdom (in particular in coelenterates, annelids, echinoderms, bony fishes, frogs and mammals). If more than one spermatozoon enters the egg in these animals, due to a very high concentration of the spermatozoa around the egg, or to a retarded reaction of the egg after the first spermatozoon had made contact, the development is inevitably abnormal and the embryo not viable (**pathological polyspermy**). In other groups, mainly those having yolky eggs, such as some molluscs, selachians, urodeles, reptiles and birds, several spermatozoa enter the egg as a rule **(physiological polyspermy),** but of these only one participates fully in the development of the embryo; the rest degenerate sooner or later, although in some cases (as in birds and reptiles) the nuclei of the accessory spermatozoa may undergo several abortive divisions in the cytoplasm of the egg.

3–3 THE ESSENCE OF ACTIVATION AND PARTHENOGENESIS

If the condition of the egg before fertilization is compared with that after fertilization, it becomes clear that the system is brought from a quasi stationary state to a condition characterized by a series of rapid changes. If it is not fertilized, the ripe egg remains quiescent for some time and eventually becomes subjected to degenerative processes, and in the end it becomes necrotic. If the egg is fertilized, it goes into action, as it were: the reduction divisions, if not performed before, are brought to completion; the male and female pronuclei fuse; complicated dislocations of cytoplasmic substances of the egg may take place (see section 3–5); and the egg enters into a period of rapid divisions (cleavage).

What is the physiological mechanism of this activation of the egg, and how does the spermatozoon produce the profound change in the egg's condition? The answer to this question has been sought by comparing the processes of metabolism and other physicochemical properties of the unfertilized and the fertilized eggs. The results of these studies have been far from satisfactory. In 1908 Warburg measured the oxygen consumption of unfertilized and fertilized sea urchin eggs and found that immediately after fertilization there was a sharp increase of oxygen consumption of up to 600 per cent as compared with unfertilized eggs. It would have been tempting to suggest that the primary action of the spermatozoon consists in release or activation of the oxidative enzymes of the egg, and that the ensuing increase of oxidation provides the energy necessary for the performance of the other changes in the egg, as outlined above, and for the development of the egg in general. It was soon discovered, however, that the increase of oxidation rate in eggs is by no means a general rule: although the oxygen uptake increases in sea urchins and some annelids (*Nereis*), it does not change appreciably in starfish and amphibians, and it actually decreases after fertilization in the eggs of the mollusc *Cumingia* and the annelid *Chaetopterus* (Needham, 1942, and Rothschild, 1956).

Obviously, the increase of oxidation rate cannot be considered as the "key reaction" of fertilization. Other metabolic changes found after fertilization in some, though by no means all, animals are: (1) production of considerable amounts of acid during first several minutes after fertilization, (2) increase of permeability of the egg membrane, (3) slight shrinkage of the egg (decrease in volume), which has been mentioned earlier, (4) increased exchange of phosphate with the surrounding medium (it was found that the rate at which labeled radioactive phosphorus, P^{32}, was taken into the fertilized eggs was in some experiments up to 160 times greater than in the case of unfertilized eggs), (5) acceleration (up to 16 times) of potassium exchange between the egg and the environment, (6) diffusion of calcium out of the eggs, (7) increase in the activity of proteolytic enzymes, and (8) changes in the birefringence of the cortical layer of the egg cytoplasm, which may be interpreted to mean that the predominantly radial arrangement of the structural constituents of the cortex found in the unfertilized egg is replaced by a tangential arrangement of the cytoplasmic micellae soon after fertilization.

The chemical changes in the egg following fertilization indicate in general a trend toward a higher rate of metabolism, but this hardly adds much to already existing knowledge that the egg is "activated," and it does not give any clue as to how the spermatozoon achieves this change.

A rather logical hypothesis would be that the action of the spermatozoon rests on the introduction of some special substance into the egg which triggers off all the other reactions, of which the changes in the metabolism of the egg are visible expressions. In view of the importance of the ferti-lizin-antifertilizin reaction for the entry of the spermatozoon into the egg, it has been suggested by Lillie (1919a) that the activation of the egg is also due to the action of the same substances. The combining of the fertilizin of the egg with the antifertilizin of the sperm would then be the beginning of the changes in the physicochemical system of the egg—the key reaction of fertilization.

We shall not go into the further assumptions of the theory, as in any case it cannot be accepted for quite a number of reasons. One reason is that no plausible explanation has been given as to how the binding of fertilizin and antifertilizin is connected with the other processes involved in the activation of the egg. Antifertilizin, as we have seen, can be extracted from the spermatozoa and added to the water surrounding ripe eggs. The presence of the antifertilizin can be ascertained by the sticking together of the eggs—an exact counterpart of the agglutination of the spermatozoa by the fertilizin contained in egg water—but there is observed no trace of activation of the egg by this treatment (Metz, 1957).

A different approach to the solution of the problem of fertilization lies in attempting to imitate the action of the spermatozoon by some known agent. Now it has long been known (see p. 13) that in some animals an egg can develop without fertilization, as in the aphids, phyllopods and rotifers at some times of the year, or as in bees and wasps in which a fer-tilized egg produces a female individual and an unfertilized egg develops into a male. These are cases of natural **parthenogenesis** (virginal repro-duction). In other animals, such as most of the echinoderms and many others, the eggs under natural conditions do not develop unless they are fertilized. It has been found, however, that certain treatments of the ripe eggs may incite them to develop, and this phenomenon is known as **artificial parthenogenesis.**

A great amount of work on artificial parthenogenesis has been done with sea urchin eggs. O. Hertwig and R. Hertwig discovered that ripe sea urchin eggs may be caused to start developing by treatment with chloroform or strychnine. Later it was found that the same and even better results may be obtained by treating the eggs with a variety of substances: hypertonic or hypotonic sea water; various salts, such as the chlorides of potassium, sodium, calcium, magnesium, etc.; the weak organic acids—butyric acid, lactic acid, oleic and other fatty acids; the fat solvents—toluene, ether, alcohol, benzene and acetone; and urea and sucrose. Similar results are obtained by temperature shocks—that is, by transferring the eggs for a short time to warm (32° C.) or cold (0° to 10° C.) water; by electric induc-tion shocks; by ultraviolet light; and even by shaking the eggs in ordinary sea water (see Loeb, 1913, and Harvey, 1956). This long list of agents (it is by no means exhaustive) clearly shows that there is no one agent which can be recognized as the specific cause of the activation of the egg, the cause which determines the nature of the processes that are to take place. Obviously, factors determining the nature of the reaction of the egg are contained in the egg itself. The agents causing the artificial parthenogenesis of the egg are instrumental only as factors triggering off this reaction of the egg, and if this is so it becomes plausible that the spermatozoon itself exercises only a similar triggering action.

Something further may be deduced, however, from the array of agents causing artificial parthenogenesis. Most of the agents used are of such a nature that they **damage** the cells to a greater or lesser degree, and if applied in greater intensity, or for a longer time, they can cause the death of the cells. It is reasonable to suppose, then, that activation of the egg involves some type of sublethal damage to the egg cytoplasm. Furthermore, some of the agents used, such as fat solvents, may be expected to affect primarily the cell surface, which is known to consist partly of lipoids; and all other chemical agents may likewise primarily affect the surface of the cytoplasm, which would also not be immune from the action of physical agents.

We may therefore take one step further and suggest that the activation of the egg is connected somehow with a damage to the cortical layer of the egg cell cytoplasm. This is the more likely because (1) the reaction of the egg is started as soon as the spermatozoon comes in touch with the surface of the egg, and (2) the first detectable reactions of the egg occur in the cortical layer, and actually involve the breakdown of parts of the cortex (the cortical granules). It must be recognized, on the other hand, that even if it were proved that sublethal damage to the egg cortex is essential for activation, it would not yet be explained how this damage causes the other progressive changes in the egg. One must agree that "the activation of the egg cannot as yet be described completely in terms of the interaction of known specific egg and sperm substances." (Metz, 1957, p. 59.)

It will be interesting to conclude this section by adding some information on natural and artificial parthenogenesis in some vertebrates.

In frogs artificial parthenogenesis may be achieved by some of the methods used on echinoderm eggs, such as the use of hypertonic and hypotonic solutions, some poisons (corrosive sublimate) and electric shock. The activation achieved by these methods is incomplete, however, and the development does not go beyond abortive cleavage. A more efficient method is pricking the eggs with a fine glass needle (a method also used successfully with echinoderm eggs). However, for the full success of this method it is necessary that the needle be smeared with blood or be contaminated by cells or cellular particles from other tissues. If particles from foreign cells are thus introduced into the ripe egg cell, the cleavage is greatly improved, and a small percentage of the treated eggs may go through the whole development apparently quite normally (Bataillon, 1910). Instead of pricking the eggs with a contaminated needle, constituent parts of cells may be also introduced into the egg with a micropipette. In this way it has been possible to investigate what fractions of cell constituents are most active in causing parthenogenesis, and it has been found that "large granules" of a centrifuged tissue homogenate (this fraction includes the mitochondria) have the strongest effects, while the liquid supernatant had no more effect than did the pricking with a clean needle or a needle wetted with a buffer saline solution (Shaver, 1953). (See Table 4.)

Table 4.   Activation of Frog Eggs by Injecting Them with Tissue (Testis) Extract and Various Fractions of the Same Obtained by Centrifugation. (From Shaver, 1953.)

Preparation	% of Eggs Reaching Blastula Stage
Whole tissue extract	9.7
Large granules	35.2
Medium and small granules	19.5
Supernatant	0
Phosphate buffer	0.3
Pricked "dry"	0

So far no successful experiments have been reported in producing artificial parthenogenesis in birds, but in connection with other work it may be of interest to note that spontaneous parthenogenetic development has been observed in one representative of the class, namely in the domestic turkey. Eggs laid by females which had been isolated from males often start cleaving. The formation of germ layers similar to those in the extraembryonic parts in normal development has been recorded, but not the formation of an embryo proper (Olsen and Marsden, 1954).

In mammals the possibility of artificial parthenogenesis was discovered in connection with experiments cultivating unfertilized eggs *in vitro* which had been collected from the fallopian tubes. Extensive investigations in this field were undertaken by Pincus and his collaborators. Most of the experiments were done with rabbits' eggs (Pincus, 1936). It was found that if the eggs are simply kept for up to 48 hours in the ordinary tissue culture medium (blood plasma plus embryo extract) some of them become activated. The first sign of activation in this case is the completion of the second meiotic division and extrusion of a second polar body. Some of the eggs go even beyond that and start cleaving. Chemical treatment (with butyric acid, which has been used for activating sea urchin eggs) does not seem to give any better results. However, a temperature shock, in particular a treatment with cold, gives much better results.

To allow the activated eggs to develop further, they have been in some experiments injected into the fallopian tubes of rabbit does, which were made "pseudopregnant" by mating with a sterile buck or by injection of hormones (the luteinizing hormone). In the body of the female the development progresses further, and quite a number of embryos reached the blastocyst stage (18 per cent in one experiment in which the eggs were activated by cooling for 24 hours at 10° C.) (Chang, 1954). In two cases where the fertilization of the eggs by spermatozoa seems to have been completely excluded, the parthenogenetic embryos completed intra-uterine development, and one of the young was born alive (the other was stillborn) (Pincus, 1939). In another case the eggs were given a cold shock *in vivo* by opening the body cavity of a rabbit doe which had unfertilized eggs in its fallopian tubes, and cooling the fallopian tubes with cold water. One live young was born (Pincus and Shapiro, 1940). So far these are the only records of living mammalian young produced by parthenogenesis, and in view of the importance of the results it would be highly desirable that the experiments should be corroborated independently by other workers in this field. (See also Beatty, 1957.)

3–4   THE SPERMATOZOID IN THE EGG INTERIOR

There is some variation in different animals as to how much of the spermatozoon is taken into the interior of the egg. In many animals, notably in the mammals, the whole of the spermatozoon head, middle piece and tail penetrates into the cytoplasm and for a short time may be seen lying intact in the interior of the egg. In some animals (echinoderms), however, the tail of the spermatozoon breaks off and is left outside the vitelline membrane, and even the middle piece of the spermatozoon may be left without, so that only the head and the centrosome enter the egg (*Nereis*). That the tail of the spermatozoon often does not enter the egg gives additional proof that its functions are purely locomotive.

The information concerning the middle piece of the spermatozoon is not unequivocal. On the one hand, the middle piece appears to enter the egg cytoplasm in most cases; on the other hand, there is no definite proof that any constituents of the spermatozoon except for the nucleus and the centrosome play any active part in subsequent development. The mitochondria contained in the middle piece have been observed in some cases to scatter in the cytoplasm of the egg, but it is not known how long they maintain their existence there.

The subsequent behavior of the spermatozoon, or rather its head, is dependent on the stage of maturation (reduction divisions) which the egg has reached at the time of fertilization. In the sea urchins the eggs are shed and become fertilized after the reduction divisions have been completed and both polar bodies have been extruded from the egg. This is, however, by no means a general occurrence. In vertebrates the rule is that the egg completes its first reduction division in the ovary and reaches the metaphase stage of the second meiotic division. In this stage all further progress is arrested, ovulation takes place and the eggs may become fertilized. The second reduction division is completed and the second polar body extruded only if the egg is fertilized by a spermatozoon or activated in some other way (see pp. 62–63). In ascidians, some molluscs and annelids the egg reaches only the metaphase of the first meiotic division when it becomes ripe and is fertilized. Only then does the egg complete the first reduction division and carry out the second. Lastly, in some annelids, in nematodes and in chaetognaths the eggs are fertilized even before the beginning of meiotic division, while the oocyte nucleus is still intact.

It follows that although in animals like the sea urchins the spermatozoon nucleus may immediately proceed to join the egg nucleus, in other cases the immediate effect of the fertilization, so far as the nuclear apparatus is concerned, is the completion of the meiotic divisions, only after which may the fusion of the male and female pronuclei take place.

When the spermatozoon first penetrates into the egg cytoplasm it moves with the acrosome (or acrosomal filament) at its front. The nucleus and the centrosome are arrayed behind the acrosome in that order. Soon, however, a rotation of the nucleus and the centrosome can be observed in many, though not in all, animals, the centrosome coming ahead of the nucleus, and the nucleus turning through 180°, so that its original posterior end turns forward. The other parts of the spermatozoon, if still discernible by this stage, lose connection with the nucleus and the centrosome. Both the nucleus and the centrosome change their appearance. The nucleus, which is now referred to as the **male pronucleus,** starts swelling, and the chromatin, which is very closely packed in the spermatozoon, becomes again finely granular. By imbibition of fluid from the surrounding cytoplasm the pronucleus becomes vesicular. The centrosome at the same time becomes surrounded by an aster, similar to that of the centrosome in the early stages of an ordinary mitosis. While these changes are going on, the sperm nucleus, together with the centrosome, moves through the egg cytoplasm toward the area where the fusion with the nucleus of the egg, the **female pronucleus,** is to take place. This area is generally near the center in holoblastic eggs having a fairly small amount of yolk, but in telolecithal eggs it is in the center of the active cytoplasm at the animal pole of the egg. As the sperm head moves inwards it may be accompanied by some cortical and subcortical cytoplasm. If the latter is heavily pigmented, as in amphibian eggs, the trajectory of the sperm head may be marked by pigment granules trailing along its path. This is sometimes referred to as the penetration or copulation path (Fig. 34).

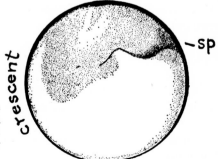

Fig. 34. Penetration path of the spermatozoon (*sp*), seen in a section through a fertilized egg of a frog, and its relation to the gray crescent. (After Schultze, from Morgan, 1927.)

The female pronucleus also has to traverse a greater or lesser way before it reaches the male pronucleus. At the beginning of its migration the female pronucleus is invariably at the surface of the egg, where the second meiotic division has been taking place, since it is only after the completion of the meiotic division that the nucleus of the egg may fuse with the nucleus of the spermatozoon. The haploid nucleus of the egg, after the completion of the second meiotic division, is often in the form of several vesicles known as the **karyomeres.** These fuse together to form the female pronucleus, which swells and increases in volume as it approaches the male pronucleus. In the last stage before they meet, the male and female pronucleus may become indistinguishable.

The actual fusion of the male and female pronuclei may differ in detail. In some animals the two pronuclei actually fuse together; that is, the nuclear membranes become broken at the point of contact and the contents of the nuclei unite into one mass surrounded by a common nuclear membrane. At the approach of the first cleavage of the fertilized egg of sea urchins and vertebrates the nuclear membrane dissolves and the chromosomes of maternal and paternal origin become arranged on the equator of the achromatic spindle. In other cases, however, the male and female pronuclei do not fuse as such, but the nuclear membranes in both dissolve and the chromosomes become released. In the meantime the centrosome has divided in two and an achromatic spindle has been formed to which the chromosomes derived from the male and the female pronucleus become attached. Only after the completion of the first division of the fertilized egg do the paternal and maternal chromosomes become enclosed by common nuclear membranes in the nuclei of the two daughter cells into which the egg has become divided (*Ascaris,* some molluscs, and annelids).

In both types the chromosomes of the maternal and paternal set retain, of course, their individuality. Lastly, in some animals, of which the copepod *Cyclops* is a well-known example, the paternal and maternal nuclear components remain separate even after cleavage has started, so that each blastomere has a double nucleus consisting of two parts lying side by side, but each surrounded by its own nuclear membrane.

A closer union of the homologous chromosomes takes place much later in preparation for meiosis in the gonads of the new individual and also in cases of somatic conjugation of chromosomes, as in the salivary gland chromosomes of Drosophila.

3–5   CHANGES IN THE ORGANIZATION OF THE EGG CYTOPLASM CAUSED BY FERTILIZATION

In addition to activating the egg and providing an opportunity for amphimixis, the penetration of the spermatozoon into the egg (or the parthenogenetic activation of the egg) causes in many animals, perhaps in all, far-reaching displacements of the cytoplasmic constituents of the egg. As a result of this, the distribution of various cytoplasmic substances and inclusions in the egg

at the beginning of cleavage may be very considerably different in arrangement from that in unfertilized egg, and even qualitatively new areas may
sometimes appear. It will be evident later that these changes in the organization of the egg at fertilization may be of a most profound importance for
further development of the fertilized egg.

One result of the extrusion of the cortical granules is that a large part of
the original outer egg cell surface becomes replaced by the inner surfaces
which surrounded the cortical granules and now are everted onto the exterior
(Afzelius, 1956). In view of the importance of the cell surfaces for morphogenetic processes (see further, sections 4–7 and 5–8), this change in the
composition of the surface layer of the egg may have a considerable significance, though at present it has not been actually shown in what degree the
changes in the physiological properties of the fertilized egg, as compared
with the egg before fertilization, are due to the replacement of part of its
outer surface.

Displacements of cytoplasmic substances may be observed best in those
eggs in which easily distinguishable kinds of cytoplasm have been evolved
during the growth and maturation of the oocyte.

We have seen that in the mature egg of the ascidian *Cynthia partita* three
distinguishable types of cytoplasm are already present. The yolky cytoplasm
fills most of the interior of the egg: this cytoplasm is slaty gray in the living
egg. Toward the animal pole lies a mass of clear cytoplasm which is derived
largely from the nuclear sap released when the nuclear membrane dissolves
in preparation for the first reduction division. Lastly, the surface of the egg
is covered by a layer of cortical cytoplasm containing yellow granules. The

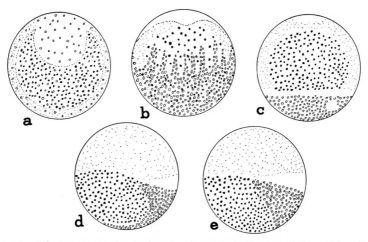

Fig. 35. Displacements of cytoplasmic substances in the egg of the ascidian *Cynthia
partita* following fertilization. *a,* Ripe unfertilized egg; *b,* egg immediately after entry
of spermatozoon; *c, d,* two successive stages of the egg at the time when ♂ pronucleus
moves to meet ♀ pronucleus; *e,* egg just before first division. Yolk granules are represented by black dots, yellow granules (mitochondria) by circles. (After Conklin, from
Huxley and de Beer, 1934, and Morgan, 1927.)

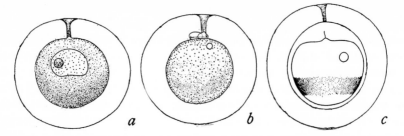

Fig. 36. Displacement of pigment in the egg of the sea urchin *Paracentrotus lividus* following fertilization. *a,* Oocyte; *b,* ripe unfertilized egg; *c,* fertilized egg. (After Boveri, from Morgan, 1927.)

presence of the clear cytoplasm at the animal pole stresses the polar organization of the egg, but there is no trace of a bilateral organization: all meridians of the egg are exactly alike. The spermatozoon penetrates into the egg at any meridian and nearer to the vegetal pole.

As soon as the spermatozoon enters the egg, the yellow cortical cytoplasm falls into violent commotion (Fig. 35). First of all, the yellow cytoplasm begins to stream down along the surface of the egg toward the vegetal pole, and for a time it accumulates as a cap on the vegetal pole. As the spermatozoon pronucleus penetrates deeper into the cytoplasm on its course toward the egg pronucleus, the yellow cytoplasm reverses its movement and streams upward but this time on one side of the egg only, namely, on the side where the spermatozoon had entered the egg. Shortly before the first cleavage the yellow cytoplasm takes up a position just below the equator of the egg, forming there a crescentic area which later gives rise to mesoderm and may be called the **mesodermal crescent.** At the same time a crescent of light gray cytoplasm appears subequatorially on the opposite side of the egg. The origin of this cytoplasmic substance has not been traced, but it is found to give rise later to the notochord, and may therefore be denoted as the **notochordal crescent.** Thus the cytoplasmic displacements following fertilization not only bring some kinds of cytoplasm to more strictly defined, restricted areas, but also in this case give the egg a distinct bilateral structure (Conklin, 1905).

Displacements of cytoplasmic substances follow fertilization in other animals also. In the sea urchin *Paracentrotus lividus* the mature egg contains red granules in the cortical cytoplasm, evenly distributed over its whole surface. After fertilization, streaming movements of the surface layer bring the pigmented granules into a subequatorial zone, leaving the whole of the animal hemisphere clear of pigment and also removing pigment from a smaller area at the vegetal pole (Fig. 36).

The structure of the amphibian egg at the time of fertilization was described on p. 35. Only a few minutes after the sperm has penetrated the egg, the surface layer of protoplasm starts contracting toward the animal pole of the egg, and in the course of this contraction the pigmented protoplasm covers up the gap in the pigment layer at the animal pole, where the first polar body has been given off. Considerably later (10 minutes after insemination of a *Rana temporaria* egg at 18° C.) a much more extensive movement of the cytoplasm begins. This involves practically the whole cortical layer of the egg, which is rotated in respect to the internal mass of cytoplasm (Ancel and Vintemberger, 1948). At the future dorsal side of the egg the cortical cytoplasm moves upward toward the animal pole, and at the opposite, future ventral, side of the egg the cortical cytoplasm moves downward toward the vegetal pole (Fig. 37). As the edge of intensely pigmented cortical cytoplasm rises above the equator on the dorsal side, it reveals the deeper lying ring of marginal cytoplasm. The marginal cytoplasm, although bearing some pigment, is much lighter colored than the cortical cytoplasm. The result is that there appears on the dorsal side of the egg a subequatorial, lightly pigmented, crescentic area, known as the **gray crescent.**

The gray crescent presents an arrangement of cytoplasmic constituents which had not occurred in the egg previously. Due to the sliding upward of the cortical layer, not only is the heavily pigmented surface protoplasm removed from this area, but the cortical layer of the vegetal field is drawn over it, with some of the yolky cytoplasm of that part of the egg superimposed over the marginal cytoplasm. The superimposed layer is sufficiently thin not to conceal the pigment of the deeper lying marginal cytoplasm. On the ventral side of the egg the reverse occurs; the marginal cytoplasm is partly covered by the layer of pigment.

As a result of these displacements the egg acquires a very well-defined bilateral symmetry, with the median plane bisecting the gray crescent. It is in this condition that the cytoplasm of the egg is found at the beginning of cleavage.

The new distribution of cytoplasmic substances of the egg is immediately followed by changes in the physiological properties of the surface layer of the egg. Before the appearance of the gray crescent the permeability of the cortical cytoplasm for vital dyes is the same over the whole of the surface of the

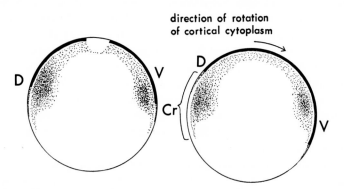

Fig. 37.   Diagrammatic median sections showing displacement of the cortical cytoplasm in the frog's egg following fertilization, with the formation of the gray crescent (*Cr*). *D,* Dorsal surface; *V,* ventral surface of the egg. (According to Banki, 1929, and Ancel and Vintemberger, 1948.)

egg, but after the rotation of the cortical cytoplasm differences become discernible. The permeability to vital dies becomes decreased around the animal pole and on the ventral surface of the egg, but in the gray crescent area this decrease does not take place. The permeability at the gray crescent thus becomes relatively higher than at other parts of the egg's surface (Dollander, 1953). Further physiological distinctions follow in later stages, and at the approach of gastrulation the gray crescent takes the lead in the formation of the blastopore, as will be explained in section 5–4.

There is some experimental evidence that factors external to the egg may decide on which side of the amphibian egg the layer of cortical cytoplasm moves upward, and on which side, therefore, the gray crescent will develop. In 1885 W. Roux attempted to control the point of entry of the spermatozoon into the frog egg by applying a sperm solution on one side of the egg only, and he found that the gray crescent appeared on the side opposite to that of the point of entry into the egg. This was confirmed by some more recent work (Ancel and Vintemberger, 1948). The arrangement of the cytoplasmic substances in the egg of an ascidian follows, as we have seen, the same course. However, if the frog's egg is stimulated to parthenogenetic development by pricking, the gray crescent appears without any relation to the point at which the egg was pricked. In the egg of the sturgeon there appears after fertilization and before the beginning of cleavage a "gray crescent" very similar to that of the amphibian egg and bearing the same relationship to the later formation of the blastopore. In the sturgeon, however, the spermatozoon enters the egg through a micropyle exactly at the animal pole; it can therefore in no case determine the position of the gray crescent or the plane of symmetry of the embryo (Ginsburg, 1953).

From this it is obvious that the direction of the rotation of the cortex of the frog's egg, and the development of the gray crescent in a frog's or in a sturgeon's egg, can be determined by some special properties of the cytoplasm of the egg and independently of the entry of the sperm. What these properties are is not fully understood. In this connection it is noteworthy that if the unfertilized egg of a frog is allowed to rotate freely, it takes up a position with its main axis not strictly vertical but slightly oblique. The part of the equator which is raised in this position is that part where the gray crescent will later develop and which corresponds later to the dorsal side of the embryo. The fact that the dorsal side tends to be raised and that the opposite ventral side sinks down indicates that the cytoplasm on the dorsal side has a slightly lower specific gravity. This may mean that there is less yolk on this side (though of course the greatest concentration of yolk is toward the vegetal pole of the egg). It is doubtful, however, whether the amount of yolk is the decisive factor in determining the position of the gray crescent.

It has been mentioned before (p. 35) that the unfertilized egg of a frog may show some traces of bilateral symmetry: the ring-shaped zone of marginal cytoplasm has been found to be slightly thicker on one side, and this is the side that may well be the one on which the gray crescent develops in a parthenogenetically activated egg.

On the other hand, the possibility of determining the plane of bilateral symmetry of the egg by the point of entry of the spermatozoid shows that the inherent differences in the egg may be overridden by an external factor. In other words, any part of the marginal zone which lies opposite the point of entry of the spermatozoon may be transformed into the gray crescent. In respect to the development of its bilateral symmetry, the egg appears to have a very considerable amount of flexibility.

CHAPTER 4

Cleavage

One of the peculiarities of sexual reproduction in animals is that the complex multicellular body of the offspring originates from a single cell—the fertilized egg. It is necessary, therefore, that the single cell be transformed into a multicellular body. This transformation takes place at the very beginning of development and is attained by means of a number of cell divisions following in rapid succession. This series of cell divisions is known as the process of **cleavage.**

Cleavage can be characterized as that period of development in which:

1. The unicellular fertilized egg is transformed by consecutive mitotic divisions into a multicellular complex.
2. No growth occurs.
3. The general shape of the embryo does not change, except for the formation of a cavity in the interior—the blastocoele.
4. There is no qualitative change in the chemical composition of the egg,

though a transformation of the food reserves into active cytoplasm and of cytoplasmic substances into nuclear substance does take place.

5. The constituent parts of the cytoplasm of the egg are not displaced to any great extent and remain on the whole in the same positions as in the egg at the beginning of cleavage.

6. The ratio of nucleus to cytoplasm, very low at the beginning of cleavage, is, at the end, brought to the level found in ordinary somatic cells.

4-1   PECULIARITIES OF CELL DIVISIONS IN CLEAVAGE

The cleavage of the fertilized egg is initiated by the division of the nucleus (the synkaryon), and, as a general rule, the division of the nucleus is followed by the division of the cytoplasm, so that the egg cell divides into two daughter cells. The daughter cells are termed the **cleavage cells** or the **blastomeres.** The first two blastomeres divide again, thus producing four blastomeres, then 8, 16, 32 and so on. The first cleavages tend to occur simultaneously in all the blastomeres, but sooner or later the synchronization is lost and the blastomeres divide at different times, independently of one another.

The division of the blastomeres is essentially a typical mitosis, and the chromosomes have the appearance and structure of somatic chromosomes. There is, however, one very important difference between the cell divisions in cleavage, and the cell divisions in later stages of development and in the adult organism. In the later stages and in the adult the cell division is intimately connected with growth. After each division the daughter cells grow, and when they are approximately doubled in size they divide again. The cells thus maintain an average size in every type of tissue. During cleavage this is not so. The consecutive divisions of the blastomeres are not separated by periods of growth. A blastomere does not increase in size before the next division begins. Consequently, with each division the resulting blastomeres are only half the original size. Thus cleavage begins with one very large cell and ends with a great number of cells, each of which is no longer very much larger than the tissue cells of the adult animal. Indeed, at the end of cleavage, the cells are usually even smaller than most of the differentiated cells of an adult animal, because cellular differentiation is often accompanied by an increase in the size of an individual cell.

The nuclei of the early cleavage cells are considerably larger than they are in ordinary somatic cells of the same animal. This is, however, mainly due to the presence of larger amounts of nuclear sap and not to chromosomal material. The amount of nucleic acid in individual nuclei may be determined by using the property of these acids of absorbing ultraviolet light at a wavelength of 2,600 Ångströms. If a very narrow beam of light of this wavelength is passed through a nucleus in a microscopic preparation, the light absorbed by the nucleus may be measured and the actual content of the nucleic acid may then be calculated. Alternatively, the nuclei may be stained in the preparation by the Feulgen reagent which is specific for deoxyribonucleic acid, and the absorption of light by the fuchsin fixed in the nucleus may be taken as an estimate of the nucleic acid present.

Since the chromosomal material in the cleavage cells is being doubled at each mitotic division, one would expect the amounts of deoxyribonucleic acid to vary from a diploid amount in cells that had just undergone mitosis, to a tetraploid amount in cells ready to divide again. This has actually been found in the embryo of the sea urchin *Lytechinus,* from the one-cell stage (zygote nucleus) to the larval stage (McMaster, 1955). (See Table 5.)

Table 5.   Mean Amounts of Deoxyribonucleic Acid per Nucleus in Arbitrary Units in Early Development of the Sea Urchin *Lytechinus variegatus.* (From McMaster, 1955.)

Stage	DNA	Condition of Nucleus
1-cell	1.11 ± .06	near 4n
2-cell	1.02 ± .04	"
4-cell	1.09 ± .04	"
8-cell	1.08 ± .07	"
16-cell micromeres	0.82 ± .06	2n < 4n
mesomeres	1.04 ± .04	near 4n
macromeres	1.02 ± .04	"
28-cell micromeres	1.10 ± .05	"
mesomeres	0.95 ± .06	"
macromeres	1.01 ± .03	"
blastula ectoderm	0.68 ± .04	2n < 4n
gastrula ectoderm	0.60 ± .03	2n +
early pluteus ectoderm	0.62 ± .03	"
late pluteus ectoderm	0.60 ± .03	"

As an example we may also quote results obtained by measurement of the deoxyribonucleic acid content in nuclei during fertilization and cleavage in an annelid, *Chaetopterus* (Alfert and Swift, 1953). (See Table 6.)

Table 6.   Nucleic Acid Content in Nuclei during Gametogenesis and Cleavage in *Chaetopterus* (Deoxyribonucleic Acid in Arbitrary Units— Average from Several Determinations)

		Condition of Nucleus	
1st polar body	127 ± 3	2n	
Spermatozoid	61 ± 1	1n	
Cleavage, interphase	210 ± 9	2n ———— 4n	
Cleavage, prophase	263 ± 10	4n	
Cleavage, telophase	124 ± 3	2n	

In spite of the greater volume of the cleavage nuclei, these are small in

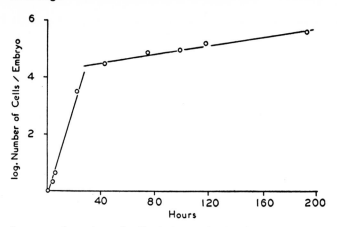

Fig. 38.   Increase of number of cells during early development of the frog's egg.
(From Sze, 1953.)

proportion to the volume of the cells. In a mature sea urchin egg before fertilization the ratio $\dfrac{\text{volume of the nucleus}}{\text{volume of cytoplasm}}$ is 1/550 (Brachet, 1950b). At the end of cleavage (in the blastula stage) the same ratio is 1/6.

If the numbers of cells in a developing embryo at various stages are counted, the change in the rate of reproduction of cells between the period of cleavage and the later development becomes very obvious. In Fig. 38 the logarithm of the number of cells in a frog embryo is plotted against time, and it is seen that the curve shows a very distinct bend at about 40 hours, which is just at the end of cleavage (between the blastula stage and the beginning of gastrulation), if development has been proceeding at 15° C. Before this time the rate of increase is far more rapid than afterwards (Sze, 1953).

The rhythm of cleavage as measured by the time interval between two consecutive divisions is not quite the same in different animals. In the goldfish, divisions follow each other continuously at rather regular intervals of about 20 minutes. The interval is nearer one hour in the case of the frog, though it depends very much on temperature. The eggs of mammals cleave very much more slowly, about 10 to 12 hours elapsing between consecutive cell divisions in the mouse at 37° C (Kuhl, 1941).

4–2   CHEMICAL CHANGES DURING CLEAVAGE

Although there is no growth during the period of cleavage, certain limited chemical transformations are going on. These may be classed under two headings. One is the utilization of yolk or other food reserves. This consumption is necessary to cover the energy requirements of the cleaving egg. But it has also been observed that the amount of active cytoplasm increases throughout the period of cleavage. This can best be seen in microscopic sections, since it is difficult to differentiate chemically between active cytoplasm and reserve protein. Furthermore, the respiration increases steadily throughout the period of cleavage, and this is generally attributed to an increase in the amount of active cytoplasm (Boell, 1945; Weber and Boell, 1955). The second type of change that is observed during cleavage is a steady increase of nuclear material at the expense of cytoplasm. The number of nuclei is of course doubled with every new division of the blastomeres, and this increase is accompanied by an increase of nuclear substance, which involves an increase of deoxyribonucleic acid—the amount per nucleus of the latter remaining constant (McMaster, 1955) (see Table 5).

The increase of the deoxyribonucleic acid, at least during the earlier phases of development, must be at the expense of some other materials contained in the egg. There are several possible sources of such materials. In the first place one should mention the nucleic acids present in the cytoplasm of the eggs. In the sea urchin eggs there is a large amount of ribonucleic acid in the egg cytoplasm, and this gradually disappears later in development (Brachet, 1950b). It has been reported that in the frog egg there is also a considerable amount of deoxyribonucleic acid in the cytoplasm (Sze, 1953), and it has been suggested that the cytoplasmic nucleic acids may be utilized for the rapid increase of the chromosomal nucleic acid during cleavage. To check the origin of the deoxyribonucleic acid in cleavage cells, developing sea urchin eggs were kept in sea water containing one of the possible precursors of nucleic acids, the amino acid glycine, which can be used for the synthesis of purine groups in the nucleic acid molecule. The glycine was "labeled" by one of its carbon atoms being radioactive (C^{14}). It was found that the radioactive carbon atoms were incorporated in large amounts into the deoxyribonucleic acid, bypassing the ribonucleic acid. It is therefore obvious that the building up of the chromosomal nucleic acid (which is deoxyribonucleic acid) does not necessarily proceed at the expense of the cytoplasmic nucleic acids already present; instead, the chromosomal nucleic acid may be synthesized anew from simpler substances contained in the cells (Abrams, 1951).

The yolky eggs of birds, reptiles and also of insects contain very little nucleic acid, and the synthesis of deoxyribonucleic acid proceeds directly from the foodstuffs contained in the yolk. In mammals nucleic acids are built up from substances taken up by the embryo from without (Brachet, 1950b).

The changes taking place during cleavage seem to be predominantly quantitative, no new substances appearing during this phase.

4–3   PATTERNS OF CLEAVAGE

The way in which the egg is subdivided into the daughter blastomeres is usually very regular. The plane of the first division is as a rule vertical; it passes through the main axis of the egg. The plane of the second division is also vertical and passes through the main axis, but it is at right angles to the first plane of cleavage. The result is that the first four blastomeres all lie

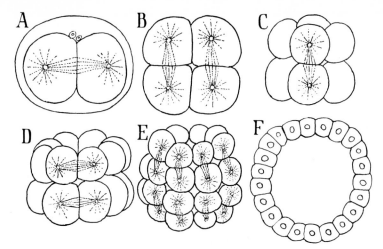

Fig. 39.   Radial cleavage with almost equal size of blastomeres in sea cucumber
Synapta digitata. (After Selenka, from Korschelt, 1936.)

side by side. The plane of the third division is at right angles to the first two
planes and to the main axis of the egg. It is therefore horizontal or parallel to
the equator of the egg. Of the eight blastomeres four lie on top of the other
four, the first four comprising the animal hemisphere of the egg, the second
the vegetal hemisphere.

If each of the blastomeres of the upper tier lie over the corresponding
blastomeres of the lower tier, the pattern of the blastomeres is radially
symmetrical (Fig. 39). This is called the **radial type** of cleavage. In many

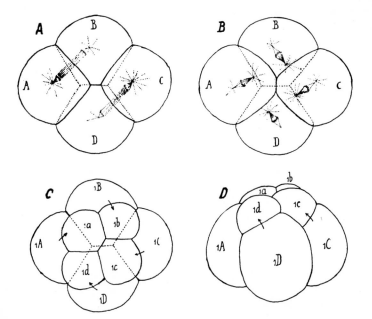

Fig. 40.   Spiral cleavage in the mollusc *Trochus.* (After Robert, from Korschelt, 1936.)

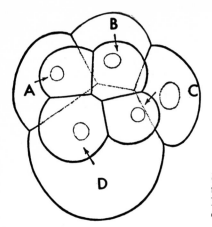

Fig. 41. Spiral cleavage in the mollusc *Unio* with blastomere *D* distinctly larger than the blastomeres *A, B, C.* (After Lillie, from Kellicott, General Embryology, 1914.)

animals, however, the upper tier of blastomeres may be shifted with respect to the lower tier of blastomeres, and the radially symmetrical pattern becomes distorted in various degrees. The distortion may be due sometimes to individual variation, but there are certain groups of animals in which distortion always takes place and is due to a specific structure of the egg.

In the annelids, molluscs, nemerteans and some of the planarians (the Polycladida) all of the blastomeres of the upper tier are shifted in the same direction in relation to the blastomeres of the lower tier, so that they come to lie not over the corresponding vegetal blastomeres, but over the junction between each two of the vegetal blastomeres (Fig. 40). This comes about not as a result of secondary shifting of the blastomeres but because of oblique positions of the mitotic spindles, so that right from the start the two daughter cells do not lie above one another. The four spindles during the third cleavage are arranged in a sort of spiral. This type of cleavage is therefore called the **spiral type** of cleavage.

The turn of the spiral as seen from above may be in a clockwise direction or in a counterclockwise direction. In the first case the cleavage is called **dextral;** in the second case it is called **sinistral.** Since the cleavage planes are at right angles to the spindles, they also deviate from the horizontal position found in radial cleavage, and each cleavage plane is inclined at a certain angle. The spiral arrangement of the mitotic spindles can be traced even in the first two divisions of the egg; the spindles are oblique and not vertical as in radial cleavage. However, the resulting shifts in the position of the blastomeres are not so obvious as after the third cleavage. During the subsequent cleavages the spindles continue to be oblique, but the direction of spiraling changes in each subsequent division. Dextral spiraling alternates with sinistral so that the spindle of each subsequent cleavage is approximately at right angles to the previous one.

It will be noticed that the type of cleavage of the egg as a whole, whether dextral or sinistral, depends on the direction of spiraling occurring during the third division of the egg.

Peculiarities of the cleavage pattern can also be introduced by differences in the size of the blastomeres. Of the four blastomeres in the four-cell stage of eggs having a spiral type of cleavage, it is often found that one blastomere is larger than the other three (Fig. 41). This allows us to distinguish the individual blastomeres. The four first blastomeres are denoted by the letters *A, B, C, D,* the letters going in a clockwise direction (if the egg is viewed from the animal pole), and the largest blastomere being denoted by the letter *D.* In some animals having otherwise an approximately radial type of cleavage, two of the first four blastomeres may be larger than the other two, thus establishing a plane of bilateral symmetry in the developing embryo. Subsequent cleavages may make the bilateral arrangement of the blastomeres still more obvious (as in tunicates and in nematodes, although in a different way). The resulting type of cleavage is referred to as the **bilateral type** (Fig. 59).

The yolk, which is always present in the egg at the beginning of cleavage in greater or lesser quantities, exerts a very far reaching effect on the process of cleavage. Every mitosis involves movements of the cell components: the chromosomes, parts of the cytoplasm constituting the achromatic figure, the mitochondria and the surface layer of the cell, the activity of which along the equator of the maternal cells leads to the eventual separation of the daughter cells. During these movements the yolk granules or platelets behave entirely passively and are passively distributed between the daughter blastomeres. When the yolk granules or platelets become very abundant, they tend to retard and even inhibit the process of cleavage. As a result, the blastomeres which are richer in yolk tend to divide at a slower rate and consequently remain larger than those which have less yolk. The yolk in the uncleaved egg is more concentrated towards the vegetal pole of the

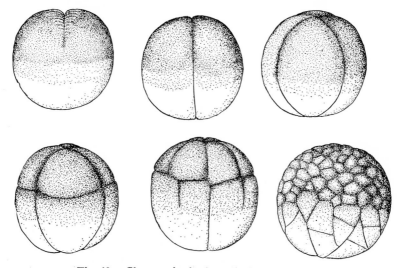

Fig. 42. Cleavage in the frog, semidiagrammatic.

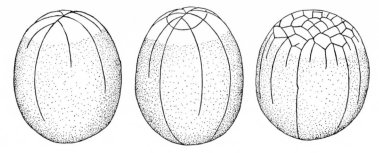

Fig. 43. Cleavage in the ganoid fish *Amia*. (After Whitman and Eyklesheimer, from Korschelt, 1936.)

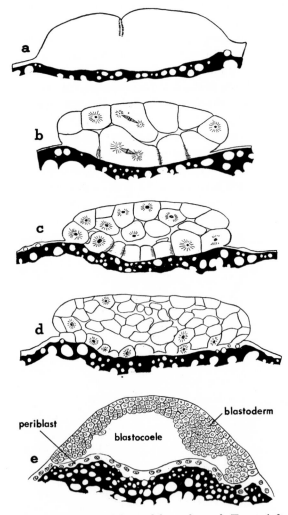

Fig. 44. Cleavage and blastula of bony fish. *a, b, c, d,* Trout (after Kopsch). *e, Muraena* (after Bocke). (From Brachet, 1935.)

egg. It is therefore at the vegetal pole of the egg that the cleavage is most retarded by the presence of yolk, and where the blastomeres are of the largest size.

A very good example of the effect of the yolk on cleavage is given by the frog's egg (Fig. 42). The influence of the yolk on cleavage may be detected even during the first division of the fertilized egg. During the anaphase of the mitotic division on the surface of the egg appears a furrow which is to separate the two daughter blastomeres from one another. This furrow, however, does not appear simultaneously all around the circumference of the egg, but at first only at the animal pole of the egg, where there is less yolk. (It has been indicated above that the first cleavage plane is vertical, and therefore passes through the animal and vegetal poles of the egg.) Only gradually is the cleavage furrow prolonged along the meridians of the egg, until, cutting through the mass of yolk-laden cytoplasm, it eventually reaches the vegetal pole and thus completes the separation of the two first blastomeres. The same process is repeated during the second cleavage. During the third cleavage, when the plane of separation of the daughter blastomeres is horizontal, the furrow appears simultaneously over the whole circumference of the egg, for it meets with an equal resistance from yolk everywhere.

A further accumulation of the yolk at the vegetal pole of the egg causes still greater delay in the cell fission at this pole, so that the cleavage becomes

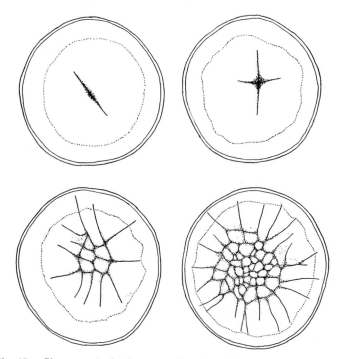

Fig. 45. Cleavage of a hen's egg, surface view. (From Patterson, 1910.)

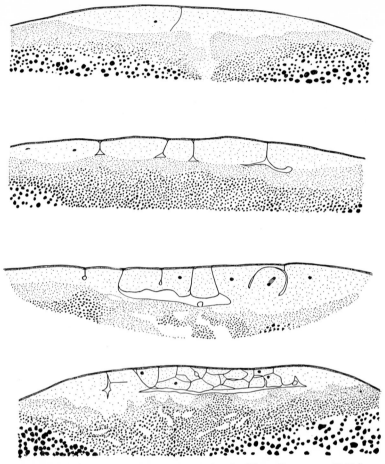

Fig. 46. Cleavage of a hen's egg (sections). (From Patterson, 1910.)

inhibited more and more. This can be traced very well in a series of various ganoid fishes, whose eggs possess an increasing amount of yolk. In *Acipenser* the cleavage is complete, as in the amphibians, but the difference between the micromeres of the animal hemisphere and the macromeres of the vegetal hemisphere is much sharper than in amphibians. In *Amia* (Fig. 43) cleavage starts at the animal pole, and the cleavage furrows reach the vegetal pole, but they are so retarded that subsequent divisions begin at the animal pole before the preceding furrows cut through the yolk at the vegetal pole. In *Lepidosteus* the cleavage starts at the animal pole as in *Amia,* but the cleavage furrows never reach the vegetal pole, so that the vegetal hemisphere of the egg remains uncleaved. This results in what is called **incomplete cleavage.** The eggs which are completely divided into blastomeres are called **holoblastic,** those with incomplete cleavage **meroblastic.** As a result of incomplete cleavage the egg is divided into a number of separate blastomeres and a residue which is an undivided mass of cytoplasm with numerous nuclei scattered in it.

In eggs in which the yolk is segregated from the active cytoplasm (elasmobranchs, bony fishes, birds and reptiles), the cleavage is incomplete right from the start (Figs. 44, 45 and 46). Cleavage in these cases proceeds only in the superficial layer of cytoplasm. At first, all the cleavage planes are vertical, and all the blastomeres lie in one plane only. The cleavage furrows separate the daughter blastomeres from each other but not from the yolk, so that the central blastomeres are continuous with the yolk at their lower ends, and the blastomeres lying on the circumference are, in addition, continuous with the uncleaved cytoplasm at their outer edges. As the nuclei at the edge divide, more and more cells become cut off to join the ones lying in the center, but the new blastomeres are also in continuity with the uncleaved yolk underneath.

In a later stage of cleavage the blastomeres of the central area become separated from the underlying yolk in one of two ways: either slits may appear beneath the nucleated parts of the cells, or else the cell divisions may occur with horizontal (tangential) planes of fission. In the latter case one of the daughter cells, the upper one, becomes completely separated from its neighbors, while the lower blastomere retains the connection with the yolk mass. The marginal cells, which remain continuous with each other around their outer edges, are also continuous with the mass of yolk and hence with the lower cells resulting from tangential divisions. All these blastomeres eventually lose even those furrows which partially separated them from one another, and fuse into a continuous syncytium with numerous nuclei but no indication of individual cells.

The cytoplasmic cap on the animal pole of the egg is now, in its central part, subdivided into a number of "free" blastomeres, while around the outer margin of the mass of blastomeres, and underneath, closely adhering to the yolk, is the syncytial layer which is called the **periblast.** The periblast is not destined to participate directly in the formation of the embryonic body, but it is supposed to be of some importance in breaking down the yolk and making it available for the growing embryo.

Another type of incomplete cleavage is found in centrolecithal eggs. At the beginning of cleavage the nucleus (synkaryon) lies in the center of the egg surrounded by a small amount of cytoplasm. The mitotic division of the nucleus starts in this position, but at first it is not followed by a division of cytoplasm (Fig. 47). As a result, a number of nuclei are formed all embedded in the undivided central mass of cytoplasm. After several divisions have taken place (the actual number of divisions varies in different animals), the nuclei start moving away from the center of the egg. Each nucleus, surrounded by a small portion of the original central cytoplasm, travels outward toward the surface of the egg. The central mass of cytoplasm is thus broken up and disappears. When the nuclei reach the surface of the egg, the cytoplasm surrounding them fuses with the superficial layer of cyto-

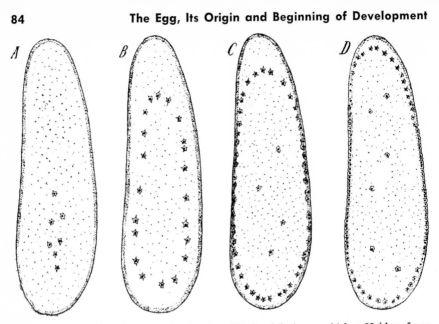

Fig. 47. Superficial cleavage of a beetle's (*Hydrophilus*) egg. (After Heider, from Korschelt, 1936.)

plasm. The surface layer of the embryo is then a syncytium with numerous nuclei embedded in an undivided layer of cytoplasm. In the next stage the cytoplasm becomes subdivided, by furrows going inward from the surface, into as many sections as there are nuclei. The sections can now be called cells, even though these are at first still connected to the yolk mass. Sooner or later the cells become completely separated from the yolk. The latter persists as a compact mass until it is gradually used up as a food reserve by the developing embryo. This type of cleavage is called **superficial cleavage.** It is the usual type of cleavage in insects and many other arthropods.

4–4   MORULA AND BLASTULA

The blastomeres in the early cleavage stages tend to assume a spherical shape like that of the egg before cleavage. Their mutual pressure flattens the surfaces of the blastomeres in contact with each other, but the free surfaces of each blastomere remain spherical, unless these outer surfaces are also compressed by the vitelline membrane. The whole embryo acquires, in this stage, a characteristic appearance reminiscent of a mulberry (Fig. 39, *e*). Because of this superficial similarity the embryo in this stage is called a **morula** (Latin for mulberry).

The arrangement of the blastomeres in the morula stage may vary in the different groups of the animal kingdom. In coelenterates it is often a massive structure, blastomeres filling all the space that had been occupied by the uncleaved egg. Some of the blastomeres then lie externally, others in the interior. (Some embryologists apply the name morula to this type of embryo only.) More often, as the egg undergoes cleavage, the blastomeres become arranged in one layer, so that all the blastomeres participate in the external surface of the embryo. In this case a cavity soon appears between the blastomeres, which at first may be represented just by narrow crevices between the blastomeres, but which gradually increases as the cleavage goes on. This cavity is called the **blastocoele.**

As the cleavage proceeds, the adhesion of the blastomeres to each other increases, and they arrange themselves into a true epithelium. In cases where a cavity has been forming in the interior of the embryo, the epithelial layer completely encloses this cavity, and the embryo becomes a hollow sphere, the walls of which consist of an epithelial layer of cells. Such an embryo is called a **blastula.** The layer of cells is called the **blastoderm,** and the cavity is the blastocoele already mentioned (Fig. 48, *a*).

In oligolecithal eggs with complete cleavage (echinoderms, amphioxus) the blastomeres at the end of cleavage are not of exactly equal size, the blastomeres near the vegetal pole being slightly larger than those on the animal pole. When the blastula is formed, the cells arrange themselves into a simple cylindrical epithelium enclosing the blastocoele. Because the vegetal cells are larger than the animal cells, the blastoderm is not of an equal thickness throughout; at the vegetal pole the epithelium is thicker, at the animal pole it is thinner. Thus the polarity of the egg persists in the polarity of the blastula.

Animals with a larger amount of yolk, such as the frog, show a difference

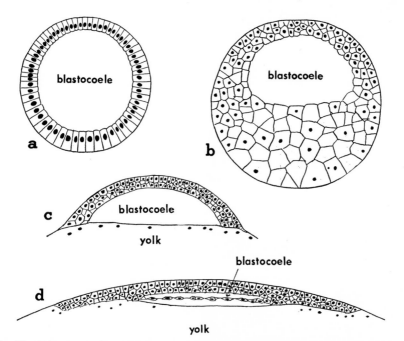

Fig. 48.   Diagrammatic comparison of blastulae of an echinoderm (*a*), a frog (*b*), a bony fish (*c*) and a bird (*d*).

in the size of the cells of the blastula that may be very considerable, and the blastula still further departs from the simple form of a hollow sphere. The cells here are also arranged into a layer surrounding the cavity in the

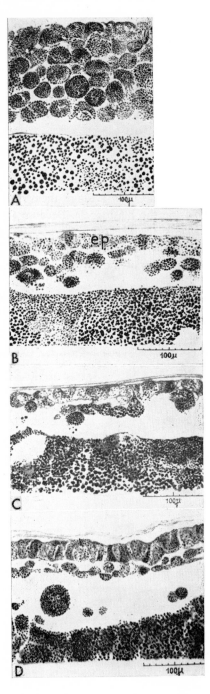

Fig. 49. Four stages in the formation of the epiblast and the hypoblast from the cleavage cells of the duck embryo. *A,* Late cleavage, all cells in one undifferentiated layer. *B,* Superficial cells join closer to form epiblast (*ep*). Underneath this there remain numbers of loose cells. *C* and *D,* Loose cells join to form hypoblast. (From Pasteels, 1945.)

interior, but the layer is of very uneven thickness. At the vegetal pole the layer of cells is very much thicker than at the animal pole, and the blastocoele is consequently distinctly eccentric, nearer to the animal pole of the embryo (Fig. 48, *b*). Furthermore, the blastoderm is no longer a simple cylindrical epithelium but is two or more cells thick. The cells in the interior are rather loosely connected to each other, but at the external surface of the blastula the cells adhere to one another very firmly because of the presence of a cementing substance joining the surfaces of adjoining cells in a narrow zone just underneath the surface of the blastoderm (Fig. 102).

A process corresponding to blastula formation occurs also in animals whose eggs have incomplete cleavage. In a bony fish or a shark the early blastomeres tend to round themselves off, thus showing that they are only loosely bound together. Later the blastomeres adhere to each other more firmly and so become converted into an epithelium. Again, as in amphibians, the superficial cells are firmly joined to one another while the cells in the interior may remain loosely connected until a later stage. The epithelium, however, cannot have the form of a sphere. Just as the cleavage is restricted to the cap of cytoplasm on the animal pole of the egg, so also the blastoderm is developed only in the same polar region. The blastoderm therefore assumes the shape of a disc lying on the animal pole. The disc, which is called the **blastodisc,** is more or less convex and encloses, between itself and the uncleaved residue of the egg, a cavity representing the blastocoele (Fig. 48, *c*).

The earlier stages of the discoidal cleavage in the eggs of reptiles and birds resemble essentially the cleavage in the meroblastic eggs of the fishes. At the time when the blastoderm is being formed, however, a very essential difference becomes apparent. In fishes, when the loosely connected blastomeres become converted into an epithelial layer, there is only one layer formed. The same is found in tortoises, but in most reptiles, especially lizards and snakes, and in birds, only some of the cells are incorporated into the superficial layer, which we will call the **epiblast.** A small number of the blastomeres remain underneath the superficial layer, lying loose in the space between the latter and the yolk. These cells subsequently unite themselves into a thin layer of flat epithelium, called the **hypoblast,** separated from the surface layer by a cavity (Fig. 49). This lower layer of cells has caused many controversies among embryologists as to its significance and homologies. It has recently been shown (Pasteels, 1945 and 1957) that this layer corresponds to the cells lying in the floor of the blastocoele on the vegetal side of the blastocoele cavity in animals such as amphibians. The cavity between the superficial layer and this deeper layer in birds and reptiles corresponds therefore to the blastocoele of amphibians and fishes (Fig. 48, *d*).

There is a further difference in that the cavity in birds and reptiles does not appear under the whole of the blastoderm but only under the central part of it. No cavity is formed under the layer of blastomeres in the periphery of the blastodisc. The blastodisc is thus subdivided into two parts: the central part, called the **area pellucida,** and the peripheral part, the **area opaca.** The names indicate that the central part appears more transparent in the living embryo, owing to the presence of the cavity under the layer of blastomeres; the peripheral area is opaque because the blastomeres rest directly on the yolk.

Only the area pellucida furnishes material for the formation of the body of the embryo. The cells of the area opaca are concerned with the breakdown of the underlying yolk and thus indirectly supply the embryo with foodstuffs.

In centrolecithal eggs having a superficial cleavage (insects), there is no cavity comparable to the blastocoele. Nevertheless, the formation of the epithelium on the surface of the egg, after the nuclei have migrated to the exterior, can be compared to the formation of the blastula. The layer of cells thus formed on the surface of the embryo is the blastoderm. Instead of surrounding a cavity, the blastoderm envelops the mass of uncleaved yolk. We may also compare this stage to an embryo whose blastocoele has been filled with yolk.

Up to the blastula stage the developing embryo preserves the same general shape as the uncleaved egg. So far the results achieved are the subdivision of the single cell into a multiplicity of cells and the formation of the blastocoele. In addition, the substances contained in the egg remain very much in the same position as before. The yolk remains near the vegetal pole. In pigmented eggs, such as those of amphibians, the pigment remains as before, more or less restricted to the upper hemisphere of the egg. Only a slight intermingling of cytoplasm seems to be produced by the cleavage furrows cutting through the substance of the egg.

It has been pointed out that during cleavage no qualitative changes can be discovered in the chemical composition of the developing embryo. No new substances, either chemically defined or microscopically detectable, have been found to appear during cleavage. It is conceivable, however, that the substances present in the egg may be redistributed in some way during cleavage and that such a redistribution may be essentially important for further development.

In this connection we shall examine first of all whether the numerous nuclei produced during the mitotic divisions of the egg are all alike in their properties, or whether any differences may be discovered among them.

4–5   THE NUCLEI OF CLEAVAGE CELLS

In his "germ plasm theory" A. Weismann presented a hypothesis to explain both heredity and the ontogenetic development of organisms. According to Weismann (1904), every distinct part of an organism (animal or plant) is represented in the sex cell by a separate particle: a **determinant.** The sum total of determinants would represent the parts of the adult organism with all their peculiarities. The complete set of determinants would be handed down from generation to generation, which would account for hereditary transmission of characters. The determinants, according to Weismann, are localized in the chromosomes of the nucleus, just as are the genes of modern genetics. However, there is this difference, that the genes are not supposed to represent *parts* of the organism but rather properties which may sometimes be discernible in all the parts of the body.

During the cleavage of the egg, the various determinants, according to Weismann, become segregated into different cells. The blastomeres would get only some of the determinants, namely, those which correspond to the fate of each blastomere. The successive segregation of the determinants would eventually result in each cell's having only determinants of one kind, and then nothing would be left to the cell but to differentiate in a specific way in accordance with the determinants present. Only the cells having the sex cells among their descendants, Weismann held, would preserve the complete set of determinants, since these would be necessary for directing the development of the next generation.

Even though Weismann's conception of the properties of determinants is not tenable from the genetic viewpoint, it is still important to know whether the difference in the fate of the blastomeres and parts of the embryo may be attributed to differences in the nuclei of the cleavage cells. Experiments are now known that give convincing information on this point.

Our knowledge of the development of complete embryos from one of the

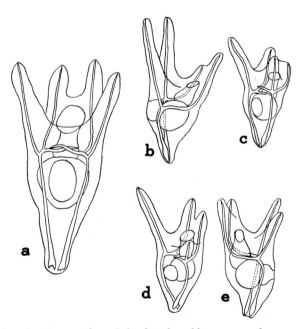

Fig. 50.   Results of separation of the first four blastomeres of a sea urchin's egg. *a,* Normal pluteus. *b, c, d* and *e,* Plutei of normal structure but diminished size, each developed from one of the blastomeres of the four-cell stage. All drawn to the same scale. (From Hörstadius and Wolsky, 1936.)

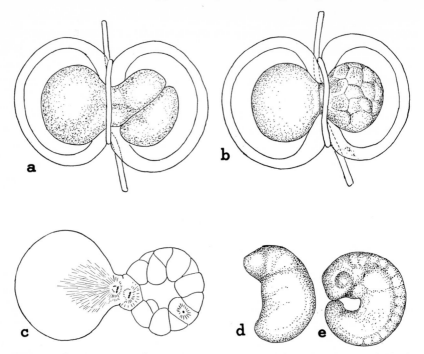

Fig. 51. Spemann's experiment of delayed supply of the nucleus to one half of a newt's egg. *a,* Beginning of cleavage in nucleated half. *b,* One of the nuclei has penetrated into the uncleaved half, and a cell boundary appears between this and the half which originally contained the zygote nucleus. *c,* Same stage in section. *d,* Embryo from the half with delayed nuclear supply. *e,* Embryo from the half which contained the zygote nucleus. (From Spemann, 1938.)

two daughter blastomeres of the egg (as a result of either separating the two first blastomeres or killing one of the first two blastomeres—see section 1–3) contradicts Weismann's hypothesis about the segregation of determinants during cleavage. The evidence becomes still more conclusive because it has been found that not only are the first two blastomeres capable of developing into complete embryos, but that the blastomeres in later cleavage stages sometimes possess the same ability. In the case of sea urchins one of the first four blastomeres, and even occasionally one of the first eight blastomeres, may develop into a whole embryo with all the normal parts but reduced in size (see Fig. 50) (Hörstadius and Wolsky, 1936).

However, the method of isolating blastomeres does not permit one to test the properties of nuclei of later generations of cells. In the four-cell stage the quantity of cytoplasm contained in one blastomere may already be too small for development to take place in an approximately normal fashion. This is true in increasing degree as cleavage proceeds and the individual blastomeres become smaller and smaller.

To push the investigation further, a different method has been devised. A most elegant experiment in this field was carried out by Spemann (1928).

Spemann constricted fertilized eggs of the newt *Triturus* (*Triton*) into two halves with a fine hair, just as they were about to begin to cleave. The constriction was not carried out completely, so that the two halves were still connected to each other by a narrow bridge of cytoplasm. The nucleus of the fertilized egg lay in one half, and the other half consisted of cytoplasm only. When the egg nucleus began to divide, the cleavage was at first restricted to that half of the egg which contained the nucleus (Fig. 51). This half divided into two, four, eight cells, and so on, while the nonnucleated half remained uncleaved. At about the stage of 16 blastomeres, one of the daughter nuclei, now much smaller than at the beginning of cleavage, passed through the cytoplasmic bridge into the half of the egg which had hitherto no nucleus (Fig. 51, *c*). Forthwith this half also began to cleave. After both halves of the egg were thus supplied with nuclei, Spemann drew the hair loop tighter and separated the two halves of the egg completely from each other. They were now allowed to develop into embryos. In a number of cases two completely normal embryos developed from the two halves of the egg as a result of this experiment.

Of the two embryos, each started by having one half of the egg cytoplasm, but as to the nucleus they were in a very different position. While one of the embryos possessed 15/16 or even 31/32 of all the nuclear material of the egg, the other got only 1/16 or 1/32 of the nuclear material. This proves conclusively that even in the 32-cell stage every nucleus has a complete set of hereditary factors necessary for the achievement of normal development. All the nuclei in this stage are completely equivalent to one another, and to the nucleus of the fertilized egg. The hypothesis of an unequal division of the hereditary substance of the nucleus, of the segregation of determinants or genes to the different cleavage cells, is thus disproved. It is now assumed that every cell of the metazoan body has a

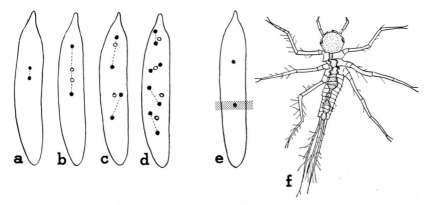

Fig. 52.   *a, b, c,* and *d,* Early cleavage in the dragonfly *Platycnemis pennipes.* Hollow circles indicate the previous position of the nuclei in each case. *e,* Results of killing of one of the first two cleavage nuclei by ultraviolet irradiation. Irradiated area is stippled. *f,* Larva which developed from an egg treated as in *e.* (Adapted from Seidel, 1932.)

complete set of nuclear factors necessary for development (a complete set of genes, in the terminology of modern genetics).

The experiment on the delayed nuclear supply to one half of an amphibian egg has been corroborated by experiments on the eggs of other animals. It will be useful to relate here a corresponding experiment carried out on a very different kind of animal: the dragonfly *Platycnemis pennipes* (Seidel, 1932). In the dragonfly the cleavage is incomplete, and only the nucleus divides at first, the cytoplasm remaining uncleaved. The egg is elongated, and after the first division the daughter nuclei move, one into the anterior and the other into the posterior half of the egg. When eight nuclei are available, these are spaced along the length of the egg. By further divisions nuclei are provided for all cells in the respective regions (Fig. 52, *a–d*). In the stage when two nuclei are present, either of them may be killed by a short exposure to a narrow beam of ultraviolet light, which does not damage the cytoplasm to any great extent (Fig. 52, *e*). The remaining nucleus continues to divide, and its daughter nuclei are distributed to all parts of the egg instead of supplying only one half of it. Completely normal embryos develop, no matter which of the two nuclei is allowed to survive (Fig. 52, *f*). The two nuclei prove to be completely equivalent for development, although normally they would have supplied very different parts of the embryo.

The methods used in the above experiments in testing the properties of the cleavage nuclei are of necessity confined to the earlier stages of cleavage. At present a more universal method is available which allows the investigation to be extended to nuclei of cells of much more advanced embryos, and possibly it may ultimately be applied even to fully differentiated cells of an adult organism. This is the method used for transplantation of nuclei. The transplantation of nuclei from one cell to another was first carried out successfully on *Amoeba,* and the method was then applied to test the properties of nuclei in developing frog embryos (Briggs and King, 1952, 1953 and 1957; King and Briggs, 1954 and 1956). The method, as applied to the frog embryo, consists essentially in taking the nucleus of any cell from an embryo during the stages from blastula to tailbud, and injecting it into an enucleated uncleaved egg.

The egg receiving the nucleus must be specially prepared. The ripe eggs are removed from the oviducts and activated by pricking with a glass needle (see section 3–3). The egg nucleus then approaches the surface of the cytoplasm in preparation for the second maturation division and is removed by a second prick with a glass needle at the exact spot where the nucleus is located. Next, a cell of an advanced embryo is separated from its neighbors and sucked into the tube of an injection pipette. The diameter of the pipette is smaller than that of the cell and as a result the surface of the cell is broken. The contents of the pipette, consisting of the nucleus and the debris of the cytoplasm, are injected deep into the enucleated egg. When the pipette is withdrawn, the egg cytoplasm tends to escape through the hole in the egg membranes, forming an extraovate protrusion. This has to be cut off by a pair of glass needles to prevent further loss of egg substance (Fig. 53).

As a result of these procedures up to 80 per cent (King and Briggs, 1956) of the eggs operated on start cleaving and producing numerous cells, the nuclei of which are derived from the injected nucleus, and the cytoplasm of which is from the enucleated egg (the small amount of cytoplasm injected with the nucleus, comprising less than 1/40,000 of the volume of the egg cytoplasm, may be ignored). Not all eggs which start cleaving develop normally later, but at least a small proportion do so and may go through all the stages of embryonic and postembryonic development up to metamorphosis (Fig. 54). In other cases the development is arrested or becomes abnormal in various stages. These cases of abnormal development do not concern us in the present instance, but the cases of complete development show that the nuclei of embryos in advanced stages, particularly in the late blastula and the early gastrula stages, can provide everything necessary for the differentiation of all types of cells of a normal embryo, whereas they would have normally participated in only one type of differentiation. In the late blastula, the stage used for some of these nuclear transplantations, there are 8000 to 16,000 cells, which means that about 13 to 14 divisions (or generations) of the original nucleus of the fertilized

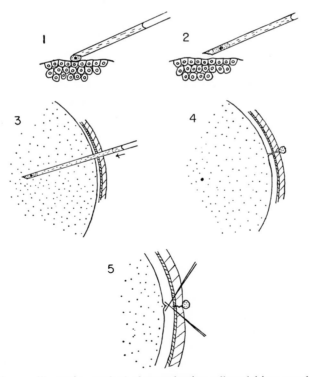

Fig. 53.   Diagram illustrating method of transplanting cell nuclei into enucleated eggs.
(From Briggs and King, 1953.)

Fig. 54. Metamorphosing tadpole developed from an egg with transplanted nucleus. (From King and Briggs, 1954.)

egg have been performed without diminishing the power of the nucleus to support every type of differentiation provided for by the specific genotype. Nuclei of cells of an early gastrula show the same properties, although the subdivision of the embryo into areas with different roles in development is well on the way at that stage, as will be shown in subsequent sections (Chapter 5).

4–6 DISTRIBUTION OF CYTOPLASMIC SUBSTANCES OF THE EGG DURING CLEAVAGE

The fact that the nuclei of the developing embryo are fully equivalent to each other does not allow us to ascribe to the nucleus the origin of differences in the fate of the cells during development of the embryo. What is then the cause of the observed differences in the developmental behavior of cells? By exclusion, we may suppose that the cause of the differences lies in the cytoplasm, and that the cells produced during cleavage do not receive the same cytoplasmic substances. There is direct experimental support in favor of this supposition.

Curiously enough, a further elaboration of Spemann's experiment of delayed nuclear supply to one half of the *Triturus* egg can be brought to bear on this problem (Spemann, 1928).

It has already been mentioned that pigment is unequally distributed over the surface of amphibian eggs (sections 2–5 and 3–5). The animal hemisphere of the egg is more or less heavily pigmented; the region around the vegetal pole is poor in pigment or completely white; the equatorial, or rather subequatorial, marginal zone is more lightly pigmented than the animal hemisphere but more darkly pigmented than the vegetal region. On one side of the egg the marginal zone is broader than on the other, and this region is known as the **gray crescent** (from the color of this region in the common European frogs). The gray crescent is known to correspond to the **dorsal** region of the embryo and also indicates the position of the dorsal lip of the blastopore (see section 5–4).

In the egg of the newt the pigment is not very abundant, but the gray crescent is readily distinguishable in the uncleaved egg. It is therefore possible to determine the position of the ligature in respect to the gray crescent (Fig. 55). If the ligature subdivides the gray crescent cytoplasm equally between the two halves of the egg, two more or less normally developed embryos are found to develop (Fig. 55, *b, e*). If the ligature comes to lie so that all the gray crescent cytoplasm is contained in one half of the egg, whereas the other half does not get any of it, then a complete embryo is developed only from the first half. The second half remains highly abnormal: it does not develop a nervous system, a notochord or segmented muscle; the only parts that can be discerned are epidermis, unsegmented (lateral plate) mesoderm and yolk endoderm. All these are parts of the ventral region of the body, thus justifying the name "belly" (Bauchstück) given by Spemann to the defective embryos (Fig. 55, *a, c, d*).

Now it is found that the fate of the egg half, whether it develops into a complete embryo or into a "belly," does not depend on which of the two halves had retained the egg nucleus originally and which was delayed in receiving a nucleus. A half with a delayed nuclear supply will develop into a complete embryo if it contained the gray crescent cytoplasm. On the other hand, the egg half without gray crescent cytoplasm will become a "belly" even if it contained the egg nucleus right from the beginning. The difference in the time of nuclear supply and the amount of nuclear material received is completely overridden by differences in the cytoplasmic composition of the two egg halves.

The importance of cytoplasmic substances in the egg for the differentiation of parts of the embryo is further proved by a number of experiments

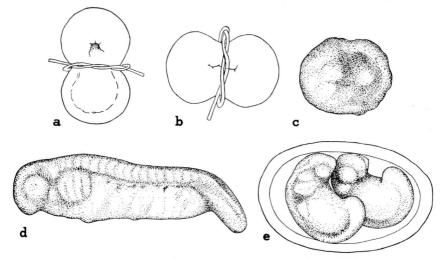

Fig. 55.   Constriction of a newt's egg in the frontal (*a*) and medial (*b*) plane. After frontal constriction the ventral half developed into a "belly" (*c*), the dorsal half into a complete embryo (*d*). After medial constriction (*b*) both halves developed into complete embryos (*e*). (From Spemann, 1938.)

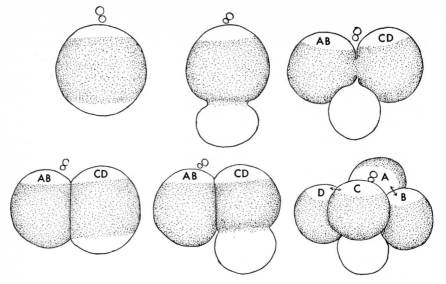

Fig. 56. Cleavage of the mollusc *Dentalium*. (From Wilson, 1904.)

carried out on the eggs of various animals. The following will be chosen as examples:

In the egg of the mollusc *Dentalium* (Wilson, 1904) there may be distinguished three layers of cytoplasm: a layer of clear cytoplasm at the animal pole of the egg, a broad layer of granular cytoplasm in the equatorial zone (actually forming the bulk of the egg substance), and a second layer of clear cytoplasm at the vegetal pole of the egg (Figs. 20 and 56). When the egg begins to cleave, the clear cytoplasm at the vegetal pole is pushed out in the form of a **polar lobe.** As the division of the egg draws to its conclusion, the polar lobe is rounded off and remains connected to the rest of the egg only by a narrow cytoplasmic bridge. When the division is completed, the polar lobe is found to be connected with one of the daughter cells, namely, cell *CD* of Figure 56 (the cleavage of *Dentalium* is spiral), and at the end of mitosis it is fused with this cell. In this way cell *CD* gets all the three types of cytoplasm, while the other daughter cell, cell *AB*, gets a half of the animal pole's clear cytoplasm, a half of the granular cytoplasm, but no vegetal cytoplasm. Cell *CD* is consequently slightly larger than cell *AB*.

The formation of the polar lobe is repeated when the second cleavage begins, that is, when the egg divides into four blastomeres. Of the two daughter cells of blastomere *CD*, the polar lobe passes to blastomere *D*, while blastomere *C* receives only the clear cytoplasm of the animal pole and the granular cytoplasm.

The two blastomeres of the two-cell stage and the four blastomeres of the four-cell stage are therefore not equivalent as to their cytoplasmic composition, only one blastomere (*CD* or *D* respectively) containing the whole of the vegetal polar lobe material. The behavior of the blastomeres is also found to be different. If the first two blastomeres are separated and allowed to develop, blastomere *CD* is found to produce a complete larva (a **trochophore** in this case), though of diminished size. Blastomere *AB* also develops into a larva, but the larva is defective in some respects, the most significant defect being that it completely lacks the rudiment of the mesoderm normally contained in the posterior or "post-trochal" part of the larva (Fig. 57).

If the blastomeres are isolated in the four-cell stage, they are again found to differ in their development. Only blastomere *D,* which possesses the clear cytoplasm of the vegetal pole, develops into a complete, though smaller, trochophore larva; the other three blastomeres develop into defective larvae lacking the mesoderm rudiment. It has been concluded, therefore, that the clear cytoplasm of the vegetal pole is necessary for the development of the mesoderm.

This conclusion is further supported by the following experiment:

At the time of the first or second cleavage the polar lobe containing the clear cytoplasm of the vegetal pole can be nipped off. The remainder then contains all the clear cytoplasm of the animal pole and the granular cytoplasm. It also contains all the nuclear material but no vegetal polar lobe material. The result is that the larva developed from such an egg is defective and lacks the mesoderm rudiment. This is a very clear proof that it is the cytoplasm and not the nucleus that makes blastomere *D* different from the three others and able to produce a certain differentiation (the mesoderm rudiment) which the other blastomeres are incapable of producing.

The egg of the oligochaete *Tubifex* contains polar plasms that are very similar to those of *Dentalium.* There is a cap of transparent cytoplasm at the animal pole of the egg—the animal polar plasm. A broad equatorial

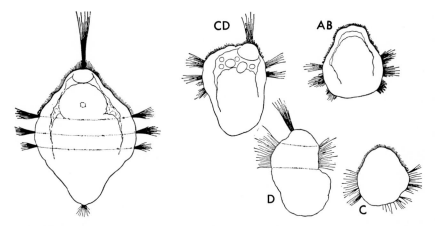

Fig. 57.   Larvae of *Dentalium* developing from a whole egg (left) and from separated blastomeres (right). The letters indicate from which blastomere each larva has developed. (From Wilson, 1904.)

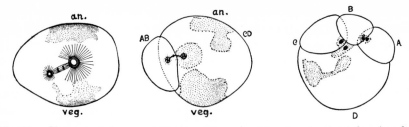

Fig. 58. Cleavage of the egg of *Tubifex*. The animal (*an.*) and vegetal (*veg.*) polar plasms are indicated by stippling. (After Penners, from Morgan, 1927.)

zone of the egg consists of granular cytoplasm. At the vegetal pole there is again a cap of transparent cytoplasm—the vegetal polar plasm. No polar lobe is formed, but nevertheless, both polar plasms are passed to blastomere *CD* during the first cleavage of the egg, and to blastomere *D* during the second cleavage (Fig. 58). The potentialities of the individual blastomeres may be tested by killing the other blastomeres by means of a localized exposure to ultraviolet rays (Penners, 1925). The survival of blastomere *D* in this animal is essential not only for the development of mesoderm but also for the development of the nervous system and the segmentation of the body. If blastomere *D* is killed, the remainder develops into ectoderm and endoderm, but no further differentiation is produced. If blastomeres *A, B* and *C* are killed, the remaining blastomere *D* may develop into a complete embryo. This it does the more easily because it is very much bigger than the other blastomeres.

That the difference between blastomere *D* and the other blastomeres is due to the presence of the vegetal polar plasm is shown by the following experiment (Penners, 1924). If the fertilized egg is exposed for a short time to an excessively high temperature, the first division of the egg is abnormal: the vegetal polar plasm, instead of passing to one of the daughter blastomeres, is equally divided between the two. The result is that both the first two blastomeres have the properties of the *CD* blastomere. Each of them is now capable of developing into a normal embryo with segmented mesoderm and a nervous system. In fact, if the egg is allowed to develop without further interruption, each of the first two blastomeres gives rise to a complete set of organs (segmented mesoderm, nervous system), so that a twinning from one egg is the result. The twins may be fused with each other to a varying degree. The importance of cytoplasmic substances for the development of various structures in the larvae has also been investigated in polychetes (Costello, 1945, 1948).

In the egg of the ascidian *Cynthia partita* no less than four cytoplasmic substances may be distinguished because of differences in the color of these substances (Conklin, 1905). In the egg at the beginning of cleavage, the animal hemisphere consists of clear transparent cytoplasm. The vegetal pole of the egg is taken up by a slaty gray cytoplasm, rich in yolk. In between, slightly below the equator, may be discerned two crescentic areas on oppo-

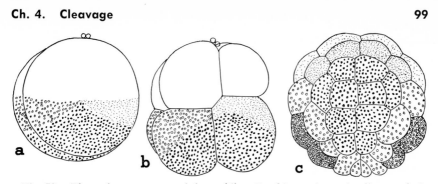

Fig. 59.   Three cleavage stages of the ascidian *Cynthia partita*. *a*, 2-cell stage. *b*, 8-cell stage. *c*, 64-cell stage. Large black dots represent slaty gray cytoplasm; loosely arranged circles represent yellow cytoplasm; closely spaced circles represent muscle cells; heavy stippling represents presumptive notochord; light stippling represents presumptive neural plate. (After Conklin, 1905, redrawn from Korschelt, 1936.)

site sides of the egg. One crescentic area consists of light gray cytoplasm, and the other consists of yellow cytoplasm, the yellow color due to the presence of yellow granules (Fig. 59).

During cleavage the four kinds of cytoplasm are distributed to different cells. Later the cells containing clear cytoplasm give rise to ectoderm; the cells containing slaty gray cytoplasm become endoderm; the cells containing yellow cytoplasm develop into mesoderm; and the cells containing light gray cytoplasm become neural system and notochord. The assumption may be made, therefore, that the development of the parts of the embryo enumerated is caused by the kind of cytoplasm that is contained in the cells. This can actually be proved by centrifuging the eggs before they have

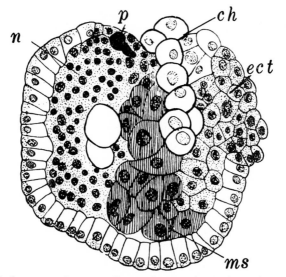

Fig. 60.   Displacement of organ rudiments in an embryo of *Cynthia partita* as a result of the centrifugation of the egg. *n*, Neural system; *p*, eye pigment; *ch*, notochord; *ect*, ectoderm; *ms*, muscle cells. (From Conklin, 1931.)

begun cleaving (Conklin, 1931). In this way the cytoplasmic substances within the egg may be displaced, and when the eggs begin to cleave, the various kinds of cytoplasm are held in abnormal positions when the substance of the egg is subdivided into cells. An embryo may develop from a centrifuged egg, possessing recognizable tissues and rudiments of organs, but these are arranged in a chaotic way. As the distinctions between the various cytoplasmic substances can still be discerned, it is discovered that each organ and tissue develops in that position where the respective cytoplasmic substance came to lie as the result of centrifugation (Fig. 60). The results of these experiments support the assumption made above about the decisive part played by the cytoplasmic substances in differentiation.

4–7   ROLE OF THE EGG CORTEX

Distribution of visible materials in the egg and in the embryo during cleavage is, however, not always of crucial importance for the localization of parts in the developing embryo. Some substances and cellular inclusions may be displaced without disturbing the normal segregation of the embryo into its subordinate parts. This can most conveniently be done by centrifugation of the uncleaved eggs. Centrifuging the eggs of most animals for a few minutes with moderate speeds, giving an acceleration of several thousand times gravity, is sufficient to rearrange various cellular inclusions in the interior of the egg according to their specific gravity.

After sufficiently strong centrifugation the eggs become stratified and show at least three typical layers. At the centripetal pole there is usually an accumulation of fat or lipoid droplets, which are the lightest constituents of the egg cytoplasm. There follows a layer of hyaline cytoplasm, which is the ground substance of the egg. The nucleus, or asters with chromosomes (if the centrifuged egg was in meiosis or mitosis), are also found in the hyaline layer. The yolk, as the most dense and heavy constituent of the egg, accumulates at the centripetal pole.

In the eggs of some animals the vegetal pole is so much heavier, owing to the presence of yolk, that it becomes oriented centrifugally during centrifugation. In this case the yolk is not displaced from its normal site at the vegetal pole but is only more concentrated. To displace the yolk in these cases the eggs have to be fixed in the desired position, so that they cannot freely rotate. This can be done sometimes by sucking them into narrow capillaries or by embedding the eggs in gelatin prior to centrifugation.

If the main axis of the embryo lies at random to the centrifugal force, as often happens, cellular inclusions will be dislocated to different parts in individual eggs. Figure 61 shows the results of centrifuging eggs of the sea urchin *Arbacia*. The red pigment granules present in these eggs are concentrated at the centrifugal end of the egg (Fig. 61, *a*). Some scattering of the granules occurs after the centrifuging is stopped and before cleavage progresses sufficiently to prevent further redistribution of the granules by subdividing the egg into blastomeres. It now becomes evident that the granules are concentrated in different positions in respect to the egg axis: near the vegetal pole (Fig. 61, *b*), near the animal pole (Fig. 61, *c*), or toward one side (Fig. 61, *d*). Independently of the position of the granules, the invagination of the blastopore occurs at the vegetal pole, so that the region of the blastopore will contain the pigment granules in some embryos but not in others (Fig. 61, *e, f*) (Morgan, 1927).

It has often been found that the pattern of cleavage may be highly independent of the distribution of cytoplasmic substances inside the egg. We have already described the cleavage of the mollusc *Dentalium,* in which a polar lobe appears at the vegetal pole during the first and second divisions of the egg and contains the cytoplasm necessary for the development of the mesoderm in the larva. A similar polar lobe is observed during cleavage of another mollusc, *Ilyanassa.* The polar lobe in this species is normally filled with yolky cytoplasm, while at the animal pole the egg cytoplasm is fairly free from yolk. Eggs of *Ilyanassa* have been centrifuged "in reverse," that is, with the vegetal pole fixed in position facing the axis of the centrifuge. As a result the heavy yolk is thrown into the animal hemisphere (still marked by the position of the polar bodies), and the hyaline cytoplasm and lipoid droplets are concentrated at the vegetal pole. Nevertheless, the polar lobe appears at the vegetal pole when the egg starts cleaving, although the lobe now contains mainly hyaline cytoplasm and lipoid instead of the yolk granules (Fig. 62). It is obvious that the formation of the polar lobe

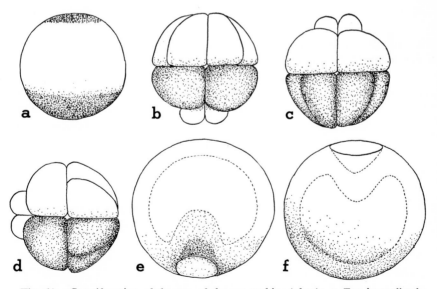

Fig. 61.   Centrifugation of the egg of the sea urchin *Arbacia. a,* Egg immediately after centrifugation: red pigment (stippling) thrown to centrifugal pole, oil droplets assembled at the centripetal pole. Drawings show position of red pigment in respect to the main axis during cleavage (*b, c, d*) and at the beginning of gastrulation (*e, f*). (From Morgan, 1927.)

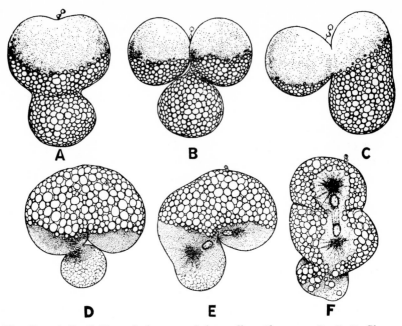

Fig. 62. *A, B, C,* Normal cleavage of the mollusc *Ilyanassa. D, E, F,* Cleavage stages after centrifugation "in the reverse," with the heavy yolk displaced towards the animal pole. (From Morgan, 1927.)

is not dependent on the yolk normally located at the vegetal pole, but on something that is not displaced by the centrifugal force (Morgan, 1927).

What can this something be? There are two possibilities. The first is that there exists in the cytoplasm some fixed network with sufficiently broad meshes to allow for the free movement of yolk granules and other inclusions without itself being torn or distorted. The existence of such a network cannot, however, be supported by any observation on the physical state or ultramicroscopic structure of cytoplasm. There remains the other alternative that the fixed system which is not displaced by centrifugation is the cortical layer of cytoplasm, or the **cortex** of the egg. This is in conformity with direct observation; the cortical granules (see section 2–5) are not displaced by centrifugation (Harvey, E. B., 1946). The layer of cytoplasm in which they are embedded is therefore sufficiently viscous to resist the forces usually generated in centrifugation experiments.

Since the immovable cortex of the egg appears to determine the point at which invagination begins in centrifuged *Arbacia* eggs, as well as to determine the position of the vegetal polar lobe in *Ilyanassa,* the further suggestion may be made that the cortex is the actual carrier of the polarity of the egg, or that the polarity of the egg is ingrained in its cortex. If this were correct, the distribution of the substances in the interior of the egg might be expected to be controlled or determined by the egg cortex. This suggestion finds support in some further results of centrifugation experiments, namely, the fact that cell constituents tend to return to their normal positions after the cessation of centrifuging. A scattering or mixing up of the strata into which the egg contents had been arranged by centrifugation could be a result of random movement of particles. But this explanation does not apply to cases in which after centrifugation certain particles not only move out from the position to which they were brought by the centrifugal force but take up a very definite location in the egg.

We have seen that in the egg of the ascidian *Cynthia* different kinds of egg cytoplasm may be displaced by centrifugation. Immediately after centrifugation the eggs show a very clear stratification, as can be seen in Fig. 63, *a*. The yellow granules, which go into the formation of mesoderm (see p. 99), are displaced to the centripetal pole; the yolk is displaced to the centrifugal pole; and the hyaline cytoplasm remains as a layer in between.

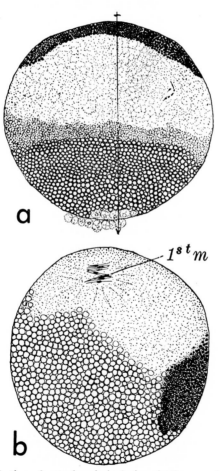

Fig. 63. Redistribution of cytoplasmic inclusions in the egg of *Cynthia partita* following centrifugation. *a*, Egg immediately after centrifugation; *b*, centrifuged egg 22 hours later (the yellow granules, shown black, have taken up a subequatorial position on one side of the egg). *1st m*, Spindle of first meiotic division. (From Conklin, 1931.)

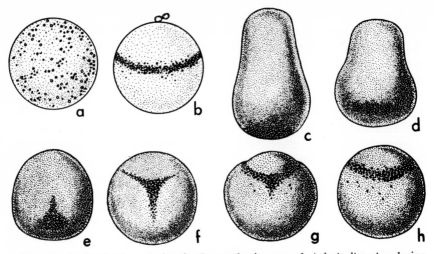

Fig. 64. Redistribution of vitamin C granules in eggs of *Aplysia limacina* during maturation and following centrifugation. *a*, Immature egg (oocyte); *b*, mature egg; *c*, egg immediately after centrifugation. *d–h*, Consecutive stages of recovery of the centrifuged egg, with the vitamin C granules returning to their normal position. (After Peltrera, from Raven, 1958.)

When the eggs are left to themselves after centrifugation, the egg substances start flowing and rearranging themselves in the interior of the egg. If the cell divisions set in sufficiently soon, this rearrangement is stopped by partitions appearing between the cleavage cells, and the result is the formation of the abnormal embryos dealt with earlier. If, however, the eggs are centrifuged well in advance of the beginning of cleavage, or if the first divisions of the egg are delayed, the redistribution of the egg contents may proceed so far that normal conditions are attained. Figure 63, *b* shows the end result of such a redistribution: the yolk is at the vegetal pole, the animal hemisphere is filled with hyaline cytoplasm, and the yellow granules take up a subequatorial area on one side of the egg, corresponding to the mesodermal "yellow crescent" of normal development (Conklin, 1931).

Another and perhaps even more impressive example is that of the egg of the mollusc *Aplysia limacina*. In the eggs of this animal there are to be found granules containing ascorbic acid (vitamin C), probably connected with the Golgi bodies. In immature oocytes the ascorbic acid granules are uniformly distributed throughout the egg cytoplasm. During maturation the granules accumulate in a ring lying inside the cortical cytoplasm and somewhat above the equator. By centrifugation the granules are concentrated at the centrifugal pole (Fig. 64), but after cessation of centrifugation the ascorbic acid granules soon start moving and again take up their normal position as a supra-equatorial ring (Peltrera, 1940).

The most plausible explanation of the last two experiments is that the displaced cytoplasmic particles tend to return to the proximity of certain regions of the egg cortex which had remained in their respective positions all the time the egg was being centrifuged. It would follow that in normal development the cytoplasmic substances in the egg are distributed in relation to local differences in the egg cortex, and that it is the egg cortex, therefore, that foreshadows in some way the pattern of the future development of the embryo.

4–8 THE MORPHOGENETIC GRADIENTS IN THE EGG CYTOPLASM

While we have learned that the cytoplasm of the egg, and in particular its cortex, starts the chain of reactions which eventually leads to the differentiation of parts of the embryo, it should not be imagined that the structure of the future embryo is rigidly determined by local differences in the cytoplasm of the egg. The local peculiarities of the egg cytoplasm are only some of the factors necessary for the formation of organ rudiments. That this is so can be shown by examining the development of the sea urchin, for instance.

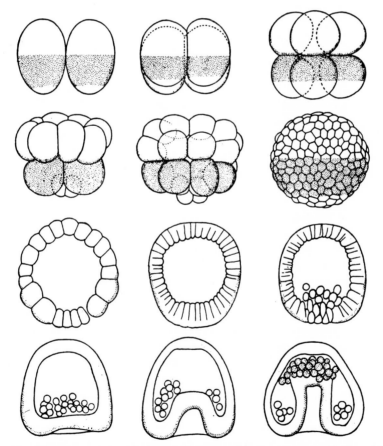

Fig. 65. Cleavage and gastrulation of the sea urchin *Paracentrotus lividus*. (After Boveri, from Spemann, 1936.)

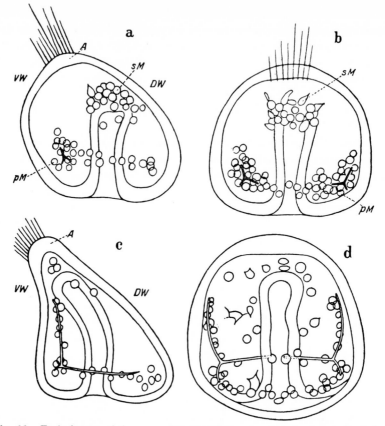

Fig. 66. Early larvae of the sea urchin *Parechinus* showing development of skeleton. *pM,* Primary mesenchyme; *sM,* secondary mesenchyme; *A,* apical tuft of ciliae; *VW* and *DW,* ventral and dorsal body walls. (After Schmidt, from Spemann, 1936.)

In the 16-cell stage of the sea urchin the blastomeres are of three different sizes. The animal hemisphere consists of eight blastomeres of medium size, the **mesomeres,** which are destined to produce most of the ectoderm of the larva. The vegetal hemisphere consists of four very large blastomeres, the **macromeres,** and of four very small blastomeres, the **micromeres,** which lie at the vegetal pole of the egg. The macromeres contain some more material for the ectoderm and all the material for the endoderm. In a subsequent cleavage the future ectoderm and endoderm are segregated from each other into an upper tier of macromeres and a lower tier of macromeres. The upper tier of macromeres lying immediately under the equator of the embryo contains the ectodermal material; the lower tier of macromeres, lying nearer to the vegetal pole, contains the endodermal material. The micromeres develop into mesenchyme, which later produces the larval skeleton consisting of calcareous spicules (Hörstadius, 1935). In the sea urchin *Paracentrotus lividus* the cytoplasm of the macromeres possesses a surface layer of red pigment granules, making the macromeres easily distinguishable from the other cleavage cells. The red pigment is already present in the egg at the beginning of cleavage as a broad subequatorial zone (Fig. 65).

In the blastula stage the cytoplasmic substances are found in the same arrangement as at the beginning of cleavage. Subsequently, the descendants of the micromeres migrate into the blastocoele where they develop the skeletal spicules; the descendants of the lower tier of macromeres invaginate, forming a pocketlike cavity—the gut; and the ectoderm produces a ciliary band, serving for locomotion, and a tuft of rigid cilia at the former animal pole. The ectoderm also sinks inward to produce a stomodeum, coming into communication with the endodermal gut at its anterior end, while the original opening of the pocketlike invagination (the blastopore) becomes the anal opening (Fig. 66).

The embryo of the sea urchin develops into a larva called a **pluteus** (Fig. 28).

If the blastomeres of the sea urchin are separated in the two-cell stage or in the four-cell stage, each of them develops into a complete pluteus of diminished size. This may be related to the fact that the first two cleavage planes are meridional, passing through the main axis of the egg. All of the first four blastomeres therefore get equal portions of the three cytoplasmic regions of the egg (ectodermal, endodermal and mesenchymal). A different result is observed if the egg is separated into the animal and vegetal halves after the third cleavage, the third cleavage plane being equatorial. In this case both halves produce, as a rule, defective embryos (Fig. 67). The animal half differentiates as an ectodermal vesicle; it does not produce a gut. Even purely ectodermal structures are abnormally differentiated (i.e., the ciliary band is not developed; the tuft of long cilia on the animal pole, on the other hand, develops excessively, growing over a much greater surface than it does normally) (Fig. 67, a). The vegetal half is differentiated into an ovoid embryo with a disproportionately large endodermal gut but without a mouth. There may be a few irregular skeletal spicules but no arms and no ciliary band (Fig. 67, b).

In later cleavage stages it is possible to cut the morula transversely below

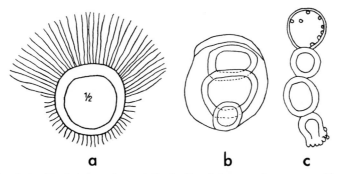

a   b   c

Fig. 67.   Animalized (a) and vegetalized (b, c) larvae of a sea urchin. (After Hörstadius, from Lehmann, 1945.)

the equator. In such cases the vegetal part develops still more abnormally: it produces a large endodermal gut and a small ectodermal vesicle (Fig. 67, c). The gut does not lie inside the ectodermal vesicle but is evaginated to the exterior (turned inside out), due to an interference with the normal processes of gastrulation. This phenomenon is known as **exogastrulation.** The mesenchyme cells migrate into the interior, but they usually produce no skeletal spicules or only very small and abnormal ones (Hörstadius, 1928, 1935).

The result of this experiment is obviously due to the fact that each of the two halves lacks some parts contained in the other half. However, each half does not simply produce what would normally have been the fate of the respective part. Instead, the differentiation of each half seems to be "exaggerated" as compared with its normal fate. This is especially clear in the case of the increased animal tuft of cilia.

The same "exaggeration" of the ectodermal or of the endodermal differentiations can be achieved by exposing the developing eggs to certain chemical substances, even without removing any parts of the egg or embryo. If the fertilized sea urchin egg is exposed to sea water containing some lithium salts in solution, the embryo develops just as if it were only the isolated vegetal half. The gut is increased and tends to exogastrulate instead of invaginating toward the interior (Herbst, 1893). The skeleton is absent or abnormal; the ectoderm is represented by an epithelial vesicle and fails to differentiate further. The increase in the size of the gut occurs at the expense of the ectoderm, and in extreme cases most of the embryo differentiates as an enormous everted gut, with the ectodermal vesicle reduced to a tiny appendage. The opposite effect is achieved if before fertilization the egg is exposed to artificial sea water lacking calcium ions but with sodium thiocyanate (NaSCN) added to it. In this case the gut is diminished or completely absent, the ciliary bands in the ectoderm fail to develop, and the tuft of stiff cilia at the animal pole is increased in size (Lindahl, 1933).

It appears that all these phenomena may be accounted for by assuming that there exist in the sea urchin's egg two factors or principles which are mutually antagonistic yet interact with each other at the same time, and that the normal development is dependent on a certain equilibrium between the two principles. Each has its center of activity at one of the poles of the egg. The activity diminishes away from the center, producing a **gradient** of activity. The two gradients of activity decline in opposite directions: the one from the animal pole, the other from the vegetal pole. Taking into account this type of distribution of activities, the factors or principles have themselves been called **gradients.** The two gradients are therefore the **animal gradient,** with a center of activity at the animal pole, and the **vegetal gradient,** with a center of activity at the vegetal pole.

According to this theory (originally suggested by Boveri in 1910, and developed in application to the sea urchin egg by Runnström, 1928, and Hörstadius, 1928) normal development depends on the presence of both gradients and an equilibrium between them. If the animal gradient is weakened or suppressed, the vegetal gradient becomes preponderant and the embryo is **vegetalized,** i.e., it develops to excess parts pertaining to the vegetal gradient, such as the gut. If the vegetal gradient is weakened or suppressed, the animal gradient becomes preponderant and the embryo is **animalized,** i.e., it develops to excess parts pertaining to the animal gradient, such as the tuft of stiff cilia at the animal pole. Other structures of the embryo, such as the ciliary bands and the skeleton, can only develop if both gradients are active, and the development is the more nearly normal the more the two gradients approach a correct equilibrium.

The equilibrium between the two gradients may be upset in various ways. The animal gradient may be weakened, with concomitant vegetalization of the embryo, by removing its center of activity (the blastomeres at the animal pole of the egg) or by the action of lithium salts. The vegetal gradient may be weakened by removing its center of activity (the vegetal part of the egg) or by the action of sodium thiocyanate.

The gradient concept aptly covers the result of various experiments on sea urchin's eggs. In addition we shall consider the following experiments. It is possible to separate from each other the three groups of blastomeres in the 16-cell stage—the mesomeres, the macromeres and the micromeres—and then recombine them at will (Hörstadius, 1935). In isolation the three groups develop as follows:

Isolated mesomeres develop into a vesicle with a tuft of cilia (as mentioned before), due to a preponderance of the animal gradient and animalization.

Isolated macromeres plus micromeres develop into an extremely vegetalized embryo with evaginated gut, the result of a preponderance of the vegetal gradient.

Mesomeres plus macromeres develop into a practically normal embryo; the macromeres bear a sufficiently strong vegetal gradient to counterbalance the animal gradient. Mesenchyme and skeleton develop in such embryos in spite of the absence of micromeres.

Micromeres alone are not capable of any development, since they do not keep together but fall apart.

Mesomeres (ectoderm) plus micromeres (mesenchyme) develop into a complete and more or less normal pluteus. This combination is especially illuminating since the gut of such embryos develops in spite of the absence of the macromeres which should normally have supplied the material for the gut. However, the embryo possesses the two gradients, the animal gradient borne by the mesomeres and the vegetal gradient borne by the micromeres, and the possession of the two gradients seems to create the necessary conditions for normal development.

What has been said is sufficient to show that the presence of different cytoplasmic substances in the egg does not necessarily mean that there is a direct relation between the substances in the cytoplasm of the egg and certain specific organs, in the sense that the cells containing the particular kind of cytoplasm develop directly into the corresponding organ. In the last experiment quoted the absence of the cytoplasm with red pigment did not prevent the development of the gut, because the necessary conditions for the gut development were provided for in another way.

Physicochemical Nature of the Animal-Vegetal Gradient System in Sea Urchin Eggs. The existence of animal and vegetal gradients is proved not only by the patterns of morphogenetic processes occurring after certain interventions in the normal development of the egg, but also by the direct demonstration of the gradients as peculiar physicochemical states of the cells of the developing embryo. One way of proving their existence is to study the reduction of dyes by the embryo under conditions of anaerobiosis. Sea urchin

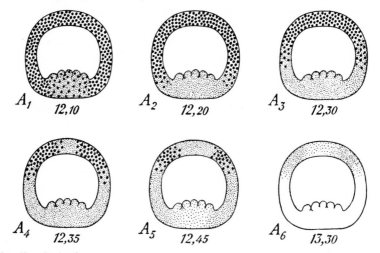

Fig. 68. Reduction of Janus green in a late blastula of a sea urchin. Large dots represent blue color; small dots, red; no dots show a complete fading of the stain. Figures indicate time of observation. (From Hörstadius, 1952.)

embryos (or embryos of other animals for that matter), slightly stained with the vital dye Janus green (diethylsafraninazodimethylaniline), are placed in a small chamber sealed off with petroleum jelly. After some time all the free oxygen contained in the chamber is used up as result of the respiration of the embryos. After that the embryos respire using Janus green as an acceptor of hydrogen, which is first reduced to a red dye, diethylsafranin, and further to a colorless substance, leucosafranin. The light grayish blue color of Janus green first changes into red, and then, as a second step in the reduction of the dye, the color disappears completely.

The reduction of Janus green, shown by the color change, does not occur simultaneously in the whole embryo but follows a very characteristic sequence. In the late blastula and early gastrula of the sea urchin the change in color is first noticeable at the vegetal pole, at the point where the primary mesenchyme is given off (Fig. 68). Then the color change spreads out, involving the whole vegetal hemisphere, and reaches the equator. At this stage a spot of changed color appears at the animal pole and also gradually increases, so that the bluish color remains only in the form of a ring lying well above the equator in the animal hemisphere. Even this ring eventually disappears. The red color now begins to fade in the same sequence, starting first from the vegetal pole and then from the animal pole (Child, 1936; Hörstadius, 1952).

The order in which the dye is reduced in the embryo shows such a nice correspondence to the postulated animal and vegetal gradients, that this alone would justify our mentioning these experiments. However, the connection goes much further. Where either the animal or vegetal gradient is suppressed the corresponding gradient of reduction also disappears. In isolated animal halves of sea urchins' eggs the animal tendencies of development are predominant, and it is found that such animalized embryos start reducing Janus green at the animal pole only; there is no center of reduction at the vegetal pole (Fig. 69). In isolated vegetal halves the animal tendencies of development are suppressed and the vegetal tendencies are supreme. Correspondingly, the only center of reduction is at the vegetal pole; no center of reduction appears at the animal pole (Fig. 70). Embryos animalized or vegetalized chemically show reduction gradients similar to isolated animal and vegetal halves of the egg (Hörstadius, 1955).

We have seen that the micromeres are carriers of the highest point of the vegetal gradient, and they preserve this property after transplantation. Accordingly, the micromeres can also serve as a center of a reduction gradient. If a group of four micromeres is implanted laterally, and the embryo is tested

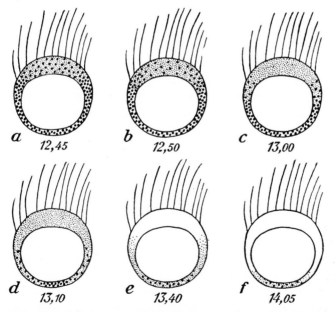

Fig. 69. Reduction of Janus green in animalized sea urchin embryo (isolated animal half). (From Hörstadius, 1952.)

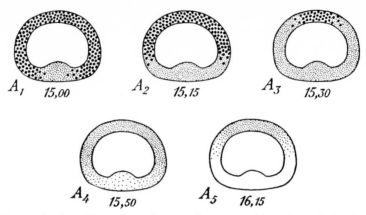

Fig. 70.   Reduction of Janus green in vegetalized sea urchin embryo (isolated vegetal half). (From Hörstadius, 1952.)

for reduction of Janus green, it can be seen that in addition to the two normal gradients, one from the vegetal pole and one from the animal pole, a third gradient appears, having the implanted micromeres as its center, but spreading out from these to the adjacent cells. In short, every modification of the gradient system of the embryo that is postulated from the morphogenetic behavior of the embryo is reflected in the gradients of reduction of Janus green (Hörstadius, 1952).

The animal and vegetal gradients can thus be considered to be definite metabolic processes or systems of metabolic reactions which involve oxidation and which have their points of highest intensity at the animal and vegetal pole respectively. The nature of these reactions is probably very complex and is as yet not fully understood, but some indications concerning these reactions may be deduced from a comparison of the chemical agents which may cause vegetalization or animalization.

Apart from the use of lithium ions, vegetalization may be caused by sodium azide (Child, 1948) and dinitrophenol. Both of these substances belong to a group of enzyme poisons; the azide is known to inactivate the cytochrome oxidase system, and the dinitrophenol disturbs respiration by peventing oxidative phosphorylation, that is, the formation of energy-rich bonds between phosphoric acid and adenosine diphosphate (with the formation of adenosine triphosphate). This might mean that vegetalization is based essentially on a disturbance of oxidative processes in the embryo (Lindahl, 1936) and, perhaps even more specifically, a disturbance of the processes of phosphorylation (Hörstadius, 1953b).

That the action of lithium ions is along the same lines may be deduced from a number of observations of which we shall single out the following. Lithium salt treatment suppresses the rise of oxygen consumption which occurs normally at the beginning of gastrulation, that is, at the same time as the morphological effects of vegetalization begin to be apparent (Lindahl, 1936). Furthermore, inorganic phosphate accumulates in the sea water during development of lithium ion–treated embryos — again a hint that these embryos are incapable of utilizing the energy of oxidation for phosphorylation and synthesis of adenosine triphosphate.

The whole sequence of reactions involved in vegetalization must, of course, be far more complicated. This we can conclude from the fact that the relative amounts of various amino acids change as a result of vegetalization in lithium ion–treated embryos (Gustafson and Hjelte, 1951), and this, of course, means that the structure of the proteins is modified by the treatment. Furthermore, if a modification of the respiratory system of the embryo is the essential feature of vegetalization, it remains to be discovered why the processes of morphogenesis at the animal pole of the embryo (essentially the ectodermal organs) are more severely affected than those of endodermal parts developing at the vegetal pole.

In this connection it may prove important that the distribution of mitochondria in a sea urchin blastula and early gastrula conforms to the animal-vegetal gradient; i.e., mitochondria are found in larger numbers in cells near to the animal pole and in smaller numbers toward the vegetal pole. Furthermore, the gradient of mitochondrial distribution closely follows the changes in the whole system of the embryo produced by animalization and vegetalization. For instance, the numbers of mitochondria in the animal hemisphere are decreased by lithium salt treatment or by implantation of micromeres (Lenicque, Hörstadius and Gustafson, 1953). Now it is known that the mitochondria are the main sites of oxidative phosphorylation, which seems to be the reaction most affected by vegetalizing agents. It may then be that the mitochondria are the immediate site of action of these agents.

As to animalizing agents, the first of those mentioned above, sodium thiocyanate, has been found to be rather unreliable. Some batches of eggs treated by this chemical do not react at all. In other batches individual embryos become animalized to greatly varying degrees. Subsequently, many other substances were found whose animalizing action is much more predictable. These belong to several distinct groups:

1. Some metals: zinc, mercury.
2. Some proteolytic enzymes: trypsin, chymotrypsin.
3. Many acidic dyes: in particular some possessing sulfonic (HSO_3) groups in their molecules, such as Evans blue, chlorazol sky blue, trypan blue, Congo red, and others possessing carboxyl groups (COOH), such as uranin and rose bengal.
4. Some other sulfonated organic compounds: germanin and others.
5. Some anionic (acid) detergents.

Animalization may be achieved by extreme dilutions of a chemical; for instance, 1:50,000 in the case of Evans blue.

It seems that the common property of most if not all the above agents is their ability to attack proteins or to form compounds with proteins, especially with basic proteins. In the case of the very large number of sulfonated and carboxylated dyestuffs it is fairly certain that their acidic groups become attached to the functional side groups of protein molecules. The two metals mentioned above, zinc and mercury, also become easily bonded to side groups of protein molecules. The case of mercury is very remarkable: mercuric chloride, a powerful poison used therefore as a fixative for proteins, causes animalization when applied for a very short time in great dilutions (1:90,000). It is probable that by blocking the active side groups of the protein molecules the agents referred to "immobilize" these molecules, preventing them from interacting normally with other cell constituents. Even the steric configurations of protein molecules may thus be altered, since these are based on lateral bonding between molecules and their parts. A far-reaching change of the properties of proteins may thus be brought about (Lallier, 1956 and 1957).

In the case of animalizing agents we are in the fortunate position that their point of attack can be clearly demonstrated. The dyes which have been used for this purpose stain the cells of the embryo, and it can be seen that the first cells to be stained are the primary mesenchyme cells at the vegetal pole. With higher concentrations of the dye or with a longer duration of the treatment the adjoining presumptive endodermal cells show the staining too. The zinc taken up into the cells of embryos may be made visible because it gives a pink coloration with dithizone. In embryos first treated with zinc and later immersed in a dithizone solution, the pink color is detectable in the same position as in the case of animalizing dyes (Lallier, 1956 and 1957). It is thus quite clear that animalization is due to damage to the center of the vegetal gradient.

Considering what has been said about the mechanism of action of the vegetalizing and animalizing agents, we may conclude that neither animalization nor vegetalization is ever achieved by a stimulation of one of the gradients, but always by a depression of the opposite gradient (or of the metabolic system which is responsible for the gradient). The increased development of any of the structures is then brought about by the alternative morphogenetic system expanding when the one is reduced in vigor. The competitive relation between the two gradients is thus demonstrated in a most convincing way.

4–9   MANIFESTATION OF MATERNAL GENES DURING
         THE EARLY STAGES OF DEVELOPMENT

In the previous sections evidence has been presented that: (1) The differences between the various parts of the embryo are not due to differences in their nuclei but to differences of the cytoplasm. From this it may be inferred that (2) the nuclei of the cleavage cells and the genes contained therein do not control the earliest stages of development of the animal egg.

This does not mean, however, that genetic factors have nothing to do with the early stages of development. The genetic factors, the genes, do play a part in this period but in a very special way. It is not the genes contained in the nuclei of the blastomeres, but the genes in the cells of the maternal body that determine the peculiarities of the egg and its early development. The best known example of this type of gene action is concerned with the inheritance of coiling in gastropod molluscs.

In the snails, as a rule, the shell is coiled spirally, and the internal organization of the animal shows a corresponding dislocation, some of the organs (heart, kidney and gills) being twisted around through an angle of nearly 180°. The direction of coiling of the shell is clockwise, if viewed from the apex of the shell. This type of coiling is called **dextral.** As an exception, the coiling may be counterclockwise, or **sinistral.** Individuals having a sinistral coiling of the shell have the internal organs dislocated in the opposite direction from that in dextral individuals. In short, dextral and sinistral individuals comport themselves in every detail of organization as mirror images of each other.

The eggs of gastropod molluscs show a spiral type of cleavage (see section 4–3), and it has been noted that the type of cleavage of the egg has a definite relation to the type of coiling of the adult. If the cleavage is dextral, that is, if the cleavage spindles show a clockwise spiraling, the adult snail also has a dextrally coiled shell. If the cleavage is sinistral, the coiling of the shell is likewise sinistral (Fig. 71). The connection between the type of cleavage and the coiling of the shell is established through the position of blastomere D in respect to the other blastomeres. Since it is from blastomere D that the mesoderm develops, its situation is reflected in the position of the internal organs, and the coiling of the shell is one of the secondary expressions of this asymmetry.

There are some species of gastropods in which all the individuals are sinistral, but the main interest attaches to a species in which sinistral individuals occur as a mutation among a population of normal dextral animals. Such

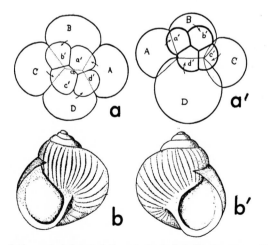

Fig. 71. Correlation of sinistral (a) and dextral (a') cleavage with sinistral (b) and dextral (b') coiling of the shell in gastropods. (After Conklin, from Morgan, 1927.)

a mutant was discovered in the fresh-water snail *Limnaea peregra* (Boycott, Diver, Garstang, and Turner, 1930). Breeding and cross breeding of dextral and sinistral snails showed that the difference between the two forms is dependent on a pair of allelomorphic genes, the gene for sinistrality being recessive (1), and the gene for the normal dextral coiling being dominant (L). The two genes are inherited according to Mendelian laws, but the action of any genetic combination is manifested only in the next generation after the one in which a given genotype is found.

Thus, if the eggs of a sinistral individual are fertilized by the sperm of a dextral individual, the eggs cleave sinistrally, and all the snails of this F_1 generation show a sinistral coiling of the shell. The genes of the sperm do not manifest themselves, although the genotype of the F_1 generation is Ll.

If a second generation (F_2) is bred from such sinistral individuals, it is all dextral, instead of showing segregation as would be expected in normal Mendelian inheritance. In fact, segregation does take place in the F_2 generation so far as the genes are concerned, but the new genic combinations fail to manifest themselves, since the coiling is determined by the genotype of the mother. The genotype being Ll, the gene for dextrality dominates and is responsible for the exclusively dextral coiling of the second generation. Only in the third generation (F_3) does segregation in the proportion of 3:1 become apparent, and then not as segregation between individuals in each brood, but as a segregation of broods. Each brood, that is, the offspring of an individual of the F_2 generation, shows a coiling that is determined by the maternal genotype. Since the individuals of the F_2 generation had the genotypes 1 LL, 2 Ll, 1 ll, three quarters of them, on the average, produce eggs developing into dextral snails, and one quarter produce eggs developing into sinistral individuals. Of the dextral broods, one third breed true, giving only dextral offspring, and two thirds show further segregation among the broods of the next generation.

It is easy to understand that the results of a reciprocal cross, that is, of the fertilization of the eggs of a dextral individual by the sperm of a sinistral individual, will lead to a somewhat different type of pedigree: the F_1 generation will be all dextral (with genotype Ll) and the F_2 generation again all dextral (with genotypes LL, Ll, lL and ll). The F_3 generation will show segregation among broods, just as in the cross examined first.

The whole case becomes clear if it is realized that the type of cleavage depends on the organization of the egg which is established before the maturation divisions of the oocyte nucleus. The type of cleavage is therefore under the influence of the genotype of the parent (the mother). The haploid state of the egg nucleus continues for only a very short time and cannot materially affect the organization of the egg. The sperm enters the egg after this organization is already established. Similarly, we should expect that the elaboration of cytoplasmic organ-forming substances is under the control of maternal genes, even though this cannot be proved because no mutants are known which produce a difference in these substances.

It is conceivable that maternal genes might produce conditions in the egg leading to morphogenetic processes which take place at a later stage of development, without visibly modifying the cleavage pattern. Several cases of such an influence are actually known.

In the meal moth *Ephestia kühniella* the dark pigmentation of the eyes, both in the adult moth and in the caterpillar, depends on the presence of a gene *A*, which controls the production of one of the precursors of the eye pigment, kynurenin (see also Chapter 17). In the allelomorph with recessive gene *a*, the ability to produce kynurenin is lost, and the eyes of the caterpillars and moths contain no dark pigment. In crosses between the normal moth and the *a* mutant, the pigmentation of the larval eyes is determined by the genotype of the mother, exactly as is the case with the coiling of the shell in *Limnaea peregra*. In a back cross between a heterozygous *Aa* female and a homozygous recessive *aa* male, 50 per cent of the offspring have the genotype *aa*; nevertheless, all the caterpillars that emerge from the eggs have dark pigment in their eyes. Only later, in the fourth instar, a difference becomes apparent between the *Aa* caterpillars and the *aa* caterpillars: in the *Aa* caterpillars the eye pigmentation persists, and the moths have black eyes; in the *aa* caterpillars the dark pigment gradually fades and eventually disappears altogether—the adult moths lack the dark pigment in their eyes.

The explanation of this is that the eggs which will have the genotype *aa* after fertilization will accumulate a certain amount of kynurenin while still in the ovaries. Before maturation the genotype of all the oocytes is of course

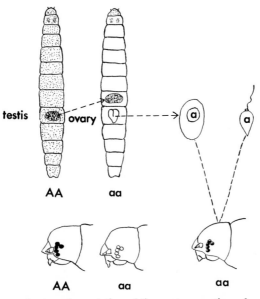

Fig. 72.   Effect on the eye pigmentation of the next generation when the testis from an *AA* caterpillar of the meal moth *Ephestia kühniella* is transplanted onto an *aa* caterpillar. Eye pigmentation of untreated *AA* and *aa* caterpillars is shown for comparison. (Modified after Kühn, 1941.)

the same as the genotype of the somatic cells of the mother, Aa in this case, and the presence of the A gene in the oocyte accounts for the elaboration of the eye pigment precursor. Besides, kynurenin is diffusible, and when produced by some cells of the mother, it may diffuse into other cells and so augment their supply. This latter circumstance has been cited as a direct proof of the mechanism by means of which the maternal genotype can influence the color of eyes of caterpillars (Kühn, 1936). Organs (testis, brain) from a normal caterpiller (AA) have been transplanted into caterpillars of the aa mutant. When the caterpillars metamorphosed into moths, the treated females were paired with aa males. The offspring from such a mating consisted of individuals with a purely recessive (aa) genotype. Nevertheless, the young caterpillars had pigmented eyes, just as if they had developed from eggs of a female bearing the gene A (Fig. 72). Since the gene A was actually absent both from the somatic cells of the mother and from the cells of the offspring, the only explanation for the experimental results is as follows: the precursor of the pigment, the kynurenin, was elaborated by the transplanted testis or brain, and by diffusion it penetrated into the oocytes of the female and was stored there. In the developing caterpillar the kynurenin was further processed into the eye pigment.

Kynurenin is of course not the only substance produced under the influence of the maternal genes while the egg is developing in the ovary. Very many if not all of the cytoplasmic substances in the egg, mentioned in section 2–5, must have a similar origin. The cytoplasmic substances determining the properties of blastomeres during cleavage may thus be an after-effect of the action of genes at a previous stage of the reproductive cycle.

PART III

Gastrulation and the Formation of the Primary Organ Rudiments

The process of gastrulation is one of displacement of parts of the early embryo. As a result the endodermal and mesodermal organ rudiments are removed from the surface of the embryo, where the presumptive material for these rudiments is to be found in the blastula stage, and brought into the interior of the embryo, where the respective organs are found in the differentiated animal. At the same time the single layer of cells, the blastoderm, gives rise to three germinal layers—the ectoderm, the endoderm and the mesoderm.

The most prominent feature of gastrulation is thus:
1. A rearrangement of cells of the embryo by means of **morphogenetic movements,**

and at the same time
2. The rhythm of cellular divisions is slowed down.
3. Growth, if any, is insignificant.
4. The type of metabolism changes; oxidation is intensified.
5. The nuclei become more active in controlling the activities of the embryonic cells. The influence of the paternal chromosomes becomes evident during gastrulation.

CHAPTER **5**

Gastrulation

5–1 FATE MAPS

A correct interpretation of gastrulation is impossible without a knowledge of the position which the presumptive germinal layers occupy in the blastula. This position may be ascertained in various ways. A chart, showing the fate of each part of an early embryo, in particular a blastula, is called a **fate map.**

In tracing the fate of the various parts of the blastoderm, it is sometimes possible to make use of the peculiarities of the cytoplasm in various parts of the egg, such as the presence of pigment granules, etc. In the developing amphibian egg, for instance, one may trace in which part of the differentiated embryo the black pigment comes to lie. Originally this pigment is restricted to the animal hemisphere of the egg. However, peculiarities of pigmentation are seldom sufficient to make it possible to reconstruct the fate map in any great detail. Recourse must be made to artificial marking of parts of the blasto-derm. A satisfactory method of marking was devised for this purpose by Vogt (1925). The method consists in soaking a piece of agar in a vital dye (Nile blue sulfate, neutral red, Bismarck brown) and then applying the piece of agar to the surface of the embryo in the necessary position. The dye diffuses from the agar, and in a matter of minutes the cells of the embryo to which the agar has been applied take up sufficient dye to produce a stain on the surface of the embryo. This marking can be done without removing the vitelline membrane, since it is permeable to the vital dyes, and so the embryo continues to develop normally. The presence of the stain does not change the normal development of the embryo, and the position of the stained cells in the differentiated embryo clearly shows the fate of the stained area. It has been established by trial that the vital dye remains, on the whole, restricted to the cells which had originally taken up the dye and to their descendants. The diffusion of the stain, if the staining has been done correctly, is negligible, and does not interfere with interpreting the results. Several stain marks may be made on the surface of the same embryo, using different colors (red, blue, brown). In this way one experiment may disclose the fate of many parts of the early embryo at the same time.

The vital stain marking method was in the first place applied to the reconstruction of the fate map in the amphibian embryo, and the original investigations have been subsequently checked by many embryologists. It is most advantageous, therefore, that the fate map of an amphibian embryo, such as the embryo of a newt (*Triturus*) or axolotl (*Ambystoma*), should be dealt with first in this description.

The Fate Map of an Amphibian Embryo. The whole surface of the blastula of an amphibian may be roughly divided into three main regions (Vogt, 1925): (1) a large area on and around the animal pole, (2) an intermediate zone, also known as the **marginal zone,** going all around the equator of the blastula, (3) the area on and around the vegetal pole (Fig. 73). These three main regions coincide approximately with areas which differ in their pigmentation. The animal region is all deeply pigmented. The marginal zone, which is much broader on one side of the embryo, is pig-

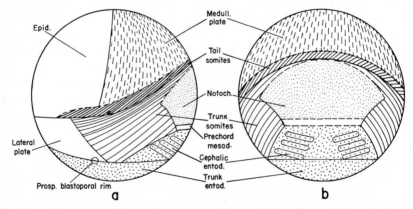

Fig. 73. Fate maps of the early gastrula of the axolotl. *a,* Lateral view; *b,* dorsal view. (After Pasteels, from Willier, Weiss and Hamburger, 1955.)

mented but not so deeply as the region around the animal pole. The vegetal region is more or less devoid of pigment.

Inside each region we find areas corresponding to the future organs of the animal. The animal region consists of two main areas: the area whose fate is to develop into the nervous system of the embryo, and the area which is to become the skin epidermis of the embryo. The material for the sense organs is also contained in these two areas. Inside the nervous system area a small sub-area may be traced which is to participate in the formation of the eyes of the embryo; inside the epidermis area the material for the nose and the ears and the ectodermal part of the mouth may similarly be traced. In the intermediate or marginal zone we find the material for the notochord occupying a large area on the dorsal side of the blastula. This is followed by an area lying nearer to the vegetal pole and containing the material for the prechordal connective tissue; this area is known as the prechordal plate. Further down, toward the vegetal pole but still inside the marginal zone, lies the material for the anterior parts of the alimentary canal: the endodermal lining of the mouth, gill region and pharynx. The parts of the marginal zone on both sides of the notochordal area are taken up by the material for the segmental muscles of the body. The lateral and ventral parts of the marginal zone give rise to the mesodermal lining of the body cavity, the kidneys, etc. (ventrolateral mesoderm). The vegetal region is composed of cells which are later found in the midgut and hindgut.

It will be noticed that the areas destined to develop into the organs of the mid-dorsal part of the animal lie on one side of the blastula, on that side where the marginal zone is broadest. The central part of the marginal zone is taken up by the notochord area. Nearer to the animal pole is situated the area of the neural system. This side of the blastula corresponds therefore to the dorsal side of the embryo. Similarly, parts of the head (eye, nose, ears) develop from areas near the animal pole of the blastula, which therefore corresponds to the anterior end of the embryo. Since the materials of the egg have not been displaced to any great extent during the cleavage, it may be inferred that the animal pole of the fertilized egg corresponds to the anterior end of the embryo. Where the marginal zone is the broadest is the dorsal side; the opposite side may be considered as ventral, and the vegetal pole as the posterior. We find, however, that the foregut area (pharynx, part of the epithelium of the mouth) is also situated on what we have agreed to call the dorsal side of the egg. The explanation of this fact will be found later (section 5–4).

On the whole, however, the location of the areas destined to develop into most of the organs does not seem to have anything in common with the position of the same organs in the adult animal. Organs which later are situated in the interior of the animal's body, such as the notochord, the gut or the brain, are represented by areas laid out on the surface of the blastula. The cells destined to cover the whole of the animal's body, such as the epidermis of the skin, occupy only a limited area on the surface of the blastula. It is obvious that a far reaching displacement or reshuffle of the parts of the blastoderm must take place before every cell can arrive at its final position.

We shall now leave the amphibians and consider the fate maps of other animals, confining ourselves, however, to the chordates.

Conditions in Acrania.   In *Amphioxus* the egg is too small to be marked by vital stains after the method of Vogt, but there are differences in the various regions of the egg cytoplasm which permit us to trace these regions into the later stages of development and thus to reconstruct a fate map, at least in rough outlines.

At the beginning of cleavage three regions can be distinguished in the *Amphioxus* egg (Conklin, 1932). Near the vegetal pole there is found a mass of cytoplasm which contains the greatest amount of yolk (although yolk in this case is not abundant and the yolk granules are relatively very

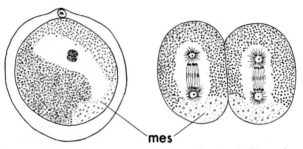

Fig. 74.   Medial section of uncleaved egg of *Amphioxus* (left), and an equatorial section of the two-cell stage (right), showing the position of the mesodermal crescent. (After Conklin, 1932.)

small). The animal hemisphere of the egg consists of cytoplasm that has less yolk and is in consequence more transparent. On one side of the egg, in a position roughly corresponding to the marginal zone of the amphibian egg, there is a special type of cytoplasm. This cytoplasm does not contain much yolk, but it can be distinguished from the animal cytoplasm by its ability to be deeply stained by basic dyes. The mass of cytoplasm of this kind has a crescentic shape, the attenuated ends of the crescent being drawn out along the equator of the egg about halfway round (Fig. 74).

During the period of cleavage the three regions become subdivided into blastomeres without the cytoplasmic substances having been displaced to any great extent. The distinctions which could be traced in the cytoplasm of the egg now become accentuated by further distinctions in the size and shape of the blastomeres. The vegetal material is now contained in a number of rather large cells taking up the position on and around the vegetal pole of the blastula. The animal hemisphere is made up of cells containing the clear cytoplasm. The cells are columnar and form a very closely packed columnar epithelium. The cells containing the basophilic cytoplasm are clearly discernible even as to shape. They are the smallest cells in the blastula, smaller even than the animal cells, and they are rather loosely packed, the external surfaces bulging out, as is usually found in the earlier cleavage stages (Fig. 75).

The fate of the three regions is the following: The clear cytoplasm that later becomes the animal hemisphere of the blastula develops mainly into skin epidermis. The granular cytoplasm, which takes up the region around the vegetal pole of the blastula, develops into the lining of the alimentary canal. The crescent of basophilic cytoplasm is the material which gives rise to the muscles and the lining of the body cavity.

It is thus seen that there is a general similarity in the fate maps of *Amphioxus* and amphibians. The crescent of basophilic cytoplasm corresponds to part of the marginal zone of amphibians. The marginal zone in *Amphioxus* cannot be traced all around the equator of the egg by using

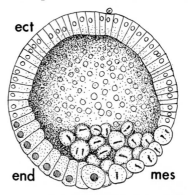

Fig. 75.   Blastula of *Amphioxus* showing distinction between the cells of the mesodermal crescent (*mes*) and the cells of the presumptive endoderm (*end*) and ectoderm (*ect*). (After Conklin, 1932.)

only the landmarks presented by differences in the cytoplasm. By tracing the development of the embryo in later stages, it is fully evident, however, that in the *Amphioxus* blastula the material for the notochord and the neural system occupies a position near the equator of the egg on the side opposite to the one taken up by the muscle area. This makes the similarity to the amphibians practically complete. There is also a striking similarity to the distribution of different kinds of cytoplasm in the ascidian embryo (see section 3–5).

Of all the other holoblastic eggs in lower vertebrates (cyclostomes, ganoids, lungfishes) the fate map has been investigated only in the lampreys. It has been found that the fate map is essentially the same as in the amphibians (Pasteels, 1940).

In meroblastic vertebrates the fate map must necessarily be different, owing to the presence of the uncleaved yolk. Nevertheless, a distinct similarity can be discovered in the position of the various areas relative to one another. The fate maps have been worked out for elasmobranchs, bony fishes, reptiles and birds.

Fate Map in a Bony Fish (Fig. 76).   In fishes the areas destined to develop into the organs of the dorsal region of the animal are concentrated toward one edge of the blastodisc which can be denoted as posterior. At the edge of the disc lies the area for the alimentary canal; in front of the

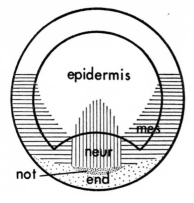

Fig. 76. Fate map of a bony fish *Fundulus,* after Oppenheimer. *neur,* Presumptive neural system; *mes,* presumptive mesoderm; *end,* presumptive endoderm; *not,* presumptive notochord. (After Oppenheimer, 1947.)

area for the alimentary canal lies the area for the notochord; and still further forward, toward the center of the blastodisc, lies the area for the neural system. The material for the mesodermal muscles and the lining of the body cavity is situated on both sides of the gut and notochord areas and is drawn out along the edge of the blastodisc. In *Fundulus* the mesoderm area is interrupted in the anterior part of the blastodisc, and here lies the presumptive material for the skin epithelium (Oppenheimer, 1947). In *Salmo,* however, the mesodermal area reaches all around the circumference of the blastodisc (Pasteels, 1940).

The Fate Map of the Avian Blastula.   The fate map in birds presents a new feature as compared with the fate map in fishes. It has already been indicated that the blastoderm in birds consists of two parts: the area pellucida and the area opaca. The area opaca is completely excluded from the formation of the body of the embryo, and the fate map, in so far as the parts of the embryo are concerned, is restricted to the area pellucida. Furthermore, since the blastoderm is bilaminar, consisting of epiblast and hypoblast, fate maps have to be made out for each of the two layers separately. The fate map, as described below, is the outcome of numerous experiments carried out by different investigators. Some of these have consisted of marking the blastoderm by vital staining (Wetzel, 1929; Pasteels, 1940), others of marking with carbon particles (Spratt, 1946). The marking was done either *in situ* in the egg after a window was cut in the shell, or on blastoderms cultivated *in vitro.* We shall first deal with the fate map for the epiblast (Fig. 77).

Roughly the anterior half of the epiblast of the area pellucida contains material for the skin epithelium and also for the embryonic membranes (see section 7–1). Almost in the middle of the area pellucida, or slightly anterior to the middle, lies the presumptive neural plate in the form of a transverse crescentic area, broadest in the middle and tapering to the sides. Behind the neural plate area, in the middle, lies the presumptive notochord; neither of these areas reaches far out to the sides. The area containing material for segmented muscle is situated in the midline behind the notochordal

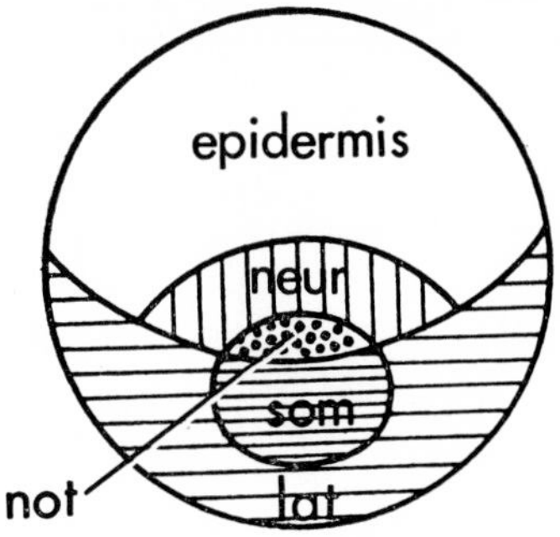

Fig. 77. Fate map of bird. *neur,* Presumptive neural system; *not,* presumptive notochord; *som,* presumptive somites; *lat,* presumptive lateral mesoderm. (From Waddington, 1952.)

area. Most of the posterior half of the area pellucida is taken up by the material producing the mesodermal lining of the body cavity (the ventro-lateral mesoderm) and by the extraembryonic mesoderm, which will not participate in the formation of the embryo (see Pasteels, 1940; Spratt, 1946; Waddington, 1952).

As we shall see later (section 5–7), the epiblast also contains some presumptive endoderm. The cells which go into the formation of the endoderm, however, are few in number and apparently are not concentrated as a separate area but are derived from a fairly large field along the midline, mainly in the posterior part of the area pellucida.

The hypoblast is less accessible for local marking, since it is covered by the epiblast, but it has been possible to mark it with carbon particles when the blastoderm is excised from the egg and cultivated *in vitro* upside down (Fraser, R., 1954). In this way it was found that some of the hypoblast cells contribute to the formation of the notochord. Most of the hypoblast, however, is presumptive endoderm, though only a small part of it will become incorporated into the gut of the embryo. The remainder is destined to become the lining of the yolk sac (see further, section 7–1). It must be noted that owing to difficulties of work with the avian blastoderm, the boundaries of the various areas in the fate map should be considered as being only approximate (Rudnick, 1948).

5–2   GASTRULATION IN *AMPHIOXUS*

In the preceding section it has been made amply evident that in all vertebrates the position of the organ-forming areas in the blastula stage differs greatly from the position of the respective organs in the adult animal. The various parts of the embryo must be moved about, displaced from their original locations, if they are to arrive at their final position. This displacement of parts takes place in the process of **gastrulation.** This will first be described as it occurs in *Amphioxus,* then in the amphibians, and lastly in the meroblastic eggs of the fishes and birds.

We have seen that in the blastula of *Amphioxus* the following areas can be distinguished: (1) the endodermal material, which lies on and around the vegetal pole; (2) the marginal zone, which is made up of the mesodermal

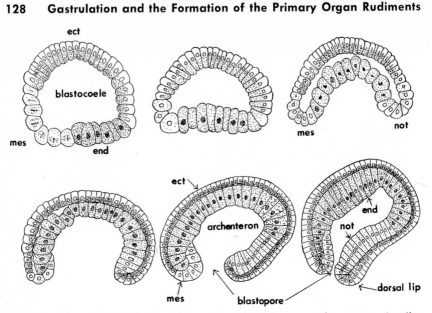

Fig. 78. Gastrulation of *Amphioxus*. A series of consecutive stages (median sections). (After Conklin, 1932.)

crescent on the one side and the area for the notochord on the other side; (3) the animal hemisphere, which consists of cells of the presumptive skin epidermis and the presumptive nervous system, the latter lying above the presumptive notochord.

Gastrulation is initiated when the blastoderm at the vegetal pole becomes flat and subsequently bends inward, so that the whole embryo, instead of being spherical, becomes converted into a cup-shaped structure with a large cavity in open communication with the exterior on the side that was originally the vegetal pole of the embryo. The cup has a double wall, an external and an internal one, the latter lining the newly formed cavity. The external and internal epithelial layers are continuous with each other over the rim of the cup-shaped embryo. In this stage there is still a space between the external and the internal wall representing the remnants of what was the blastocoele of the blastula (Fig. 78).

The external lining consists of presumptive epidermis and presumptive nervous system. In other words, it consists of parts which have been classified as ectoderm. The internal lining consists mainly of the presumptive gut material, that is, of endoderm. The presumptive material of the notochord and the mesodermal crescent at first lie on the rim of the cup, but very soon they shift inward so as to occupy a position on the internal wall of the cup. In this way the endoderm, the mesoderm and the notochord disappear from the surface of the embryo into the interior where they belong. The external surface of the embryo now consists of ectoderm. The embryo in this stage of development is called a **gastrula.** The movements of infolding or inward bending of the endoderm and mesoderm are known as **invagination.** The cavity arising through the invagination of the endoderm and mesoderm is called the **primary gut** or **archenteron.** The opening of the archenteron to the exterior is called the **blastopore.** At the same time the blastopore denotes the pathway by which the endoderm and mesoderm pass into the interior of the embryo. The rim which surrounds the blastopore of the cup-shaped embryo may be likened to a mouth, so that it is usual to speak of the **lips** of the blastopore. We may distinguish the dorsal lip, the ventral lip and the lateral lips of the blastopore, respectively.

The blastopore is very broad in the initial stage of gastrulation, but soon the lips of the blastopore begin to contract, so that the opening which leads into the archenteron becomes smaller and is eventually reduced to an insignificant fraction of the original orifice. This contraction of the lips of the blastopore is connected with the disappearance of the mesodermal crescent material and the presumptive notochord from the rim of the cup-shaped embryo. As more material is shifted into the interior of the gastrula the remnants of the blastocoele become completely obliterated by the two walls of the embryo coming into contact with each other.

As the presumptive notochord and the mesodermal crescent shift into the interior of the gastrula, they also change their position relative to each other (Fig. 79). In the blastula these two areas lie on opposite sides of the embryo. Now the lateral horns of the mesodermal crescent converge toward the dorsal side of the embryo and come to lie on both sides of the presumptive notochord. In the next stage that follows after the contraction of the rim of the blastopore, the embryo becomes elongated in an anteroposterior direction, all the various presumptive areas participating in this elongation. The elongation of the notochordal and the mesodermal material brings them into still closer contact with each other, the notochordal material shifting backward, in between the two horns of the mesodermal crescent material. The result of these movements is that the notochordal material becomes stretched into a longitudinal band of cells lying medially in the dorsal inner wall of the gastrula and flanked on both sides by bands of mesodermal cells similarly stretched in a longitudinal direction. The remainder

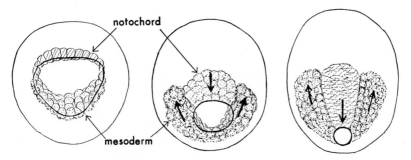

Fig. 79. Change in the relative position of the presumptive mesoderm and the presumptive notochord during closure of the blastopore in amphioxus. Diagrammatic. (After Conklin, 1932.)

of the lateral, ventral and anterior parts of the inner wall of the gastrula consists of endodermal cells.

The external wall of the gastrula similarly takes part in the elongation of the embryo. One of the results of this is that the presumptive material of the nervous system becomes stretched into a longitudinal band of cells lying mediodorsally over the notochordal material but being somewhat broader than the latter.

5–3 FORMATION OF THE PRIMARY ORGAN RUDIMENTS IN *AMPHIOXUS*

Immediately after the various presumptive organ rudiments have taken up their position in the inside and on the surface of the gastrula, the next phase of development sets in. The sheets of epithelium which were representative of the various parts of the future animal become broken up into discrete cell masses of diverse shape, which can be called the **primary organ rudiments.** The term "primary" indicates that the structures in question are not final; they are actually complex in nature, and a further subdivision of the cell masses, or at least of most of them, is necessary before every single organ and structure of the adult animal appears as such.

In the interior of the embryo the presumptive materials of the notochord, the mesoderm and the gut become separated from each other by crevices appearing along the boundary lines of each (Fig. 80). The strip of cells lying mid-dorsally rounds itself off forthwith and becomes a cylindrical cord

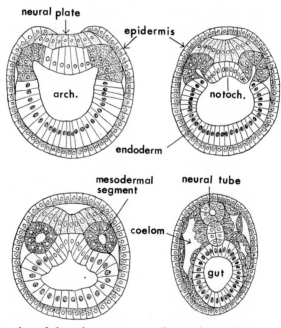

Fig. 80. Formation of the primary organ rudiments in *Amphioxus*. Transverse sections. (After Hatchek, from Korschelt, 1936.)

of cells, the **notochord,** which still differs from the same organ of the adult *Amphioxus* in the structure of its cells.

As the presumptive material of the mesoderm becomes separated both from the notochord and the endoderm, it breaks up into a series of roughly cuboidal masses of cells lying on each side, one behind the other along the length of the animal's body. These blocks of mesodermal cells are called the **mesodermal segments.** In connection with their formation one detail should be noted. Just before the mesodermal cells become separated from the endoderm and notochord, a longitudinal groove appears on the inner surface of the mesodermal bands, that is, on the surface toward the archenteron. The groove becomes drawn out into each of the mesodermal segments, so that a pocketlike invagination is formed in each. At the same time crevices cut in between adjoining segments from the outside, separating them from each other (Fig. 81). When the mesodermal segments become eventually separated from the endoderm and notochord, the pocketlike invaginations in each become completely closed off from the cavity of the archenteron, and become small cavities inside the mesodermal segments. In a later stage of development, these cavities expand and become the secondary body cavity or **coelom** of the adult animal. The derivation of the cavities of the mesodermal segments from the archenteric cavity is distinct in the case of the most anterior segments, but the posterior mesodermal segments are solid masses of cells at first and acquire cavities later as a result of the separation of the cells in the middle.

When the notochord and the mesodermal segments dissociate themselves from the endodermal material, the free edges of the endoderm approximate each other and fuse along the dorsal midline. The endoderm thus becomes converted into a closed sac. The cavity of this sac becomes the cavity of the alimentary canal.

In the ectoderm the presumptive nervous system material becomes separated from the surrounding presumptive epidermis in the form of an elongated plate, the **neural plate.** The neural plate sinks below the level of the remainder of the ectoderm and is then covered by the free edges of the epidermal epithelium (Fig. 80). The neural plate forthwith rolls itself into a tube, the lateral edges folding themselves upward and fusing along the

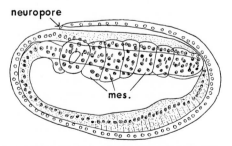

Fig. 81.   Embryo of *Amphioxus* in lateral view showing mesodermal segments (*mes*) overlying the notochord and neural plate. (After Conklin, 1932.)

midline. The neural plate is then transformed into the **neural tube,** which becomes the spinal cord of the animal. The neural tube does not close completely at the anterior end but leaves an opening, the **neuropore,** which remains patent until the later stages of development.

The epidermal epithelium which covers the neural plate by shifting over its surface is derived from the sides and from the area posterior to the neural plate. Because the hind end of the neural plate borders on the blastopore, the posterior portion of epidermis covering the neural plate is derived from the region below (ventral to) the blastopore. As the epidermis shifts forward over the surface of the neural plate, it also passes over the blastopore. The blastopore becomes cut off from the exterior and opens into a space lined by the walls of the neural tube. The canal thus formed, which connects the blastopore, and therefore the archenteron, with the cavity of the neural tube, is called the **neurenteric canal.** The canal persists for only a short time, until the cavity of the neural tube (later becoming the central canal of the spinal cord) becomes completely separated from the cavity of the archenteron, which in turn becomes that of the alimentary canal. The cavity of the alimentary canal later acquires communication with the exterior by means of the oral and anal openings which break through the body wall, but this does not take place until a very much later stage of development.

After the formation of the spinal cord, the notochord, the gut and the mesodermal segments, the early embryo of *Amphioxus* possesses most of the essential features of the organization of chordate animals. The features which are still lacking are: a mouth, an anus and the gill clefts. The gill clefts, as well as the mouth, are developed in all chordates at a comparatively late stage. We will see that a stage of development in which an embryo possesses a spinal cord, notochord, gut and the segmented mesoderm recurs with great tenacity in all vertebrates, and that the subsequent stages of development in all classes of vertebrates are much more similar than are the processes of cleavage and gastrulation.

5–4   GASTRULATION IN AMPHIBIANS

The purpose which gastrulation must achieve in the amphibians is the same as in *Amphioxus:* the areas of the blastoderm which are destined to become gut, notochord, muscle, etc., have to be brought into the interior of the embryo. However, in the amphibians this cannot be done by the bending inward of the vegetal region of the blastoderm, as in *Amphioxus,* because the vegetal wall of the blastula is far too thick and overladen with yolk, and is therefore rather passive in its behavior. The processes leading to the disappearance of the material for the internal organs from the surface are therefore carried out mainly by the more active cells of the marginal zone. A correct understanding of amphibian gastrulation has been achieved by the method of localized vital staining—by applying marks of vital stain on the surface of the embryo and tracing the movements of the stained areas in the course of gastrulation (Vogt, 1929).

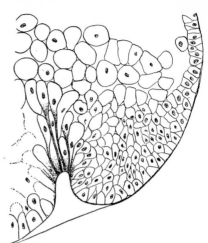

Fig. 82. Median section through the blastopore region of an early gastrula of a newt, showing the cells at the bottom of the pit streaming into the interior. (After Vogt, 1929.)

The first trace of gastrulation which can be observed in an egg of a newt or a frog is the formation of a depression or groove on the dorsal side of the embryo, at the boundary between the gray crescent (see p. 69) and the vegetal region. If the embryo is studied in sections in this stage, one may observe that the cells at the bottom of the groove are streaming into the interior of the embryo toward the blastocoele cavity (Fig. 82). In the next phase of gastrulation the groove begins to spread out into a transverse furrow, and the lateral ends of the furrow are prolonged all along the boundary between the marginal zone and the vegetal region until they meet at the opposite, ventral side of the embryo, thus encircling the vegetal region (Fig. 83). The groove and, later, the furrow represent the blastopore. Both are produced by the invagination of the endodermal and mesodermal material into the interior of the embryo.

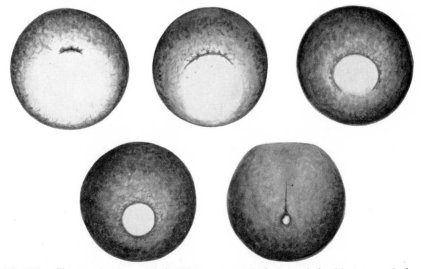

Fig. 83. Changes in shape of the blastopore and closure of the blastopore during gastrulation in a frog.

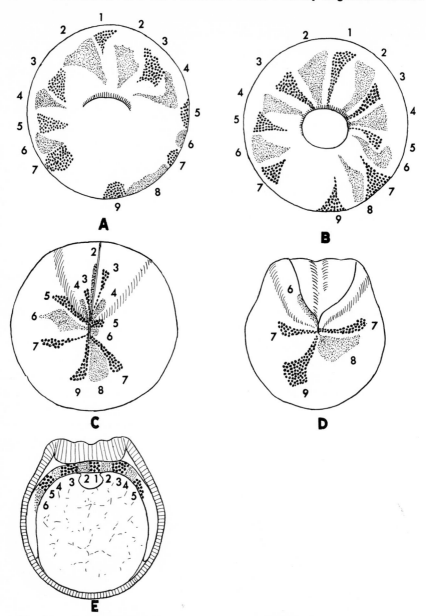

Fig. 84. Experiment of local vital staining on a gastrula of *Triturus*. The marks were made around the circumference of the marginal zone. *A, B, C, D*, Surface views; *E*, transverse section to show position of the stained areas in the roof of the archenteron. (From Vogt, 1929.)

The invagination in this case is not quite similar to the invagination in amphioxus. The original groove may be said to have been the result of the bending inward of a portion of the blastoderm and the formation of a pocketlike depression. Once the depression has been formed, however, the further invagination can better be described as the rolling of the superficial cells over the rim of the blastopore and into its interior while new portions of the blastoderm approach the rim in their stead. While the superficial material is rolling over the rim of the blastopore, the rim itself does

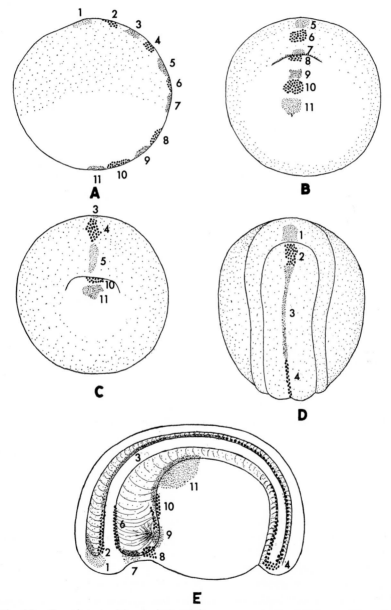

Fig. 85. Experiment of local vital staining on a gastrula of *Triturus*. The marks were made along the dorsal median surface. *A, B, C, D,* Surface views; *E,* the embryo dissected in the medial plane to show the position of the stained areas in the interior. (From Vogt, 1929.)

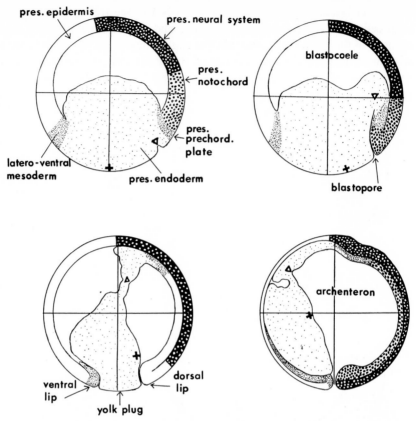

Fig. 86. Diagrammatic median sections of amphibian embryos from the beginning to the completion of gastrulation. The triangle marks the cells which start invagination. The cross marks the position of the vegetal pole of the blastula. (After Vogt, 1929.)

not remain stationary but gradually shifts over the surface of the vegetal region, moving away from the animal pole and toward the vegetal pole. By the time the blastopore has the form of a ring, the cells of the vegetal region are still to be seen filling the space enclosed by the edge of the blastopore. These cells are then called the **yolk plug.** The rim of the blastopore, however, continues to contract and at last covers the yolk plug altogether. In this way the material of the vegetal region disappears eventually into the interior of the embryo. When the blastopore contracts to such a degree that the yolk plug disappears from view, the blastopore is said to be "closed." This is not quite exact, however, since a narrow canal leading into the interior still persists.

If the blastoderm above the rim of the blastopore is marked with spots of vital stain, it can easily be seen that the stained areas stretch toward the lips of the blastopore, approach its rim, roll over the edge and disappear inside (Fig. 84). Once inside, the stained material does not come to rest but continues its movements in the interior of the embryo, but this time it moves away from the blastopore in the opposite direction from that which it followed while the stained material was still on the surface of the embryo. It can also be seen that the movement of the superficial material is most rapid and extensive on the dorsal meridian of the embryo. More material is invaginated over the dorsal lip of the blastopore than over the lateral lips and least over the ventral lip of the blastopore. This accounts for the different breadth of the marginal zone around the circumference of the egg. In fact, the upper limit of the marginal zone is none other than the limit to which the blastoderm is invaginated during the

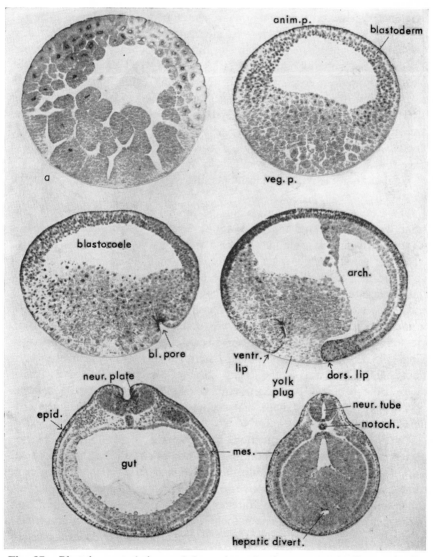

Fig. 87. Blastula, gastrulation and formation of primary organ rudiments in the frog. *a,* Late cleavage stage showing difference in the size of the blastomeres.

process of gastrulation. The vegetal region is similarly that part of the blastoderm which is enclosed by the rim of the blastopore. The animal region is the part of the blastoderm that does not pass into the interior by way of the blastopore. The extent of the invagination can be clearly seen if a series of vital stain spots is made along the mid-dorsal meridian of the embryo, as shown in Fig. 85.

On the dorsal side of the embryo, where the groove had first appeared and where the streaming of the cells into the interior of the embryo is the most active, a cavity is soon formed leading from the groove on the surface into the interior of the embryo. This cavity is lined on all sides by the invaginated cells and represents the archenteron. The archenteron is at first a narrow slitlike cavity, but later, as more material becomes invaginated, it expands at its anterior end and in so doing encroaches on the blastocoele. The latter is eventually obliterated (Figs. 86 and 87). While the rim of the blastopore moves over the surface of the vegetal region, the vegetal region is drawn into the interior and at the same time caused to rotate, so that after the end of gastrulation it comes to lie in the ventral part of the archenteron, its originally exterior surface facing its cavity. The opposite, dorsal wall of the archenteron consists of cells of the marginal zone which have rolled into the interior over the dorsal lip of the blastopore.

We can now trace the displacements and movements of each area of the blastula individually.

The Ectoderm.   The material of the animal region including the presumptive epidermis and the presumptive nervous system greatly increases its surface during gastrulation, because at the end of gastrulation it has to cover the whole embryo after the mesoderm and the endoderm have disappeared inside. The **expansion** of the ectoderm is an active process, and the increase of surface area goes on at the expense of a thinning out of the epithelial layer. The presumptive epidermis expands in all directions, but in the case of the presumptive nervous system the expansion is mainly in the longitudinal direction, i.e., toward the blastopore. In the transverse direction, on the other hand, the presumptive nervous system area contracts, and the material of the lateral horns of the crescentic area is drawn in toward the dorsal side of the embryo. As a result the whole nervous system area changes its shape and becomes oval, elongated in an anteroposterior direction.

The Notochord.   The notochord rolls over the dorsal lip of the blastopore into the interior of the embryo and becomes stretched along the dorsal side of the archenteron, forming the mid-dorsal strip of the archenteron roof. As it does so, the presumptive notochord undergoes a very considerable elongation in the longitudinal direction and a corresponding contraction in the transverse direction. The notochordal material becomes concentrated on the dorsal side of the embryo, just as is the presumptive neural system but even to a much greater degree.

The Prechordal Plate.   The prechordal plate, which in the blastula lies just below the presumptive notochord, invaginates by rolling over the dorsal lip of the blastopore and becomes a part of the archenteron roof in front of the anterior end of the notochordal material.

The Mesoderm.   Of all parts of the blastula, the mesodermal area undergoes the most complicated movements. Most of the mesoderm invaginates into the interior by rolling over the lateral and ventral lips of the blastopore. Once inside the embryo the mesoderm moves in an anterior direction as a sheet of cells, penetrating between the ectoderm on the outside and the endoderm on the inside. The mesoderm in Urodela detaches itself from the endoderm and moves forward between the ectoderm and the endoderm, having a free edge anteriorly, but preserving an uninterrupted connection with the notochordal material on the dorsal side of the embryo (Fig. 88). The notochordal and the mesodermal material in this stage are in the form of one continuous epithelial sheet, the **chordo-mesodermal mantle.** In the Anura, the mesoderm does not split off from the adjoining endoderm until gastrulation is nearly finished. The result is, however, the same: the formation of the chordo-mesodermal mantle lying between the ectoderm and the endoderm. As the mesoderm moves from the posterior end of the embryo (represented by the blastopore) toward the anterior end, there remains at the anterior end a region which the mesodermal mantle has not yet reached. This region, which is free of mesoderm, diminishes as gastrulation proceeds, but does not disappear completely. It is in this mesoderm-free region at the anterior end of the embryo that the mouth is later formed.

The concentration toward the dorsal side of the embryo, noted in respect to the notochordal and nervous system material, is also very distinct in the case of the mesodermal mantle. The movement of the mesoderm inside the embryo does not follow the same path as on the surface. If the trajectories of all parts of the presumptive mesoderm are traced, it is seen

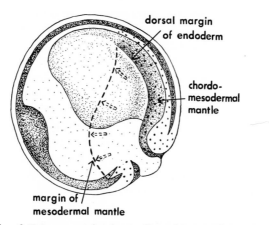

Fig. 88.   Gastrula of *Triturus* cut in the median plane to show separation of the chordomesodermal mantle from the endoderm. (After Vogt, from Hamburger, 1947.)

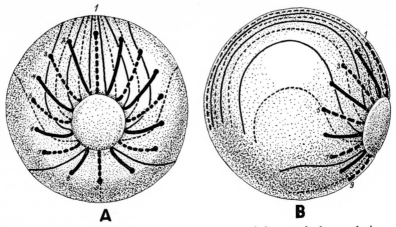

Fig. 89.   Trajectories of the movements of parts of the marginal zone during gastrulation in amphibians. Thick lines show movements of cells on the surface of the embryo; thin lines show movement of invaginated cells. (From Vogt, 1929.)

that they converge toward the dorsal side where they represent the movement of the mesoderm after its invagination (Fig. 89). As a result, the mesodermal material becomes concentrated toward the dorsal side. The mesodermal layer is thickest in the roof of the archenteron, where the mesoderm adjoins the notochord; it is thinned out in the lateral part and still more so in the ventral part of the embryo. The mesoderm continues to invaginate by rolling over the rim of the blastopore even after the endoderm has come to lie inside, and the yolk plug has disappeared from the surface. The invagination of the mesoderm may therefore be considered as retarded as compared with the invagination of the endoderm. The degree of retardation is not the same in different animals. It is the least in frogs, greater in the urodeles, and greatest in the lamprey (whose gastrulation is, otherwise, very similar to the gastrulation of the amphibians). As result of the retardation, the presumptive mesoderm of the tail region and sometimes also the presumptive mesoderm of the posterior trunk region may still be on the surface of the embryo when the blastopore is "closed."

The Endoderm.   The presumptive endoderm is found, in the blastula, partly in the marginal zone and partly in the vegetal region. The two parts of the presumptive endoderm invaginate in different manners: the part lying in the marginal zone is mainly absorbed into the original pitlike invagination of the blastopore; the vegetal region disappears from the surface when it becomes covered up by the contracting rim of the blastopore.

The blastopore first appears as a pit in the endodermal area, between the marginal zone endoderm and the vegetal endoderm. The endodermal cells lying at the bottom of the pit are later found in the duodenal region of the embryo. In the early gastrula the presumptive endoderm of the pharyngeal and oral region lies on the anterior slope of the pit, and is therefore invaginated as part of the dorsal lip of the blastopore. This material later forms

the most anterior part of the advancing archenteron. In the later stages of
gastrulation the oral and pharyngeal endoderm expands so as to form the
spacious **foregut,** whose lateral, ventral and anterior walls then consist of
a rather thin layer of endoderm. Only part of the dorsal wall of the foregut
is taken up by the prechordal plate and the anterior tip of the notochord.

The endoderm of the vegetal region passes into the interior of the embryo
more or less passively and there comes to lie in the floor of the archenteron.
It is not pushed forward as far as the marginal zone endoderm, but remains
confined to the middle and posterior parts. The layer of endoderm in the
floor of the archenteron is very thick, and the cavity of the archenteron is
therefore reduced posteriorly to a rather narrow canal underneath the
chordo-mesodermal mantle. This narrow part of the archenteron later
becomes the **midgut** of the embryo. The lateral endodermal walls of the
midgut are much thinner, and dorsally they end at a free edge after the
mesoderm has been split off from the endoderm.

5–5   FORMATION OF THE PRIMARY ORGAN RUDIMENTS IN AMPHIBIANS

At the time of the closure of the blastopore, the separation of the organ
rudiments in amphibians has gone somewhat further than in *Amphioxus;*
the chordo-mesodermal mantle has, to a great extent, separated itself from
the endoderm, and the mesoderm has attained its definitive position between
the endoderm and the ectoderm. To make the separation complete the
anterior end of the notochord becomes split at its anterior end from the
prechordal plate, and the prechordal plate itself separates from the adjoining endoderm. The endoderm forthwith closes under the prechordal plate.
The foregut is then surrounded on all sides by endodermal cells. In the
midgut region the free edges of the endoderm approach each other in the
midline and fuse, thus completely enclosing the midgut. The chordo-mesodermal mantle becomes subdivided into its two components: the notochord
in the middle and the mesoderm on both sides. As the notochord is sepa-

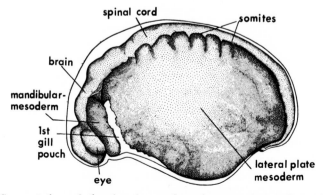

Fig. 90. Segmentation of the dorsal mesoderm in an embryo of a salamander
(*Ambystoma punctatum*). (Modified from Adelmann, 1932.)

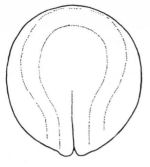

Fig. 91.   Neurula of a frog. (After Spemann, 1938.)

rated from the mesoderm, the strand of notochordal cells becomes converted into a cylindrical body, round in cross section. The mesoderm, at the same time, becomes subdivided into a series of mesodermal segments. Contrary to what is found in *Amphioxus,* however, not the whole of the mesoderm becomes segmented, but only its dorsal part, the part adjoining the notochord. The mesodermal layer has been thickened here owing to the convergence of the mesoderm toward the dorsal side of the embryo during gastrulation. This thickened part now becomes segmented by a series of transverse crevices cutting into the mass of mesodermal cells and separating them (Fig. 90). The thinned out lateral and ventral parts of the mesodermal mantle on each side do not become subdivided into segments, and these parts of the mesoderm are known as the **lateral plates.** The mesodermal segments in amphibians are not fully homologous to the primary mesodermal segments of *Amphioxus;* they are given the name of **somites.** Each somite is separated from the adjoining somites but remains connected to the dorsal edge of the lateral plate. The strand of cells connecting the somite to the lateral plate is known as the **stalk** of the somite, and these cells, as will be shown later, have a special part to play in the development of the embryo.

At about the same time as the segmentation of the mesoderm in its dorsal part takes place, the lateral plate mesoderm becomes split into two layers, an external layer applied to the ectoderm known as the **parietal layer,** and an internal layer applied to the endoderm the **visceral layer.** The narrow cavity between the two layers is the **coelomic cavity** or **coelom.** In later stages the cavity expands and becomes the body cavity of the adult animal (Fig. 185). Small cavities also appear in the somites, but these are obliterated later and disappear without a trace. There is no connection between the coelomic cavities in the mesoderm of the amphibians and the archenteric cavity.

The presumptive nervous system has been traced previously to the stage when it forms an approximately oval shaped area on the dorsal side of the embryo. The neural area covers the areas of the notochord and the somites in the middle and posterior parts of the embryo and the prechordal plate in front. The anterior end of the neural area is underlain by a part of the endodermal lining of the foregut. After the closure of the blastopore the presumptive area of the nervous system becomes separated from the epidermis in the form of the **neural plate.** The ectodermal epithelium of the neural plate becomes thickened by the further concentration of the epithelium toward the dorsal side of the embryo. At the same time the cells of the neural plate change in shape; they become elongated and arrange themselves into a columnar epithelium. Thus they are different from the cells of the epidermis which remain more or less flat and arranged as a stratified epithelium usually two cells thick.

Superficially the neural plate becomes visible because of a concentration of pigment at the edges of the plate. Soon, however, the edges of the neural plate become thickened and raised above the general level as **neural folds** (Fig. 91). A shallow groove may be seen at the same time along the midline of the neural plate, separating it into right and left halves. The neural plate continues to contract in a transverse direction, especially in its posterior parts. The neural folds become higher, so that eventually the neural plate is converted into a longitudinal depression, bordered laterally and in front by the neural folds. Subsequently the folds make contact in the midline and fuse, beginning from the point corresponding to the occipital region of the embryo and progressing forward and backward. In this way the neural plate is transformed into the neural tube (Fig. 92). The cavity of the neural tube is broadest in the anterior part of the tube, and this part later develops into the brain with the brain cavities. In the posterior part of the neural tube the cavity is much narrower. It will later become the central canal of the spinal cord. Sometimes the central canal is not found in the posterior part of the neural tube at the time of the closure of the neural folds, and in that case the tube is hollowed out later by the separation of the cells in the middle of the organ.

The formation and closure of the neural plate is known as **neurulation,** and the embryo, in the stages when it possesses a neural plate, is called a **neurula.**

After the neural folds have fused in the midline the neural tube sepa-

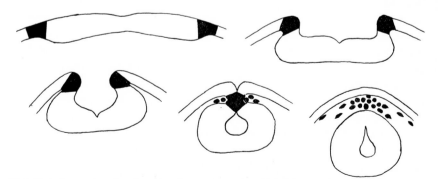

Fig. 92.   Stages in the formation of the neural plate and neural tube in amphibians. Transverse sections (diagrammatic). Neural crest cells shown in black.

rates itself completely from the overlying epidermis. The free edges of the epidermis fuse, so that the epidermis becomes continuous over the back of the embryo. A certain number of cells, however, do not become included either in the neural tube or in the epidermis. These cells can be traced back to strips running all along the crest of the neural folds (Fig. 92). After the neural tube has become separated from the epidermis these cells are found as an irregular flattened mass between the neural tube and the overlying epidermis. This mass of cells is the **neural crest.** The neural crest may be continuous across the midline at first, but very soon the mass of cells separates into a right and a left half, lying dorsolaterally to the neural tube. The neural crest cells have been found to play a very special part in the development of the embryo, as will be indicated later (section 10–3).

During neurulation the embryo already begins to stretch in length. At the same time it is flattened from side to side and also becomes smaller in a dorsoventral direction. The volume of the embryo does not change appreciably. The stretching is greatest in the posterior part of the embryo. As the neural tube takes part in this general stretching, it becomes still more attenuated in its posterior parts, so that the difference between the presumptive brain and the presumptive spinal cord is increased. The part of the embryo above the blastopore becomes elongated beyond the blastopore and this elongation becomes the rudiment of the tail. It is known as the **tailbud.** The gill pouches are formed as lateral outpushings of the foregut, and in later development these give rise to the gill clefts. The mouth later breaks through at the anterior end of the foregut. The blastopore persists as the anus.

5–6 GASTRULATION IN FISHES

In the bony fishes the invagination of the mesoderm and the endoderm proceeds around the circumference of the blastodisc. There is no inpushing of epithelial layers, however, and no archenteron is formed. The endoderm, which was located around the posterior half of the circumference of the blastodisc in the blastula stage, shifts into the interior along the surface of the yolk (or, to be more exact, over the surface of the periblast) (see section 4–3). As the cells move inward they are also concentrated toward the midline of the embryo. They eventually form a thin layer or plate underneath the blastodisc (Fig. 93). The mesoderm invaginates in the same way at first, but later the invagination is continued as a process of rolling over the edge of the blastodisc. Although no archenteron is formed, the edge of the blastodisc must be recognized as the blastopore, because it is the pathway by which the endoderm and the mesoderm pass into the interior of the embryo. The concentration of the material toward the dorsal side of the embryo is, in the case of the mesoderm, even more pronounced than in the case of the endoderm. The mesodermal cells begin to converge toward the midline even before they invaginate. By applying vital staining it has been ascertained that the mesodermal material lying around the rim of the blastopore (the edge of the blastodisc) streams along the edge of the blastodisc toward the posterior end, where the body of the embryo begins to be formed. Once invaginated, the mesodermal cells move obliquely forward and toward the midline, converging in the midline where the axial organs of the embryo will be formed.

The presumptive notochordal area, which is elongated transversely in the blastula, is, during the process of invagination, concentrated toward the midline and stretched in a longitudinal direction so that eventually it is converted into a narrow strip of cells lying in the midline of the embryo, above the endodermal plate and under the ectoderm. The notochord as well as the prechordal plate invaginates by rolling over the posterior edge of the blastodisc, which corresponds, of course, to the dorsal lip of the amphibian blastopore. Notochord, prechordal plate and mesoderm are continuous with each other, forming a chordo-mesodermal mantle. The neural system area is stretched toward the posterior edge of the blastodisc (to the dorsal lip of the blastopore), and thus it takes the place of those parts which had disappeared from the surface. At the same time, the presumptive nervous system area concentrates toward the midline and becomes a strip of tissue lying over the presumptive notochord, but reaching considerably further forward from the posterior edge of the blastodisc.

It is a peculiarity of the fishes that the formation of the primary organ rudiments in the anterior part of the embryo begins long before the invagination has been accomplished in the posterior part (Fig. 94). The first to appear is the anterior end of the neural plate, corresponding to the prechordal part of the brain. Next in succession appears the neural plate of the hindbrain region and of the anterior trunk region. At the same time the notochord starts separating from the mesoderm, with which it had previously been

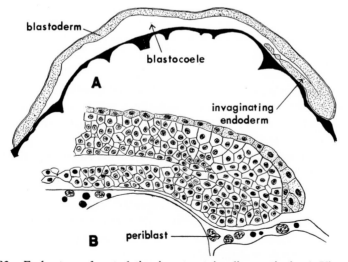

Fig. 93.   Early stage of gastrulation in a trout (median section). *A,* View of the whole blastodisc under low magnification; *B,* part adjoining the blastopore (posterior edge of blastodisc) at higher magnification.

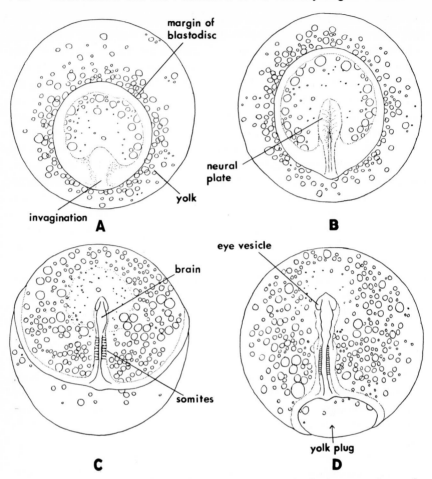

Fig. 94. Stages of development of the trout. *A*, Blastodisc in the stage of gastrulation; *B*, neural plate stage; *C*, formation of the brain, eye vesicles, and somites; *D*, overgrowth of yolk almost completed.

united into a continuous layer. The mesodermal mantle had already been thickened all along the presumptive notochord, owing to the concentration toward the midline during gastrulation. These thickened parts of the mesoderm adjoining the notochord are now subdivided into segments—the somites. The remainder of the mesodermal mantle stays unsegmented and represents the lateral plate mesoderm. The endoderm had been separated from the mesoderm in a previous stage. The endodermal plate now concentrates toward the midline and consequently becomes thicker, but the cavity of the alimentary canal does not appear until a very much later stage. When it does appear, it is formed by a separation of the cells along the middle of the endodermal cord of cells.

The process of neurulation in the bony fishes differs considerably from that in amphibians. Although a neural plate is found, the neural folds are only slightly indicated. The neural plate does not roll into a tube, but narrows gradually, at the same time sinking deeper and deeper into the underlying tissues. Eventually it separates from the epidermis which becomes continuous over the dorsal surface of the embryo. No cavity appears in the nervous system rudiment while it is being separated from the remainder of the ectoderm. The ventricles of the brain and the central canal of the spinal cord are formed later by the separation of cells in the middle of the brain and spinal cord.

While the invagination of the endoderm and mesoderm and later the formation of the primary organ rudiments of the anterior part of the body is going on, the blastodisc as a whole is spreading to cover an ever-increasing part of the yolk. The rim of the blastodisc shifts over the surface of the yolk in a centripetal direction. This movement of the blastodisc rim may be compared to the movement of the rim of the amphibian blastopore away from the animal pole and toward the vegetal pole of the embryo. Eventually the rim of the blastodisc converges at or near the opposite side of the yolk, and the opening closes by constriction of the rim. We have seen that the rim of the blastodisc corresponds to the lips of the blastopore. The process described just now is therefore equivalent to the closure of the blastopore in the amphibians and amphioxus. Incidentally it ends up by having all the yolk covered by the blastodisc. Just before the blastopore closes a yolk plug may be seen projecting from the blastopore. This yolk plug looks very much like the yolk plug in the amphibians, but it differs from it in so far as it consists only of nonsegmented yolk, whereas the yolk plug in the amphibians consists of cells of the vegetal region. At the time when the lips of the blastopore move over the surface of the yolk, they do not contain presumptive endoderm except at the posterior edge of the blastodisc. The endoderm therefore does not form a continuous sheet all around the yolk, which in the lateral and ventral parts of the embryo is in direct contact with the lateral plate mesoderm.

Gastrulation and the formation of the primary organ rudiments in elasmobranchs follows the same pattern as in the bony fishes—with the difference, however, that an archenteron is formed at the posterior edge of the blastodisc by the invagination of the presumptive notochord and endoderm. The rudiment of the nervous system is similarly not solid, but the neural plate is rolled into a tube having a distinct cavity right from the start.

5–7 GASTRULATION IN BIRDS

The organ-forming part of the blastoderm in the birds takes up only a limited area inside the blastodisc (the area pellucida). The invagination cannot therefore proceed on the edge of the blastodisc. Instead, the invagination takes place in a special region inside the area pellucida. This region is a narrow strip of blastoderm starting from the posterior edge of the area pellucida and running forward along the midline of the embryo for about three fifths of the entire length of the area pellucida. This strip of blastoderm

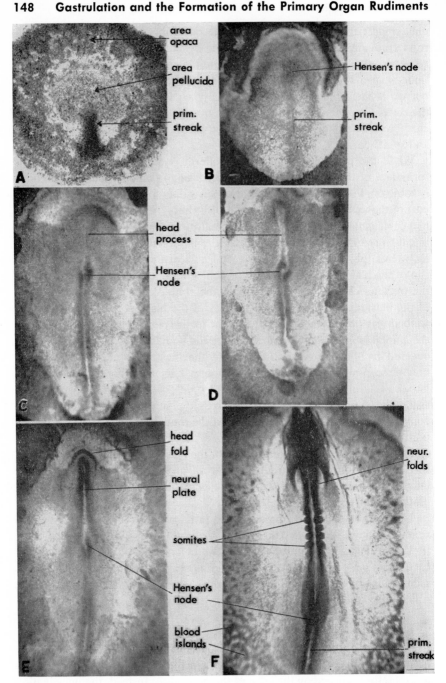

Fig. 95. Gastrulation and formation of primary organ rudiments in a chick embryo (surface views). *A,* Short primitive streak (stage 3) (from Hamburger and Hamilton, 1951); *B,* definitive primitive streak (stage 4); *C,* head process stage (stage 5); *D,* retreating primitive streak (stage 6); *E,* neural plate (stage 6 +); *F,* closure of neural tube and formation of somites (stage 9 —). (Numbers of stages after Hamburger and Hamilton, 1951.)

becomes thickened during the stages of gastrulation and is known as the **primitive streak.** Along the middle of the primitive streak, when it is fully developed, there runs a narrow furrow, the **primitive groove.** At the anterior end of the primitive streak there is a thickening, the **primitive knot** or **Hensen's node** (Fig. 95, *B*). The center of Hensen's node is excavated to form a funnel-shaped depression. In some birds (the duck for instance) this depression is continued forward into a narrow canal which will be referred to later.

The thickening of the blastoderm is brought about by a convergence of the surface layer of the blastoderm, the epiblast, toward the midline in the posterior half of the area pellucida. As the cells nearest the midline become concentrated to form the early primitive streak, the parts of the epiblast lying further out laterally and anterolaterally swing in a curve backward and inward to take the place of parts of the blastoderm shifting toward the midline (Fig. 96). The primitive streak first becomes visible in the hindmost part of the area pellucida—the **short primitive streak** (Fig. 95, *A*).

The primitive streak now elongates by the concentration of more and more material from the sides toward the midline in front of the original short primitive streak. The early primitive streak is broad and its edges are rather vaguely indicated. In later stages it contracts in a transverse direction, becomes narrower and quite sharply delimited. This is the **definitive primitive streak** (Fig. 95, *B*). In the process of this transformation the primitive streak elongates and its anterior end is pushed even further forward, though the amount of this forward thrust has been a matter of controversy among embryologists (see Waddington, 1952).

While the movements in the epiblast lead to the formation of the primitive streak, there is some indication that the cells of the lower layer of the blastoderm, the hypoblast, also do not remain quiescent, but shift in an anterior direction. The hypoblast first becomes a continuous layer in the posterior part of the blastoderm, and is only later found to be present in the anterior half. It is possible that the forward shift of the hypoblast contributes toward

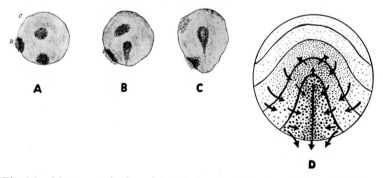

A   **B**   **C**

D

Fig. 96. Movements in the epiblast during the formation of the primitive streak in a chick embryo as shown by displacement of vitally stained areas (*A, B, C*). *D,* Diagram of movements based on this and similar experiments. (From Wetzel, 1929.)

the completion of this layer anteriorly. It has been suggested, however, that the polarity of the hypoblast, which shows itself, among other things, in its directed movements (forward), may have an even more profound signifi-cance (see p. 173).

The movements in the blastoderm leading to the formation of the primi-tive streak may be called pregastrulation movements, to distinguish them from the gastrulation movements proper.

At the stage of the short primitive streak the cells of the blastoderm al-ready begin to invaginate into the space between the epiblast and the hypo-blast (see section 4-4). The process of invagination is different from all that has been seen in the lancet fish, in the amphibians and in the bony fishes. There is no infolding of the epithelial layer of the blastoderm; neither does a sheet of cells shift from the surface to the interior, as in bony fishes. In the birds the cells seem to be moving downward singly, even if there are many cells moving in the same direction. The regular epithelial arrangement of cells found in the epiblast becomes lost (Fig. 98, *a*). This type of gastrulation movement bears the name of **immigration.**

Very soon the migrating cells reach the hypoblast and establish an intimate contact with the cells of this layer. Henceforward the whole of the primitive streak is a mass of moving cells. The direction of movement is mainly downward from the surface of the blastoderm toward the hypoblast, but the mass of migrating cells also spreads out sideways and forward from the anterior end of the primitive streak. By marking cells of the hypo-blast underneath the primitive streak with carbon particles it was found that these cells also become a part of the shifting migrating mass, and some of the cells originally belonging to the hypoblast move into the midst of the primitive streak cells and are carried with them, especially in an anterior direction (Fraser, 1954).

Although the cells of the chick blastoderm move individually, their movements are obviously directed by common causes, and therefore the whole mass of cells moves in a coordinated fashion. The formation of de-pressions on the surface of the blastoderm, the furrow along the midline of the primitive streak and the funnel-shaped depression in the center of Hensen's node are due to this mass movement of cells from the surface of the blastoderm into the interior (just as in the case of a vortex appearing on the surface of water).

As the cells of the epiblast migrate into the interior, whole areas of the blastoderm disappear from the surface. They are replaced, however, by the adjoining areas moving toward the midline, and taking their place in the primitive streak. The newly arrived cells in their turn migrate down into the interior. Thus the primitive streak persists, although the cells of which it is made up do not stay in the same place but are replaced all the time. The first areas to start invaginating are the presumptive prechordal plate, the notochord, and the presumptive lateral mesoderm. The prechordal plate soon disappears from the surface through Hensen's node and spreads under the epiblast in front of the primitive streak.

The presumptive notochord cells disappear at a fairly early stage from the surface of the blastoderm and become concentrated in the definitive primitive streak in the deeper parts of Hensen's node. After the prechordal plate has spread out in front of the anterior end of the primitive streak, the presumptive notochord cells start moving as a dense mass in the mid-line straight forward from Hensen's node underneath the surface of the epiblast (Fig. 97).

The mass of notochordal cells can be distinguished from the surrounding (mainly mesodermal) cells, and is called the **head process**, or, more correctly, the **notochordal process** (Fig. 95, *C*). The narrow canal mentioned above and found in some birds (e.g. the duck) penetrates into this notochordal process. Since the formation of the canal is due to the movement of the invaginating cells of the gastrula, the cavity of the canal must be recognized as corresponding to a part of the archenteron, even though the lining of the canal consists exclusively of presumptive notochordal cells. There is no other cavity in the development of birds that could be considered as a homologue of the archenteron. The invagination of the mesoderm or the endoderm does not lead to the appearance of a cavity. We have seen above that in other vertebrates the archenteron may also be completely absent (see gastrulation in the bony fishes).

Presumptive lateral plate mesoderm is contained throughout gastrulation in the posterior part of the primitive streak, and from this position cells of the presumptive lateral plate mesoderm migrate downward, outward and forward. Indeed, no other kinds of mesodermal cells are ever found in the posterior half of the primitive streak. This part of the primitive streak can therefore be compared with the lateroventral and ventral lips of the blastopore of a frog.

The area of presumptive somites lies in the anterior part of the primitive streak. From this part of the streak, therefore, the somite material invaginates into the interior. After having passed into the interior, the cells of the presumptive somites migrate outward and forward and become distributed in a strip on each side of the notochordal process. The presumptive lateral mesoderm cells move still further outward and are later found as sheets of cells on each side of the embryo as a lateral continuation of the dorsal mesoderm.

After its first appearance the primitive streak increases somewhat in length, mainly due to a longitudinal stretching of its posterior half. The

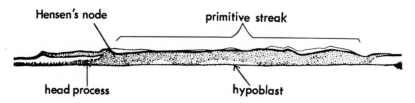

Fig. 97.   Longitudinal medial section of a chick embryo in the head process stage.
(After Wetzel, 1929.)

adjoining part of the area pellucida is also involved in the stretching, so that the area pellucida loses its circular shape and becomes more or less pear-shaped, the attenuated end directed posteriorly. The elongation of the primitive streak is, however, only temporary. As the cells destined to become notochord, mesoderm, and so on migrate into the interior of the embryo, the material of which the primitive streak consists becomes gradually exhausted. The influx of cells from the sides becomes retarded, and can no longer compensate for the expenditure of cells due to immigration. The whole primitive streak begins to shrink, the anterior end receding backward, while the posterior end remains more or less stationary (Fig 95, *D, E*). It has been ascertained by marking with carbon particles that Hensen's node is carried bodily backward during this recession of the primitive streak (Spratt, 1947). The presumptive notochord cells contained in Hensen's node one after another continue migrating downward and forward, so that the notochord process is prolonged backward owing to continual apposition of new cells. In the same way the strip of somite mesoderm and the sheets of lateral plate mesoderm are continued backward from parts of the primitive streak which retain their properties although the whole structure is reduced in scale. By the end of gastrulation the primitive streak has shrunk almost to nothing (Fig. 95, *F*); the residue becomes partly incorporated into the tailbud which is formed at the posterior end of the body of the embryo and partly into the cloacal region of the embryo.

As the primitive streak recedes backward, the neural system area, which lies just in front of Hensen's node and even forms the most anterior half of the node itself, stretches backward, while the lateral horns of the area are drawn in toward the midline. As a result, a strip of presumptive neural plate is left in the wake of the receding primitive streak. A part of the neural plate area is also found in front of the place where Hensen's node was located originally. This part will become the prechordal part of the brain.

The primitive streak, being the path by way of which the presumptive internal organs are invaginated into the interior of the embryo, must be considered as the blastopore, even if it does not lead into an archenteron. We have met with this situation in the bony fishes where, notwithstanding the fact that no archenteron is to be found, the edge of the blastodisc was recognized as the blastopore. The recession and disappearance of the primitive streak corresponds therefore to the closure of the blastopore. The remnant of the blastopore, as in the amphibians, is to be found in association with the anus (or cloaca).

We have seen that the primitive streak, while it is active, is a mass of cells, continuous with both the epiblast and the hypoblast, as well as with the sheets of mesodermal cells migrating in between the two (into the blastocoele) (Fig. 98, *a*). In the part of the embryo from which the primitive streak has receded, the continuity of all three layers no longer occurs. The sheet of mesodermal cells, together with the notochordal process, is separated from the overlying epithelium, which no longer contains pre-

sumptive mesoderm or endoderm and from now on is pure ectoderm. Likewise the chordo-mesodermal mantle has been split off from the underlying
layer of cells, the former hypoblast (Fig. 98, b). This underlying layer is,
however, no longer identical with the original hypoblast. Cells from the
primitive streak, originally situated in the surface layer of the blastoderm
(in the epiblast), have found their way into the lower layer. These are
the cells of the presumptive alimentary canal (Hunt, 1937). The lower layer
of cells from now on will be referred to as the endoderm. It is seen that in

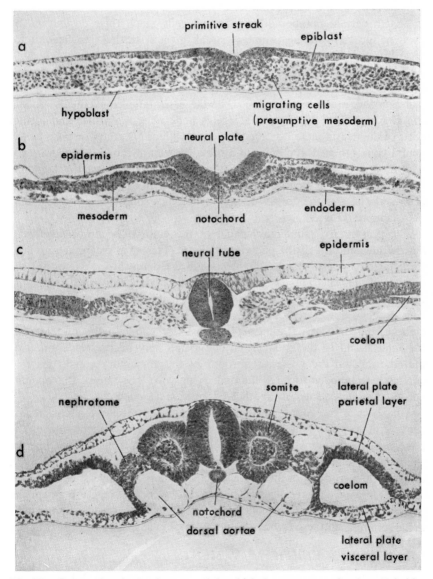

Fig. 98. Stages of early development of the chick (transverse sections). *a*, Primitive
streak; *b*, neural plate stage; *c*, neural tube stage; *d*, primary organ rudiments.

birds the endoderm has a double origin: some of the cells are derived from the hypoblast, a layer corresponding to the floor of the blastocoele in the amphibians. These cells are used only to a limited extent for the formation of the actual alimentary canal of the embryo; most of them become the endodermal lining of the **yolk sac** (see section 7–1). Other endodermal cells are derived from the primitive streak and join up with the first kind after a process of invagination. These cells remain in the median strip of the endodermal layer, in that area of it which is actually used up for the formation of the walls of the alimentary canal.

That a certain number of endodermal cells reach their destination by invaginating through the primitive streak gives a further justification for regarding the primitive streak as the blastopore.

The neurulation and the formation of the other primary organ rudiments proceeds in an anteroposterior direction, just as in the fishes. The neural plate appears in the brain region while the gastrulation movements are still in full swing (Fig. 95, *E*). As Hensen's node recedes, further and further parts of the neural plate become differentiated, and the anterior parts of the neural plate proceed to close into a tube, the neural tube. The neural tube is formed in a very typical way, with a large cavity enclosed by the walls of the tube. The notochord becomes separated from the adjoining sheets of mesoderm, and the dorsal mesoderm becomes subdivided into segments —the somites (Fig. 98, *d*).

The presumptive alimentary canal of the embryo is represented by a narrow median part of the endodermal layer. This strip of epithelium later forms a fold, with the apex of the fold directed **upward,** i.e., toward the notochord. The margins of the fold then approximate to each other and fuse, enclosing a cavity which will be the cavity of the alimentary canal. This process, however, does not take place along the whole length of the embryo, but only at its anterior and posterior ends. In the middle part, although a fold of the endodermal epithelium may be present connecting the anterior and the posterior portions of the alimentary canal, this fold continues to be open towards the underlying yolk. The fate of this opening will be considered later in another connection (section 12–1).

The sheet of mesoderm invaginated through the primitive streak at first consists of loosely lying cells. At the stage when the segmentation of the mesoderm begins, or shortly before, the mesodermal cells reunite into an epithelial arrangement. After the somites and the lateral plates have been formed, the mesoderm of the lateral plate splits into two layers: the external or parietal layer, and the internal or visceral layer. The cavity between the two layers is the coelom (Fig. 98, *c, d*). Small coelomic cavities appear also in the somites, but these cavities, as in the amphibians, soon disappear again.

5–8   MORPHOGENETIC MOVEMENTS

Gastrulation is essentially a process of movement of parts of the embryo. These movements are very different from the movements of parts of an adult animal. Whereas the movement of parts of an adult are usually of a reversible nature, the gastrulation movements are irreversible; each part remains in the position into which it has been brought by the preceding movement. As the result of the movements the structure of the embryo is changed, or, in other words, new structural elements are created, such as the archenteron, the neural tube and the notochord, in place of the simple layer of cells found in the blastula stage. The movements have created new shapes, new forms. They have been therefore designated as the **morphogenetic movements** (Vogt, 1925).

The morphogenetic movements appear to be movements of large parts of the whole embryo, which stretch, fold, contract or expand. The question arises, how are these movements achieved? They cannot be ascribed to contractility in the narrow sense of muscle contractility. Neither can they be interpreted as an ameboid movement of the embryo as a whole, because each moving part consists of numerous cells, and the movement of the whole, we should expect, would be an integrated result of the movements of the individual cells.

That the gastrulation cannot be a function of the embryo as a whole has been proved by investigating isolated parts of the young gastrula. We have seen that the presumptive ectoderm contributes to gastrulation by expanding its surface. The expansion is an active process depending on the presumptive ectoderm itself. If large pieces of the animal region of an amphibian blastula or early gastrula are cut out and cultivated in a suitable medium (Holtfreter's saline solution, a modification of Ringer's solution) the presumptive ectoderm rounds itself up into a vesicle, and later the epithelium of this vesicle increases its surface greatly and is thrown into a series of irregular folds as it does so. Also, if presumptive ectoderm is combined with cells of presumptive endoderm and mesoderm in such a proportion that the ectoderm is far in excess of the endodermal and mesodermal parts which it has to cover, the ectoderm tends to be thrown into folds (Spemann, 1921). This shows that the expansion of the ectoderm is active and proceeds independently of the other movements involved in gastrulation. The increase of the surface of the ectoderm is due partly to a flattening of the cells of the epithelial layer and partly to a rearrangement of cells, those of the lower layers becoming intercalated among the cells of the upper layers, so that the epithelium eventually becomes two cells thick.

Similarly, the movements of invagination can be performed by parts of the blastoderm independently of their surroundings. The most suitable method of testing this is to transplant small pieces of the dorsal lip of the blastopore into some other part of the embryo, into the animal region or into the ventral part of the marginal zone. The transplanted piece will invaginate and form an archenteric cavity which may be completely independent of the archenteric cavity of the host embryo. The mechanism by means of which the invagination is achieved is the change of shape and the movements of the cells of the marginal zone. If cells of a columnar

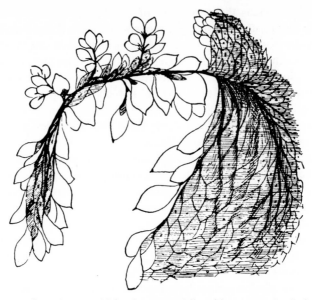

Fig. 99.   A fraction of an amphibian blastoporal lip with groups of cells held together by the "surface coat." (From Holtfreter, 1943a.)

epithelium contract at their external ends and expand at their internal ends, so that instead of being prismatic they become pyramidal, the whole epithelial layer must inevitably change its shape and become bent in toward the interior. This is what actually takes place in the blastopore region; the external parts of the cells here become attenuated, and the opposite ends of the cells expand. The cells acquire a bottle shape with the neck of the bottle keeping the cells in touch with the surface, while the bodies of the cells are drawn away from the surface. The area occupied by a given group of cells on the surface of the embryo contracts to a small fraction of what it had been originally (Fig. 82). The change in the shape of the cells is not due to a lateral compression from neighboring cells, but is performed by the cells themselves. This is best seen if the cells are isolated by tearing the embryo to pieces: the cells preserve their shape and, what is more, the infolding of the surface layer may even be facilitated by releasing a piece of it from its connections with the surrounding parts (Fig. 99) (Holtfreter, 1943a).

However, invagination cannot always be interpreted as an infolding of an epithelial layer. It often takes the form of immigration of individual cells into the blastocoele. In this case the process involves two components. One component is the loosening of the ties existing between the adjoining cells of an epithelial layer. This process is due to a change in the properties of the surface layer of cytoplasm in the cells. When epithelial cells are brought into contact, they stick together and stay together. When cells of the mesenchyme touch other mesenchyme cells they do not usually adhere to each other. If the cells are moving they are not checked in their movements by contacts with other similar cells and draw apart as easily as they come together. Cells in the process of invagination by immigration have the same properties. There must, however, be still another factor involved in the process of immigration of cells: a factor ensuring that the cells withdraw from the exterior of the embryo and move into the interior.

This factor has been demonstrated experimentally in the following way (see Townes and Holtfreter, 1955). The cells of amphibian embryos placed in water, which has been made alkaline (*p*H 9.6 to 9.8) by addition of caustic potash, lose cohesion and the embryo or parts of it treated with alkali disaggregate into a mass of disconnected single cells. Masses of cells prepared in this fashion from different embryonic tissues (gastrula ectoderm, mesoderm or endoderm, neurula epidermis, neural plate and others) were then mixed together in various combinations and returned to a fresh solution of *p*H 8.0, in which the cells return to their normal state and again join together (Fig. 100). After the original clump of cells became solid, the cells began to sort themselves out according to their properties. The ectodermal and epidermal cells combined with endodermal, mesodermal or neural plate cells always crawled out to the exterior. The mesodermal and neural plate cells combined either with epidermis or endoderm crawled inward and formed masses inside the aggregate. The masses of neural plate cells became hollowed out later and formed structures resembling brain vesicles. Masses of mesodermal cells arranged themselves around "coelomic cavities" (see p. 142). If epidermis, mesoderm and endoderm were present in an aggregate, the epidermis cells concentrated on the exterior, and mesodermal cells took up position between the epidermis and the endoderm. At a later stage cells of different nature, while adhering to their own kind, split off from cells belonging to a different type, forming cavities between them (Fig. 101). In this way arrangements of cells similar to those produced at the end of normal gastrulation and neurulation are brought about by the activities of individual cells beginning from a completely abnormal starting point. The behavior of the cells under these conditions—and by inference also in normal development—is presumably due to their **selective affinities** (Holtfreter, 1939a, Weiss, 1947). Depending on their special properties, a cell may either adhere to another

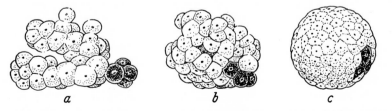

a   *b*   *c*

Fig. 100.   Separated cells of presumptive ectoderm and endoderm joining together into a spherical mass. (From Holtfreter, 1939b.)

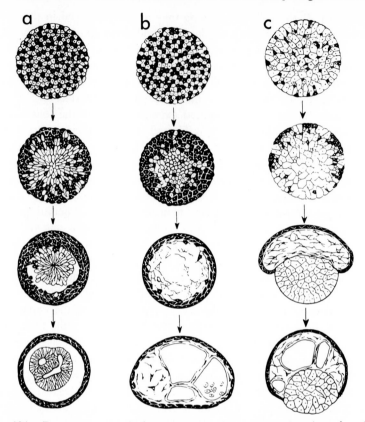

Fig. 101. Rearrangement of disaggregated and reaggregated embryonic cells. *a,* Combined epidermal and neural plate cells. *b,* Combined epidermal and mesodermal cells. *c,* Combined epidermal, mesodermal and endodermal cells. (From Townes and Holtfreter, 1955.)

cell and stay in contact with it, or move away from a position in which the surroundings are not of such a nature that they keep the cell bound (Fig. 101).

In the case of the infolding of epithelial sheets, an attraction of the invaginating cells to the interior seems to play a part, besides the change of shape of the cells. The invaginating cells exercise a certain amount of pull on the surface of the embryo, especially when the archenteric invagination is being formed. Once inside, the invaginating cells push away from the rim of the blastopore with a force that can be measured. By inserting an obstacle in the form of a steel ball in the way of the invaginating chordomesoderm of an amphibian, and by measuring the force which was necessary to keep the ball in place (an electromagnet was used for this purpose), it was possible to measure the pressure developed by the migrating masses of cells (Waddington, 1939a). This pressure was found to be small, but measurable; it was 0.34 mg/mm².

Change of arrangement of cells, change of their shape and active ameboid movement of cells are therefore the equivalent of morphogenetic movements on the cellular level. There are, however, further factors present which serve to coordinate the activities of individual cells. As has been mentioned in section 4–4, the cells at the surface of the blastoderm are more firmly held together than the cells which do not reach to the surface. It has been claimed (Holtfreter, 1943a) that the adhesion of cells in the surface layer is due to the presence of an extracellular continuous cuticular membrane, the "surface coat," to which all the cells reaching the surface are firmly connected. With the aid of the electron microscope it can be shown (Fig. 102) (Balinsky, unpublished) that no such membrane exists, but that the cells reaching the surface adhere to each other very closely just at their distal ends, and are probably joined by some cementing substance, which is not easily broken. During the gastrulation movements in amphibians the surface layer of cells remains intact, so that the cells of the epithelia move together, whether the movement be a general expansion, a concentration, or invagination at the lip of the blastopore (Holtfreter, 1943a). The result is that the cells move about as if borne by a common force.

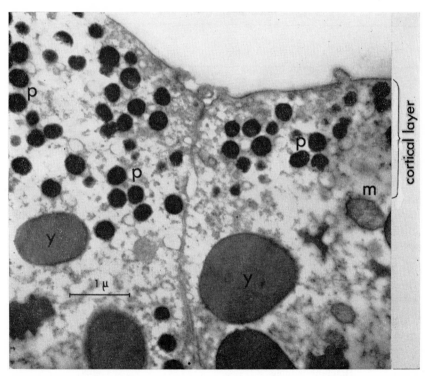

Fig. 102. Surface of the blastoderm in an early gastrula of the frog *Phrynobatrachus natalensis.* The electron micrograph shows the junction of two cells and the cell boundary reaching right to the external surface. *Y,* Yolk; *lp,* lipochondria; *m,* mitochondria; *p,* pigment granules.

5–9   METABOLISM DURING GASTRULATION

Throughout gastrulation the volume of the embryo does not change appreciably. Every expansion in one direction occurs at the expense of a contraction in another direction or directions. What has been said about the absence of growth during cleavage applies in the same way to the period of gastrulation. Division of cells by mitosis continues, however, throughout gastrulation, and this means that there is an increase of nuclear material at the expense of the cytoplasmic substances. Breakdown and assimilation of reserve materials are also going on, but here a new feature is observed that makes the metabolism of the gastrula different from the metabolism of a blastula.

It could be expected that the morphogenetic movements during gastrulation would cause an increased expenditure of energy and consequently an increased oxidation. This is what is actually found: the oxygen consumption during gastrulation shows a further increase as compared with the cleavage stages and with the blastula. The oxygen consumption of the frog embryo is shown in Table 7.

Table 7.   Oxygen Consumption during Cleavage and Gastrulation in the Frog (after Cohen from Barth and Barth, 1954)

Stage	Oxygen Consumption (in Microliters per Milligram of dry Weight per Hour)
Cleavage (stage 6 +)	0.054
Early blastula (stage 8 +)	0.069
Late blastula (stage 9)	0.108
Mid-gastrula (stage 11 −)	0.141
Late gastrula (stage 12 +)	0.162

A similar sharp increase in total oxygen consumption is also observed in sea urchin eggs.

Not only the amount of oxidation changes, but the nature of the oxidative processes changes at the same time. In the cleavage and blastula stages the respiratory quotient of a frog's egg is low; with the beginning of gastrulation it rises and becomes very near unity and remains so until hatching, as shown for the frog embryo in Table 8.

Table 8.   Respiratory Quotient (R.Q.) during Early Stages of Development in the Frog (after Brachet, 1950b)

Stage	R.Q.
Morula	0.66
Advanced blastula	0.70
Gastrula	1.03
Neurula	0.98
Hatching tadpole	0.97

When the respiratory quotient is equal to unity, it indicates that the substances broken down for respiration are carbohydrates. A low respiratory quotient is usually considered to be due to oxidation of proteins. However, we have seen (section 2–7) that there is very little if any breakdown of proteins in the early stages of development of both the sea urchin and the frog, but that carbohydrates, and in particular glycogen, decrease in quantity quite considerably. It has also been found that if amphibian eggs are kept under conditions of anaerobiosis (in an atmosphere of nitrogen), they produce lactic acid starting with the stages of early cleavage—a clear indication of carbohydrate breakdown. If the substrate for oxidation (or fermentation) is the same during cleavage and during gastrulation, there must be some qualitative difference in the processes of oxidation occurring between these two stages. The nature of this difference is not clearly understood.

The number of cells in the embryo continues to increase during gastrulation, and this means that, just as happened during the period of cleavage, further amounts of deoxyribonucleic acid are being synthesized. The protein turnover after the beginning of gastrulation seems in addition to be essentially different from what had been going on during cleavage. There is no doubt that during gastrulation cytoplasmic proteins are being synthesized at the expense of the yolk. The new proteins are qualitatively different from those of the yolk, and this manifests itself among other ways in a changing proportion of amino acids. In a careful study of the amino acid composition of the sea urchin egg it has been found (Gustafson and Hjelte, 1951) that the relative amounts of the various amino acids (18 of these could be traced) remain practically unchanged during the first ten hours of development up to the stage of the mesenchyme blastula. With the beginning of gastrulation, although the total nitrogen of the embryo remains at the same level, some of the amino acids decrease in amount, others increase, and these changes continue throughout the period of development up to the pluteus stage (no observations were made on later stages). The changes in the constitution of the proteins have their counterpart in the changes in the quantities of some enzymes, which is perhaps another way of saying the same thing, as enzymes are proteins. The enzymes investigated fall into two groups (Gustafson and Hasselberg, 1951). Some remain unchanged in activity throughout the development up to the pluteus stage. Others, in particular the enzymes apyrase, succinic-dehydrogenase, malic dehydrogenase and glutaminase, while remaining at the same level up to late blastula stage, show a clear increase beginning with the early gastrula stage.

The appearance at the gastrula stage of proteins which are qualitatively different from those present in the egg has been proved in sea urchins and in amphibians by applying the extremely sensitive methods of immunology (see also section 14–3). It has been found that the gastrula contains antigens, capable of causing the formation of antibodies, which were not present before (besides containing antigens already present in the egg at the beginning of development) (Perlman and Gustafson, 1948; Clayton, 1951).

CHAPTER 6

Determination of the Primary Organ Rudiments

The experiments of local vital staining and similar methods, allowing the experimenter to reconstruct fate maps, do not give any information as to whether or not the cells in the various areas actually have different potentialities for development. The only information that these experiments give is about the eventual destination of each part of the early embryo in normal development. Certain distinctions between cells belonging to the various areas are apparent, such as differences in yolk content, pigmentation, and so on. There is, however, no *a priori* reason to believe that such differences are actually linked together with the future development of each part. In the previous pages it has been indicated that even if isolated from its normal environment each area of the blastula tends to perform morphogenetic movements that roughly correspond to the part which it would play in normal gastrulation. This again does not prove in itself that the type of differentiation which is to be expected from each presumptive area is already fixed in an early stage.

To find out whether the presumptive areas of a blastula are actually preformed for a specific part in the future development, methods other than observing normal development are necessary.

One of the methods used is the method of **transplantation,** already mentioned in another connection (section 5–8). Small pieces may be cut out of embryos in various stages of development and inserted into suitably prepared wounds of the same or another embryo. In the case of transplantation of a piece of the same embryo to another place the transplantation is said to be **autoplastic.** If the transplantation is from one individual to another of the same species, the transplantation is **homoplastic.** If the transplantation is to an individual of another species belonging to the same genus, the transplantation is **heteroplastic.** A transplantation to an individual more distantly related than species of one genus is called **xenoplastic.** The animal (embryo) from which a part is taken is referred to as the **donor,** the animal to which the part is transplanted is called the **host.**

In adult animals, especially highly organized animals such as vertebrates, transplantation is not easy, only autoplastic and homoplastic transplantations usually being successful. In embryos, however, and in lower invertebrates (such as the coelenterates) the grafts may heal successfully even after xenoplastic transplantations. Successful transplantations have been carried out between embryos of frogs and salamanders, and mammals and birds.

It is usually important to know later which tissues and cells are derived from the graft and which from the host. The distinction is easy if donor and host are sufficiently different from each other. The cells may be told apart by their size, staining properties, and other factors. Sometimes differences may be found even between cells belonging to closely related animals. Differences in pigmentation may be very useful. The eggs of *Triturus (Triton) cristatus,* for instance, are entirely devoid of pigment, whereas eggs of other species of *Triturus* (*T. taeniatus, T. alpestris*) have fine granules of black pigment in their cytoplasm. These pigment granules are distributed to all the cells during cleavage. If tissues of *T. taeniatus* and *T. cristatus* are joined in transplantation, the cells of *T. taeniatus* can be distinguished both macroscopically and in sections by the presence of pigment granules. If no natural differences can be discovered between the two animals (embryos) used for transplantation, artificial differences may be introduced by staining one of the embryos with vital dyes. The presence or absence of the dye will indicate the position of the graft. In some cases the vital staining may be preserved in sections of the operated embryo by special methods.

6–1   DETERMINATION OF THE NEURAL PLATE

Heteroplastic transplantation between *Triturus cristatus* and *Triturus taeniatus* embryos has been used to investigate the determination of the presumptive ectodermal areas—the epidermis area and the neural system area (Spemann, 1921). A small piece of the epidermis area of an early gastrula was transplanted into the neural system area of another embryo in the same stage of development (Fig. 103). The result was that the

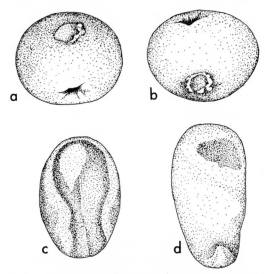

Fig. 103. Exchange of presumptive epidermis and presumptive neural plate in the early gastrula stage of two *Triturus taeniatus* embryos differing slightly in density of pigmentation. *a, b,* Embryos immediately after the operation. *c, d,* Embryos operated on in the neural plate stage. In *c* the graft forms part of the neural plate. In *d* the graft forms part of the anteroventral epidermis. (From Spemann, 1938.)

grafted material developed in conformity with its new surroundings and became first a part of the neural plate of the host (Fig. 103, *c*) and later a part of the neural tube (Fig. 104). There was every reason to conclude that it was differentiating as nervous tissue. In the case of a reverse transplantation, presumptive neural material differentiates into skin epidermis in conformity with its new surrounding (Fig. 103, *d*). Presumptive ectoderm (either from the epidermis area or from the neural system area) was transplanted also into the marginal zone of an early gastrula. In this case the graft was drawn into the blastopore in the course of gastrulation, and was later found in different places in the interior of the host. In every case the graft differentiated in conformity with its surroundings and took part in the development of various organs of the host: the notochord, the somites, lateral plate mesoderm, kidney tubules, wall of the gut (Mangold, 1923).

It is thus evident that the fate of presumptive nervous system and presumptive epidermis is not fixed at the time of the stages used for the above experiments. Besides having a definite normal fate—called the **prospective significance**—the parts tested possess abilities to develop in various other ways, under experimental conditions. This ability of the parts of an early embryo to develop in more than one way is called the **prospective potency** of these parts.

The prospective potency of the neural area is therefore shown to include not only epidermis, but also mesodermal and endodermal tissues. The prospective potency of the epidermis area includes nervous system and mesodermal and endodermal tissues as well. The two prospective potencies are practically identical, in spite of the different prospective significance.

The potencies of the marginal zone and the vegetal region could not be tested as easily as those of the presumptive ectoderm because they both tended to invaginate into the interior from every position in which they were placed. However, in exceptional cases parts of the transplanted marginal zone may remain on the surface at the end of gastrulation, and then they also conform to their surroundings in their development and differentiate as epidermis or neural system.

An entirely different result is observed if the transplantation of pieces of ectoderm is performed at the end of the period of gastrulation. A transplanted piece of presumptive neural system area will differentiate as

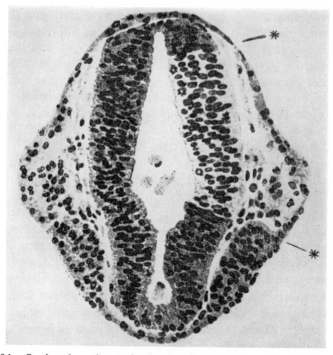

Fig. 104. Section through anterior head region of a *Triturus taeniatus* embryo in which part of the brain (shown by asterisks) is developed from grafted *Triturus cristatus* presumptive epidermis. (From Spemann, 1938.)

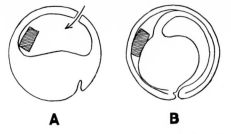

A   **B**

Fig. 105. Diagram of operation. *A,* Inserting a graft into the blastocoele of an amphibian early gastrula. *B,* Position of the graft after completion of gastrulation.

brain or spinal cord in whatever part of the embryo it has been placed. Usually in this case the transplanted tissue sinks from the surface of the embryo—this would correspond to neurulation in normal development— and develops into a vesicle with thickened walls. Sometimes definite parts of the neural system can be recognized, such as one or other of the brain vesicles or an eye (Spemann, 1919). The presumptive epidermis of a completed gastrula has likewise lost the ability to differentiate as nervous system. If it is transplanted into the nervous system area it does not conform to its surroundings and may interfere with the closing of the neural tube, but even if it does get enclosed inside the neural tube it differentiates into epidermis and not into nervous tissue.

Obviously some change has occurred between the early gastrula and the completion of gastrulation. The prospective potencies of presumptive neural tissue and epidermis have been narrowed down and are now the same as their prospective significance, so far as these experiments go. This narrowing down of the prospective potency, which is equivalent to the fixing of the fate of a part of the embryo, is known as **determination.** After the process of determination has taken place the respective parts are said to be **determined.** The experiments described so far show that the neural system as a whole is determined by the end of gastrulation. They do not show, however, whether *every part* of the neural system is determined at the same time. The determination of parts inside the primary organ rudiments will be dealt with later (Chapters 10, 11 and 12).

The experiments described above also show that the determination of parts of the ectoderm does not take place from causes inherent in the ectoderm itself. The differentiation of parts of the ectoderm is dependent on the surroundings in which the ectodermal cells find themselves. Further experiments proved that the condition which is decisive in determining the development of the neural plate is contact of the ectoderm with the roof of the archenteron, i.e., with the sheet of cells representing mainly the presumptive notochord and somites, which shifts forward underneath the ectoderm during gastrulation. Parts of ectoderm in contact with the archenteron roof differentiate into neural plate and nervous tissue. There are two ways in which parts of the archenteron roof can be brought into contact with ectoderm of the epidermis area which normally would not develop into neural plate. One way is to place the tissue whose influence it is desired to test into the blastocoele of a blastula or early gastrula through a slit cut in the animal region of the embryo. The wound heals very easily and quickly, and later as the archenteron of the host encroaches on the blasto-coele, the graft is wedged in between the invaginated endoderm + meso-derm and the ectoderm (Fig. 105). Another way is to transplant a piece of the dorsal lip of the blastopore into the marginal zone of another embryo in a suitable stage (blastula or early gastrula) and allow the graft to invagi-nate into the interior. In both cases the graft comes to lie under the ecto-derm, and the development of the overlying ectoderm is changed in such a way that it develops as nervous tissue.

One of the first experiments of the latter kind was performed by Spe-mann's pupil, Hilde Mangold (Spemann and H. Mangold, 1924). Hilde Mangold transplanted heteroplastically (from *Triturus cristatus* to *Triturus taeniatus*) a piece of the dorsal lip of the blastopore of an early gastrula. The graft was placed near the lateral lip of the blastopore of the host embryo (also an early gastrula), and subsequently most of the graft in-vaginated into the interior, leaving, however, a narrow strip of tissue on the surface. When the host embryo developed further, it was found that an additional whole system of organs appeared on the side where the graft had been placed. The additional organs together comprised an almost com-plete secondary embryo. This secondary embryo lacked the anterior part of the head. The posterior part of the head was present, however, as indi-cated by a pair of ear rudiments. There was a neural tube flanked by somites and a tailbud at the posterior end (Fig. 106). A microscopic

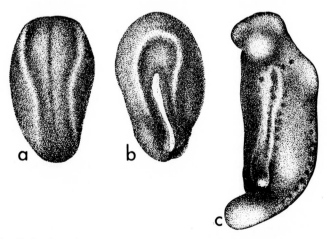

Fig. 106.   Induction of secondary embryo in *Triturus* by means of a transplanted piece of blastopore lip. *a, b,* Embryo operated on in the neurula stage. *a,* Dorsal view with host neural plate. *b,* Lateral view with induced neural plate. *c,* The induced sec-ondary embryo on the side of the host in the tailbud stage. (After Spemann and H. Mangold, from Spemann, 1938.)

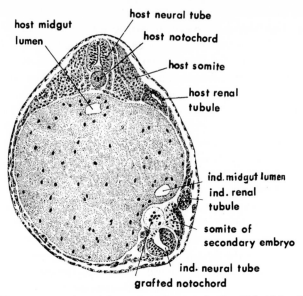

Fig. 107.   Transverse section of the same embryo as in Fig. 106. (After Spemann and
H. Mangold, from Spemann, 1938.)

examination revealed the presence of a notochord, kidney tubules and an
additional lumen in the endoderm representing the gut of the secondary
embryo. All the parts of the secondary embryo were found to lie at about
the same level as corresponding parts of the host. Since heteroplastic transplantation was used in this experiment, it was possible to determine what
parts of the secondary embryo were developed from the graft, and which
were developed from the cells of the host (Fig. 107). It was found that
the notochord of the secondary embryo consisted exclusively of graft cells,
the somites consisted partly of graft and partly of host cells. A small number of graft cells were present in the neural tube; these were certainly the
cells that had not invaginated. The bulk of the neural tube, part of the
somites, the kidney tubules and the ear rudiments of the secondary embryo
consisted of host cells. The additional lumen in the endoderm was also
surrounded by host cells. All these latter parts would not have developed
if the graft had not been present. It follows that their development was
due to some influence of the graft. This influence, causing the development
of certain organs of the embryo, is called **embryonic induction.** The part
which is the source of the influence is called the **inductor.**

From the experiment described above it may be concluded that the
induction of the neural system is due to the activity of the underlying tissues, the presumptive notochord and the presumptive somites in particular.
The issue is somewhat complicated by the participation of both graft and
host cells in some of the organs of the secondary embryo. This was later
shown not to be essential: in other experiments the whole of the induced
neural tube was developed exclusively from host tissue. It was further found

that anterior parts of the brain could also be induced by the roof of the archenteron, so that the induced nervous system may be complete.

In special experiments it could be ascertained which parts of the gastrula are capable of inducing the neural system. Pieces were taken from all parts of the early gastrula and each slipped into the blastocoele of an embryo in the early gastrula stage. Only grafts taken from the dorsal lip of the blastopore and the adjoining parts of the marginal zone were found to be able to induce. The area capable of induction coincided, in fact, almost exactly with the presumptive areas of notochord, somites and prechordal plate (Bautzmann, 1926).

Likewise it has been ascertained what embryonic tissues are able to react to the induction by developing as nervous system. The tissue in question is the ectoderm or presumptive ectoderm of the gastrula. The reactive ability is highest in the early gastrula stage, it is still high in the mid-gastrula stage, begins to decline in the late gastrula and fades away with the beginning of neurulation. This can be shown by transplanting the inductor underneath the ectoderm in successive stages, and the result may be expressed as a percentage of successful inductions (Machemer, 1932).

In later stages not only does the percentage of inductions diminish, but the volume and degree of differentiation is also reduced. Complete neural systems or large brain vesicles may be induced by transplantation of inductors in the early gastrula stage, but in the early neurula inductions are very weak; they may consist of only a few cells differentiating as nervous tissue. The presumptive ectoderm of the blastula is not able to react to the inductive stimulus; if the inductor is transplanted in the blastula stage, the ectoderm reacts by the formation of a neural plate only after the gastrulation is completed. In all cases the induced neural system develops simultaneously with the neural system of the host. All this shows that the reacting cells must be in a particular state to be able to differentiate into nervous system under the influence of the inductor. This particular state of reactivity is referred to as **competence.** The competence for neural induction is restricted to the ectoderm and is present during a short period only. If during this period the presumptive ectoderm is not stimulated to differentiate to neural tissue, it differentiates as epidermis. No special stimulation (induction) is necessary to cause the latter differentiation, though it has been found that certain conditions must be satisfied if the epidermis is to develop progressively and produce normal skin epithelium. The condition is the presence of underlying connective tissue. Without underlying connective tissue epidermis very quickly degenerates: the cells lose their polarity, they do not remain arranged as an epithelial layer, but form a spongy mass consisting of irregular strands of cells and eventually they perish.

6–2   SPEMANN'S PRIMARY ORGANIZER

Hilde Mangold's experiment brought out the fact that the transplanted dorsal lip of the blastopore exercises its influence not only on the ectoderm by inducing the neural plate, but also on the mesoderm and endoderm. This has also been corroborated by further experiments. Extensive inductions can be produced in the mesoderm by transplanting pieces of the dorsal lip of the blastopore. The organs developed from the graft are usually supplemented by parts produced from the host tissue as a result of the induction, so that a more or less complete whole is developed. Together with the parts developed from ectoderm a secondary embryo of various degrees of completeness arises. All the parts of the secondary embryo may be in harmonious relationship to each other, both in size and position. It is this ability of the dorsal lip of the blastopore to cause, when transplanted, the development of a complete whole that led to the dorsal lip of the blastopore being called **organizer,** or **primary organizer** (Spemann, 1938). The organizer of the gastrula is that region of the egg that can be distinguished in the stages before the beginning of cleavage, as the **gray crescent.** We can therefore link up the phenomena occurring during gastrulation with factors present at the beginning of development; i.e., we can trace the dorsal lip of the blastopore with its peculiarities to a specific part of the cytoplasm of the fertilized egg. The gastrulation movements and the determination of parts of the embryo during and after the gastrulation are thus foreshadowed by the organization of the egg before cleavage.

The inductions produced by a transplanted organizer or its part may vary in degree of complexity. When the induction is weak and the quantity of induced neural tissue is small, it is sometimes impossible to compare the induced tissue with any part of the normal nervous system. If, however, the inductor used is powerful, and the reaction of the competent tissue good, large organs or groups of organs are developed. In this case the induced parts can be recognized as representing certain organs of the normal embryo: specific parts of the brain, such as forebrain, midbrain and medulla, the spinal cord, eyes, nose rudiments, ear vesicles, somites, pronephric tubules and tailbuds. The various parts are always in a more or less harmonious relationship, so that parts belonging to the head are not mixed at random with parts belonging to the posterior trunk and tail region. In other words, the inductions are regionally specific.

The regional specificity may be imposed on the induced organs by the inductor. The anterior part of the archenteric roof induces head organs, and may be called a head inductor, the posterior part of the archenteric roof induces trunk organs and tailbuds and may be called the **trunk** or **spino-caudal** inductor. The head inductor may be further differentiated into the **archencephalic** inductor, inducing forebrain, eyes and nose rudiments, and the **deuterencephalic** inductor, inducing hindbrain and ear vesicles (Lehmann, 1945).

During gastrulation, the anterior part of the archenteric roof invaginates over the dorsal lip of the blastopore earlier than the posterior part of the archenteric roof. The dorsal blastopore lip of the early gastrula therefore contains the archencephalic and deuterencephalic organizer (inductor), and the dorsal blastopore lip of the late gastrula contains the spino-caudal organizer (inductor). The inductions produced by the dorsal lip of the blastopore taken from the early and the late gastrula differ in accordance with expectation: the first tends to produce head organs, and the second tends to produce trunk and tail organs (Spemann, 1931).

In the case illustrated in Fig. 106 the induction could be classified as combined deuterencephalic and spino-caudal, but lacking archencephalic parts.

Although the "primary organizer" was first discovered in urodele amphibians, it was soon found that the dorsal lip of the blastopore and the roof of the archenteron of other vertebrates has the same function in development. In particular it was found that the chordo-mesoderm in all vertebrates induces the nervous system and the sense organs.

In frogs the induction of a secondary embryo can be performed by the dorsal lip of the blastopore transplanted into the blastocoele of a young gastrula, in very much the same way as in newts and salamanders (see Fig. 108) (Schotté, 1930). The same applies to the cyclostomes, in particular to the lampreys, whose cleavage is holoblastic and whose blastula and gastrula resemble those of amphibians (see Fig. 109, *a*) (Bytinski-Salz, 1937; Yamada, 1938). In bony fishes inductions of secondary well-developed embryos were produced by transplanting the posterior edge of the blastodisc, which corresponds to the dorsal lip of the blastopore, into the blastocoele of another embryo (Fig. 109, *b, c*) (Oppenheimer, 1936, in *Fundulus* and *Perca*) or by transplanting the invaginated chordo-mesoderm and endoderm (Luther, 1935, in *Salmo*). Neural inductions were also obtained by transplanting the dorsal lip of the blastopore in the sturgeon (Ginsburg and Dettlaff, 1944). (See also Oppenheimer, 1947.)

In the birds the existence of the primary organizer was established by experiments by Waddington and his collaborators. The inducing part was found to be the anterior half of the primitive streak, which for other reasons we have recognized as corresponding to the lips of the blastopore in

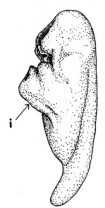

Fig. 108. Induction of a secondary embryo (*i*) by means of a grafted primary organizer in a frog. (After Spemann and Schotté, 1932.)

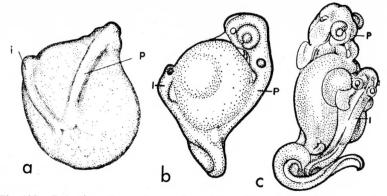

Fig. 109. Induction of secondary embryos by grafting of the primary organizer in the lamprey (*a*) and the perch (*b, c*). *p*, Primary embryo; *i*, induced secondary embryo. (*a*, From Yamada, 1938; *b, c*, from Oppenheimer, 1936.)

amphibians. To test the inducing ability of the primitive streak, whole blastoderms were removed from the egg in early gastrulation or pregastrulation stages and cultivated *in vitro* on a blood plasma clot. Parts of the primitive streak from another embryo were then inserted between the epiblast and hypoblast. Very good inductions of secondary embryos were obtained in this way (Fig. 110) (Waddington, 1933; Waddington and Schmidt, 1933; see also Waddington, 1952). Hensen's node does not differ in inducing ability from the part of the primitive streak immediately

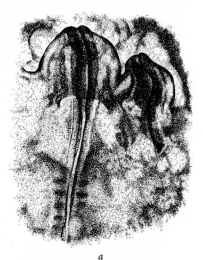

a

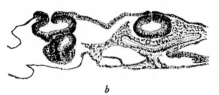

b

Fig. 110. Induction of a secondary embryo by means of a grafted primitive streak in a bird. *a*, Surface view; the secondary embryo is on the right. *b*, Section through the same embryo, showing the host axial system (right) and the induced neural tube lying above the neural tube and mesoderm developed from the graft (left). (After Waddington and Schmidt, from Waddington, 1952.)

following, but the posterior third of the primitive streak cannot induce neural differentiation. We have seen that cells migrating through the posterior part of the primitive streak give rise to lateral plate mesoderm (p. 151). This part of the streak corresponds therefore to the lateroventral lips of the blastopore of an amphibian gastrula, so the lack of inducing ability of this part is not surprising. In later stages neural induction may be caused by transplantation of the head process—the notochord rudiment.

Much light on the relation of the hypoblast to the epiblast of the avian blastoderm has been thrown by experiments performed by Waddington and his collaborators (Waddington, 1933, 1952). After excision of the whole blastoderm it is possible to separate the hypoblast from the epiblast, to join them together so that the anteroposterior (longitudinal) axis of the epiblast is placed at an angle of 90 to 180 degrees to the anteroposterior axis of the hypoblast. The preparation is then cultivated *in vitro* on a blood plasma clot. It was found that as a result of the discrepancy in the orientation of the epiblast and hypoblast, the primitive streak developing in the epiblast became curved, so as to coincide in part with the orientation of the hypoblast. When the rotation of the epiblast was 108 degrees, a completely new primitive streak appeared in some cases, with its anteroposterior axis orientated in accord with the position of the hypoblast, and thus pointing in the opposite direction to the one in which it should have been pointing normally. The hypoblast is thus capable of imparting its polarity to the epiblast, and causing the complicated system of movements in the epiblast described above (section 5–7).

The primitive streak in birds is thus dependent on the underlying hypoblast for its formation. This process of induction is something for which no analogue has been found so far in other vertebrates.

No experiments on the primary organizer have been done in reptiles, but one would expect that the archenteron has the same inducing activity in this class as in other vertebrates.

Because of technical difficulties only very inadequate information is available on the primary organizer and neural induction in mammals. A successful neural induction has been performed in a rabbit embryo by cultivating the early blastodisc on a plasma clot and implanting the primitive streak of the chick as an inductor (Fig. 111) (Waddington, 1934).

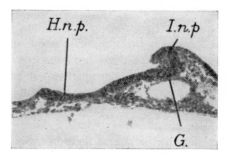

Fig. 111. Induction of a neural plate in a rabbit embryo by means of a grafted piece of primitive streak of the chick. The induced neural plate (*I.n.p.*) is joined to the host neural plate (*H.n.p.*). *G*, Cyst developed from the graft. (From Waddington, 1934.)

The experiment proves that the tissues of a mammalian gastrula have the competence for neural induction. It is practically certain that the primitive streak and the cells migrating from the streak (chordo-mesoderm) are the source of the inducing stimuli. This suggestion may be supported by the fact that the anterior end of a somewhat later rabbit embryo (with two pairs of somites) induced a neural plate in a chick embryo when placed under a chick blastoderm (Waddington, 1936).

We will now consider the nature of the inducing stimuli. Most of the work has been done with amphibian material as the reacting system.

6–3   ANALYSIS OF THE NATURE OF INDUCTION

The induction of the neural plate by the underlying chordo-mesoderm suggests the problem of analyzing the nature of the influence exerted by the inductor on the reacting ectoderm. Spemann gave the active region the name of "organizer," meaning that this was the part which organizes the process of development. He wanted to suggest that the term "organizer" was not merely a metaphor, and to say that the action of the organizer "is not a common chemical reaction, but that these processes of development, like all vital processes, are comparable, in the way they are connected, to nothing we know so much as to those vital processes of which we have the most intimate knowledge, viz., the psychical ones" (Spemann, 1938, p. 372).

This opinion of Spemann's has not, however, been borne out by subsequent experiments, or at least it has not been supported in its original form. It has been found that the nature of the inducing agent can be analyzed by physicochemical methods.

A priori it could be suggested that the inducing agent, that is the influence exercised by the roof of the archenteron which causes the ectoderm to be differentiated as neural tissue, may be of either a physical, a chemical or a vital nature.

A physical factor which might be responsible is the *contact* between the roof of the archenteron and the ectoderm. In induction of the neural plate by the archenteron the contact between inducing and reacting parts is a very intimate one. In the living state it is difficult to separate the roof of the archenteron from the presumptive ectoderm in mid- and late gastrula stages without damaging the cells. The inducing stimulus does not seem to be able to pass from the archenteron roof to the reacting ectoderm otherwise than by means of this intimate connection. Putting a thin layer of cellophane between inducing and reacting cells effectively suppresses induction. An attempt has been made to insert a porous membrane between inductor and reacting material. If chemical substances were liberated by the inductor into the intercellular fluid and subsequently taken up by the reacting cells, such substances should be able to pass through the pores of the membrane, provided the size of the molecules was not too great in relation to the size of the pores. So far it has not been possible to prove that neural induction can be performed through a porous membrane, using membranes with pores 50 mμ and 3 to 4 mμ (Brachet, 1950a).

The suggestion that contact may have something to do with the induction, nevertheless, has been ruled out experimentally because it was found that not every kind of tissue when placed underneath the ectoderm of the gastrula causes the latter to differentiate as neural tissue. No induction occurs, for instance, if a piece of gastrula ectoderm is placed in the blastocoele so that it comes to lie underneath the host ectoderm. The same is true if the graft consists of gastrula endoderm.

To determine whether some vital activity of the organizer is essential for neural plate induction the attempt was made to find out whether the organizer can exercise its action only in the living state. The organizer—the dorsal lip of the blastopore—was killed by various means—by boiling, by treating it with alcohol or petrol ether, by freezing or by desiccation—and it was then implanted into a living embryo in an appropriate stage of development (the early gastrula stage). It was found that a killed organizer can still induce (Bautzmann, Holtfreter, Spemann and Mangold, 1932). The vital activity of the organizer is therefore not essential for induction. These experiments lead to the conclusion that the roof of the archenteron produces its effect by liberating some chemical substance which is the immediate cause of induction. It is plausible that such a substance could be liberated even from dead tissue.

Experiments with Abnormal Inductors and Induction by Substances of Known Chemical Composition. The question arises then as to the nature of such a chemical substance, and also whether such a substance can emanate only from the roof of the archenteron, or whether perhaps other tissues can also produce a similar substance. It has been shown experimentally that the second alternative is the correct one. First of all, it was found that the neural plate, once it has begun to differentiate, may be a source of an inductive stimulus for neural differentiation (Mangold and Spemann, 1927). The inductive ability is not restricted to the early stages of neural development, but is also present in the tissues which are derived from the neural plate, i.e., in the tissues of the brain and spinal cord in the larva and in the adult animal. If the roof of the archenteron is the normal inductor of the neural plate, the nervous tissue may be called an abnormal inductor of the same. The notochord and the dorsal mesoderm (the muscles derived from the somites) also preserve their ability to induce long after the time when the neural plate induction takes place in the normal embryo. It has been found, furthermore, that almost any tissue of the adult animal can induce a neural plate and other organ rudiments besides, if it is placed under the presumptive ectoderm of an embryo in the early gastrula stage. Liver, kidney and muscles have been found to be very good inductors, but gut, skin and various other tissues have also been found to be effective (Holtfreter, 1934b).

It is of special interest that the inductions can be performed not only by tissues of the same or closely related species, but also by tissues of various animals, even of animals belonging to different phyla of the animal kingdom. Using the newt as the host, the tissues of the following animals were found to be effective as inductors: *Hydra,* insects, fishes, reptiles, birds, mammals (mouse, guinea pig, man). This lack of specificity applied to both normal and abnormal inductors. The conclusion from these experiments must be either that the inductive substance is very widely distributed in animal tissues, or that various substances may act as inducing agents.

As the normal inductor can act after being killed, so the abnormal inductors can induce after being killed by boiling or by immersion in alcohol or ether. A rather remarkable discovery made in this connection is that some tissues can induce better when they have been killed than if they are used in the fresh state. Some tissues that do not induce in the living state may induce when dead. This is the case with parts of the amphibian gastrula and blastula other than the dorsal blastopore lip region. Presumptive ectoderm and endoderm if killed by boiling or in other ways and implanted into the blastocoele will induce neural plates (Holtfreter, 1934a). Whether the inducing agent in this case is the same as the one emerging from the archenteron roof, cannot be affirmed without further proof.

The next stage was obviously to try to isolate the active substance from the inducing tissues. Waddington, Needham and collaborators (1934) succeeded in preparing ether extracts from inducing tissues (amphibian embryos, adult newts, mammalian liver) which were capable of inducing a neural plate. The solubility of the inducing substance in ether suggests that it is of a fatty or lipoid nature. Further purification of the extract showed that the unsaponifiable fraction retained the inductive ability. Thus the inductive substance should be identified as belonging to the steroid group. Forthwith inductions were attempted by introducing into the embryo various pure substances of the steroid group. The substance to be tested was, for this purpose, emulsified in egg albumen, the albumen was then coagulated, and a piece of the coagulum implanted in the blastocoele of a young gastrula in the usual way.

Quite a number of the substances tested—some of them also known by their carcinogenic (cancer-causing) action—induced very clear neural plates and neural tubes. Only small quantities of the steroids were necessary to produce induction; the carcinogen $1:2:5:6$-dibenzanthracene-endo-α-β-succinate can induce large and well-developed neural tubes if only 0.0125 microgram of this substance is introduced into the embryo. Weak neural inductions, however, may be obtained by as little as 0.000125 microgram of the inducing substance. This would correspond to a concentration of 0.025 milligram per kilogram wet weight (Shen, 1939). Waddington, Needham and collaborators concluded from this that the natural inducing substance is a steroid. It was soon found, however, that the steroids are not the only substances that can cause neural induction, neither are fat solvents the only means of extracting the active substance from inducing tissues. One of

Spemann's collaborators, Wehmeier (1934) prepared an active extract by prolonged boiling of muscle in water. The extract was made into a jelly by adding agar, and pieces of the jelly were implanted into the blastocoele. Very large and well-developed neural plates were induced in this way (Fig. 112). The active principle can also be extracted from the inducing tissues with other solvents such as a 0.14 M sodium chloride solution or ammonium acetate.

The latter extraction experiments are further corroborated by the implantation of various chemically defined substances besides the substances of the steroid group. Fischer, Wehmeier and collaborators (1935) found that a number of weak organic acids are good inductors, as for instance muscle adenylic acid, thymonucleic acid, dihydroxystearic acid, linolenic acid and stearic acid.

A chemical substance that has attracted attention as possibly the active principle of the "organizer" is the cytoplasmic ribonucleic acid. The amount of ribonucleic acid in the region around the dorsal lip of the blastopore and in the dorsal lip itself is higher than in the lateroventral parts of the blastula and in the vegetal region. As the cells of the dorsal lip of the blastopore invaginate and become the roof of the archenteron, the amount of ribonucleic acid in these cells diminishes. In the cells of the presumptive

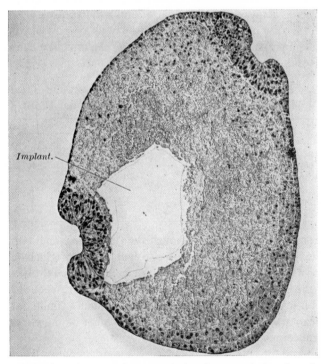

Fig. 112.   Induction of a secondary neural plate in an axolotl embryo by means of an implanted piece of agar containing substances extracted from muscle by boiling the muscle in water. (From Wehmeier, 1934.)

neural plate the amount of ribonucleic acid increases. The ribonucleic acid content in the neural plate and neural tube is higher than in the epidermis (Brachet, 1947). Thus the behavior of the ribonucleic acid during the process of induction corresponds to what we would expect if it was the inducing substance.

Special experiments have been carried out to bring further evidence that ribonucleic acid is the acting principle in induction. One method used consisted in destroying the nucleic acid in the tissue or preparation used as an inductor, and testing whether the inducing power would disappear after the treatment. In an early experiment it was claimed that dead tissues lose their inducing ability after treatment with the enzyme ribonuclease which decomposes ribonucleic acid (Brachet, 1947). It was found later, however, that the result was probably due to an impurity of the enzyme used, which contained besides ribonuclease some proteolytic enzymes, and the latter were responsible for loss of inducing power. When pure preparations of nucleases were used, they did not appreciably change the inducing activity of tissues or extracts. In one experiment, for instance, a very active preparation was obtained from a liver homogenate by ultracentrifugation. The microsome fraction was used, which contains most of the cytoplasmic ribonucleic acid. The experimental lot of this preparation was incubated for 3.5 hours with ribonuclease. After ribonuclease treatment the substance was precipitated, and its inducing ability (as well as that of untreated controls) was tested by placing pieces of precipitate between two layers of gastrula presumptive ectoderm. The preparations were cultivated as explants in a saline solution. Table 9 gives the results of the experiment.

Table 9.   Inducing Effect of Liver Ribonucleoprotein Treated with Ribonuclease (after Hayashi from Yamada, 1958)

	Control	Experimental
Ribonucleic acid content in mg. of RNA phosphorus per mg. protein nitrogen	148.6	0.84
Number of explants	49	52
Total neural induction	100%	100%
Archencephalic inductions	14%	15%
Deuterencephalic inductions	82%	77%
Atypical inductions	18%	17%

From Table 9 it is seen that although the ribonucleic acid content of the inductor was reduced to less than 1 per cent of the original quantity, the inducing power of the preparation was the same as in controls, not only in respect of the total number of inductions, but also in respect of the type of induction observed. This result is supported by many similar experiments, and it appears therefore that ribonucleic acid cannot be the essential part in the chemical composition of an inductor.

In contrast to the action of ribonuclease, the treatment of tissues or tissue extracts with proteolytic enzymes, pepsin, trypsin or chymotrypsin, destroys their ability to induce.

Figure 113 shows in diagrammatic form the results of treating a liver extract with trypsin for varying times (Hayashi, 1958). It was found that the preparation completely lost its ability to induce neural structures after treatment for 120 minutes and only caused weak atypical inductions. Figure 114, *a* and *b*, shows the induction in the control compared with the lack of neural induction after treatment of the liver extract with trypsin (120 minutes). There seems to be no doubt that the integrity of the proteins is essential for inductions by tissues and tissue extracts.

The general conclusion from the above experiments is that the inducing principle in abnormal inductors is of protein nature. Unfortunately this does not prove that the same substance acts in normal development.

The quest after the "normal" inductor of the neural plate has been further confused by the discovery (Waddington, Needham and Brachet, 1936) that isolated gastrula ectoderm may be stimulated to produce neural tissue (in the form of neural tubes) by keeping it in a weak solution of the dye methylene blue, which obviously has nothing to do with the development of the normal embryo. Lastly it was found (Holtfreter, 1947c) that isolated presumptive ectoderm may be caused to develop into nervous tissue by a short exposure to a saline solution of a composition which is not quite favorable to the cells. This is the case if the standard solution is more acidic or more alkaline than it should normally be—if the *p*H is lower than 5.0 or higher than 9.2

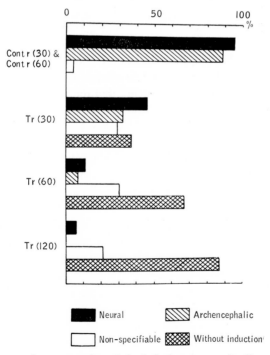

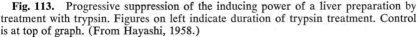

Fig. 113. Progressive suppression of the inducing power of a liver preparation by treatment with trypsin. Figures on left indicate duration of trypsin treatment. Control is at top of graph. (From Hayashi, 1958.)

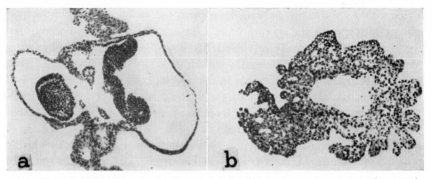

Fig. 114. Results of implanting into a vesicle of *Triturus pyrrhogaster* early gastrula blastoderm of a pentose-nucleoprotein liver preparation without enzyme treatment (*a*) and after treatment with trypsin for 120 minutes (*b*). There is archencephalic induction in *a*, lack of neural induction in *b*. (From Hayashi, 1958.)

Continued development in strongly acid or alkaline solutions is of course impossible; embryonic tissue exposed to these solutions very soon shows signs of damage. First of all the connections between the cells are broken; embryonic epithelia fall apart. The cytoplasm of the cells treated with acids or alkalis shows signs of liquefaction, noticeable because of an increased Brownian movement in the interior of the cells, and the amount of the hyaline cytoplasm at the periphery of the cells increases. If the acid or alkaline treatment is continued, cytolysis sets in, the cell membrane breaks down and the cells disintegrate. (See Fig. 115.) If the duration of the treatment is short, and the cells are soon transferred to a normal solution, the disintegration of the cells is checked. For a time they may show very active ameboid movement, but later the cells tend to clump together—reaggregate—and may resume normal development except that the end result of this development is changed: neural differentiation is observed in a great proportion of the aggregates. Complete dispersal of the tissues into individual cells is not necessary for the change in their differentiation. The main effect appears to be a sublethal damage (cytolysis) to the cells from which they can recover. Some agents such as urea causing damage to the cells without changing the *p*H of the medium may have the same effect as acids and alkalis (Karasaki, 1957).

The neural induction by cytolyzing agents is obviously a completely different way of controlling differentiation from the normal processes of induction by the roof of the archenteron. The action of the roof of the archenteron does not cause a sharp swing of the *p*H to the acid or the alkaline side, neither does it cause cytolysis in the overlying ectoderm. Even if it could be assumed that some specific inducing substances pass from the roof of the archenteron (or from pieces of adult tissues) into the ectoderm and thus change it qualitatively, no similar mechanism can be at work when solutions of ammonia or hydrochloric acid are used in cytolysis experiments. It has been suggested that the cytolysis "unmasks" some substance present in the ectodermal cells themselves, and that this substance, once released or activated, changes the presumptive epidermal cells into neural cells. Whether this interpretation is correct or not, it makes us regard with extreme caution any experiment in which neural differentiation is caused by treatment with various chemical substances. These substances may have nothing to do with normal induction, but cause neural differentiation through their toxic action which causes sublethal cytolysis of cells.

Whereas treatment of presumptive ectoderm with ammonia transforms it into nervous tissue, a similar treatment of presumptive mesoderm from the ventral lip area, which normally develops into lateral plate mesoderm and

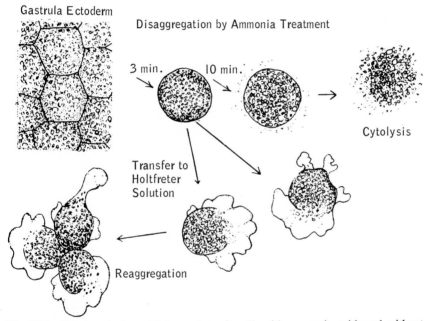

Fig. 115.   Treatment of amphibian embryonic cells with ammonia, with and without return of the cells to a normal medium. (From Karasaki, 1957.)

blood rudiments, can cause it to differentiate into notochord and somites (Yamada, 1950). In both cases structures pertaining to the dorsal side of the embryo (neural plate in the case of ectoderm, notochord and somites in the case of mesoderm) are produced from material which otherwise would develop into lateral or ventral parts (epidermis, lateral plate mesoderm). The transformation has therefore been denoted as "dorsalization" (Yamada, 1950). The term dorsalization would apply to normal induction, and not only to that produced by abnormal inductors or cytolyzing agents.

Physiological Properties of the Organizer.   A different approach to the problem would be to investigate what peculiarities of the cells in the dorsal lip of the blastopore are not found in other parts of the embryo, and what changes, if any, can be discovered in these cells at the time when they are inducing a neural plate.

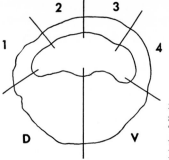

Fig. 116. Diagram showing the pieces into which frog gastrulae were cut to study respiration of the different parts. The dorsal side, with the blastopore, is on the left. (After Sze, from Barth and Barth, 1954.)

Many attempts have been made to analyze the physicochemical and physiological properties of the dorsal blastopore lip material, as compared with other parts of the embryo.

One of the first facts discovered in this connection was that the amount of glycogen becomes considerably diminished in the invaginating cells of the dorsal lip of the blastopore. This was first discovered by histochemical methods, by using a specific stain for glycogen on sections of gastrulating embryos (Woerdeman, 1933). Later, by methods of chemical analysis the exact amounts of glycogen consumed were determined; it was found that in the dorsal lip of the blastopore 31 per cent of the glycogen is lost during gastrulation, whereas in other parts of the embryo only from 1 per cent to 9 per cent is lost in the same time (see Brachet, 1950b).

Rapid breakdown of glycogen in the dorsal lip suggests a particularly active respiration in this area. Direct measurements of respiration, however, showed that the dorsal lip region is by no means the part of the gastrula which respires at the highest rate. To make measurements of respiration, frog gastrulae were cut into several regions, as indicated in Fig. 116. Two big pieces were made of the vegetal hemisphere (a dorsal and a ventral one) and four pieces of the marginal zone and the animal hemisphere (pieces 1 to 4) starting with the dorsal lip of the blastopore. The oxygen consumption was determined for each piece as well as the dry weight, total nitrogen and extractable nitrogen—the latter, as the reader will remember (see section 2–7) is equivalent to the nitrogen of the active cytoplasm (total nitrogen less the nitrogen contained in the yolk).

When the oxygen consumption is related to dry weight of the fragments, it is seen that there is a great difference in the oxygen consumption of various parts of the embryo (see Table 10).

The highest oxygen consumption, if referred to dry weight or to total nitrogen, that is to the whole mass of the embryonic tissue, is found at the animal pole of the gastrula, and the lowest oxygen consumption is at the vegetal pole. The dorsal side, including the dorsal lip, has a distinctly higher oxygen consumption than the ventral side. This would account for the difference in the breakdown of glycogen as between the dorsal and the ventral lip (see above). We must realize, however, that not all parts of the cells of a

Table 10. Oxygen Consumption in Different Parts of the Frog Gastrula in Microliters of Oxygen Consumed per Hour per Unit of Reference (from Barth and Barth, 1953)

Per Unit of	Vegetal dorsal	Dorsal lip piece 1	Animal pole piece 2	piece 3	Ventral lip piece 4	Vegetal ventral
Dry weight	1.2	3.7	4.4	3.9	2.7	0.9
Total nitrogen	12.0	37.0	53.0	46.0	30.0	8.2
Extractable nitrogen (nitrogen of active cyto- plasm	11.0	12.0	13.0	13.0	13.0	11.0

frog gastrula respire; the yolk presumably does not respire at all whereas it contributes both to the dry weight and to the total nitrogen of parts of the embryo.

It would be desirable to eliminate the yolk from calculations of embryonic respiration, and this was done by calculating the oxygen consumption per unit of extractable nitrogen (= nitrogen of active cytoplasm). The data are given in the last line of Table 10 and they show that equal amounts of active cytoplasm have practically the same respiration rate in all parts of the gastrula. In other words, the differences in oxygen consumption observed are due to different amounts of active cytoplasm in relation to yolk, that is, to the gradient of yolk distribution (dealt with in section 2–4), and not to a local specifically higher rate. Studies of respiration have thus failed to reveal a qualitative distinction of cytoplasm of the "organizer region" or to give an indication about a peculiar type of metabolism that would account for its particular role in development.

Emission of Inducing Substances by the Natural Inductors. We have seen that extracts of inducing substances made by chemical methods cannot be entirely relied on since a variety of substances may produce inductions, partly through the channel of sublethal cytolysis. It was therefore a great advance when it was discovered, by Niu and Twitty, that tissues capable of induction themselves release inducing substances into the surrounding medium. It was only necessary to find a means of accumulating these substances in sufficient concentrations, and to discover the best way to apply them to cells competent to react to the induction. The first part of the problem was solved by cultivating inductor tissues, such as the dorsal lip of the amphibian blastopore, parts of the neural plate, rudiments of the notochord and somites, in small quantities of saline by the hanging drop method (a drop of culture medium with the tissue to be cultivated is placed on a coverslip, and this is inverted over a hollow on a slide; the preparation is then sealed to prevent desiccation). The saline solution used was a specially modified one, now known as the "Niu and Twitty solution" (see Niu and Twitty, 1953). The inducing substances accumulate in the medium gradually; the active concentration is reached only after a week to 10 days. The fragment of early gas-

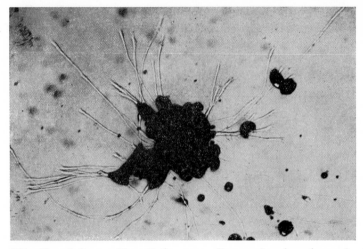

Fig. 117. Neural tissue with radiating nerve fibers and melanophores developed from a piece of young *Triturus rivularis* ectoderm cultivated in conditioned medium. (From Niu, 1956.)

trula ectoderm is then placed in the same drop of medium, and cultivated for a further one to three weeks. It was found important for the success of the experiment that the fragment of ectoderm be very small, containing something like 15 to 20 cells. If the fragment is large, it rounds off and becomes covered with the surface coat (see section 5–8) which prevents substances from without from penetrating into the cells. With a very small fragment this does not occur and there is a sufficient surface of unprotected cytoplasm into which the inducing substances may be taken.

In successful cases, and these may be up to 90 per cent in some experimental series, the cells of the ectodermal explant behave as if they were a part of a normal neural plate. Many of the cells scatter in the medium and become transformed into typical melanophores. Others remain in a clump and differentiate as nerve cells. The latter may produce nerve processes, which radiate into the surrounding medium (Fig. 117). The presence of the inductor tissue is no longer necessary for the success of the induction once the medium has been "conditioned," i.e., has accumulated the necessary amount of inducing substance. Pieces of ectoderm cultivated in drops withdrawn from the original cultures and containing no inductor cells differentiate into nerve cells and melanophores just as readily as those which are kept in a drop with a piece of inductor (Niu, 1956).

The action of "conditioned" saline must be essentially different from the shock treatment by media with very high or very low pH which was referred to above. It was found by direct measurement that the saline after cultivation in it of inductor tissues is still very nearly neutral (it had a pH of 7.8). Neither was any trace of sublethal injury or cytolysis apparent. The action was obviously due to the presence of a specific substance or substances in the medium, and there appears to be no reason why these substances may not be the "natural" inducing substances, that is, the substances through the mediation of which the neural plate and other organs are induced in normal development. An explanation also suggests itself as to why in previous experiments a direct contact between inducing and reacting tissues was found to be necessary. The reason probably is that if an intercellular space of any appreciable volume, filled by any kind of medium, intervenes between the two, a "conditioning" of this medium, or the accumulation there of a sufficient concentration of inducing substance, would be reached only long after the competence in the reacting tissue would have passed. The situation is quite different when the acting and reacting cells are in immediate contact, or, to be more exact, are separated only by the ordinary intercellular gap of about 80 Å $(0.008 \ \mu)$, which could be quickly saturated with the inducing substance.

It is now possible, by analysis and later by fractionation of the "conditioned" saline medium, to determine what is the chemical nature of the inducing substance or substances. The analysis is by no means complete, but the following has so far been ascertained. The conditioned medium contains a macromolecular substance or substances of the nature of nucleoproteins. The nucleic acid component is mainly ribonucleic acid, though small amounts of deoxyribonucleic acid have also been found. So far it has not been possible to determine with certainty whether the inducing action is due to the nucleic acid component or the protein component. In some experiments (Niu, 1956) addition of ribonuclease to the "conditioned" medium reduced its inducing activity but did not abolish it, whereas addition of trypsin or chymotrypsin prevented induction completely—a result very similar to that obtained with extracts of adult tissues (pp. 178–179).

Three further points may be noted here: first, the duration of the treatment of the competent ectoderm. The minimum time of this treatment was found to be 24 hours. After this the treated ectoderm cells may be removed from the "conditioned" medium to ordinary saline, where they proceed to differentiate into nerve cells or chromatophores. Twenty-four hours, incidentally, is also the time which was found to be necessary for contact between normal inductor and reacting tissues, in order for the induction to have full effect (*cf.* Toivonen, 1958).

The second point of interest is that the substances released by the inductors into the surrounding medium appear to change with the age of the inductor, or with the duration of its sojourn in the medium. The reaction consisting of nerve cell and melanophore differentiation is observed when the inductor had been cultivated in the medium for seven to ten days. If the inductor itself differentiates into muscle cells, which happens especially if somites are used, or in some cases of explantation of parts of the neural plate (*cf.* data on the fate of parts of the neural plate in section 10–1), and if the cultivation is extended to 12 to 16 days, the reaction of the ectoderm becomes different: up to 73 per cent of the inductions lead to differentiation of myoblasts in the cultivated ectoderm. In this way a "mesodermal induction" (see section 6–4) could be reproduced *in vitro*. The presence of sub-

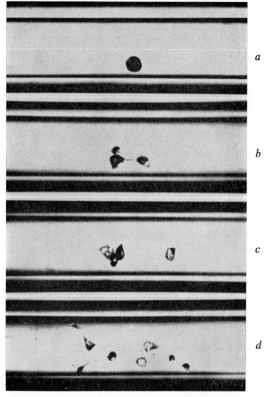

Fig. 118.   Single ectodermal cell cultivated in a glass capillary tube filled with conditioned medium. Four photographs of the tube were taken on the 2nd (*a*), 4th (*b*), 5th (*c*) and 7th (*d*) days after explantation. *a*, The single cell which has been placed in the capillary tube. *b*, Three cells have been produced. *c*, Four cells have been produced. *d*, Eight cells have been produced by division of the original single cell. (From Niu, 1956.)

stances inducing muscle may have been a direct result of the particular type of differentiation going on in the inductor, the induction tending here to assimilate the reacting material to the inductor itself.

The third point that emerges from Twitty and Niu's work is that the ectoderm cells need not be in their typical association to be able to react to the inductor. Even in the standard experiment in which very small clumps of cells (15 to 20 cells, as indicated above) were used, the arrangement of cells as in a normal blastoderm was lost. It was found possible, however, to expose individual, single, cells to the inducing stimulus. For this purpose single ectodermal cells were sucked into capillary tubes filled with "conditioned" saline. The cells seemed to thrive well, underwent several divisions and showed signs of transformation into promelanophores (Fig. 118) (Niu, 1956). From this it follows that the reaction to an inducing substance consists in the change of properties of individual cells in the first instance, in the creation of *new kinds of cells*. Once the properties of the cells are changed their behavior is changed, and the formation of the multicellular organ rudiment, such as the neural plate, is the outcome of the new properties of all the cells comprising the rudiment. In the case of the neural plate the new properties of the cells cause them to form a thicker layer than the layer of epidermis, to make the edges of the layer curve upward, and so on. These processes will be dealt with in Chapter 9 and more specifically in section 9–3.

Do Inducing Substances Penetrate into Reacting Cells?   The last question which may be considered in connection with the physicochemical nature of the processes of induction is about the way in which the inducing substances control the differentiation of cells. In principle at least two possibilities may be envisaged. First, it may be assumed that the inducing substances penetrate into the cells and by interfering in the metabolic mechanisms in the interior of the cells change their physicochemical composition. Although this mechanism of action seems to be fairly plausible, it is not self-evident. It may be suggested (see Weiss, 1947) that the inducing substances, although not penetrating into the interior of the cells, become attached to the surface of the cells, forming bonds with some types of molecules present in the cytoplasm of the reacting cell. As this bonding may be selective, it may be expected to cause a shift in the equilibrium of the cell sufficient to lead its differentiation into a new channel.

Radioactive tracers have been used to learn whether in the process of embryonic induction substances pass from the inducing tissue into the reacting tissue. The general plan of such an experiment is as follows. Parts of one embryo are cultivated for some time in a solution containing substances which carry a "marked" atom—a radioactive isotope of carbon or sulfur. Labelled amino acids, methionine S^{35} and glycine C^{14}, and a nucleic acid precursor, orotic acid C^{14}, were used in some experiments (Ficq, 1954; Sirlin, Brahma and Waddington, 1956; Waddington and Mulherkar, 1957). These compounds are taken up into the tissues of the embryo. Next, inducing parts—roof of the archenteron or parts of brain rudiments—are excised from the embryo containing radioactive substances and transplanted into a normal embryo where the graft may induce neural plates or other structures from the host tissue. The embryo is then fixed and cut into sections, and the sections are coated with a photographic emulsion. The electrons emitted by radioactive atoms cause a darkening of the emulsion, and it is thus possible to discover where the radioactive atoms were situated at the time of fixation.

It has been found that the radioactive substances do not remain restricted to the cells of the grafts, but that they become fairly widely dispersed in the host embryo. A high radioactivity is shown by induced neural plates, which may be the result of the passage of inducing substances from the graft into the host tissue. However, the host neural plates and tubes are also strongly radioactive, and a lesser radioactivity can be discovered in the mesodermal and endodermal tissues of the host. These results are consistent with the assumption that the radioactive atoms are carried around in the host tissue by simple diffusion and that their concentration in the neural plates is simply due to a higher metabolic turnover in the neural rudiments, whether they were induced by the graft or not. The experiments thus do not give a proof of the passage of inducing substances from the inductor into the reacting tissue. Neither do they give support to the "surface action" theory, as no concentrations of radioactive substances at the surfaces of the induced structures have been recorded.

In a new approach, anuran (*Rana pipiens*) chordo-mesoderm was used to induce neural plates in newt (*Taricha torosa*) ectoderm. After four to five days' cultivation the neural plates were tested for the presence of *Rana* antigens. It was found that the extracts of the tested tissue gave a slight positive reaction with rabbit serum immunized against *Rana* embryo homogenates. It would seem that frog proteins had actually passed into the newt ectoderm (Rounds and Flickinger, 1958). Unfortunately there is no means of knowing whether the proteins detected by the serological reaction were actually those which caused the induction.

6–4   GRADIENTS IN THE DETERMINATION OF THE PRIMARY ORGAN RUDIMENTS OF VERTEBRATES

The Dorsoventral and Anteroposterior Gradients.   The phenomenon of regional specificity, referred to above, raises the question whether the gray crescent does not contain after all two or even more different specific substances, responsible for the development of the head and the trunk regions (or the archencephalic, deuterencephalic and spinocaudal regions), with their separate inductors or organizers. Certain results achieved with abnormal inductors have been adduced in favor of this supposition.

When different tissues of various adult animals are tested as inductors, it is found that they do not all act exactly alike: some of them act preponderantly as head inductors, others as trunk inductors. Some induce only ectodermal parts (neural structures, sense organs), others also induce mesodermal organs and tissues (see Toivonen, 1949, 1950, 1953). For instance the liver of the guinea pig, treated with alcohol, acts as an archencephalic inductor, that is, it induces large brain vesicles, bearing a resemblance to the telencephalon, diencephalon and mesencephalon (see p. 251), sometimes with eyes; it also induces noses and balancers (see p. 283). The kidney of the adder is a rare case of predominantly deuterencephalic inductor; it induces brain parts resembling the hindbrain and also ear vesicles. The kidney of the guinea pig treated with alcohol is a spinocaudal inductor; it induces mainly spinal cord, notochord, bands of muscle segments arranged as they are normally in the trunk region of the embryo and often also complete tails with fin folds around their edges. Lastly, alcohol-treated bone marrow of the guinea pig induces almost exclusively mesodermal parts of the trunk and tail: notochord, rows of somites, nephric tubules and limb-buds.

The inducing properties of tissues may change depending on the type of treatment. In particular it was found that heating (or boiling) the tissues reduces their ability to induce mesodermal and spinocaudal structures, whereas archencephalic inductions are still easily obtained. This and similar results have led to the conclusion that the regional nature of inductions produced by adult tissues is the result of the interaction of two factors contained in various proportions in the different tissues (Yamada, 1950; Nieuwkoop and others, 1952; Toivonen and Saxén, 1955; Toivonen, 1958). One factor is the "neuralizing agent" of Toivonen and Saxén, called "dorsalizing agent" by Yamada and "activating agent" by Nieuwkoop and others. When it is present alone it causes archencephalic inductions. It is the active principle of the alcohol-treated liver. The second principle is Toivonen and Saxén's "mesodermalizing agent" which is the same as "caudalizing agent" of Yamada and the "transforming agent" of Nieuwkoop and others. If present alone, this second factor induces only mesodermal parts: notochord, muscle, kidney, limb-bud (see Fig. 119). This is the active principle of the alcohol-treated guinea pig bone marrow. A deuterencephalic induction is the result of the presence of a small amount of the mesodermalizing factor in addition to the neuralizing factor. A large amount of the mesodermalizing factor added to the neuralizing factor produces a spinocaudal induction.

That deuterencephalic structures are a result of a certain balance between neuralizing and mesodermalizing inducing substances has been shown by the following elegant experiment (Toivonen and Saxén, 1955). Two pellets, one prepared from guinea pig liver and another prepared from guinea pig bone marrow, were implanted simultaneously into the blastocoele of a young gastrula (Fig. 120). The liver preparation alone would have induced exclusively archencephalic structures, the bone marrow alone induces trunk

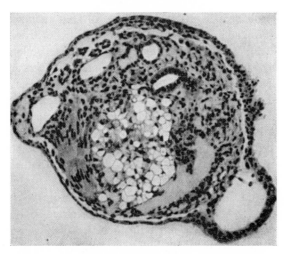

Fig. 119. A purely mesodermal induction (notochord, somites, pronephros) in isolated ectoderm caused by the non-nucleoprotein fraction of bone marrow. (From Yamada, 1958.)

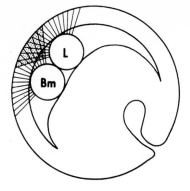

Fig. 120. Simultaneous implantation of an archencephalic (liver—*L*) and a mesodermalizing (bone marrow—*Bm*) inductor into a newt's gastrula. (From Toivonen and Saxén, 1955.)

mesoderm. Together, however, they induced organs belonging to all levels, including deuterencephalic and spinocaudal structures: medulla, ear vesicles and spinal cord.

The two factors are actually two substances with different chemical properties. The neuralizing substance is thermostable and soluble in organic solvents. The mesodermalizing substance is insoluble in organic solvents and highly thermolabile. The result of this latter property is that spinocaudal and deuterencephalic inductors on heat treatment induce archencephalic structures; in the fresh state they contain both substances but the mesodermalizing substance is destroyed by heat, whereas the neuralizing substance remains unchanged (at least after a short heat treatment—prolonged heat treatment inactivates the neuralizing substance as well). By graded heat treatment of a tissue its inducing properties may be changed gradually, as the ratio between the mesodermalizing and the neuralizing substances changes in favor of the latter (see Fig. 121).

It has been a matter of controversy for some time whether the neuralizing agent and the mesodermalizing agent are two distinct chemical substances, or whether they might be, after all, two aspects of the action of one and the same substance, or two states of one substance, a more labile state (the mesodermalizing agent) and a more stable state (the neuralizing agent). Recent experiments have now resolved this problem. It has been found that the neuralizing substance of the guinea pig liver can be isolated in the form of a ribonucleoprotein (Hayashi, 1956), either by sedimentation with streptomycin sulfate, or by ultracentrifugation as a result of which the microsomal fraction, containing the cytoplasmic nucleoprotein, shows the strongest archencephalic inductive action. The mesodermalizing substance of guinea pig bone marrow on the other hand is not sedimented by streptomycin sulfate and is not contained in the microsome fraction after ultracentrifugation (Yamada, 1958). The mesodermalizing substance is therefore a protein which does not tend to be coupled with nucleic acid. The two agents are thus distinctly different and separate substances, one possibly bound to the microsomes, the other not. There is no contradiction between the statement that the neuralizing substance can be obtained in the

form of a ribonucleoprotein, and our previous finding that the active principle of the inducing substance is a protein and not a nucleic acid: if
the active nucleoprotein prepared from liver is treated with ribonuclease the
inducing action is retained, if it is treated with proteolytic enzymes the
preparation loses its inducing ability (Hayashi, 1955 and 1958).

It may be of some interest that the neural inductions produced in ectodermal explants by cytolyzing agents are all of archencephalic type. The
cytolyzing agents can thus simulate the action of the neuralizing substance
but not that of the mesodermalizing substance.

The results obtained with abnormal inductors have been used to interpret
normal development. It has been assumed (Toivonen and Saxén, 1955)
that the same two substances are active as inducing substances of the archenteron roof, and that the regional differentiations in the normal embryo
are controlled by a balance between the two substances distributed in the
form of gradients. The gradient of the neuralizing substance is highest mid-
dorsally, and declines toward the lateral and ventral parts of the embryo.
The mesodermalizing substance is most highly concentrated at the posterior
dorsal end of the embryo and forms a declining gradient both in the anterior
direction and in a lateral direction. This distribution of substances as
originally contained in the archenteron roof is transmitted to the overlying
ectoderm.

The anterior part of the archenteron roof (including prechordal plate)
emits only the neuralizing substance, and induces the archencephalon. No
notochord and no somites are developed at this level. At a slightly posterior
level the notochord and somites are already present. There is some admixture of the mesodermalizing substance to the neuralizing substance, and as

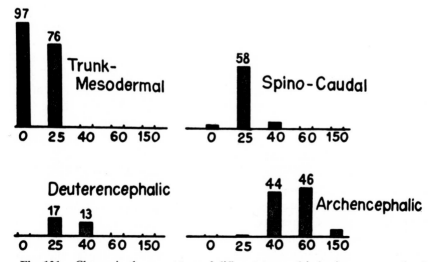

Fig. 121. Change in the percentage of different types of inductions as a result of
treating the inductor (slices of bone marrow) with heat for varying times (seconds)
indicated below the lines on the diagram. (From Yamada, 1958.)

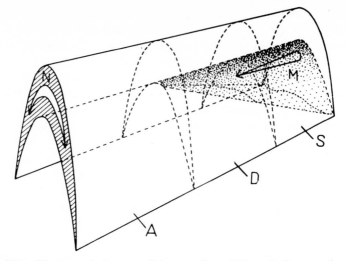

Fig. 122. Diagram of the neuralizing gradient (*N*) and the mesodermalizing gradient (*M*). *A*, Archencephalic level; *D*, deuterencephalic level; *S*, spinocaudal level. (From Toivonen and Saxén, 1955.)

a result deuterencephalic structures (medulla and ear vesicles) are induced in the ectoderm. Still further back large amounts of the mesodermalizing substance are available, mesoderm develops into notochord and large masses of muscle, and the neural structure induced is a spinal cord (see Fig. 122). It will be seen that each level of the embryo will have a different concentration of the neuralizing and mesodermalizing agents.

The concept of two mutually permeating gradients reminds us at once of the animal and vegetal gradients of the sea urchin egg, though the relationships of the gradients to the main axes of the embryo (and egg) are different.

To give further confirmation to the theory of the interaction of the neuralizing and mesodermalizing gradients, it would be necessary to show that the neuralizing and mesodermalizing agents are actually present in the normal embryo. In support of this it is indicated that the "trunk organizer" (see Spemann, 1931) (p. 171), the non-invaginated part of the dorsal lip of the blastopore of a middle and late gastrula, if treated with a heat shock (one hour at 30 to 38° C.) becomes an archencephalic inductor ("head organizer") (Takaya, 1955). We will remember also that in experiments of "conditioning" of the medium in which tissues of the archenteron roof had been cultivated (Niu, 1956) (p. 185), a neuralizing substance becomes accumulated at first (inducing nerve cells and mesectodermal melanophores), and this is later superseded by a mesodermalizing substance which induces the differentiation of myoblasts.

Apart from experiments on the isolation of chemical substances, gradients, in the form of grading or fading out of certain substances and properties, are very obvious in the early development of amphibia and other vertebrates. The dorsoventral gradient (the gradient of the neuralizing substance) can be traced both in the ectodermal and in the mesodermal layer. In the ectoderm the highest level of the gradient is associated with the development of the neural plate and neural tube. The next highest level of the gradient causes the differentiation of the neural crest. There is ample evidence that the development of the neural crest is the result of a weak induction of the same nature as the induction of neural tissue. When neural plates are induced, the cells on the periphery of the region exposed to the inductor, where the inductive stimulus is fading away, develop as neural crest. Sometimes, if the inductor is too weak to induce a neural plate, neural crest cells can still be induced. It is usually easy to notice the presence of neural crest cells on the site of induction, because some of them differentiate as pigment cells. Possibly a still lower level of the gradient is responsible for the differentiation of the various placodes including nose and ear rudiments, the rudiments of cranial ganglia and of the lateral line sense organ system (see section 10–4).

Where the gradient of the neuralizing agent fades away ectoderm differentiates as skin epidermis. In the chordo-mesodermal mantle the dorso-ventral gradient may be held to be responsible for the segregation of the notochord and the mesoderm in the gastrula of an amphibian. Those parts of the marginal zone in which the specific cytoplasmic substance is most highly concentrated become notochord, and those parts in which the concentration of the substance is lower become mesoderm (the somites). Those parts, lastly, in which the concentration of the cytoplasmic substance is minimal become unsegmented lateral-plate mesoderm (Yamada, 1937, also Dalcq and Pasteels, 1937). The gradient concept does not contradict our previous statement that the notochord may induce somites and other mesodermal parts. This may be due to the presumptive notochord, as the highest level in the gradient system, establishing a new gradient in the surroundings into which it has been transplanted, either by means of direct diffusion of the specific substance from higher levels of concentration into the surroundings where the concentration is lower, or by some other means not involving the diffusion of a specific substance.

Besides the gradient which has its highest level in the dorsomedian strip of tissues diminishing towards both sides there is a second gradient, with a center of highest concentration at the anterior end of the embryo lessening in a posterior direction. This anteroposterior gradient may be considered to be another manifestation of the caudo-cranial mesodermalizing gradient, its reverse side as it were. This gradient is responsible for the differentiation of parts of the nervous system, the most anterior part of the head being the part where the gradient is at its highest.

Control of Development by Influencing the Gradient System.   The gradient unites the parts of a developing embryo into one whole, into one morphogenetic system. Any factor that affects the gradient will therefore affect the whole morphogenetic system, no matter how simple or even elementary the factor itself may be. It is possible, for instance, to affect a

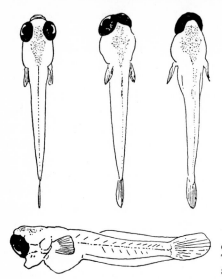

Fig. 123. Cyclopia in fish larvae, caused by magnesium chloride treatment. Top left, normal control. (After Stockard, from Huxley and de Beer, 1934.)

morphogenetic system by depressing the level of the gradient at its highest point. One way of achieving this is to expose the embryo to certain chemical substances such as magnesium chloride or lithium chloride. Another way of depressing the high level of the gradient is to remove part of the archenteron roof underlying the anterior end of the nervous system. The result in both cases is about the same: the most anterior parts of the nervous system fail to develop normally. The defects can best be traced in the structure of the eyes. The injured embryos develop a defect known as

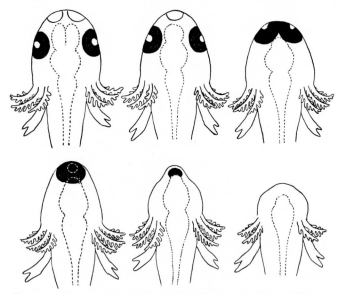

Fig. 124.   Increasing degrees of cyclopic deficiencies in newt larvae. (Diagrammatic.) Upper left, normal larva.

cyclopia, that is, the appearance of one median eye instead of two lateral ones (Fig. 123). The median eye is really the result of fusion in the midline of the two eye-forming areas, and every intermediate state may be found in embryos in which the injury has not been very severe. If very large parts of the underlying chordo-mesoderm have been removed, or if the action of the chemical agent has been very strong, the defects in the structure of the eyes are still greater—the single eye being reduced in size and failing to develop altogether in the extreme cases. The eyes are not the only organs affected in cyclopic embryos; other parts of the head are changed and reduced more or less in correspondence with the defects of the eyes (see Adelmann, 1936). The parts which may be involved are the brain, the nose, the ear, the mouth, the gill clefts. The defects spread from the front end backward as the injury is more and more pronounced. The nose rudiments and the forebrain are the first to be affected, so that the olfactory sacs become unpaired and finally disappear. Then follow the eyes and the mouth,

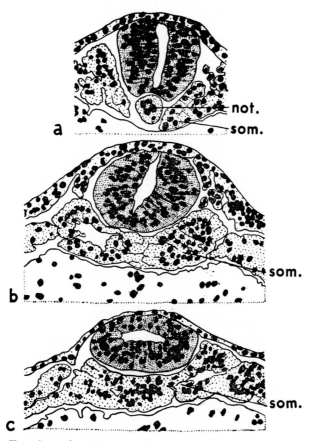

Fig. 125.   Transformation of presumptive notochord in somite mesoderm by means of the action of lithium in *Triturus*. *a,* Untreated control; *b, c,* lithium-treated embryos. *not.,* Notochord; *som.,* somites. (From Lehmann, 1945.)

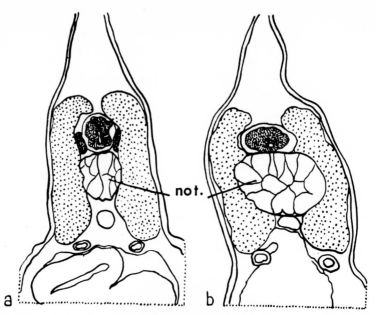

Fig. 126. Raising the dorsoventral gradient in a frog embryo with thiocyanate. *a,* Normal embryo; *b,* embryo with increased notochord after thiocyanate treatment. (After Ranzi and Tamini, from Lehmann, 1945.)

the mouth becoming narrower and also disappearing. After the eye and mouth follow the posterior parts of the brain, the gill clefts and the ear vesicles. The gill clefts fuse in the midventral line, before disappearing one by one. In extreme cases practically the whole head is lacking (Fig. 124).

It must be understood that the complete presumptive ectoderm remains in place during the above experiments; none of it is removed, neither do any of its parts become necrotic. All cells are there, but their development is changed as a result of an interference with the gradient system. The experiment with the lithium chloride or magnesium chloride brings out a further important phenomenon: although the whole embryo may be exposed to the chemical substance, it is only the highest parts of the gradient system that are affected. This is a very common result, namely that the highest point of a gradient is more easily damaged than the other parts. In fact a gradient can often be discovered by exposing the embryo to any mild injurious factor, such as weak poisons, abnormal temperature or ultraviolet radiation. The effect will be observed first, and sometimes only, at the highest point of the gradient if the intensity of the injurious factor has been chosen correctly. In vertebrates in stages following gastrulation such a sensitive region is invariably the anterior part of the head.

Lithium can also affect the gradient in the chordo-mesodermal system. Treatment of the embryo during gastrulation with a weak solution of lithium chloride suppresses the development of the notochord, which has been postulated as representing the center of highest activity of the mesodermal gradient. If the development of the notochord is suppressed, the presumptive notochord cells differentiate according to the next highest level of the gradient and develop into somite tissue (Fig. 125). The right and left rows of somites are then continuous with each other across the midline, underneath the neural tube (Lehmann, 1937, 1945).

It has been reported that the opposite effect may be produced by treating frog embryos in the blastula stage with a solution of sodium thiocyanate (NaSCN). The result is the development of embryos in which the notochord is much larger and thicker than in control animals (Fig. 126). This means that a greater than normal area has shown a differentiation characteristic of the highest level of the gradient. This may be called raising the level of the gradient (Ranzi and Tamini, 1939).

Raising the dorsoventral gradient in the ectoderm should lead to an increase in the size of the neural system. In fact embryos with excessively broad and massive brains have been observed after sodium thiocyanate treatment in fishes (see Huxley and de Beer, 1934) and frogs (Ranzi, 1957) (Fig. 127). It should be mentioned, however, that the above results have been contradicted. It was suggested that the number of cells in the notochord is not increased by thiocyanate treatment, and that the embryos thus treated are stunted, which naturally leads to an increase of the notochord in cross section. The same applies obviously to the thickening of the neural system. Furthermore it was claimed that in the above experiments not sufficient attention was given to the variability of the amphibian embryos (Ogi, 1957). The last word in this field has not yet been said.

Cyclopia is occasionally observed in natural conditions both in man and in domestic animals (see Adelmann, 1936). It is therefore very significant that the same effect can be produced experimentally. In this way we get

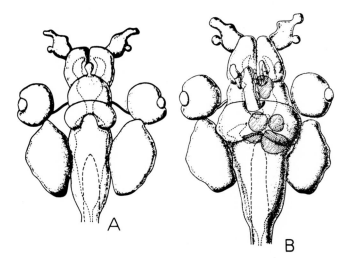

Fig. 127.   Reconstruction of the brain and sensory organs in a normal frog tadpole (*A*) and in a tadpole after treatment with thiocyanate (*B*). (From Ranzi, 1957.)

an inkling of what goes on behind the scenes when a cyclopic monster is born. There must have been some injury to the gradient system concerned with the formation of the head in the embryo. The injurious effect may be produced by both hereditary and environmental factors. The cyclopia or similar defects may be the result of some toxicosis of the mother occurring at an early stage of pregnancy. This is known to be the case if a woman contracts the illness known as German measles (rubella) during the early weeks of pregnancy. The illness is not a serious one for the mother, but the toxin poisons the embryo, producing defects of the anterior part of the head (see Glathaar and Töndury, 1950).

On the other hand, a gene has been found in guinea pigs (Wright and Wagner, 1934) which produces, in varying degrees, defects of a cyclopic nature, from underdevelopment of the nose and mouth to almost complete absence of the head.

CHAPTER **7**

Embryonic Adaptations

In this connection we will consider special organs or parts of the embryo which are not precursors of any of the organs of the adult or the larva, but serve to satisfy the requirements of the embryo in respect to food or oxygen supply or to its protection.

The yolk in the egg cannot be considered as part of the embryo in this sense, although it fulfills the purpose of supplying nourishment for the embryo. In animals, however, in which the amount of yolk is very large, special organs may be evolved to store the yolk and utilize it.

In the meroblastic eggs of fishes the yolk is originally covered only by a thin membrane of cytoplasm which is continuous with the cytoplasmic cap at the animal pole at the beginning of cleavage. During gastrulation (section 5–6) the edge of the blastodisc, now becoming the rim of the blastopore, spreads over the surface of the yolk and eventually encloses the yolk altogether. The cell layers spreading over the yolk are the ectoderm, the mesoderm and the periblast—the syncytial layer formed on the surface of the

199

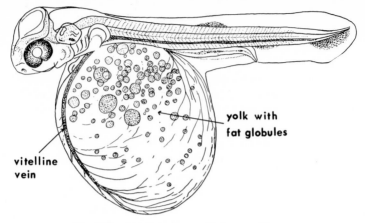

Fig. 128.   Trout embryo with large yolk sac.

yolk underneath the blastoderm. The endoderm in the bony fishes does not follow the movements of the blastopore rim, so that the yolk is not enclosed by the endodermal layer.

In the meantime the organs of the embryo are developed, and the body of the embryo raises itself away from the surface of the mass of yolk. Thus is introduced a partial separation between the body of the embryo proper and the **yolk sac** enclosing the mass of yolk. The broad connection between the body of the embryo and the yolk sac becomes constricted later, so that the yolk sac remains connected to the embryo proper by the stalk of the yolk sac (Fig. 128). A system of blood vessels is developed in the walls of the yolk sac, entering the heart by means of a pair of omphalo-mesenteric veins, which collect blood from several vitelline veins. The yolk is broken down mainly due to the activity of the periblast, and so it becomes transportable through the blood stream to the embryo. As the yolk is used up, the yolk sac diminishes in size, while the embryo grows. Hatching takes place in the meantime, the embryo becoming a free-swimming larva. The yolk sac is eventually retracted into the ventral body wall of the larva.

7–1   THE EXTRAEMBRYONIC STRUCTURES IN REPTILES AND BIRDS

In reptiles and birds as in fishes, the yolk is eventually enclosed into a yolk sac, but the way this is done differs from that found in fishes. In reptiles and birds the gastrulation movements do not take place on the edge of the blastodisc but inside it. The yolk is covered by the spreading of the edge of the blastodisc, and this process is not connected, therefore, in any way with gastrulation. The periphery of the blastodisc in reptiles and birds is differentiated as the area opaca, and it is the external edge of this that spreads over the surface of the yolk and eventually covers it. The cells of which the area opaca consists, beginning with the part nearest the embryo and progressing outward, split into three layers which are continuous with the germinal layers developed in the area pellucida: the ectoderm, the mesoderm and the endoderm, the latter adhering to the surface of the yolk.

During the second and third days of incubation in the case of the chick, a network of blood vessels is developed in the inner part of the area opaca, which from this time onward becomes the **area vasculosa.** The outer part of the area opaca becomes the **area vitellina.** The development of the blood vessels in the area vasculosa is intimately connected with the differentiation of the first blood cells. This happens in the following way. First of all, groups of densely packed mesodermal cells appear in the area opaca all around the sides and the posterior edge of the area pellucida. These groups of cells are known as the **blood islands** (Fig. 129). Next the cells of the blood islands become differentiated into two kinds: the cells on the periphery join up to form a thin epithelial layer—the endothelium of the future blood vessels; the central cells, on the contrary, become separated from each other and are differentiated as blood corpuscles. Thus from the beginning the blood corpuscles are lying inside the blood vessels. Meanwhile the adjacent blood vessels establish communications among themselves and the whole is transformed into a very irregular network. All the vessels of the area vasculosa are joined together on the periphery by the terminal sinus, which is the boundary line between the area vasculosa and the area vitellina (Fig. 130). The network of the area vasculosa becomes prolonged into the area pellucida and eventually establishes connection with the embryo proper. A connection with the blood system of the embryo is formed at two points: with the venous system, by means of right and left omphalo-mesenteric veins, joining in an unpaired ductus venosus which enters the sinus

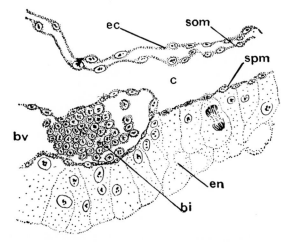

Fig. 129.   Section through area vasculosa of a chick embryo, showing a blood island. *ec,* Extraembryonic epidermis; *som,* parietal layer of mesoderm; *spm,* visceral layer of mesoderm; *en,* endoderm; *bi,* blood island; *bv,* blood vessel. (From Wieman, 1949.)

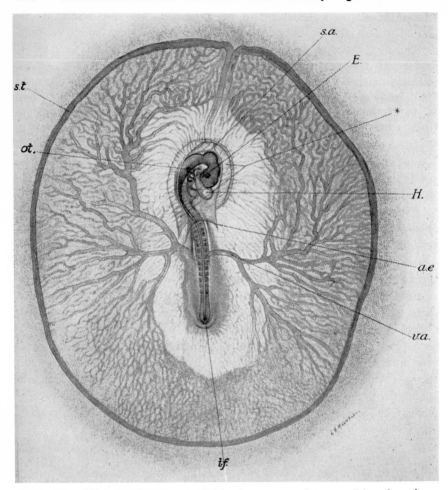

Fig. 130. Chicken embryo surrounded by the area vasculosa. *a.e.*, Edge of amnion; *E.*, eye; *H.*, heart; *ot.*, ear vesicle; *s.a.*, suture along the line of fusion of amniotic folds; *s.t.*, terminal sinus; *t.f.*, tail fold; *v.a.*, vitelline artery. The asterisk indicates a depression in which lies the head of the embryo. (From Kerr, 1919.)

venosus of the heart, and with the arterial system by means of right and left vitelline arteries, which branch off from the dorsal aorta.

By the end of the second day of incubation the heart of the embryo begins to beat and the blood can circulate through the network of the yolk sac. The wall of the yolk sac develops outgrowths on its inner surface, i.e., on the surface facing the yolk. The outgrowths, supplied by blood vessels from the area vasculosa network, penetrate deep into the yolk and thus facilitate its absorption. The yolk, however, is not completely absorbed during embryonic life. Shortly before hatching the yolk sac is retracted into the abdominal cavity of the embryo, and the walls of the abdominal cavity close behind it.

As in the fishes, the body of the embryo in birds and reptiles becomes separated from the yolk sac. This is achieved by the formation of folds which appear all around the body of the embryo. The folds involve all three germ layers and are directed downward and inward, undercutting the body of the embryo proper. These folds are known as the **body folds.** Although the body folds eventually surround the whole embryo, the various sections of the fold do not appear simultaneously. The first to appear is the part of the fold just in front of the head. This section, which may be referred to as the **head fold,** undercuts the head and anterior part of the trunk of the embryo, so that these parts project freely over the surface of the yolk sac. The lateral and the posterior parts of the body fold develop soon after, the latter undercutting the tail end and the posterior part of the trunk, which now also project freely over the surface of the yolk sac, though this posterior part of the embryo is initially much shorter than the anterior part, separated from the yolk by the head fold. The body folds gradually contract underneath the embryo, and eventually only a rather narrow stalk, the **umbilical cord,** connects the embryo with the yolk sac and other extraembryonic parts. The cord includes all the germ layers: an outer ectodermal lining, a double sheath of mesoderm and an endodermal canal connecting the cavity of the gut with the cavity of the yolk sac. In addition it contains the blood vessels going out to the yolk sac and other structures which will be referred to later.

Besides the yolk sac, the reptiles and birds develop three other extraembryonic organs, known as the **embryonic membranes.** These are the **amnion,** the **chorion** and the **allantois.**

The amnion and the chorion are developed together as upwardly projecting folds, the **amniotic folds,** appearing on the area pellucida just outside the body folds and eventually closing over the dorsal surface of the embryo (Fig. 131). The rudiment of the amnion and chorion first appears as a transverse fold anterior to the head of the embryo. The fold bends backward over the anterior end of the head and covers it as with a hood. Subsequently the lateral ends of the fold are prolonged backward along both sides of the embryo. The lateral folds approximate each other over the body of the embryo and fuse from the front end backward, so that more and more of the embryo becomes covered by the folds. Eventually a fold also develops behind the embryo, and the free edges of the folds fuse, thus completely enclosing the body of the embryo in a cavity—the **amniotic cavity.** The amniotic cavity is at first a very narrow slit between the embryo and the inner wall of the amniotic fold, but soon a fluid is secreted into the cavity, which distends it, so that the embryo floats freely in the cavity, connected to the extraembryonic parts only by the umbilical cord.

When the amniotic fold is first formed in front of the head of the embryo, it consists only of a part of the extraembryonic ectodermal layer. The lateral folds, however, from the start consist of the ectoderm and the somatic layer of the extraembryonic mesoderm. Mesoderm also penetrates secondarily

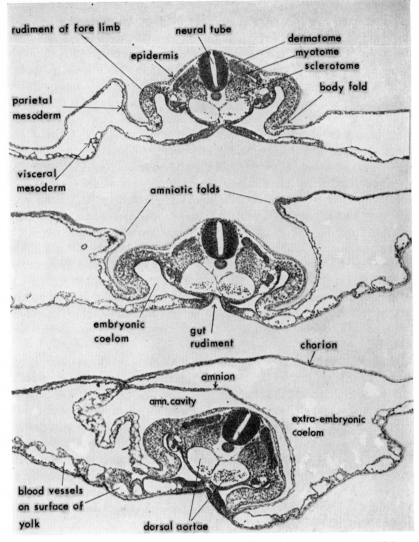

Fig. 131. Three stages in the development of the amniotic cavity in the chick.
(Transverse sections.)

into the anterior part of the fold, the composition of the fold therefore becoming uniform throughout. The amniotic fold has two surfaces running out into an acute edge—the inner surface, facing the body of the embryo, and the outer surface, facing away from the embryo. The fusion of the amniotic folds of opposite sides occurs at the expense of their acute edges. The surface of the fold facing the embryo now lines the amniotic cavity, and becomes the **amnion.** The surface of the fold facing away from the embryo, after the fusion of the folds, becomes continuous over the dorsal surface of the embryo (and over the amniotic cavity) and is now called the **chorion.** The chorion is, of course, continuous with the epithelial layers covering the surface of the yolk sac. From what has been said of the composition of the amniotic folds it is clear that the amnion consists of a layer of extraembryonic ectoderm on the inside (facing the amniotic cavity) and a layer of extraembryonic somatic mesoderm on the outside. The chorion consists of a layer of extraembryonic ectoderm on the outside and a layer of extraembryonic mesoderm on the inside. In between the amnion and the chorion is the extraembryonic coelomic cavity, which is continuous with the coelomic cavity in the embryo proper.

The advantages that the embryo obtains from the development of amnion, chorion and the amniotic cavity are clear enough: the embryo becomes immersed in a container filled with fluid and thus can accomplish its development in a fluid medium, although the egg is "on dry land." It would appear therefore that the primary function of the amnion is to protect the embryo from the danger of desiccation. Besides this the fluid of the amniotic cavity is an efficient shock absorber, and protects the embryo from possible concussions. Lastly, it isolates the embryo from the shell of the egg, and thus protects it from adhesion to the shell or from friction against it. At the same time the formation of the amniotic cavity has a slightly negative effect: it removes the embryo away from the surface of the egg and thus away from the source whence it could obtain oxygen.

The third embryonic membrane, the **allantois,** is very different in nature from the first two membranes. Originally the allantois is none other than

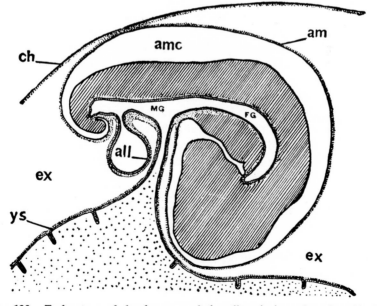

Fig. 132. Early stage of development of the allantois in a four-day-old chick embryo. *all,* Allantois; *am,* amnion; *amc,* amniotic cavity; *ch,* chorion; *ex,* extraembryonic coelom; *FG,* foregut; *MG,* midgut; *ys,* yolk sac. (From Wieman, 1949.)

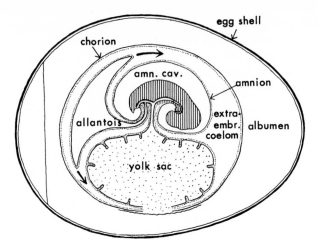

Fig. 133. Later stage of development of the allantois in a chick embryo (diagrammatic), showing expansion of the allantois (indicated by arrows) in the extraembryonic coelom.

a urinary bladder. A cleidoic egg, as that of a bird or reptile, has no means of disposing of the waste products of protein breakdown by removing them from the egg. A solution was found to the problem of waste disposal by excreting the protein breakdown products in the form of uric acid, which is deposited as water-insoluble crystals and stored inside the egg up to the time of hatching. The storage place is the urinary bladder. As the storage has to go on throughout the time of incubation, the urinary bladder has to increase to enormous proportions.

The allantois appears as a ventral outgrowth of the endodermal hindgut. It thus corresponds exactly in nature to the urinary bladder of the amphibia. The outgrowth consists of endoderm with a layer of visceral mesoderm lining it from the outside (Fig. 132). The allantois, however, grows very rapidly, and soon penetrates into the extraembryonic coelom, into the space between the yolk sac, the amnion and the chorion. The distal part of the allantois expands, and remains connected with the hindgut of the embryo by means of a narrow neck. When the body folds contract, separating the embryo from the extraembryonic parts, the neck of the allantois is enclosed together with the stalk of the yolk sac in the umbilical cord. The distal part of the allantois becomes flattened and penetrates between the amnion and the yolk sac on the one side, and the chorion on the other side (Fig. 133). By the middle of the incubation period the allantois spreads over the complete circumference of the egg underneath the chorion. Thus the allantois comes into a position where it can take up a second function in addition to that of serving as a reservoir for storing uric acid.

The new function of the allantois is to supply the embryo with oxygen. A network of blood vessels develops on the external surface of the allantois, and this network is in communication with the embryo proper by means of blood vessels running along the stalk of the allantois and through the cord. The blood flows to the allantois from the dorsal aorta through a right and a left umbilical artery. These arteries leave the dorsal aorta at a point which is much more caudal than the starting point of the vitelline arteries. The returning blood flows to the heart through a pair of umbilical veins. These veins originally enter the right and left ducts of Cuvier respectively. Soon, however, this arrangement of blood vessels is changed. The right umbilical artery and the right umbilical vein disappear, and the left umbilical vein acquires a new central connection; it joins up with the left hepatic vein, whereupon the channel to the duct of Cuvier is closed, and the whole blood flow from the allantois passes into the left hepatic vein.

The allantoic circulation continues until the chick breaks the eggshell and begins to breathe the surrounding air. Then the umbilical vessels close, the circulation ceases, and the allantois dries up and becomes separated from the body of the embryo.

7–2   THE EARLY DEVELOPMENT OF MAMMALS

The development of the mammals is greatly modified as compared with that of other vertebrates, and this is due to adaptations found in this class. There is good evidence that the mammals are derived from ancestors which were very closely related to the early reptiles or which can actually be classified as reptiles. As such, they must have had large yolk-laden eggs, and their ontogenesis must have been very like that of reptiles. At some stage in their evolution some of the mammals became viviparous, the embryo receiving an adequate supply of nutrition from the mother while it was retained in the uterus, and the yolk supply of the egg became superfluous. The yolk supply consequently became progressively reduced and eventually disappeared altogether. In this respect, however, there is a distinct gradation in the three subdivisions of the class.

In the subclass Prototheria the eggs are laid and they develop outside the maternal body (*Ornithorhynchus*), or, though developing in the maternal body, they are not supplied with nourishment (*Echidna*), this adaptation not going further than the state of ovoviviparity (section 2–7). The eggs therefore have a large amount of yolk, they are meroblastic, and the subsequent development follows essentially a reptilian pattern (Flynn and Hill, 1939).

In the subclass Metatheria (the marsupials) the developing embryos receive nourishment from the mother in the uterus, although the adaptations for this have not been pushed as far forward as in the next group, and the young are born poorly developed. The yolk here has already become superfluous, and its quantity in the egg is insignificant. Even so, the yolk is not used by the embryo as in the reptiles and birds, but the mass of yolk is ejected at the beginning of cleavage as the egg becomes divided into the first two blastomeres (Hill, 1918).

In the most advanced subclass of the mammals, the Eutheria (placental mammals), the egg is devoid of yolk right from the start.

Cleavage, Blastocyst and Development of Germinal Layers. With the disappearance of the yolk, the mammalian eggs have reverted to complete cleavage, but the subsequent development bears ample evidence of the former presence of yolk, and in many respects the morphogenetic processes resemble the processes found in meroblastic eggs with a discoidal type of cleavage.

The eggs of mammals are fertilized in the region of the funnel of the oviducts or in the uppermost part of the oviduct. Through the action of the ciliated epithelium of the duct the fertilized eggs travel slowly down the oviduct, and the cleavage of the egg, or at least the first stages of cleavage,

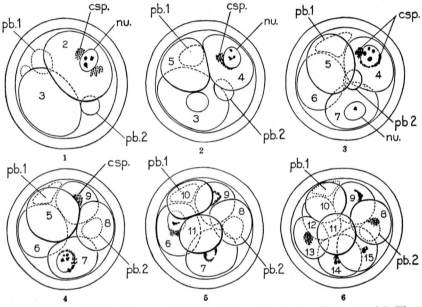

Fig. 134. Cleavage of the egg of a monkey *Macacus rhesus*. *pb.1* and *pb.2*, First and second polar bodies; *nu.,* nucleus; *csp.,* centrosphere material. (From Lewis and Hartmann, 1933.)

take place during the sojourn of the eggs in the oviduct. Although the cleavage is complete, it is by no means as regular as in the oligolecithal eggs of the invertebrates and lower chordates. The synchronization of mitoses in the blastomeres is lost very early. Even the first two blastomeres cleave at different rates, consequently a three-cell stage is found, and subsequently stages with five, six, seven blastomeres, and so forth (Fig. 134).

The course of the cleavage process varies quite considerably in different mammals. In the marsupials and a few insectivores, as for instance, *Hemicentetes,* the blastomeres become arranged in the form of a hollow sphere. In the African elephant shrew (*Elephantulus*), at the four-cell stage the blastomeres already surround and enclose a cleavage cavity (blastocoele). In the majority of mammals, however, the result of cleavage is a solid mass of cells, a **morula,** in which some cells are superficial and others lie inside, completely cut off from the surface by the enveloping cells (Fig. 135, *A*). The sorting out of the superficial and the internal cells of the morula has been shown, at least in some mammals, to have some relation to the arrangement of the cytoplasmic substances in the egg.

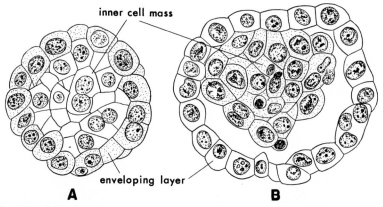

Fig. 135.   Morula (*A*) and early blastocyst (*B*) of a bat, showing differentiation into an inner cell mass and the enveloping layer. (After van Benden, from Brachet, A., 1935.)

The cleavage of the rat egg may be taken as an example. In this animal as also in some others (see section 2–5), the subcortical cytoplasm on one side of the egg is basophilic owing to the presence of numerous granules containing ribonucleic acid. The subcortical cytoplasm of the opposite hemisphere of the egg is vacuolated. The axis indicated by these two kinds of cytoplasm is approximately at right angles to the animal-vegetal axis of the egg (see p. 37). The cleavage planes bear no definite relationship to the cytoplasmic areas of the egg, and the two kinds of cortical cytoplasm are segregated into the blastomeres at random (Fig. 136). In the eight-cell

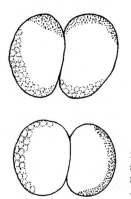

Fig. 136.   Random distribution of different kinds of cytoplasm at the first division of the egg of the rat. Stippled area represents basophilic cytoplasm. (From Jones-Seaton, 1950.)

stage the blastomeres are all arranged in one plane (Fig. 137). By this time the basophilic and the vacuolated cytoplasms are distributed to different blastomeres. Now an important displacement of the blastomeres takes

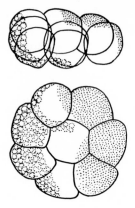

Fig. 137. Cleavage of the rat's egg: eight-cell stage. (From Jones-Seaton, 1950.)

place; the blastomeres containing vacuolated cytoplasm spread out around the blastomeres with basophilic cytoplasm. The latter sink inward and are eventually completely surrounded by the former (Jones-Seaton, 1950).

In the rhesus monkey the embryo early becomes subdivided into a mass of cells lying in the interior and an enveloping layer of cells. The origin of the two kinds of cells has not been traced in the primates. A subdivision into a layer of superficial cells and into a group of cells lying in the interior also takes place later in those mammals, in which the cleavage cells are at first arranged as a hollow sphere.

The internal and enveloping cells of early mammalian embryos diverge further and further in their physiological properties and later have a very different function and fate. The cells lying in the interior, known as the **inner cell mass,** have a more basophilic cytoplasm to start with; the basophilia greatly increases during the subsequent stages, indicating a rapid synthesis of ribonucleic acid in the cytoplasm. By contrast the cells of the **enveloping layer** appear pale if they are stained with basic dyes, in particular with dyes which reveal the presence of ribonucleic acid (pyronine, toluidine blue). The cells of the inner cell mass also give a positive reaction for alkaline phosphatase, whereas the reaction for alkaline phosphatase in the enveloping layer remains negative (Dalcq, 1954).

Sooner or later a cavity appears inside the compact mass of cells of the morula (if it was not already present in the cleavage stage). The cavity is formed of creases which appear between the inner cell mass and the cells of the enveloping layer. Fluid is imbibed into this cavity, so that it increases, the whole embryo becoming bloated to the same degree. The enveloping layer becomes lifted off the inner cell mass on most of its inner surface, remaining attached on one side only. This side corresponds later to the dorsal side of the embryo (Figs. 135, *B,* and 138).

A mammalian embryo in this stage is called a **blastocyst.** The cavity of the blastocyst may be compared to the blastocoele, but the embryo as a whole differs essentially from a blastula since its cells are already differentiated into two types: the inner cell mass and the cells of the enveloping layer.

A further step in the progressive differentiation in the mammalian blastocyst is the appearance of a layer of very flat cells on the interior surface of the inner cell mass, that is, on the surface facing the cavity (the blastocoele). This layer of flat cells corresponds to the lower layer of cells of the chick blastoderm—the **hypoblast** (Fig. 138). The cells represent the presumptive endoderm, or at least a part of it.

The origin of the hypoblast (endoderm) cells is a subject of some controversy. The more generally accepted view is that the endoderm cells are split off from the inner cell mass. There are indications, however, that the endoderm cells in some mammals such as elephant shrew (van der Horst, 1942) or even in all mammals (Dalcq, 1954) are derived from the enveloping layer of cells. Some of these, near the edge of the inner cell mass, migrate inward along the internal surface of the inner cell mass and arrange themselves in a continuous layer, as stated above (Fig. 139). Unfortunately, the methods of local vital staining or marking of cells could not as yet be applied to the study of the migration of cells which give rise to the hypoblast in mammals. Failing this, the conclusion has to be reached on the grounds of differences in the staining reactions of the various cells. It is important to

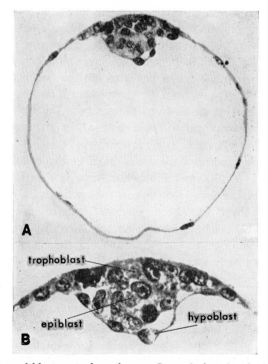

Fig. 138. Advanced blastocyst of monkey. *a,* General view (section); *b,* dorsal part of same under higher magnification. (From Heuser and Streeter, 1941.)

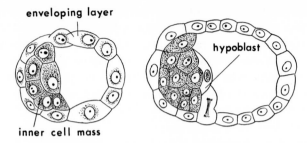

Fig. 139.   Origin of hypoblast (endoderm) in a rat blastocyst. (After Dalcq, 1954.)

note, therefore, that the early endoderm cells in their lack of cytoplasmic basophilia and in their negative reaction for alkaline phosphatase differ from the cells of the inner cell mass and resemble the cells of the enveloping layer (Dalcq, 1954), thus making their origin from the latter fairly plausible.

If we recognize the layer of flat cells (endoderm of some authors) as the hypoblast, the remaining inner cell mass is seen to be equivalent to the epiblast of the chick blastoderm. As to the enveloping layer, it does not directly contribute any longer to the formation of the embryo proper. The function of the enveloping layer of cells is to establish a connection with the maternal body and to mediate between it and the embryo in supplying the latter with nourishment. This layer of cells is forthwith known as the **trophoblast.** The term trophoblast is often applied to the enveloping layer of cells from the very earliest stages of its formation. This would be a somewhat loose use of the term in view of the possibility that the endoderm (hypoblast) cells may be given off from the enveloping layer. The word *trophe* means nourishment, therefore it seems better to use the Greek term trophoblast or "nourishing" layer only from the stage where it no longer contains cells which become part of the body of the embryo. The cells at the expense of which the embryo proper will be developed, i.e. the inner cell mass and hypoblast, may be referred to as the **formative cells.**

The trophoblast corresponds in position to the chorion of the later embryos of the reptiles and birds, but there is the difference that the chorion, in its typical form, develops in conjunction with the amnion after all the germinal layers have been segregated, whereas the trophoblast precedes the germinal layers in its development. Obviously this is so because in mammalian development the connection with the maternal body must be established at an early stage. The hypoblast cells are at first found in the region of the inner cell mass only, but later they spread along the inner surface of the trophoblast and surround the internal cavity of the blastocyst, much as the endodermal cells in reptiles and birds surround the mass of uncleaved yolk. The inner cavity of the blastocyst thus acquires a resemblance to the yolk sac of the lower amniota, except that instead of being filled with yolk it is filled with fluid. In spite of this difference the cavity within the enveloping endodermal layer is referred to as the yolk sac. As the hypoblast spreads out to enclose the cavity of the yolk sac, the inner cell mass also spreads out and becomes arranged into a plate, resembling the epiblast of the blastodisc of reptiles and birds. The arrangement of the formative cells in the mammalian blastocyst is now very similar to that in the avian blastodisc prior to the appearance of the primitive streak. Again there is the difference that the blastodisc of a bird lies on the surface of the sphere of uncleaved yolk, whereas under the blastodisc of a mammal there is only the fluid filling the yolk sac. The situation in mammals, that after a cleavage which is not discoidal the formative cells should be arranged in a disc, can only mean that the ancestors of the mammals had large yolky eggs with meroblastic discoidal cleavage.

In eutherian mammals the blastodisc is primarily formed underneath the layer of trophoblast cells. In those whose development may be considered to be more primitive, as for instance in the rabbit, the ungulates, most insectivores and lemurs, the layer of trophoblast over the blastodisc **(Rauber's layer)** disappears, so that the epiblast becomes temporarily superficial (Fig. 144, *A*). In the higher mammals, however, Rauber's layer does not disappear in the embryonic area; the formative cells are therefore never exposed to the exterior (Figs. 140 and 144, *B*).

The blastodisc, consisting of an epiblast and a hypoblast, becomes quite sharply delimited from the remainder of the embryo (Fig. 140). The epiblast consists of a thick plate of columnar cells, clearly distinguished from the flatter and more irregularly arranged cells of the trophoblast. The hypoblast cells on the underside of the blastodisc become cuboidal or even columnar, and in this they differ from the cells of extraembryonic endoderm lining the internal surface of the trophoblast. The hypoblast becomes especially thick, with the cells becoming high and columnar near one edge of the blastodisc. The thickening is the prechordal plate (see p. 138), and it denotes what will be the anterior end of the embryo.

The embryo now enters the phase of gastrulation. The gastrulation and formation of germinal layers in mammals resemble what is found in birds and in reptiles. A primitive streak is formed, and a Hensen's node is seen at the anterior end of the primitive streak. The primitive streak is, however, much shorter than in birds, and, when fully developed, does not surpass half the length of the blastodisc, being confined to its posterior part. The cells of the primitive streak migrate downward and sideways between the epiblast and the hypoblast (Fig. 142). Some of the migrating cells appear to join the hypoblast and so contribute to the formation of the endoderm (Heuser and Streeter, 1941). The loose cells migrating sideways give rise to the layer of mesoderm.

The cells migrating forward from Hensen's node remain packed more closely and give rise to the "head process"—the notochordal rudiment. The notochordal rudiment in some mammals (including man) is perforated by a canal starting from Hensen's node. This is the archenteric canal which

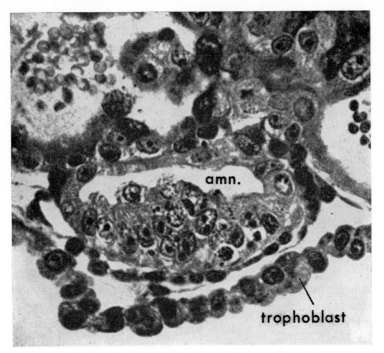

Fig. 140.   Monkey embryo. Formation of the amniotic cavity by cavitation. (From Heuser and Streeter, 1941.)

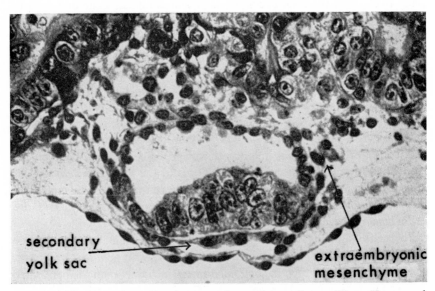

Fig. 141.   Monkey embryo. Development of secondary yolk sac. (From Heuser and Streeter, 1941.)

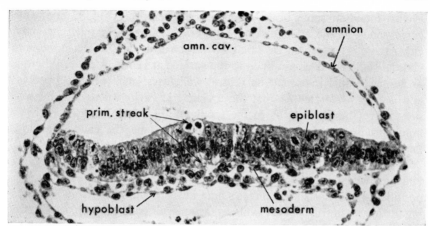

Fig. 142.   Monkey embryo in the primitive streak stage. Transverse section. (From Heuser and Streeter, 1941.)

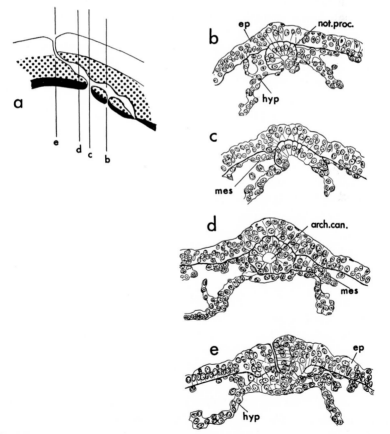

Fig. 143.   The archenteric canal (*arch. can.*) in a human embryo. *a,* Reconstructed medial section, showing the relation of the canal to epiblast (white), hypoblast (black) and mesoderm (dots). Vertical lines show the levels of sections *b* through *e. ep,* Epiblast; *hyp,* hypoblast; *mes,* mesoderm; *not. proc.,* notochordal process. (From Grosser, 1945.)

is also found in some birds, as mentioned before (p. 151), and which corresponds to the archenteron of lower vertebrates. In other mammals either there is no archenteric canal, or the canal though present in the notochordal rudiment does not open to the surface at Hensen's node. Where an archenteric canal is present its ventral wall fuses later with the hypoblast and is then perforated so that the archenteric cavity opens into the yolk sac cavity (this also occurs in the reptiles) (Fig. 143). Subsequently the notochord separates itself from the endoderm, and the endoderm closes underneath the notochord, forming again a continuous layer.

The primitive streak in mammals, as in birds, is a transient structure. Having given rise to the mesodermal layer and the notochord rudiment, the primitive streak starts shrinking, its anterior end with Hensen's node receding further and further back, while in the anterior parts of the blastodisc the germinal layers enter the next phase of development, the formation of the primary organ rudiments. The first of these to become visible is of course the neural plate. The subsequent development of the neural tube, notochord, mesodermal segments and so on will be treated together with the development of these parts in other vertebrates.

We must now turn to the fate of those parts of the blastocyst which do not participate in the formation of the embryo: to the extraembryonic parts.

Extraembryonic Parts.   Toward the end of the period of cleavage the embryo usually passes from the oviduct into the uterus. As a rule, the embryo is then in the blastocyst stage. In the uterus the next important step in the development of the mammalian egg has to take place: the embryo becomes attached to the wall of the uterus, where it is to perform its subsequent development. This attachment of the blastocyst to the uterine wall is known as **implantation.** The attachment to the uterine wall is carried out by the cells of the trophoblast. Two basically different types of attachment are to be found. In most mammals the blastocyst becomes attached to the surface of the uterine epithelium and therefore lies in the cavity of the uterus. In a smaller number of mammals—in man, but also in some of the rodents—the blastocyst penetrates deep into the uterine wall, and the development of the embryo takes place *inside* the uterine wall. The epithelium lining the cavity of the uterus at the site of implantation becomes destroyed. The destruction of the uterine epithelium is due to the activity of the trophoblast cells. The embryo is capable of imbibing the fluid filling the uterus and deriving from it nutrient substances, just as is the case in ovoviviparous dogfish, but this source of nutrition is insignificant as compared with the food supply that the embryo gets by diffusion from the maternal blood vessels, once the connection between the embryo and the uterine wall is established. The formation of this connection is called **placentation,** and it will be dealt with in the following section.

At the time of implantation the embryo is still in the blastocyst stage, and the processes of gastrulation take place only after implantation. Into roughly the same period also falls the development of the embryonic membranes. This differs to a most amazing degree (see Mossman, 1937, Goetz, 1938) in various groups of mammals, so that only the more general features of this process can be indicated here (see Fig. 144). If the blastodisc is exposed to the exterior by the disappearance of Rauber's layer in the embryonic area, the amniotic folds may develop around the embryo proper in much the same way as in reptiles and birds. After the fusion of the amniotic folds the body of the embryo becomes enclosed in the amniotic cavity. A body fold later separates the embryo from the yolk sac, with which the embryo remains connected by means of the umbilical cord. The outer wall of the amniotic fold becomes the chorion. On the periphery, in the extraembryonic area, the ectoderm of the chorion is continuous with the remaining parts of the trophoblast. The trophoblast thus becomes a part of the chorion; it actually forms the epithelial layer of the chorion over most of the extraembryonic surface of the blastocyst. The mesoderm spreading from the embryonic area may supply this part of the chorion with a connective tissue layer (Fig. 144, A).

The development of the amniotic cavity in the way described above occurs relatively late, after gastrulation and the formation of primary organ

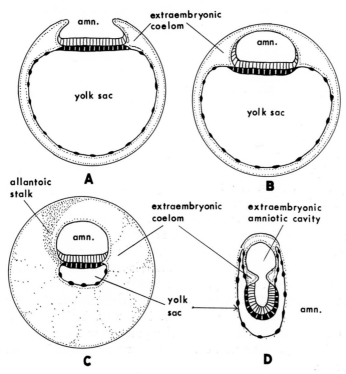

Fig. 144.   Diagrammatic comparison of the relation of the embryonic and extraembryonic parts (amnion, yolk sac) in a shrew (A), a bat (B), a human (C) and a mouse (D).

rudiments. We find this method of amnion formation in the marsupials, ungulates (pig), most insectivores and lemurs. In other mammals the formation of an amniotic cavity precedes gastrulation; the amniotic cavity is then formed by a crevice appearing between cells in the future blastodisc area. This mode of formation of the amniotic cavity is known as **cavitation.** The split may occur either between the trophoblastic layer and the inner cell mass, or inside the inner cell mass. In the second case (rodents) the amniotic cavity is at once surrounded by a double epithelial layer—by the amnion, derived from the inner cell mass, and by the trophoblast which now assumes the role of the chorion (or its epithelial component). In the first case (Fig. 140) the amniotic cavity is only covered by the trophoblast, and an epithelial lining representing the amnion proper must be developed later, either at the expense of the edges of the blastodisc (in the bats) or at the expense of cells derived from the internal surface of the trophoblast (in higher primates according to Heuser and Streeter, 1941). Further variations of amnion development have been described; references to these descriptions may be found in the papers by Mossman, 1937, and Goetz, 1938.

The development of the allantois also varies a great deal inside the class Mammalia. With the acquisition of viviparity the original function of the allantois as a urinary bladder becomes altogether lost. As the maternal organism supplies the embryo with all the necessary nutrition, so also it takes over the removal of the waste products of metabolism from the embryo. Not only does the carbon dioxide produced by the embryo diffuse into the maternal blood, but the end products of the protein metabolism also find their way into the maternal blood and are excreted by the kidneys of the pregnant mother. This of course will be possible only if the end products of metabolism are readily soluble. Uric acid, owing to its insolubility in water, was the most suitable end product of protein metabolism in the case of the cleidoic eggs of reptiles and birds, but in the mammals we find that the end product of protein metabolism is urea which, owing to its solubility, can be passed from the embryo to the mother and can thus be disposed of. As a result the storage of these waste products in the allantoic cavity is no longer necessary.

On the other hand, the function which was secondary in the case of birds and reptiles retains its importance for the mammalian embryo. The mammalian embryo, enclosed in the amniotic cavity and surrounded by the amnion and the chorion (trophoblast), is still in need of a pathway by which oxygen from outside (from the maternal organism) may be transported to its tissues. This pathway, in the higher mammals, is still supplied by the blood vessels of the allantois. Moreover, besides supplying the embryo with oxygen, the allantoic circulation now supplies the embryo with nutrient substances as well, as these can no longer be drawn from the yolk sac.

The result is that the endodermal part of the allantois (the epithelial lining of the allantoic vesicle) becomes gradually reduced in the higher mammals, while the mesodermal part, giving rise to the blood vessel system of the allantois, develops progressively. The endoderm of the allantoic stalk still fulfills a useful function as it forms a bridge growing out from the embryonic body and reaching the chorion, along which the mesoderm spreads outward and establishes a connection with the trophoblastic epithelium. The expansion of the endodermal vesicle at the outer end is, on the other hand, wholly superfluous. The development of the allantoic mesoderm, however, may proceed even in the total absence of the endodermal component (in the guinea pig).

Two special cases will be mentioned here.

In the monkeys and in man a typical blastocyst is formed with a cavity which becomes lined by hypoblast (endoderm cells). The blastocyst becomes attached to the wall of the uterus by its dorsal side, that is, by the side bearing the internal cell mass (Fig. 145).

In man the blastocyst sinks into the uterine wall, as a result of the destruction of the maternal tissues by the trophoblast, and the opening through which the blastocyst enters is closed by a clot of coagulated blood plasm and later covered up by a sheet of regenerated uterine epithelium. In the monkey the blastocyst does not enter the uterine epithelium, and, instead, becomes attached by its ventral side to the opposite uterine wall.

The amniotic cavity is formed by cavitation as described on p. 218. The epithelium surrounding the amniotic cavity is of very unequal thickness throughout. The roof of the amniotic cavity is formed by a fairly thin layer of epithelium; this is the amniotic ectoderm. The floor of the amniotic cavity consists of a very thick plate of cells arranged in a columnar epithelium. This plate, with the underlying layer of hypoblast (endoderm) which is also thickened, constitutes the bilaminar blastodisc, inside which the

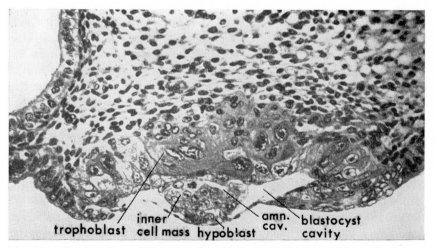

Fig. 145.   Implantation of the seven-day-old human blastocyst. (From Hertig and Rock, 1945.)

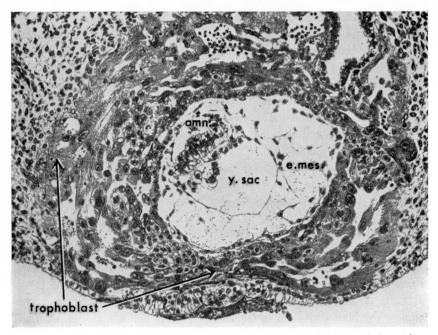

Fig. 146.   Implanted 11-day-old human embryo prior to primitive streak formation, with amniotic cavity (*amn.*), secondary yolk sac cavity (*y. sac*) and extraembryonic mesenchyme (*e.mes.*). The embryo is surrounded by the trophoblast. (From Hertig and Rock, 1941.)

primitive streak will later appear and from which the body of the embryo proper will later develop (Fig. 146).

It is common to both monkeys and man that at an early stage of development, even before the primitive streak is formed, numbers of loose mesenchyme-like cells appear between the formative parts and the trophoblast. These cells, which may be compared to extraembryonic mesenchyme, are supposed to be derived from the trophoblast (Heuser and Streeter, 1941), whereas in most mammals extraembryonic mesoderm spreads out from the blastodisc and has the same origin as the embryonic mesoderm (that is, it is derived from the primitive streak). The delay in the development of the primitive streak in the higher primates, in conjunction with the need for accelerating the development of the extraembryonic parts which are necessary to supply the embryo with food and oxygen, must have made it necessary to provide for a second, unusual source of extraembryonic mesoderm.

The extraembryonic mesoderm rapidly increases in quantity and gradually fills the cavity of the blastocyst, which was recognized as the yolk sac. A secondary, smaller yolk sac is then formed by the hypoblast cells connected to the blastodisc (see Fig. 141). The cavity of this secondary yolk sac is for some time much smaller than the amniotic cavity, but it catches up with the amniotic cavity later (at the time when the primitive streak appears). In the meantime the mass of extraembryonic mesenchyme separates to form a cavity of its own, which is, of course, the extraembryonic coelom; the mesenchyme cells arrange themselves into a mesothelium lining this coelom. The extraembryonic coelom now becomes the greatest cavity in the blastocyst and takes up most of the space inside the trophoblast. In the area adjoining the posterior end of the embryo the extraembryonic mesenchyme becomes concentrated and forms a strand leading from the blastodisc to the area of trophoblast, which enters into the formation of the placenta. The strand of

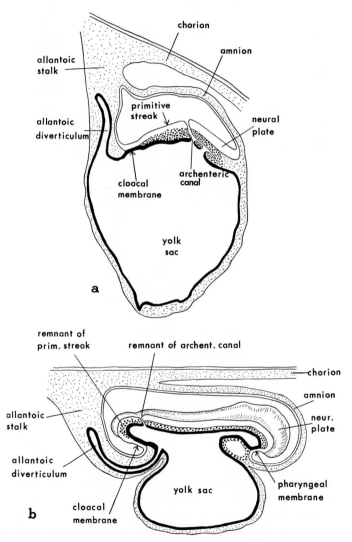

Fig. 147. Semidiagrammatic median sections of the human embryo in the primitive streak stage (*a*) and the neural plate stage (*b*). The ectoderm is white, the endoderm black, the embryonic mesoderm coarsely stippled, the extraembryonic mesoderm finely stippled. (*a,* After Jones and Brewer, 1941, and Grosser, 1945; *b,* after Grosser, 1945.)

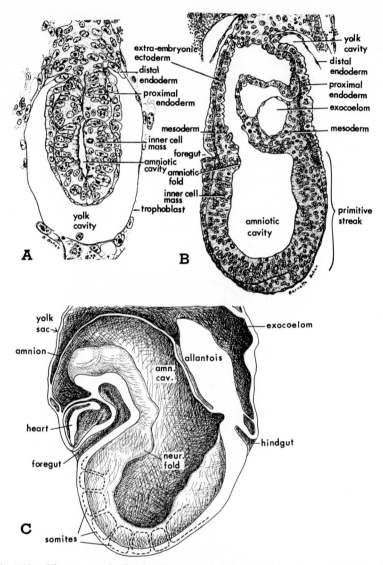

Fig. 148. Three stages in the development of the mouse. *A,* The egg cylinder stage (section); *B,* the primitive streak stage (section); *C,* neural plate stage (embryo dissected in the median plane.) (Modified from Snell, 1941.)

mesoderm known as the **connecting stalk** or **allantoic stalk** represents the mesodermal component of the allantois, and through this strand of mesoderm the allantoic circulation is established later (Fig. 147). Here again it can be noted that the necessity for an early provision of the embryo with food and oxygen supplies has brought about a short-cut in the established pattern of developmental processes. The endodermal part of the allantois—the allantoic vesicle—becomes superfluous and remains rudimentary, never reaching the placenta in monkeys and man.

The second type of modified mammalian development is that found in mice, rats and related rodents. The initial stages of cleavage and blastocyst formation in these animals do not differ essentially from that in other mammals, but after the implantation, which is interstitial as in man, the blastocyst fails to become bloated by accumulation of fluid in its cavity, and instead becomes elongated in a direction perpendicular to the surface of the uterus (Fig. 148). The embryo in this stage is known as the "egg cylinder." Both the trophoblast and the inner cell mass covered by the hypoblast (endoderm) stretch, leaving very little space between them. In due time the endoderm lines this cavity from the trophoblast side, converting it into the yolk sac. As the inner cell mass is attached to the trophoblast only at one end of the egg cylinder (the deep end), the formative cells become almost completely surrounded by the yolk sac (Fig. 148, *A*). The amniotic cavity is now formed as a rather narrow canal inside the inner cell mass. The cells of the inner cell mass form an epithelium surrounding the amniotic cavity; part of this epithelium represents the epiblast of the blastodisc, and part the amniotic ectoderm. The formative cells make up about half the lining of the amniotic cavity, that half lying nearer to the uterine cavity. The other half of the egg cylinder contains only extraembryonic parts: extraembryonic ectoderm, extraembryonic endoderm and, at a later stage, also extraembryonic mesoderm. The cavity, originally stretching throughout the length of the egg cylinder, later becomes constricted in the middle. One part of the cavity is now almost completely surrounded by formative cells (the cells of the epiblast), while only a very small part of the lining of this cavity can be recognized as being amniotic ectoderm. The other part of the original amniotic cavity will not interest us any further. The structure of the embryo at this stage may be described as resulting from an invagination of the bilaminar blastodisc into the yolk sac. The dorsal surface of the blastodisc instead of being flat as in most other mammals is concave, and the ventral side (lined by the hypoblast) is convex and surrounded by the yolk sac cavity (Fig. 148, *B*).

In this abnormal position the embryo proceeds to the formation of the primitive streak, which gives rise to the mesoderm of the embryo. The neural tube, the mesodermal somites, the heart and other parts of the embryonic body are developed in the usual way, but the body of the embryo, instead of being surrounded by the amniotic cavity, is curled round it with the dorsal side facing the cavity, while its convex ventral surface is surrounded by the yolk sac (Fig. 148, *C*). Only at a later stage, during the eighth day of intrauterine life in the case of the mouse (total duration of pregnancy—20 days), does the embryo start twisting itself out of this unusual position, starting with the head end, and eventually assumes the normal position in the amniotic cavity, with the body convex dorsally and concave ventrally. The amniotic wall becomes stretched so that the amniotic cavity surrounds the embryo. (For further details see Snell, 1941.)

7–3 PLACENTATION

The term **placenta** applies to any type of organ, built up of maternal and fetal tissues jointly, which serves for the transport of nutrient substances from the tissues of the mother into those of the embryo. Placentae are not found exclusively in mammals, but appear also in animals belonging to various groups of the animal kingdom such as in *Peripatus* (Protracheata), *Salpa* (Tunicata), *Mustelus laevis* (Elasmobranchia) and certain lizards. The nature of the tissues entering into the formation of the placenta is not the same in all cases. A prerequisite for participation in the development of the placenta is, on the one hand, the superficial position of the part in question, and, on the other, the possibility of the formation of a blood vessel network, which may undertake the transport of nutritive substances from the surface of contact with the maternal tissues to the developing embryo.

In mammals there exist two essentially different main types of placentae. In the first type, the connection is established between the uterine wall and that part of the chorion which is lined by the yolk sac with its network of vitelline blood vessels. This type of placenta is called the **chorio-vitelline placenta.** It is found in some of the marsupials (*Didelphys, Macropus*). In the second type of placenta, the connection is established by the chorion lined by the allantois, with the allantoic blood vessel system taking over the transport of substances from the mother to the embryo. This type of placenta is called the **chorio-allantoic placenta,** and it is found in all the Eutheria, as well as in some marsupials (*Parameles, Dasyurus*). Remnants of the chorio-vitelline placenta may be found either temporarily or even permanently in higher mammals, playing a subsidiary part in their placentation.

As the function of the placenta is to allow a passage of substances from the maternal to the fetal tissues, everything that accelerates this will increase the efficiency of this organ. A facilitation of the transport of substances through the **placental barrier** (the surface separating the maternal from the fetal tissues) may be achieved in two ways:

(1) By increasing the surface of the contact, and

(2) by diminishing the thickness of the layer separating the maternal from the fetal blood.

The surface of contact between maternal and fetal tissues is increased by the development of outgrowths on the surface of the chorion, known as the **villi.** The villi are fingerlike or branched outgrowths of the chorion, containing connective tissue and blood vessels in their interior (Fig. 149). They fit into corresponding depressions in the uterine wall. The villi may be scattered all over the surface of the chorion in the case of the pig, and this type of placenta is known as the **diffuse placenta.** In cattle the villi are found in groups or patches, while the rest of the chorion surface is smooth. The patches of villi are called **cotyledons,** and the placenta of this type is known as the **cotyledon placenta.** In carnivores the villi are developed in the form of a belt around the middle of their blastocyst which is more or less elliptical in shape. This is the **zonary placenta.** In man and the anthropoid apes the placenta is at first all covered with villi, but the villi continue developing only on one side, the side turned away from the lumen of the uterus, while on other parts of the chorion the villi are reduced. The functional placenta therefore has the shape of a disc, and is known as the **discoidal placenta.** A discoidal placenta has also been developed independently in the rodents (mouse, rat, rabbit and others). In the monkeys the placenta consists of two discs—a **bidiscoidal placenta.**

The thickness of the partition between the maternal and fetal blood may be diminished by the removal of some of the intervening layers of tissue. Depending on which layers have disappeared, several types of placentae may be distinguished. The names given to the various types indicate the two tissues—one maternal, the other fetal—which are in immediate contact.

In the more primitive cases the layers of tissue participating in the diffusion of substances from the mother to the embryo are the following:

1. The blood of the mother,
2. The endothelial wall of the maternal blood vessels,
3. The connective tissue around the maternal blood vessels,
4. The uterine epithelium.
5. The epithelium of the chorion,
6. Connective tissue of the chorion,
7. The endothelial wall of the blood vessels in the chorion,
8. The blood of the embryo.

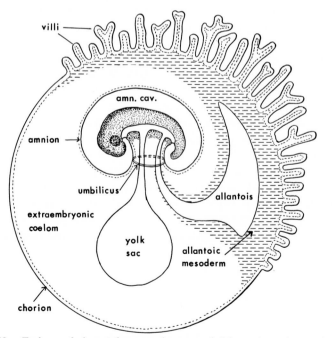

Fig. 149. Embryo of placental mammal, surrounded by embryonic membranes (diagrammatic).

If all the above-mentioned tissues are present in the placenta the chorionic epithelium is in contact with the uterine epithelium, and the placenta is designated as an **epithelio-chorial placenta.** This type of placenta is found in all marsupials and in the ungulates (horse, pig, cattle). In the case of an epithelio-chorial placenta the villi, in their growth, push in the wall of the uterus and later lie in pocketlike depressions of the uterine wall.

When the blastocyst is implanted and subsequently when the villi begin to grow, the superficial tissues of the uterine wall may be destroyed to a greater or lesser extent. If the destruction involves the uterine epithelium and the underlying connective tissue, the epithelium of the chorion may come into direct contact with the endothelial walls of the maternal capillaries. A placenta is then formed which is called an **endothelio-chorial placenta.** It is found mainly in carnivores, but also in a few other mammals.

The destruction of the maternal tissues of the uterine wall may involve the endothelium of the maternal blood vessels. The cavities of the blood vessels are then opened up, and the villi of the chorion become bathed in the maternal blood, thus facilitating the gas exchange and the diffusion of nutrient substances from the maternal blood into the blood vessels of the chorionic villi. This type of placenta, the **hemo-chorial placenta,** is found in primates, and in many insectivores, bats and rodents. Actually, the chorionic villi are surrounded by spaces (sinuses), devoid of an endothelial lining, into which maternal blood enters through the arteries of the uterus, and from which the blood flows into the uterine veins. The villi may be ramified dendritic structures or they may coalesce distally and form a more or less complicated network.

In the case of epithelio-chorial placentae, at parturition the villi can be pulled out of the pockets in which they have been embedded, and the fetal part of the placenta may be removed, leaving the surface of the uterine wall intact. There is therefore no bleeding at birth. This is not the case with other types of placentae. At parturition not only the fetal component of the placenta is shed, but a part of the uterine wall participating in the formation of the placenta is torn away as well. An open wound is left on the wall of the uterus, and hemorrhage inevitably occurs. In the latter case the placenta is said to be **deciduous,** whereas in the first case the placenta is **non-deciduous.**

The hemorrhage at parturition is normally stopped by the same mechanism as serves for the expulsion of the newborn: the contraction of the muscular wall of the uterus constricts the blood vessels and thus slows down the flow of blood, until clotting of the blood stops the hemorrhage altogether.

CHAPTER 8

The Genetic Factors during Gastrulation and Formation of Primary Organ Rudiments

There are many indirect indications that nuclear genetic factors (the genes), after having been dormant throughout the period of cleavage, begin to manifest themselves during gastrulation, and in ever-increasing measure control the processes of development from this stage onward. This does not mean that the processes taking place during cleavage are not affected by genes, but the effect of the genes is then indirect: it is an action through the medium of the initial structure of the egg as shown in the case of dextral and sinistral cleavage (section 4–9). The structure of the egg cannot of course be under the influence of the paternal genes, which enter the egg only during fertilization.

In this section we are concerned with direct interference of the genes in the processes of development, an interference in which the paternal genes as well as the maternal genes can take part.

The evidence in this field is from two sources—from the observation of cleavage without participation of nuclei, and from hybridization experiments.

Fragments of sea urchin eggs not containing the nucleus may be treated with agents which cause parthenogenetic activation of the egg (viz., hypertonic sea water). It has been found (Harvey, 1935) that the non-nucleated fragments begin to cleave and may reach the morula stage. The embryos, however, do not form a regular blastula and cannot gastrulate, which shows that the presence of nuclei is necessary for development to go beyond cleavage.

8–1 THE EVIDENCE FROM HYBRIDIZATION EXPERIMENTS

Hybridization experiments must be considered here in some greater detail. If the egg of one species is fertilized by a spermatozoon of another species, the nuclei of the two gametes may fuse and together participate in cleavage and development. Each cleavage cell would then consist of maternal cytoplasm (because the contribution of the spermatozoon to the cytoplasm of the fertilized egg is negligible), the maternal nuclear half and the paternal nuclear half. Embryos of this composition are **true hybrids.** The eggs may be treated just after fertilization, before the nuclei of the egg and the sperm have time to fuse, so as to remove the egg nucleus. This can be achieved, for instance, by puncturing the egg and allowing a small quantity of cytoplasm together with the egg nucleus to flow out of the egg. If the egg is fertilized with sperm of another species, an embryo will develop in which the cytoplasm is maternal, but the nucleus is paternal (see Baltzer, 1940). Such embryos will be referred to as **hybrid andro-merogons** (the term **merogony** is applicable to the fertilization of a part or fragment of the egg; **androgenesis** is the development of an embryo having only paternal nuclei). **Andro-merogons** are embryos in which an enucleated egg or egg fragment is fertilized by sperm of the same species. As a rule the andro-merogons are haploid.

The results of hybridization depend primarily on the relationship between the two parents. The penetration of the egg by foreign sperm is much more easily achieved than the subsequent cooperation with the maternal cytoplasm. Cases are known in which eggs were "fertilized" by sperm belonging to animals of a different class, and even of a different phylum: fertilization of the sea urchin, *Sphaerechinus,* with sperm of the sea lily, *Antedon;* fertilization of the sea urchin, *Strongylocentrotus,* with sperm of the mollusc, *Mytilus.*

In these cases the sperm penetrates into the egg and activates it but does not play any part in the subsequent development. This is tantamount to a parthenogenetic development of the egg, caused by the penetration of foreign sperm (see Morgan, 1927). If the animals are more closely related, as for instance in the case of a toad's eggs being fertilized by frog's sperm (*Bufo vulgaris* × *Rana temporaria*), the sperm nucleus fuses with the nucleus of the egg. The development (cleavage) begins, but the embryos die before the beginning of gastrulation owing to incompatibility between the sperm nucleus and the egg nucleus and cytoplasm. If the animals are still more closely related, such as species of the same genus, then the hybrids may be fully viable throughout the whole of development.

It is in the cases of rather closely related species, producing viable hybrids, that the influence of the nucleus can best be analyzed. Much of the work has been done on sea urchin development.

If two species of sea urchin, differing in the rate of cleavage, are crossed, the cleavage of the hybrid is always the same as the cleavage of the maternal species. In the sea urchin, *Dendraster eccentricus,* the egg cleaves at the rate of 29 to 30 minutes between successive divisions. In *Strongylocentrotus franciscanus* the interval between successive cleavages is 47 minutes. The hybrid has the maternal rate of cleavage. Also if fragments of the egg devoid of a nucleus are fertilized, the resulting hybrid andro-merogons cleave with a rhythm that is the same as in the maternal species (A. R. Moore, 1933).

The rate of cleavage cannot be influenced by the sperm nucleus. On the other hand the characters of the larva—the pluteus—are clearly intermediate between the two parents; at the larval stage the sperm nucleus has been able to exert its influence. In a special case the influence of the sperm nucleus has been discovered as early as the beginning of gastrulation. In the sea urchin, *Lytechinus,* the mesenchyme cells produced from the micromeres (section 4–8) migrate into the blastocoele slightly earlier than the beginning of invagination of the archenteron. In the sea urchin, *Cidaris,* the mesenchyme cells do not separate from the epithelium until after the archenteron begins to invaginate; they are consequently given off from the inner end of the archenteron. Further, the cleavage of *Lytechinus* proceeds at a much greater pace than in *Cidaris;* the blastula stage is reached after 5.5 hours in the *Lytechinus* embryos, but only after 16 hours in *Cidaris* embryos. The beginning of gastrulation is correspondingly delayed in the second species. When the *Cidaris* eggs were fertilized by *Lytechinus* sperm, the early stages of development up to the gastrula stage were performed exactly as in the maternal species, *Cidaris.* The rate of cleavage corresponded to that typical for *Cidaris.* As the gastrulation approached, the mesenchyme cells began to be separated from the blastoderm just as the invagination of the archenteron first became visible. The stage at which the mesenchyme cells migrated into the blastocoele was therefore intermediate between the two species, and this proved that at the beginning of gastrulation the paternal nucleus was already able to manifest itself (Tennent, 1914).

In hybridization experiments performed on amphibians, it is a rule without exceptions that the cleavage rate is strictly maternal. This also holds good in cases of hybrid andro-merogons. It is thus shown that the cytoplasm alone determines the rate of cleavage. The pigmentation of the egg and early developmental stages is of course always maternal, because the egg pigment is not synthesized after the egg has become mature. Differences in the pigmentation of the larvae become evident only much later when the melanophores, derived from the neural crest, become differentiated.

During the subsequent development the pigmentation depends on the nuclear factors. A hybrid andro-merogon was produced by using eggs of the black race of the axolotl (*Ambystoma mexicanum*) and sperm of the white race. In the embryo the cytoplasm was derived therefore from the black race and the nucleus from the white race. The animals were all white, thus showing that the nucleus dominated over the cytoplasm (Baltzer, 1941).

No crosses are known in amphibians in which morphological differences between the two parent species could be discovered in the early stages—the gastrulation or the neurulation stages. It is found, however, that the end of cleavage and the beginning of gastrulation is the stage when the incompatibility of the sperm nucleus and the egg nucleus and cytoplasm first manifests itself. In many hybrid combinations the hybrid goes through the early stages of cleavage more or less normally, but the development stops in the blastula or early gastrula stage (*Rana esculenta* × *Bufo vulgaris; R. esculenta* × *R. temporaria; B. vulgaris* × *R. temporaria; Triturus palmatus* × *Salamandra maculosa; R. pipiens* × *R. sylvatica; R. pipiens* × *R. clamitans* and others) (Morgan, 1927; J. A. Moore, 1946; A. B. C. Moore, 1950). If the true hybrid between two species is viable, the hybrid andro-merogon may not be. The andro-merogon embryos, even without hybridization, are always weaker than normal embryos due to the haploid state of their nuclei. However, the hybrid andro-merogons are much inferior in their capacity for development. The hybrid combination *R. pipiens* × *R. capito* is fully viable up to adult stage when the chromosomes of both parental species are present. The andro-merogon of *R. pipiens* fertilized with sperm of the same species develops into a tadpole. The hybrid andro-merogon (*R. pipiens* × *R. capito*) develops normally through cleavage and gastrulation stages, but dies in the early neurula stage (Ting, 1951). The paternal chromosomes, in the presence of maternal chromosomes, may function normally and transmit the paternal characters to the diploid hybrid, but left alone with the foreign cytoplasm they prove to be inadequate for supporting development. Further cases are known in which the andro-merogonic hybrid dies during gastrulation or shortly after (*R. pipiens* × *R. palustris; Triturus alpestris* × *T. palmatus; T. palmatus* × *T. cristatus*) (Hadorn, 1932; de Roche, 1937; A. B. C. Moore, 1950).

There is some direct evidence that in hybrids the paternal chromosomes cause the synthesis of proteins of their own species about the time of the beginning of gastrulation. The evidence was obtained by immunological methods in the following experiment. Eggs of the sea urchin *Paracentrotus lividus* were fertilized with sperm of *Psammechinus microtuberculatus*. Using the blood plasma of animals immunized against proteins of *Psammechinus microtuberculatus,* it was found that the fertilized eggs showed no trace of these specific antigens. Twenty-four hours after fertilization, however, when the hybrid embryos were in the mesenchyme blastula stage, antigens of the paternal species, *Psammechinus microtuberculatus,* could be clearly detected. As the paternal antigens could not be discovered earlier, it is evident that these antigens (predominantly proteins) had been built up in the cytoplasm, which is derived mainly from the egg, under the influence of the sperm chromosomes (Harding, Harding and Perelman, 1954). The death of non-viable hybrids and hybrid andro-merogons is preceded by an arrest of at least some of the physiological processes which characterize the onset of gastrulation, such as the increase in oxygen consumption, glycolysis, and the increase in the amount of ribonucleic acid (Brachet, 1950a); the latter is especially suggestive, as this involves a synthetic process. We have seen that besides the synthesis of ribonucleic acids the embryo during gastrulation begins synthesizing new proteins, manifesting themselves as new antigens when investigated by serological methods. It is highly probable that in cases of incompatibility between the foreign paternal nuclei and the maternal cytoplasm these synthetic processes cannot go on, or cannot go on satisfactorily, hence the arrest of development and death.

If, on the other hand, the nuclei and the cytoplasm are compatible, the synthesis of new proteins and nucleic acids (and probably also other cytoplasmic substances) goes on progressively. Since some of the newly synthesized substances are different from the ones present before, rapid changes in the constitution of the embryo are inaugurated, changes that involve not only the position and the arrangement of cells, as in the gastrulation movements, but also their inherent physiological properties. This leads to the next period of development, in which the various parts become progressively differentiated from each other: the period of **organogenesis.**

PART IV

Organogenesis

The germinal layers formed as a result of the process of gastrulation are the source of material for the development of all the organ rudiments of the embryo. The germinal layers become progessively subdivided into groups of cells which have been called **primary organ rudiments,** because some of them are composite in their nature and undergo a further subdivision until all the various parts are segregated from each other in the form of **secondary organ rudiments.**

The formation of the primary organ rudiments has been described in connection with the process of gastrulation.

CHAPTER 9

Morphogenetic Processes
Involved in Organogenesis

The way in which sheets of epithelial cells give rise to the various organ rudiments is largely the same as that by which gastrulation is achieved, that is, by means of morphogenetic movements—by changes in the position and arrangement of cells. The movements can be studied by the same methods as can those involved in gastrulation, by the application of local vital staining, for instance.

The movements are different in nature depending on whether they are performed by epithelial cells or by cells of the mesenchyme. By epithelial cells we mean cells in close formation, adhering to one another, independent of whether they are spreading in a sheet or forming a compact mass. The following classification of morphogenetic processes is a modification of the one given by Davenport (1895).

9–1 MORPHOGENETIC PROCESSES IN EPITHELIA

The changes observed in epithelial cells leading to the formation of organs are mainly of the following types:

(a) Local Thickenings of the Epithelial Layer. This is a very common method of organ formation. We have seen before that the formation of a neural plate takes place in this way. The thickening is brought about by the epithelial layer flowing, as it were, toward the region in which the organ rudiment is to appear, in this case the mid-dorsal region of the embryo. As the thickening becomes discernible, the epithelial cells may become elongated in the direction perpendicular to the surface of the epithelial layer.

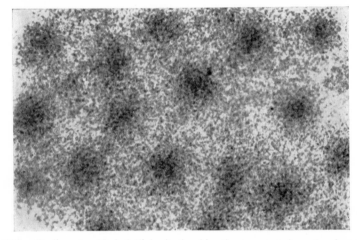

Fig. 150.   Rudiments of hair follicles in the skin of a mouse embryo (surface view).

This elongation of cells is very distinct in the neural plate. The cells become columnar, and in the case of the urodele amphibians, at least, single cells stretch right through from the external to the internal surface of the neural plate, in spite of its increased thickness. This means that the formation of the neural plate involves a rearrangement of the cells of the epithelial layer, because the ectoderm of the late gastrula in amphibians consists of at least two layers of cells, and this two-layered state is retained by the presumptive epidermis, whereas in the neural plate a single layer of columnar cells is found. It has been suggested that the elongation of individual cells of the epithelial layer in the direction perpendicular to the surface of the epithelium is the mechanism by which the thickening of the layer is brought about (Holtfreter, 1947b). This is, however, not always the case; sometimes the increase in thickness of an epithelial layer is brought about, at least in part, by a local increase in the number of cell layers.

In the case of multiple organs, more than one thickening may develop simultaneously in the same sheet of epithelium. A very clear-cut example of this may be seen in the development of the hair follicles of mammals. Before the appearance of the rudiments of the hair follicles the epidermis is a two-layered sheet of epithelium, with cells more or less flattened and of uniform thickness. Then, rather suddenly, numerous concentrations become visible, scattered at more or less regular intervals throughout the epithelial layer (Fig. 150). In each concentration the cells lie closer together, and the individual cells, especially in the lower layer of epithelium, are less flat than before, approaching the cuboidal or even the columnar shape. At a slightly later stage the number of cell layers is also increased. This concentration of cells is brought about by the cells flowing toward the respective centers, since no increase of cellular reproduction (mitoses) could be discovered in the early hair follicle rudiments (Balinsky, 1950).

(b) Separation of Epithelial Layers.   Epithelial layers or masses of cells may be subdivided into parts by the appearance of crevices between groups of cells, the cells on the opposite sides of a crevice losing connection with one another. The crevices may appear at any spot, but perhaps the most common occurrence is for the crevices to be either parallel to the surface of the epithelium or perpendicular to it. In the first case the epithelial layer is split into two layers lying on top of one another. The original crevice may be increased by the secretion of fluid into it, and may become a more or less spacious cavity. This is the case with the formation of the parietal and visceral layers of the lateral plate mesoderm and the coelomic cavity between them. An instance of crevices appearing perpendicular to the surface of the epithelium is found in the development of the mesodermal somites. The upper edge of the mesodermal mantle becomes subdivided into segments by crevices running between cells perpendicular to the surface of the epithelium and at the same time perpendicular to the main axis of the body (Fig. 90). The masses of cells thus formed (the somites) are of approximately the same size. This suggests that each somite is formed around a center of attraction whose force is limited, so that only a certain number of cells can be kept together, the cells losing connection with each other where the attraction of the center is too weak. A rhythmic pattern of differentiation is thus produced.

A splitting of epithelial masses often follows the formation of local thickenings. Thus the neural tube eventually loses its connection with the epidermis.

(c) Folding of the Epithelial Layer.   The folding of epithelial layers can occur either in the form of linear folds resulting in the formation of grooves or as approximately round depressions forming pockets. The direction of folding is, with only a few exceptions, toward the originally proximal surface of the epithelial layer. In the ectoderm and later in the epidermis the folding is therefore directed inward. In the invaginated parts of the embryo, as in the whole of the endoderm, the folding is directed outward, that is, away from the lumen of the alimentary canal.

The folding is usually foreshadowed and accompanied by a thickening of the epithelial layer. This is the case in the neural tube formation: at first the neural plate appears as a thickening of the ectoderm; next this thickened plate starts to fold inward, thus forming a neural groove and eventually rolling into a tube. This is followed by the fusion of the edges of the neural tube and its separation from the remaining epidermis (Fig. 92). If a linear fold becomes closed in and separated from the original epithelial layer, the resulting structure is a hollow tube. If a pocketlike depression becomes closed in and separated from the original epithelial layer, the result is a hollow vesicle. An example of the first formation is the neural tube and of the second formation the rudiment of the inner ear—the ear vesicle (Fig. 172). Most of the glands are developed as pocketlike invaginations of the epithelium, and this shows how common this type of morphogenetic process is. Of course the invagination of the archenteron, as found in a sea urchin or *Amphioxus,* is also an example of pocketlike infolding of an epithelial layer. What has been said about the mechanism of the invagination and the change in the shape of cells as a possible cause of the infolding applies to all infoldings in general.

(d) **Thickenings Followed by Their Excavation To Form Tubes or Vesicles.** By no means always are tubes or vesicles produced by infolding of the epithelial layer. Quite often the initial stage is a solid thickening of the epithelium (as under a). Such a solid thickening may acquire an internal cavity secondarily by the separation of cells in the middle. This cavity may or may not be connected with the external surface of the epithelium. It is very remarkable that one and the same organ in different animals may develop in different ways: as an infolding in some, and as a solid thickening becoming excavated later in others. This is the case with the development of the neural tube in various vertebrates; it is formed by infolding in the elasmobranchs, in the urodele amphibians and in the amniotes. It develops as a longitudinal thickening that later separates itself from the epidermis and acquires a central cavity in the Myxinoidea and in the bony fishes. In the Anura a sort of intermediate state is found, the groove being very shallow and the neural tube being in part solid at the beginning. Other organs such as the eye, the ear and the lens may develop either as hollow pocketlike invaginations or as solid masses of cells. The underlying mechanism of both formations must therefore be similar in nature.

If the thickening of an epithelial layer is directed outward, it takes the form of an "outgrowth." The formation of an outgrowth is probably never caused by a local increase of growth rate, that is, increased multiplication or increase in size of the cells. Much as in all other cases of organ formation, the outgrowths are caused by the concentration of cells and their rearrangement. Such outgrowths are formed on the surface of the chorion in mammals (the chorionic villi—see section 7–3), or on the gills in amphibians and fishes (the gill filaments). The outgrowths may be simple and fingerlike or may develop secondary outgrowths when the whole becomes dendritic or plumose. The tips of the outgrowths may be solid, but once they have appeared the solid outgrowths of epithelium become hollowed out, so that connective tissue and blood vessels may penetrate into the outgrowth. Blood capillaries develop in a similar way: first as solid strands of endothelial cells, pushing forward from existing blood vessels, and being hollowed out later. The cavity in this case is the lumen of the capillary. The speed with which the solid outgrowths are hollowed out differs from case to case, and no strict boundary can be drawn between these and outwardly directed folds or outpushings which are hollow right from the start.

(e) **Fusions of Previously Separated Cell Masses.** These may play a certain part in morphogenesis. For example, the edges of the neural plate fuse along the midline dorsally after the neural plate has rolled itself into a tube. Also, blood capillaries growing out from the arteries and from the veins fuse their ends to establish an open pathway for the blood circulation.

(f) **Breaking Up of Epithelial Layers To Produce Mesenchyme.** This is a very important morphogenetic process, even if it does not lead directly to the formation of organ rudiments, because it furnishes the material from which further organ rudiments may develop. An epithelial layer or a portion of it may sometimes break up into mesenchyme altogether, so that the epithelium as such disappears or a gap is left in the previously continuous layer. An example of this process is found in the development of the neural crest; a part of the ectodermal epithelium forming the edge of the neural folds splits up into disconnected cells, leaving a gap between the neural tube and the epidermis. (Sometimes, however, the neural crest cells are given off from the dorsal wall of the neural tube after its separation from the epidermis.) The other possibility is for individual cells to slip out from an epithelial layer, so that the latter remains intact though depleted of part of its cells. This process is of very common occurrence; thus the visceral layer of the lateral plate mesoderm gives off cells which, as mesenchyme, surround the alimentary canal and are later differentiated into the smooth muscle and connective tissue of the intestine.

The breaking up of epithelial layers and the migration of individual cells out of such layers are the result of a change in the physiological properties of the cells, most especially of their surface layers. The cells do not adhere to one another on contact as do epithelial cells, but glide freely, and thus they move about each independently of the other. This can be observed if such cells are cultivated *in vitro* (Holtfreter, 1947a,b).

9–2 MORPHOGENETIC PROCESSES IN MESENCHYME

The morphogenetic processes in mesenchyme naturally are somewhat different from those occurring in the epithelial cell masses. The most typical of those are reviewed here.

(a) **Aggregation of Mesenchyme into a Mass.** Such aggregations may be formed in the mesenchyme independently of epithelial parts of the embryo. Mesenchyme cells are drawn together to form densely packed masses, without, however, joining up into an epithelium. Each cell retains its ability to migrate freely, thus showing that the specific properties of the surface cytoplasm have not been changed. Such mesenchyme aggregations later differentiate as cartilage, bone or muscle tissue.

(b) Attachment of Mesenchyme to Another Body. In normal embryonic development it is often observed that mesenchyme cells become concentrated near the surface of an epithelium or around some epithelial structure such as a vesicle or tube. The aggregated mesenchyme may later develop into cartilage or bone and thus produce a skeletal capsule around or near the epithelial structure. Such skeletal capsules develop around the nasal sac and the inner ear of vertebrates. The brain case as a whole is developed in a similar way. In other cases the mesenchyme does not differentiate as skeletal tissue, but it may develop into a fibrous capsule, which is found around many internal organs (kidneys, liver, spleen).

(c) Acquisition of Epithelial Connections by Mesenchyme Cells. Mesenchyme cells are sometimes secondarily rearranged into an epithelium and become typical epithelial cells in so far as their relationships to each other are concerned. Such a transformation takes place in the mesoderm produced from the primitive streak in birds and other Amniota. The cells migrating from the streak are ameboid, but later the sheet of these migrating cells fuses into a regular epithelium—the mesodermal mantle (later subdivided into somites, lateral plates, etc.). Another case of the same kind is found in the development of blood vessels. The endothelium of blood vessels consists of cells which had been mesenchyme cells before, but which have joined together to form an epithelial layer. It is true that epithelia developed from mesenchyme differ from those which had never lost their epithelial properties. Thus the endothelia possess the capacity for phagocytosis in common with many connective tissue cells, a property which is not found in the ectodermal and endodermal epithelia of vertebrates.

9–3 FACTORS DIRECTING THE MOVEMENT OF CELLS DURING ORGAN RUDIMENT FORMATION

Some factors concerned in directing the morphogenetic movements of cells have already been mentioned. The most important factor of this kind is the one responsible for the epithelial or mesenchymal arrangement of cells and this factor operates through a change in the properties of the cell surface.

Isolated embryonic cells, kept in a suitable medium, quite generally possess the ability for movement. In the case of epithelial cells, if two or more cells come in contact, they adhere to one another and at the same time their motility is greatly reduced if it does not disappear altogether. In the case of cells of mesenchymal nature the contacts between cells do not lead to their adhering together and do not decrease their motility. Differentiation of the cells tends to immobilize them, and differentiated connective tissue cells, such as fibroblasts, do not move about if the tissue is in the normal state. Between free motility and complete immobilization there may exist an intermediary state, in which the cell constantly sends out pseudopodia, but these are soon withdrawn, and other pseudopodia are sent out in different directions, so that the cell sometimes remains in the same place in spite of very active motility (de Bruyn, 1945).

Although in epithelial masses the motility of individual cells is reduced, the sheet of epithelium can move as a whole. This movement is best observed in tissue cultures. It may be seen that the edge of an epithelial sheet becomes pushed outward, sometimes in a broad line, sometimes forming more or less narrow tongues, forcing themselves ahead of the main mass. Epithelial sheets can spread only over a solid substrate, and the same is true, with a reservation, for the mesenchyme cells. The reservation is that mesenchyme cells may travel along (and through) a network of fibers, whereas such a substrate could not support the movement of an intact epithelial layer.

From what has been said it is clear that the migration of embryonic cells can be taken for granted, and that it is the directing of this migration along definite channels and to definite destinations, so that an organized whole is gradually built up, that has to be accounted for. The properties of the substratum along which the cells migrate are among the major factors affecting the direction of migration. As has been already indicated, the cells cannot move through space filled with fluid only. In the parts of the embryo occupied by mesenchyme, however, the intercellular spaces are filled with a colloidal solution which is in part gelated, so that fibers of molecules are stretched through the space in various directions. These fibers serve as a substratum for the migration of mesenchyme cells. If the fibers of the intercellular substratum do not show any orientation, the migration of mesenchyme cells is disorganized, and the cells become scattered equally in space. If the fibers of the intercellular substratum become oriented in any direction, the mesenchyme cells migrate along the oriented bundles of fibers. This can be proved experimentally with cells in tissue cultures. The

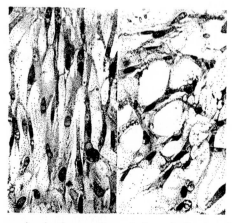

Fig. 151. Cells growing in tissue culture. Left—micelles in substrate are oriented by allowing the blood plasma to clot under stress. Right—no orientation of micelles in substrate. (After Weiss, from Kühn, 1955.)

fibers of the plasma clot on which the cells are cultivated may be caused to be stretched in a certain direction by various means, as for instance by allowing the clot to set under mechanical tension. The cells will then follow the direction of the plasma clot fibers in their migration (Fig. 151) (Weiss, 1929; Doljanski and Roulet, 1934).

If the mesenchyme cells aggregate around an epithelial vesicle, this aggregation may be the result of the orienting influence of the epithelium on the fibers of the intercellular substrate. If these fibers are stretched perpendicularly to the surface of the epithelial vesicle, the migrating mesenchyme cells will converge toward the vesicle. Once having reached the surface of the epithelium the cells of the mesenchyme may be held there by intercellular bonds, whose nature is not well understood, but whose existence can hardly be doubted (Weiss, 1947).

The influences of adjoining parts on the movements of migrating cells is not the only form of interaction of parts playing an important role in the development of organ rudiments. Influences of a nature similar to the induction of the neural plate by the roof of the archenteron have been discovered in the development of organ rudiments. Such influences may modify the movements of cells, but not directly. The direct result is a modification of the properties of cells, and as a consequence of this the movements of cells are also altered. This can be illustrated by the development of the neural crest; the neural crest differentiates together with the neural plate as a result of an induction from the roof of the archenteron. Once induced, the neural crest cells acquire new properties, they become mesenchyme cells and start migrating away from their source of origin, while both the cells of the epidermis and the cells of the neural tube remain relatively stationary.

9–4   DEVELOPMENT OF GENERAL BODY FORM. "STAGES" OF DEVELOPMENT

Changes in Body Shape.   While the various organs are being formed, the shape of the embryo as a whole undergoes far-reaching changes. During the period of organogenesis, in the case of the vertebrate embryo the main changes are:

1. Elongation of the body,
2. Formation of the tail,
3. Subdivision of the body into head and trunk,
4. Development of appendages,
5. Separation of the embryo proper from the extraembryonic parts (the latter process has already been dealt with in sections 7–1 and 7–2).

Some of the processes enumerated above also occur in invertebrates; in particular the elongation of the body occurs in annelids and arthropods. The subdivision of the body into sections such as head, thorax and abdomen is typical for insects and, with modifications, for some other arthropods.

The development of appendages is an essential feature of arthropod development. Other processes concerning the body as a whole may occur in invertebrates but have no counterpart in vertebrate development. For

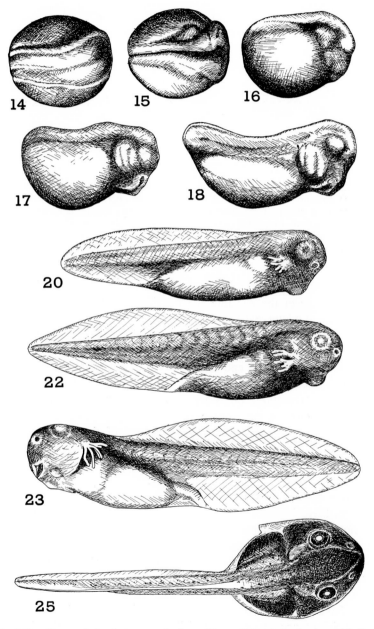

Fig. 152. Stages of development of a frog (*Rana pipiens*), beginning with the neurula stage. The numbers indicate stages of development after Pollister and Moore, 1937. (Redrawn after Rugh, 1948.)

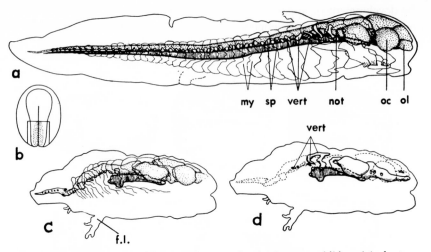

Fig. 153. Results of excision of the notochord of an amphibian (*Ambystoma punctatum*) embryo. *a,* Normal larva. *b,* Diagram of operation; stippled area of the archenteron roof removed. *c,* Operated larva; drawing shows notochord (vertical lines), neural system (stippled) and contours of muscle segments. *d,* Same larva, showing notochord and cartilaginous skeleton. *f.l.,* Forelimbs; *my,* muscle segments; *not,* notochord; *oc,* eye; *ol,* nose; *sp,* spinal ganglia; *vert,* vertebrae. (From Hörstadius, 1944.)

instance, in insects the body of the embryo undergoes a peculiar shifting from the surface of the egg into the interior of the yolk from whence it emerges again at a later stage.

In holoblastic vertebrates, of which the amphibians may serve as example, the embryo retains a spherical shape (i.e., the shape possessed by the unfertilized egg) up to the end of gastrulation and the beginning of neurulation. In the neurula stage the embryo becomes slightly elongated in an anteroposterior direction, but only after the completion of neurulation does the elongation of the embryo become really prominent. A tail rudiment, the **tailbud,** appears at the posterior end of the body and rapidly develops into an elongated appendage, but the rest of the body also stretches out, becoming at the same time flattened laterally and, to a certain extent, lower in a dorsoventral direction (Fig. 152).

Although most of the organ rudiments of the embryo are involved in this elongation, there is experimental evidence that not all are equally active. If the notochord rudiment of an amphibian embryo in the neurula stage is excised, the embryo remains stunted and does not stretch as usual (Fig. 153) (Hörstadius, 1944; Kitchin, 1949). On the other hand the notochordal rudiment will stretch and form an elongated rod even if it is cultivated *in vitro* and is not accompanied by other parts (Fig. 154) (Holtfreter, 1939b). Isolated parts of the neural system or induced brain vesicles, when no other tissues accompany them, fail to elongate. From this it appears that the notochord changes its shape actively, while the nervous system is being pulled in length by the adjacent notochord. In amniotes the elonga-

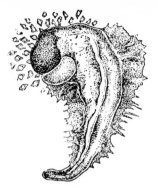

Fig. 154. Explantation of the dorsal blastopore lip of an axolotl gastrula. The notochord has developed in the form of an elongated rod in the middle of the explant. Lateral parts are mesoderm. At the anterior end is a piece of ectoderm (pigmented). (From Holtfreter, 1939b.)

tion starts in the primitive streak stage, so that the body of the embryo is already long and narrow by the time the main axial organs (neural tube, notochord, dorsal mesoderm giving rise to the somites) are laid down.

The formation of the tail rudiment will be dealt with in section 10–1.

The subdivision of the body in terrestrial vertebrates into head and trunk is largely dependent on the reduction of the branchial apparatus. As will be described in section 12–3, the system of visceral clefts and arches is fully developed in embryos of all vertebrates. In fishes and the larvae of amphibians the visceral clefts and arches persist and take up the area on the ventral side and posterior to the head. In terrestrial vertebrates the branchial apparatus loses its respiratory function and becomes reduced. As a result of this, in later embryonic stages the area of the body posterior to the head fails to grow at the same rate as other parts, thus producing a constricted section between the head and the trunk (see Figs. 155, 156). The constriction is accentuated further by:

(a) a certain amount of longitudinal stretching of the neck region, as the result of which the cervical vertebrae are as a rule somewhat longer than thoracic vertebrae (this is not true in some mammals with shortened necks such as man or the whales), and

(b) the withdrawal of the heart, originally situated in the neck region, next to the branchial clefts, into the trunk (thorax) (see Figs. 155, 156).

The development of appendages, particularly the paired limbs, will be described in section 11–3.

Normal Stages of Development. The changing appearance of the embryos, especially during organogenesis, invites the distinguishing of certain **stages** which can be referred to when it is desired to indicate how far an embryo has progressed in its development. Tables of "normal stages" have been worked out for a number of species of animals, especially those that are most often used for research.

In the latter half of the nineteenth century an ambitious work on establishing series of normal stages for a large number of animals was undertaken by Keibel and his collaborators. This work was an essential contribution to the science of comparative embryology (see p. 12) but went

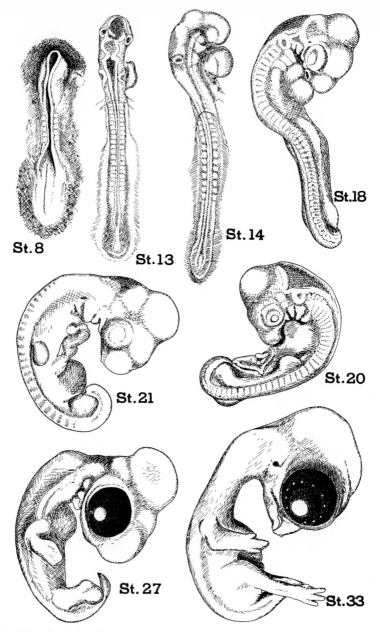

Fig. 155. Stages of development of a chick. (Stages 18 and 20 after Keibel and Abraham, from Keibel in Hertwig, 1906; stages 13, 14, 21, 27, 33 from Hamburger and Hamilton, 1951; stage 8 from Patten, 1951.)

into oblivion later, when the interest of the great majority of embryologists shifted from a descriptive to an experimental approach to the development of animals. However, it soon became evident that tables of normal stages

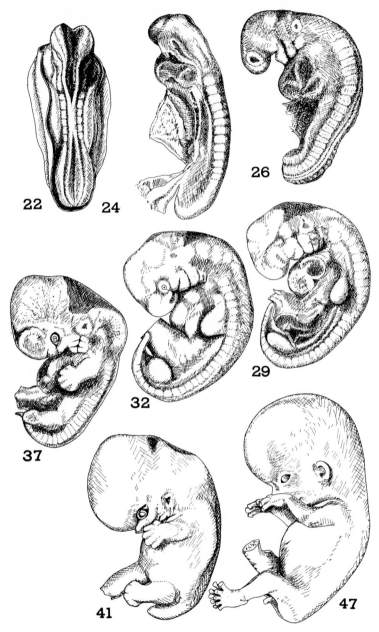

Fig. 156. Stages of development of the human embryo. Figures show approximate age in days. (22-day-old embryo, after Arey: Developmental Anatomy; 37-day-old embryo redrawn from Hamilton, Boyd and Mossman, 1947; the rest after Streeter, 1942–1951.)

were quite as important for experimental work, as it was often necessary to indicate precisely at what stage an operation or other experiment was carried out.

The table of normal stages of *Ambystoma* (= *Amblystoma*) *punctatum,* prepared by Harrison, was the first made expressly with experimental investigations in view. Harrison's table has never been published, but has been made accessible privately to many workers in the field and has been widely used. Other tables of normal stages followed; the most widely used ones are probably the stages of *Rana pipiens* by Pollister and Moore (1937) and the stages of the chick by Hamburger and Hamilton (1951). Other tables of stages may be looked up in Rugh, *Experimental Embryology* (1948). There is still no up-to-date table of normal stages of the development of the human embryo. (See, however, the series of papers by Streeter, 1942-1951.)

In compiling a series of normal stages the embryologist is confronted with the task of deciding what characteristics should be selected for distinguishing one stage from another. The characteristics that are often used are:

1. Age of the embryo,
2. Size of the embryo,
3. Morphological peculiarities of the embryo.

In human embryology in particular the first two criteria are often used. It is customary for an author to refer to embryos by age (five-week embryo, two-month-old embryo) or by size (an embryo or fetus of so many mm. crown-rump length). Both these criteria are, however, not very convenient; the age of an embryo is often not known, and, in animals other than mammals, the rate of development is dependent on the temperature of the environment to such an extent that a statement about the age of the embryo is meaningless unless the temperature at which the development has proceeded is likewise indicated. The size of the embryo is no true indication of its degree of development as the dimensions of the embryo vary to a great extent. It may be noted here that some variability in the size of embryos may be ironed out later in the course of development, which adds to the difficulty of using size as a criterion for the definition of stages.

What remains is to base the normal stages on morphological properties of the embryo, and especially on properties that can be easily ascertained by external examination of the embryo, without its fixation or dissection, that is, largely on external features. In the initial stages of development (cleavage stages) the number and size of the blastomeres may conveniently be used. During gastrulation the shape of the blastopore or its equivalent (primitive streak) may be used. Just after gastrulation the neural plate offers easily recognizable features. During early organogenesis the number of pairs of somites has often been used to define the stage of development of the embryo. The somites, though not strictly speaking "external features," can be seen on external inspection, especially in the amniotes. In still later stages the development of the appendages presents very easily distinguishable and convenient characters for the definition of normal stages.

Although morphological characters appear to be the best criteria for establishing the stage of development of an embryo, there are certain limitations even in this approach. It has been found that the development of different parts (organs) of the embryo is not always strictly coordinated in time; sometimes certain ones develop more rapidly, sometimes others. So if two embryos have certain organs (for instance the forelimbs) in exactly the same condition, they may at the same time differ in the degree of development of other organs (the nervous system or the liver, for instance). This phenomenon of **heterochrony** or unequal rate of development of parts must be always borne in mind when any tables of normal stages are being used or referred to.

CHAPTER **10**

Development of the

Ectodermal Organs

in Vertebrates*

10–1 DEVELOPMENT OF THE CENTRAL NERVOUS SYSTEM

The central nervous system of vertebrates develops from the primary rudiment—the neural tube. The origin of the neural tube has been described earlier (section 5–5). The tube when formed is of unequal diameter throughout; its anterior end is expanded, the cavity is broader and the walls are thicker than in the posterior part of the tube. This foreshadows the development of the brain from the anterior part of the neural tube and the develop-

* A more detailed description of organogenesis in vertebrates may be found in the books by Nelsen, 1953, and Witschi, 1956.

ment of the spinal cord from the posterior part. The various parts of the brain (forebrain, midbrain, etc.) are first indicated as thickenings of the wall of the neural tube, followed up, especially in the case of the cerebral hemispheres, by the development of pocketlike evaginations of the brain wall. Shallow constrictions develop early all around the neural tube, thus permitting the distinguishing of several "brain vesicles." Three such brain vesicles appear at the beginning. The most anterior brain vesicle, the **prosencephalon,** later gives rise to the **telencephalon** and the **diencephalon.** The second brain vesicle, the **mesencephalon,** is not subdivided further and develops into the midbrain. The third brain vesicle, the **rhombencephalon,** gives rise to the **metencephalon** (cerebellum) and the **myelencephalon** (medulla oblongata) (Fig. 157). The medulla oblongata becomes constricted by shallow furrows

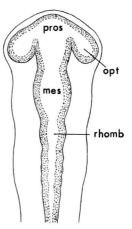

Fig. 157. Primary brain vesicles and eye rudiments of a two-day-old chick embryo. *pros,* Prosencephalon; *mes,* mesencephalon; *rhomb,* rhombencephalon; *opt,* eye vesicles.

into a number of segments, **neuromeres.** This segmentation, which is especially clear in fish embryos, is, however, only temporary and does not leave any trace in the organization of the adult brain. It is not correlated with the metameric arrangement of the cranial nerves.

Even before the prosencephalon is clearly subdivided into the telencephalon and diencephalon, a pair of saclike protrusions appear on its lateral walls. These protrusions are the rudiments of the eyes, which are thus, basically, specially differentiated parts of the brain. In this stage they are called optic vesicles. The optic vesicles become constricted off from the remainder of the prosencephalon, and the connecting **optic stalk** later forms the basis for the development of the optic nerve. The optic stalk (and the optic nerve) join that part of the brain vesicle which becomes the diencephalon. By the method of local vital staining, as well as by observation of peculiarities of pigmentation, it has been possible to trace back the cells of the optic vesicle to the open neural plate. In the neural plate the presumptive material of the optic vesicles lies far forward, close behind the transverse neural fold, and rather near to the midline (Fig. 158). In the course of development the eye rudiments are drawn out away from one another and into a more lateral position than they occupied initially. The

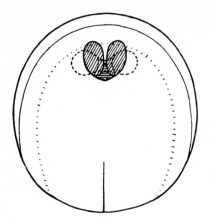

Fig. 158. Position of the presumptive eye rudiments in the neural plate of an amphibian. Shadowed areas show the position of the eyes in the very early neurula. The dashed contour shows their positions in the early neurula. (From Manchot, 1929.)

subsequent development of the optic vesicles will be dealt with in a special section.

After it has been closed, the neural tube undergoes very considerable stretching, together with the stretching of the embryo as a whole, especially in its posterior half. As this stretching is at the expense of the thickness of the tube, it tends to enhance the difference in bulk between the brain and the spinal cord.

A most peculiar transformation is suffered by the posterior part of the neural plate and tube. The neural plate here reaches right to the blastopore. When the posterior part of the neural tube elongates, it does so to a greater extent than the ventral part of the embryo. The posterior end of the neural tube is therefore carried beyond the blastopore. As its hindmost tip is attached to the blastopore, the neural tube becomes bent on itself some distance from the blastopore. The apex of the bend now becomes the tip of the tail rudiment (Fig. 159). The major part of the neural tube from its cranial end to the apex of the bend differentiates as central nervous system (brain and spinal cord). The inflected part of the tube, however, the part lying between the apex of the tail rudiment and the blastopore, differentiates as muscle of the tail region. It loses its central canal and becomes split along the midline into two lateral masses or strips of cells which forthwith shift upward, so as to lie on both sides of the notochord and spinal cord. Cranially these cell masses join up with the dorsal part of the mesodermal mantle. Together with the latter the presumptive muscle of the tail region is subdivided into muscle segments—the somites. No stalks of somites and no lateral plate are developed in the tail region, and in this respect the segmentation of the presumptive muscle derived from the neural plate differs from the segmentation of the trunk mesoderm. By local vital staining it is possible to determine exactly what part of the neural plate differentiates as caudal muscle, and what part differentiates as neural tissue. The boundary between the two parts runs straight across the neural plate at about one-sixth the distance to its posterior end (Bijtel, 1931, 1936).

It has been shown previously (in the section on gradients in amphibian development, 6–4) that the roof of the archenteron is responsible for the differentiation of the various parts of the neural plate and neural tube. The anterior part of the archenteron roof, namely the prechordal plate, induces predominantly the forebrain and eyes (archencephalic inductor), a more posterior part of the archenteron roof induces the hindbrain and associated structures (deuterencephalic inductor), and the most posterior part of the archenteron roof induces spinal cord and muscle (spino-caudal inductor). The induction of muscle by the spino-caudal inductor becomes comprehensible from what has just been stated about the fate of the posterior end of the neural plate.

A corollary to the experiments on the regional specificity of inductors is provided by the following experiments designed to test the determination of parts of the neural plate. It was desired to test whether in the neural plate, after it has already made its appearance but before it starts to differentiate into its subordinate organ rudiments, the various parts are interchangeable, or whether they are already determined for their respective destinations. For this purpose pieces of the young neural plate were cut out and reimplanted into the same or another embryo. The transplanted parts were placed in abnormal orientations, as for instance with reversed anterior and posterior ends, or they were placed in an altogether different region of the neural plate. The result was found to be different depending on whether the pieces of the neural plate were taken with or without the underlying archenteron roof. If the graft consisted of neural plate cells only, the graft differentiated in agreement with its surroundings and the original

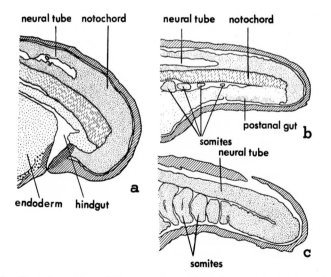

Fig. 159. Transformation of the posterior end of the neural tube into the caudal somites in an amphibian embryo. *a,* Early tailbud stage; *b,* late tailbud stage, median section; *c,* same, paramedian section. (From Bijtel, 1931.)

polarity of the graft or its place of origin did not manifest itself. Rotated grafts gave rise to parts of the brain which were in complete harmony with their surroundings (Alderman, 1935). Pieces of neural plate taken from its posterior region and transplanted into the anterior region differentiated as brain parts, instead of differentiating as spinal cord or muscle. In short, the development of the neural system went on as if nothing had happened (Umanski, 1935). The various parts of the neural plate were found not to be determined, or else their determination was not final and could be overridden by the influence of the surrounding tissues.

If, however, the neural plate material was taken together with the under-lying archenteron roof, the graft differentiated in accord with its original prospective significance. Inverted sections of the brain developed if the graft was inverted (Spemann, 1912b); the normal location and differen-tiation of the eyes was disarranged if the rotation of a piece involved the eye region of the neural plate. That this result is due to the rotation of the archenteron roof together with a portion of the neural plate is clearly proved by experiments in which only a piece of archenteron roof was rotated, while the neural plate remained in its normal position. This experiment caused a derangement in the development of the eyes (Alderman, 1938). The removal of a part of the archenteron roof underlying the eye region of the neural plate causes defects in the development of the brain and eyes; parts of the forebrain are found to be missing, and the eyes are fused into one cyclopic eye (Adelmann, 1937) (see also section 6–4).

What has been said in respect to determination of the structure of parts developing from the neural plate (of which the eyes are the most easily recognizable) may be extended to the determination of the functional mechanism developing in the brain. As will be shown later (section 16–2) the normal movements of the forelimb in salamander larvae depend on a central mechanism ("action system" of Weiss, 1955) which is located in the spinal cord, at the level of the three pairs of spinal nerves (3, 4 and 5) which supply the forelimbs. If the area of the neural plate giving rise to this segment of the spinal cord is excised and replaced by a more posterior part of the neural plate, the graft will fit into its new position and acquire the functional properties necessary for controlling the movements of the forelimbs. If a similar transplantation is carried out later, in the tailbud stage, replacing the spinal cord at the forelimb level by a more posterior section of the cord, the graft can no longer fully take over the function of the more anterior section, and the movements of the forelimbs supplied by nerves from the graft are abnormal (Detwiler, 1936). At the same stage (tailbud), however, the section of the neural tube which develops into the medulla (the rhombencephalon) may be cut out and replaced, with inver-sion of its anteroposterior axis, and not only does the graft develop into a morphologically perfect medulla, tapering from anterior end backward, but it acquires a functional polarization in harmony with the rest of the central nervous system; all the nervous responses, in particular the control of the swimming movements, of the operated larvae may be perfectly normal (Detwiler, 1949).

These experiments prove that the detailed structure of the central nervous system, on which its functional properties depend, is not determined at the time when the rudiment of the nervous system is first formed, but the peculiarities of the various parts of the brain and spinal cord are elaborated gradually throughout an extended period.

The further development of the brain and spinal cord is more or less complicated, depending on the degree of perfection that the central nervous system attains in any group of animals. Whereas in the lower vertebrates such as elasmobranch fishes and amphibians, for instance, the adult conditions do not depart greatly, in so far as the shape of the brain is concerned, from the conditions in the embryo, in higher vertebrates and especially in mammals the brain, which in the early embryo does not differ very much from the brain of an amphibian or fish embryo, changes later in a most striking way as a result of progressive development of certain parts.

The Development of the Brain.   We will first trace the features manifest in the development of the brain in all vertebrates, and then point out some of the peculiarities found in higher vertebrates, especially in mammals and in man.

As has been indicated earlier, the anterior brain vesicle, the prosencephalon, gives rise at its anterior end to the telencephalon. The latter produces in an early stage two bulges directed anterolaterally which become the

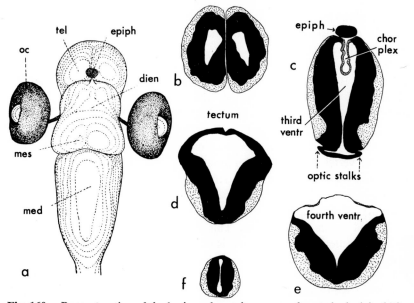

Fig. 160.   Reconstruction of the brain and eyes in a young frog tadpole (*a*). (After Spemann, from Huxley and de Beer, 1934.) Transverse sections of (*b*) telencephalon, (*c*) diencephalon, (*d*) mesencephalon, (*e*) medulla, (*f*) spinal cord. *epiph*, Epiphysis; *dien*, diencephalon; *med*, medulla; *mes*, mesencephalon; *oc*, eye; *tel*, telencephalon.

cerebral hemispheres (Fig. 160). Each bulge of the telencephalon contains a pocketlike cavity, which is an extension of the original cavity of the anterior brain vesicle (prosencephalon). The two cavities are known as the **first** and **second ventricles** of the brain. Originally they are in broad communication with the rest of the cavity of the prosencephalon, but later the channels leading into the first and second brain ventricles may become constricted. These channels are called the **foramina of Monro** (also known as the interventricular foramina).

In the lower groups of vertebrates the walls of the cerebral hemispheres are only moderately thickened, and the nerve cells remain accumulated on the inner surface of the walls, that is, on the surface facing the internal cavity.

The diencephalon in all vertebrates is remarkable in that it produces a great variety of structures with different functions, in addition to the two eye vesicles that are referred to later. The brain cavity in the region of the diencephalon remains fairly large and is known as the **third ventricle** of the brain. The cells in the brain wall become concentrated mainly in the sides of the diencephalon, while dorsally and ventrally it becomes thinned out. The greater part of the dorsal wall, or we may call it rather the roof of the diencephalon, becomes membranous and later does not contain any nerve cells at all. Instead it is richly supplied with blood vessels and becomes the **choroid plexus,** which later bulges down into the cavity of the third ventricle. The choroid plexus is the pathway by which nutrition and oxygen are brought into the ventricles of the brain. Processes of the choroid plexus may penetrate from the third ventricle into the first and second ventricles by way of the foramina of Monro.

Only the posterior section of the roof of the diencephalon retains the nervous character, but parts of it form dorsally directed outgrowths of which the most important are the parietal organ (or paraphysis) and the epiphysis. Both are formed as rather long, fingerlike outgrowths of the brain roof, the end sections of which become transformed into more or less rounded masses of cells, while the stalks become constricted and may even be interrupted later. In lower vertebrates either the epiphysis (in cyclostomes) or the parietal organ (in reptiles) become eyelike organs, but in higher vertebrates they differentiate as glandular structures. The homologies and the function of these structures have been the subject of many investigations, but in neither respect has a clear answer as yet been given. Both organs develop mid-dorsally, that is, at the site where the neural folds had fused at an earlier stage. This raises the question whether the presumptive material of each organ is contained in one of the neural folds, or in both. It has been found that at least in the case of the epiphysis there are originally two rudiments, one on each edge of the neural plate, and that after neurulation these two rudiments fuse into one single unpaired organ (van de Kamer, 1949).

The floor of the diencephalon produces in all vertebrates a funnel-like depression, the **infundibulum.** Part of the wall of the latter becomes segregated from the brain wall and fuses with a solid outgrowth from the stomodeal invagination (see section 12–2), the two together forming the **hypophysis,** the most important endocrine gland in vertebrates.

The midbrain remains a fairly simply organized part of the brain. The walls of the midbrain become thickened mainly ventrally, but the lateral walls and the roof are also fairly thick, and the latter gives rise to an important nerve center, the **tectum.** The cavity of the midbrain becomes narrow and is known as the **aqueduct of Sylvius.**

The neural tube may be fairly straight or only slightly curved at the time of its formation but in later stages it becomes bent at an angle at one or more levels. These bends are known as **flexures.** The most important flexure, and the one found consistently in all vertebrates, is that at the level of the midbrain, known as the **cephalic flexure.** Here the foremost part of the brain (the telencephalon and the diencephalon) are bent downward in front of the anterior tip of the notochord.

The rhombencephalon, as indicated before, gives rise to the metencephalon and the medulla oblongata. The cavity of the rhombencephalon expands especially anteriorly, just behind the midbrain, and becomes the **fourth ventricle.** The roof of the medulla thins out and is converted into a second choroid plexus—the **posterior choroid plexus,** similar to the one developed from the roof of the diencephalon. The future nerve cells are concentrated lateroventrally in the floor of the medulla, but separated into two masses by a median groove. This arrangement of nervous tissue gives the medulla a very characteristic appearance in cross-section (Fig. 160, *e*).

The metencephalon in a younger embryo is no more than a slightly thickened section at the anterior end of the medulla (which gives rise to the **pons varolii**) and a transverse bar in the roof of the brain just behind the mesencephalon and anterior to the choroid plexus of the fourth ventricle. The dorsal part of the metencephalon later gives rise to the **cerebellum.**

At its posterior end the medulla gradually merges into the spinal cord. The membranous part of the roof becomes narrower and eventually disappears, and the medioventral groove becomes deeper, until in the spinal cord the cavity is represented by a narrow vertical slit, separating the nervous tissue into two lateral masses, while both the floor and the roof of the cavity are thin, though not membranous as in the choroid plexus (Fig. 160, *f*).

The Human Brain.   The development of the brain in the higher vertebrates can best be illustrated by a brief description of the changes which the brain rudiment undergoes in the human embryo.

The brain of the human embryo toward the end of the first month after conception is not very different from the brain of an amphibian embryo, except that it is distinctly more elongated (Fig. 161, *a*). The eye rudiments are separated from the prosencephalon, the cephalic flexure is indicated, but there is as yet no trace of the progressive development of the hemispheres of the forebrain or of the cerebellum.

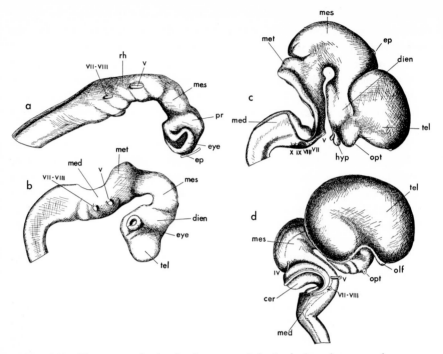

Fig. 161. Four stages in the development of the brain in a human embryo. *cer,* Cerebellum; *ep,* epiphysis; *dien,* diencephalon; *hyp,* hypophysis; *med,* medulla; *mes,* mesencephalon; *met,* metencephalon; *olf,* olfactory bulb; *opt,* optic stalk (or nerve); *pr,* prosencephalon; *rh,* rhombencephalon; *tel,* telencephalon. Roots of the cranial nerves are indicated by means of Roman numerals. (After Hochstetter, from Grosser, 1945.)

Soon after the beginning of the second month after conception the telencephalon forms a conspicuous bulge dorsally in front of the eye rudiments. The bulge is slightly bilobed; this is the first indication of the future first and second ventricles of the brain. The cephalic flexure is increased to such an extent that the brain appears to be bent on itself (Fig. 161, *b*). In addition to the cephalic flexure the brain now shows two more flexures. At the level of the anterior part of the rhombencephalon the brain is bent with the convexity facing downward; this is the **pontine flexure.** The metencephalon with the pons varolii (after which the flexure is named) lies in front of the flexure and most of the medulla remains posterior to the flexure. A third flexure, with the convexity facing dorsally (the same as the cephalic flexure), appears at the junction between the medulla and the spinal cord. This is the **cervical flexure.**

About the middle of the second month after conception the flexures of the brain become much more distinct, especially the pontine flexure. The development of the cerebellum does not make much progress, but the midbrain enlarges very considerably, and attains its largest relative size. The main advance, however, is shown by the telencephalon; the two lobes indicated previously enlarge greatly and spread out forward, upward and backward, partially covering the laterodorsal surfaces of the diencephalon (Fig. 161, c).

By the beginning of the third month after conception the hemispheres of the telencephalon constitute by far the greatest part of the brain (Fig. 161, d). They have expanded backward to such an extent that they almost completely cover the diencephalon. A broad shallow groove on the outer surface of each hemisphere (the future lateral fissure) indicates the separation of the temporal lobe of the brain. The mesencephalon is greatly expanded dorsally and forms a large mass posterior to the cerebral hemispheres.

The cerebellum is the last part of the brain to become conspicuous on inspection from the outside, since the rudiments of the cerebellum are formed originally as masses of brain tissue bulging into the fourth ventricle from the sides of the metencephalon. These masses increase and later fuse together above the cavity of the brain, and only after this does the rudiment of the cerebellum swell to the exterior in front of the fourth ventricle. This occurs toward the end of the third month after conception.

At the age of four months after conception the cerebral hemispheres have grown so large that they cover the midbrain from the sides and touch the cerebellar hemispheres, which by this time have become clearly discernible. Even at this time the surface of the cerebral hemisphere, apart from the lateral fissure, is quite smooth, but during the second half of the period of pregnancy the surface becomes wrinkled and folded, giving rise to the characteristic gyri of the human brain.

Even more important than the changes in shape and size of the various parts of the brain are the processes in the substance of the nervous tissue which lead to the establishment of the functional mechanism of the nervous system. These processes will be dealt with in the discussion on growth and differentiation (section 16–2).

10–2   DEVELOPMENT OF THE EYES

The Optic Cup.   The origin of the optic vesicles has already been described. The optic vesicles push outward until they reach the epidermis, displacing the intervening mesenchyme, so that they come into direct contact with the inner surface of the epidermis. After this the external surface of the optic vesicle flattens out and invaginates inward, so that the vesicle is transformed into a double-walled, cuplike structure—the **optic cup** (Fig. 162). The invaginated wall of the optic cup is much thicker than the remaining external wall. The first is to develop into the **retina** of the eye, the second develops into the **pigment coat** of the eye (**tapetum nigrum**). The rim of the eye cup later becomes the edge of the pupil. The cavity of the optic cup is the future posterior chamber of the eye, filled by the vitreous body. The opening of the eye cup is very large at first, but later the rims of the cup bend inward and converge, so that the opening of the

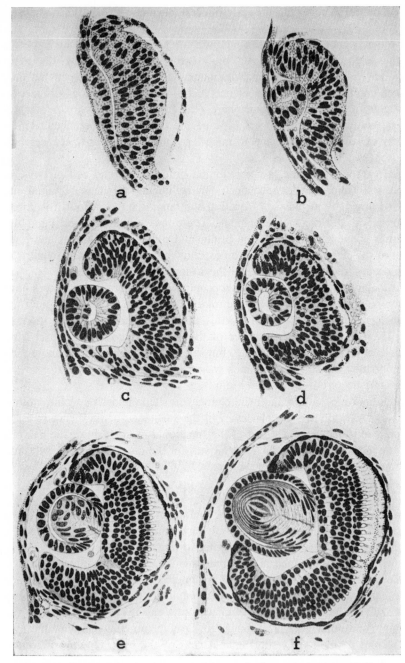

Fig. 162. Stages of the development of the eye in the axolotl. (After Rabl, from Spemann, 1938.)

pupil is constricted and reduced to its final relative dimensions. The rim of the optic cup surrounding the pupil becomes the **iris.** The constriction of the pupil does not take place equally all around the circumference of the eye, but it remains open longer on the ventral edge of the eye cup. A groove, the **choroid fissure,** remains here, cutting through the otherwise approximately circular edge of the eye cup and reaching inward as far as the optic stalk (Fig. 163). This fissure serves for the entry into the posterior chamber of the eye of a blood vessel and of mesenchyme cells which are found later in the vitreous body. The fissure normally closes during embryonic life.

The size of the optic vesicles relative to the rest of the prosencephalon may vary considerably in different vertebrates. Even in one order among vertebrates as in the frogs, it was found by direct measurement that the mass of cells used for the formation of the eye vesicles may, in different species, range from 10 per cent to 50 per cent of the volume of the prosencephalon (Balinsky, 1958). As a general rule the eye rudiments are large in bony fishes, in reptiles and birds, smaller in amphibians and relatively very small in mammalian embryos. The determination of the optic cup is due to the action of the underlying roof of the archenteron on a part of the neural plate. In the late neural plate stage the determination appears to be irrevocable, and parts of the optic vesicle cannot differentiate in any other way than by developing into retina, iris or pigmented epithelium of the eye. The eye rudiment may be excised and transplanted into any region of the embryo, and its development will continue more or less normally, producing a heterotopic eye.

A transplanted eye rudiment, although no longer capable of being transformed into other tissues, cannot develop normally unless it is surrounded by mesenchyme. Without mesenchyme in its environment the differentiation of the optic rudiment remains extremely poor (Holtfreter, 1939b; Lopashov, 1956).

The determination of the eye as a whole does not mean that all the parts of the eye rudiment are determined as well. The determination of the parts of the eye occurs much later. The eye rudiment in the neural plate stage and in the optic vesicle stage may be split into two, and each half

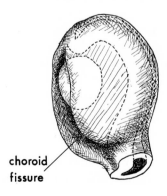

choroid fissure

Fig. 163. Eye cup, showing the position of the choroid fissure in respect to the lens and the eye stalk. (After Froriep, from Korschelt, 1936.)

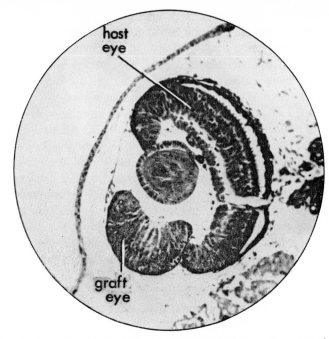

Fig. 164. A whole eye, developed from a transplanted piece of presumptive pigment coat. (From Dragomirow, 1933.)

develops into a complete eye. This can best be demonstrated if a part of the eye rudiment is transplanted. It is found then that the remaining and the transplanted part each develop into a small eye. Experiments have also been performed on the eye rudiment at the stage when the optic vesicle is being transformed into an optic cup, and the two parts of the cup, the future retina and the future pigment coat, become morphologically distinguishable. A piece of the presumptive pigmented epithelium may be excised and transplanted into the vicinity of a normal eye of another embryo, and it develops into a complete eye, consisting both of a pigment coat and a retina (Fig. 164) (Dragomirow, 1933). However, a prerequisite for this is that the piece be not too small; very small fragments of the eye rudiment usually develop into pigmented epithelium only.

The ability of a part of the organ rudiment to develop as a whole is reminiscent of a similar ability in the early stages of cleavage in some animals. The process observed in both cases is known as **self-regulation.** It is found in many organ rudiments, and, in fact, can be considered as a common property of organ rudiments. The ability to self-regulate presupposes that the parts of the rudiment (or of the egg in the early cleavage stages) are not determined.

In the case of the eye cup the absence of determination of parts can be demonstrated also in another way. If a suitable inductor is applied to the outer surface of the optic cup, that is, to the surface which normally differ-

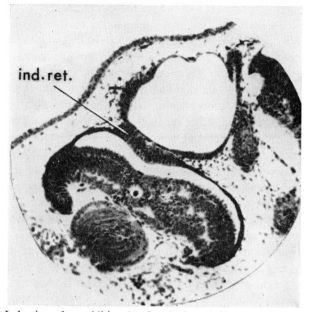

Fig. 165.   Induction of an additional retina (*ind. ret.*) in an eye by means of contact with an ear vesicle. (From Dragomirow, 1936.)

entiates into the pigmented epithelium, it may be induced to develop as retina, so that the eye has two retinas, a normal one and an additional one. The latter is never as large as the normal one. A suitable inductor for this purpose is the sensory epithelium of an ear (Fig. 165). The experiment therefore consists in the transplantation of an optic vesicle into the immediate vicinity of the ear vesicle to ensure the immediate contact of the epithelia of the two organ rudiments (Dragomirow, 1936).

With advancing development it is apparent that the rim of the optic cup becomes increasingly different from the deeper lying parts. The constriction of the pupil, referred to above, takes place at the expense of a considerable thinning out of the wall of the rim of the optic cup. The thinned out portion becomes the iris of the eye, while the remaining part, which stays considerably thicker, gives rise to the retina proper.

In the iris large amounts of pigment are deposited in the outer epithelial layer (this layer is actually a part of the pigment coat of the eye). In addition to cells carrying pigment this layer also gives rise to the smooth muscle fibers of the sphincter and dilator muscles of the iris. In the retinal layer the cells start differentiating into the sensory and nerve (or ganglion) cells. The first trace of this differentiation is seen in the arrangement of the nuclei of the cells in several layers, of which the nuclei situated in the innermost layer (nearest to the pigment coat) belong to the future rod and cone cells. The rudiments of the rods and cones appear as cytoplasmic processes on the inner ends of these cells. The remaining nuclei, arranged in two or more layers nearer to the cavity of the eye cup, give rise to the various types of intermediate and ganglion cells of the retina. Nerve processes, arising from the ganglion cells of the retina, grow out toward the brain, and the path which they take is along the stalk of the eye cup. In this way the stalk of the optic cup becomes transformed into the optic nerve. On reaching the floor of the diencephalon the nerve fibers do not enter the same side of the brain, but cross over to the opposite side and there penetrate into the wall of the diencephalon and the mesencephalon. Where the nerve fibers of the two eyes cross and bypass each other on their way to the contralateral parts of the brain arises the **optic chiasma.**

The optic cup, even when fully differentiated, is not yet a complete eye. Certain accessory structures have to be added to it to make the eye fully functional. The most important of these structures is the **lens,** which serves for the refraction of the rays of light entering the eye. The lens is not developed from the optic cup, but from the epidermal epithelium with which the optic vesicle comes in contact, as mentioned before. As the outer wall of the optic vesicle begins to invaginate to become the retinal layer of the optic cup, a thickening appears in the epithelium which is in contact with the invaginating part of the optic cup (Fig. 162, a). This thickening is the rudiment of the lens.

The Lens.   The way in which the lens rudiment is separated from the remainder of the epidermis varies in different classes of vertebrates. In birds and mammals the epidermal thickening folds in to produce a pocket which is for a short time open to the outside, and later a vesicle lying in the opening of the iris (in the pupil). In amphibians and bony fishes the thickening in the formation of which only the inner layer of epidermis takes a part is nipped off from the epidermis as a solid mass, but later the cells of this mass rearrange themselves into a vesicle. In both cases the vesicle must undergo further differentiation before the lens can function as a refracting body. This happens in such a way that the cells on the inner side of the lens vesicle elongate, become columnar at first and later are transformed into long fibers. During this transformation the nuclei of the cells degenerate and the cytoplasm becomes hard and transparent. The fibers are arranged in the lens in a very orderly way, forming the spherical or ellipsoid refracting body of the lens. Part of the lens epithelium remains unchanged and covers the sphere of fibers distally. The junction between the unchanged lens epithelium and the mass of fibers is the growth point of the lens; here the epithelial cells are continuously transformed into fibers, so that the refracting body grows by apposition of new fibers.

Between the development of the optic cup and the development of the lens there exists, in most of the vertebrates studied experimentally, a direct causal relationship; the development of the lens is dependent on an induction from the optic vesicle. As the optic cup touches the epidermis, it gives off a stimulus of some kind, which causes the epidermis cells to develop into the lens rudiments (Lewis, 1904). Any epidermal cells are able to react to the induction of the optic vesicle, and without this induction the lens does not develop at all or at least the development is defective. The dependence of the lens development on the action of the optic cup can be shown by several types of experiments. One type of experiment is to remove (excise) the eye rudiment before it can reach the epidermis. Such an operation usually leads to the absence of the lens. The other experiment is to remove the epidermis which normally would have formed the lens, and replace it with a piece of epidermis taken from another part of the body, from the head or even from the belly. In this experiment it was observed that the epidermis, if in contact with the optic vesicle, develops into a lens. A third type of experiment is to transplant the optic vesicle, without the epidermis normally covering it, under the epidermis in an abnormal position. In this case the local epidermis may be caused to develop a lens (see Mangold, 1931b).

For a long time it was believed that an intimate contact between the optic vesicle and the epidermis is indispensable for lens induction. A thin layer of cellophane inserted between the optic vesicle and the epidermis in a chick embryo completely stopped the inducing action of the optic vesicle (Mc-Keehan, 1951). Insertion of a porous membrane, which presumably allowed for the passage of macromolecules, between the eye vesicle and the epidermis in a frog embryo also precluded lens induction (de Vincentiis, 1954). However, recently experiments have been reported (McKeehan, 1958) in which a partial screening of the epidermis from the eye vesicle in a chick embryo by a thin slice of agar did not prevent the complete development of the lens. The inducing agent could thus get either through or around the agar. This would suggest that the inducing agent is a chemical substance. Apart from this, not much is known as to the nature of the stimulus responsible for the induction of the lens. It has not been possible to extract a lens-inducing substance from the optic vesicle. On the other hand, some of the "abnormal" inductors of neural plates are known to induce "free" lenses, that is, lenses without an eye. The thymus of the guinea pig seems to be especially suitable for this purpose (Toivonen, 1940 and 1945). That the stimulus in this case is exactly the same as that exercised by the optic cup remains to be proven.

There is one peculiar similarity between the induction of the neural plate and the induction of the lens; it concerns the distribution of the cytoplasmic ribonucleic acid in the components participating in the process. The eye vesicle at the time when it comes into contact with the presumptive lens epidermis contains a large amount of ribonucleic acid. At the same time the presumptive lens epidermis has little ribonucleic acid, and in this respect it is no different from the rest of the epidermis. After the contact is established, the ribonucleic acid in the cells of the eye vesicle is found to be concentrated near the outer margin of the cells, that is, where the cells touch the epidermis. Large amounts of ribonucleic acid now appear also in the presumptive lens cells, at first only at their proximal ends, where they are in contact with the eye vesicle, but later it is also found in the outer parts of the epidermal cells. In subsequent stages the ribonucleic acid content in the retinal cells decreases, while it continues to increase in the cells of the lens rudiment (Mc-

Keehan, 1956). It would be attractive to conclude that ribonucleic acid actually passes from the eye vesicle cells into the presumptive lens cells, though of course this is not the only possible interpretation of the above observation. It is probably safer to say that in the process of induction the inducing part uses up its ribonucleic acid, while the increase of ribonucleic acid content is one of the changes brought about in the part reacting to induction. This would be applicable to both the lens induction by the optic vesicle and the neural plate induction by the roof of the archenteron (*cf.* section 6–3).

The relation of the lens development and the optic cup development is complicated rather considerably by the fact that in a few amphibians (*Rana esculenta, Xenopus laevis* and others to a lesser extent) (Spemann, 1912a; Balinsky, 1951) the lens shows a certain degree of independent development ("self-differentiation") even in the absence of the optic cup, that is, when the optic cup has been previously removed. The degree of independent development varies from a tiny nodule of epidermal cells to a rather typical lens with fiber differentiation. The development is never completely normal, as lenses without eyes undergo a far-reaching degeneration once the initial stages of development have been passed. Nevertheless the independent development of lenses shows that the eye cup is not the only part which may be involved in lens development. The experiments in which free lenses have been induced by abnormal inductors (thymus) suggest that the "self-differentiation" of lenses, where it occurs, is due to some other influence on the differentiation of the epidermal cells, and that this influence is responsible, in some species, for "independent" lens development (that is, lens development which is independent of the eye cup). What these influences might be, may be concluded from some experiments in which belly ectoderm was not capable of developing a lens when transplanted just before the formation of the optic vesicle, but was able to react if transplanted in the neurula stage. It follows that to be able to react the epidermis must be in position some time before the contact with the optic vesicle is established. During this time the head mesoderm lies immediately underneath the presumptive lens epidermis, and it was concluded that the influence of the head mesoderm is the factor which prepares the epidermis for the subsequent induction emanating from the optic vesicle (Liedke, 1951, 1955). Presumably, in some species, the "preparation" goes so far that the development of the lens may start even if the eye vesicle is not present.

Our conclusion may perhaps be framed in another way: we may say that the development of the lens is a result of induction emanating from two sources, from the head mesoderm and from the eye vesicle. Normally both are necessary for successful lens development. There is some evidence that the relative importance of the two inductors may be changed by environmental factors; keeping the developing embryos at a low temperature seems to favor the induction by the head mesoderm, so that the induction by the eye vesicle becomes unnecessary (Ten Cate, 1953; Jacobson, 1955).

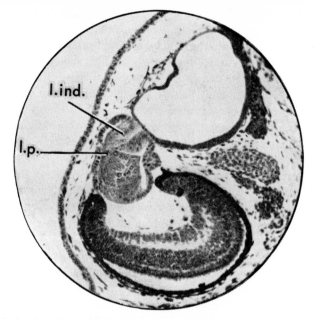

Fig. 166.   Induction of an additional mass of lens fibers (*l. ind.*) by means of contact with an ear vesicle. *l.p.,* Primary lens fibers. (From Dragomirow, 1929.)

The differentiation of the lens cells into lens fibers in normal development is caused by the same induction as the development of the lens as a whole. It has been proved that the contact with the presumptive retinal layer of the eye can induce lens fiber differentiation. The epithelial cells of the lens rudiment are all capable of fiber differentiation if they are exposed to a suitable stimulus. The contact with retinal tissue or with the sensory epithelium of an ear vesicle (Fig. 166) (compare what has been said on the induction of the retina) may cause the formation of an additional mass of lens fibers, so that under experimental conditions lenses may develop having two independent masses of fibers (Dragomirow, 1929).

Accessory Structures.   The other accessory structures of the eye which are present in all vertebrates having functional eyes are the **choroid coat,** the **sclera** and the **cornea.** The choroid coat and the sclera develop from mesenchyme, accumulating around the eyeball, in the way mesenchyme cells accumulate around many organs giving rise to their connective tissue capsules (see section 9–2). In the case of the eye, the interior layer of mesenchyme cells gives rise to a network of blood vessels surrounding the pigment epithelium. The outer layer of mesenchyme forms a fibrous capsule around the eye, which serves for its protection and for the insertion of the eye muscles. The capsule may either remain fibrous or develop cartilage or even bone (in reptiles and birds).

The cornea originates in part from mesenchyme but the epidermal epithelium also plays an essential role in its formation. The connective tissue

part of the cornea is continuous with the sclera, while the corneal epithelium
is continuous with the skin epidermis or with the epithelium of the eyelids
where such are present. Both the epithelium and the connective tissue of the
cornea become transparent, so that the rays of light may enter the eye. The
development of the cornea can easily be traced in living amphibian embryos.
Initially the epidermis covering the eye is pigmented, for the epidermal cells
contain granules of pigment derived from the egg. In the cells of the presumptive cornea these pigment granules become dissolved, and later when
the chromatophores develop in the connective tissue of the skin, the cornea
remains free of them.

The transformation of the skin into cornea is caused by an induction, the
source of which is the eyeball. This can be proved by transplanting the eyeball
heterotopically, or by replacing the normal cornea by skin from another part
of the embryo. The stimulus can be given off by both the eye cup and the
lens; if the lens alone is transplanted, the epidermis over it loses its pigment
and differentiates as cornea. If the eye is removed, the cornea does not
develop at all (Spemann, 1901; Fischel, 1919; Mangold, 1931b).

The induction of the cornea presents an interesting peculiarity as compared with the neural plate and the lens induction in amphibians. The competence to differentiate as cornea is found in the skin not only during a short
period of embryonic development, but for a long time, long after the normal
differentiation of the cornea has taken place. Also the eyeball retains its
inductive ability for a long time, probably permanently. What is more, the
persistence of the cornea is dependent on the continuous presence and influence of the eyeball. If, in a late amphibian larva or an adult, the eye is
removed, the cornea soon loses its transparency, it is invaded by chromatophores and becomes more or less normal skin. On the other hand a fully differentiated piece of skin will lose its chromatophores and become transparent
cornea if it is transplanted over the eye.

In the development of the eye induction takes place repeatedly and some
parts, after having been induced, become themselves a source of inducing
stimuli. A whole chain of inductors can thus be made out:

The roof of the archenteron induces the neural plate, and as part of the
neural plate it induces the eye cup rudiment.

The eye cup rudiment, becoming the optic vesicle, induces the lens (acting
together with head mesoderm).

The lens induces the cornea (acting together with the optic cup).

Parts developing as a result of induction, and inducing in their turn, may
be called secondary, tertiary, etc., inductors, or organizers of the second
grade, third grade, and so forth.

10–3   THE FATE OF THE NEURAL CREST CELLS

The neural crest at the time of its formation is represented by a mass of
loose cells lying dorsally to the neural tube. Almost at once after the formation of the crest, the cells of which it consists start migrating in a lateral and

ventral direction from the place of their origin. As they move, the neural crest cells form into streams, bypassing as they go certain organs (viz., the eye, the gill pouches) (Fig. 167). These streams of neural crest cells are especially conspicuous in the head and neck region, while in the trunk region the neural crest cells are more scattered, right from the start (Stone, 1926; Raven, 1931). The neural crest cells move mainly along the inner side of the epidermis, between it and the layer of mesoderm; they penetrate, however, into the interstices between the neural tube and the inner surface of the mesoderm as well. Sooner or later the most advanced neural crest cells reach the midventral line of the body. Not all of them, however, travel as far as this; rather, they become spread all along the path, some of the cells even retaining their original position dorsal to the neural tube.

The migration of the neural crest cells is a complicated process, dependent on many factors. One of the factors seems to be a repulsion of the migrating

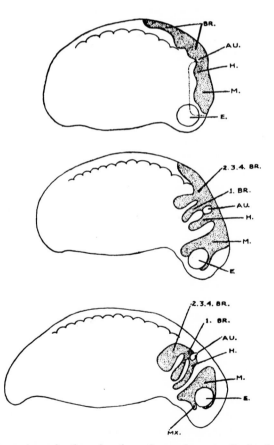

Fig. 167.   Three stages in the migration of neural crest cells in the salamander *Ambystoma punctatum. AU.,* Ear vesicle; *BR.,* branchial neural crest cells (numbers refer to particular branchial arches); *E.,* eye; *H.,* hyoid neural crest cells; *M.,* mandibular neural crest cells; *MX.,* maxillary neural crest cells. (From Stone, 1926.)

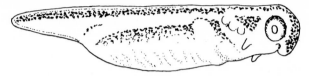

Fig. 168. Larva of the newt *Triturus cristatus,* with longitudinal stripes of pigment. (From Balinsky, 1925.)

cells from one another due to the substances produced by the metabolism of the cells. If the neural crest cells are cultivated *in vitro,* the cells spread evenly in the medium available (Twitty and Niu, 1948). In the organism, however, the even distribution of neural crest cells is disturbed by the surrounding tissues and organ rudiments which lie in the path of the migrating cells. Certain organ rudiments block the way of the migrating cells (as mentioned above) and are bypassed. Other rudiments seem to attract the neural crest cells in some way, or keep them fixed once they have reached a certain position. So the neural crest cells are held along the upper edges of the somites and also along the upper edge of the lateral plate. Later, when part of the neural crest cells differentiate into chromatophores, these accumulations of neural crest cells become conspicuous as longitudinal strips of pigment (Fig. 168) (Twitty, 1949).

The migration of neural crest cells can be observed in several different ways. The neural crest cells being ectodermal cells are, in amphibians, distinguishable from the mesoderm cells by their smaller content of yolk granules and by a greater amount of pigment derived from the egg. Another way of tracing the neural crest cells is by means of local vital staining. If the staining is applied to the neural folds, the neural crest cells derived from these can readily be seen against the background of unstained ectoderm and mesoderm. A third method is based on the ability of the neural crest cells to differentiate into melanophores. By isolating parts of the embryo in sufficiently early stages and cultivating them in suitable surroundings, it is possible to prove that without the neural crest cells no pigmentation can develop in the skin or in other organs and tissues as well. If, however, a part of the embryo is isolated and transplanted, after the migrating neural crest cells have reached this part, the pigment cells later differentiate in the graft. The actual differentiation of pigment cells occurs after the migration has been completed; while on the move the neural crest cells do not differ in their pigment content from other cells, and this of course makes it difficult to observe their migration directly. This last method of investigating the migrations of neural crest cells has been applied in mammals and birds in which the first two methods are not feasible (Rawles, 1948).

In the fishes and amphibians the pigment cells are found predominantly in the connective tissue—in that of the skin, but also in the peritoneum, in the walls of blood vessels and elsewhere. In the birds and mammals the pigment is found predominantly in the epithelial derivatives—the hairs and the feathers. Nevertheless the production of the pigment is also due to the activity of the neural crest cells. The latter penetrate into the hair and feather follicles and deposit the pigment granules in the hairs and the feathers as they grow out of the follicles. If the access of the neural crest cells to the hair and feather follicles is precluded, the hairs and feathers may develop normally, but they are completely devoid of pigment (Rawles, 1947). For further information on the role of neural crest cells in the pigmentation of feathers in birds, see also Willier (1952), "Cells, feathers and colors."

Besides the pigment cells, other types of cells are also differentiated from the neural crest cells.

The visceral skeleton is almost completely developed from neural crest cells. The visceral arches occupy approximately the same position as the streams of neural crest cells in the early embryo. The mass of neural crest cells behind the eye, that is, between the eye and the first branchial pouch, becomes the mandibular arch, the upper part of which becomes differentiated as the **quadrate,** and the lower part gives rise to the **Meckel's cartilage** or mandible proper (Fig. 169). The mass of neural crest cells between the first and second branchial pouches becomes the hyoid arch, the next becomes the first branchial arch and so forth. The mass of neural crest cells moving downward in front of the eye contributes to the formation of the anterior half of the trabeculae of the skull (Fig. 169) (Stone, 1926; Raven, 1931; Hörstadius and Sellman, 1946). The neural crest cells of the trunk region, on the other hand, do not participate in the development of skeletal tissues (Raven, 1936), both the axial skeleton (vertebral column) and the limb skeleton being derived from the mesoderm. Rather peculiarly, one element of the visceral skeleton in the amphibians, the second basibranchial, is also of mesodermal origin.

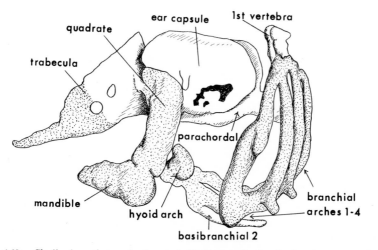

Fig. 169.   Skull of a salamander larva, indicating parts developing from neural crest material (stippled) and from mesoderm (white). (From Stone, 1926.)

The papillae of the teeth in urodele amphibians have been shown to be derived from neural crest cells (de Beer, 1947). It is highly probable that the papillae of the teeth in all other vertebrates are of the same origin.

It has been a matter of controversy whether the ganglia of the peripheral nervous system are developed from neural crest cells. At present it is accepted that the spinal ganglia are built up by neural crest cells of the trunk, but that the ganglia of the cranial nerves (V, VII, IX and X) are developed from thickenings of the epidermis—the **placodes.** Besides the spinal ganglion cells, the neural crest contributes to the development of the nervous system by supplying material for the ganglia of the sympathetic nervous system (recent confirmation by Nawar, 1956) for the sheaths of the nerves (Schwann cells) and the meninges (at least the pia mater and arachnoidea) (Piatt, 1951). Lastly, the neural crest cells are found to differentiate as subcutaneous connective tissue, although here they are joined by mesenchyme cells derived from the mesoderm.

As the neural crest cells may take such divergent paths of differentiation, the question arises whether the fate of individual crest cells is determined by environmental influences, or whether these cells already differ at the time when they leave the neural folds. Apparently both the above alternatives are partially true. If different parts of the neural fold are explanted in a culture medium (Niu, 1947) or transplanted to the side of an embryo (Hörstadius and Sellman, 1946; see also Hörstadius, 1950), different results are obtained depending on the area from which the neural fold has been taken. Pieces of cranial neural folds under these conditions produce only small numbers of melanophores but they give rise to cells which may develop into cartilage. Neural folds of the trunk region, when explanted or transplanted, give rise to very numerous melanophores, but no pro-cartilage cells are given off. On the other hand, it has been shown that cartilages develop from neural crest cells only when they are induced to do so by adjoining tissues. Pieces of cranial neural crest were cultivated in an epithelial vesicle either alone, or together with other tissues such as neural plate, notochord, foregut endoderm, midgut endoderm and lateral mesoderm. Under these conditions cartilage developed from neural crest cells only when they were cultivated together with foregut endoderm. Trunk neural crest under the same condition produced only melanophores and mesenchyme and no cartilage (Okada, E. W., 1955). It follows that

(1) only cranial neural crest is competent to produce cartilage;
(2) it can do so only under the influence of foregut endoderm (see also section 12–3).

There is some evidence (Stevens, 1954) that among the pigment cells in amphibians the two types, i.e., melanophores and guanophores, are already distinct while the cells are migrating from the site of their origin—the neural folds.

10–4 THE FATE OF THE EPIDERMIS AND THE STRUCTURES DERIVED THEREFROM

When the epidermis is first segregated from other parts of the ectoderm (neural plate, neural crest) during the process of neurulation, it is still a very complex rudiment. Most of it becomes the epidermis of the skin, but a number of other structures besides are derived from it. Some of these have been mentioned already: the lens, the cornea and the cranial ganglia.

The epidermis itself gives rise to quite a large number of special differentiations, such as various unicellular and multicellular skin glands, including the sweat glands and the sebaceous glands, the hairs, feathers and scales, and various other special structures derived from the above. The development of some of these parts falls under histogenesis rather than organogenesis, and will not be dealt with here.

In the early embryo the epidermis is a layer of epithelium. In amphibians the epithelium consists of two rows of cells: the outer **covering layer** or **periderm** and the inner so-called **sensory layer.** The latter name is given not because the layer as such has nervous functions, but because some sensory organs are derived from parts of this layer. In birds and mammals the epidermis of early embryos consists originally of one layer of cells, i.e., is one cell thick. Only at a later stage do an inner and an outer layer of cells become differentiated, the inner layer becoming the **generative** (or **Malpighian**) layer of the epidermis. The skin is then composed of the epidermis and the layer of mesenchyme partly derived from the neural crest and partly from the dermatomes (see p. 286) which gives rise to the **dermis.**

The Placodes.   Many structures derived from the epidermis make their first appearance in the form of plate-shaped thickenings of the epidermal epithelium. Such thickenings have been called **placodes.** That the ganglia of the cranial nerves are derived from placodes has been stated above. When the lens rudiment first appears as a thickening of the epidermis it bears a great similarity to the other placodes. A pair of placodes, appearing in front of the anterior end of the neural plate and probably deriving their material from the neural fold itself, develop into the olfactory sacs. A placode appearing against the side of the hindbrain invaginates and produces a vesicle which is eventually separated from the epidermis. This is the ear vesicle, the rudiment of the internal ear (the ear labyrinth). The placode from which the ear vesicle is developed is the **auditory placode.** Parts of the epidermis adjoining the auditory placode also become thickened, and from these placodes develop the **lateral line sense organs.** In the aquatic vertebrates (fishes, aquatic larvae of amphibians), the lateral line organs are distributed over the head (in several rows), and a row of the same organs stretches backward along the side of the entire body and tail. Wherever the lateral line organs are found, the cells of which they consist come from the placodes of the ear region.

In amphibians the backward migration of the lateral line organ cells could

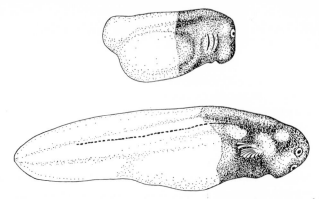

Fig. 170. Migration of cells of the lateral line rudiment, demonstrated by uniting the anterior half of an embryo of *Rana sylvatica* and the posterior half of an embryo of *Rana palustris*. (After Harrison, from Weiss, 1939.)

be demonstrated in a grafting experiment. Two embryos belonging to different species of frogs were cut transversely in halves, and the anterior half of a darkly pigmented species (*Rana sylvatica*) was grafted on the posterior half of a lightly pigmented species (*Rana palustris*). It could be subsequently observed how the darkly pigmented cells of the lateral line rudiment of the anterior half migrated out into the posterior half and along the trunk and tail (Fig. 170) (Harrison, 1904).

In the case of lateral line sense organs, the pathway of the migrating cells is dependent on the surroundings through which they migrate. If the anterior half of a frog embryo is transplanted onto another embryo whose own lateral line organ rudiment had been removed previously, the lateral line cells of the anterior half grow out into the second embryo, and once they have reached the path taken normally by the lateral line cells in their migration, they start moving along this path, even though their new direction is at an angle to the one they had been following before (Fig. 171). This shows that the path of

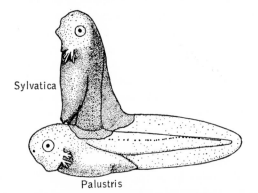

Sylvatica

Palustris

Fig. 171. Migration of cells of the lateral line rudiment in an experiment in which the anterior part of a *Rana palustris* embryo was grafted on to the back of a *Rana pipiens* embryo. (After Harrison, from Huxley and de Beer, 1934.)

migration is determined by a factor lying outside the migrating cells themselves (Harrison, 1904). It has been shown that the path of migration in this case is determined by the mesoderm.

The development of the placodes is probably always dependent on a stimulus from the tissues situated under the epidermis. This has been proved in some cases, as in the case of the lens of the eye. The ear is also dependent in its differentiation; this follows from the experiments in which the ear vesicles have been induced heterotopically as the result of the transplantation of various inductors. The ear vesicle is among the structures that are very often induced when the primary organizer is transplanted, but also can be induced in experiments with transplantation of adult tissues and of parts of the neural plate and neural tube (Guareschi, 1935; Gorbunova, 1939; Kogan, 1939). The latter experiments suggest that in normal development the ear vesicle is induced by the medulla oblongata. This, however, cannot be the sole inductor, as the ear vesicles may develop in their normal position after the medulla oblongata is removed at an early stage, either alone or together with most of the central nervous system rudiment. It is concluded, therefore, that the development of the ear vesicle is dependent on multiple induction, that similar stimuli are emitted both from the medulla oblongata and from the mesoderm developing from the roof of the archenteron (Harrison, 1935; Albaum and Nestler, 1937).

It is probable that the induction of the ear vesicle proceeds in two stages: first the presumptive ear ectoderm is acted upon by the underlying mesodermal mantle in the late gastrula and early neurula stages, and later the determination is finally stabilized by the influence of the medulla which, as the result of the closure of neural folds, comes into close contact with the epidermis in the ear region (Yntema, 1950). We have seen a similar case in the "independent" development of the lens in some species of frogs (p. 266).

A similar duplication in the sources of induction has been postulated for the nose rudiment, namely an earlier induction by mesoderm and a later induction by the forebrain. It has been claimed, for instance, that the anterior portion of the neural plate if transplanted under the epidermis on the flank may induce a nose rudiment locally. In more careful experiments it was found, however (Zwilling, 1940; Schmalhausen, 1950), that nose rudiments may develop in the absence of brain tissues, and so the alleged inductions must be due to the nose rudiment material being grafted together with the presumptive forebrain. The two rudiments are thus induced simultaneously by the underlying roof of the archenteron and initially lie very close to one another.

The Olfactory Organ.   Although the nose rudiments seem to be determined at a very early stage (late gastrula) they first become discernible morphologically after the closure of the neural tube, in the form of two thickenings of the epidermis, the **olfactory placodes,** just anterolaterally of the hemispheres of the telencephalon. The central part of each placode becomes invaginated, and the olfactory placode thus becomes converted into an olfactory sac which is open to the exterior by the external naris. Parts of the wall of the olfactory sac, especially the dorsal and lateral wall, are differentiated as olfactory epithelium. The primary sensory cells of the olfactory epithelium develop on their proximal ends nerve processes (axons) which converge to form the olfactory nerve. The olfactory nerve grows into the adjacent wall of the telencephalon, bridging the narrow gap between the olfactory organ and the brain.

In most fishes the olfactory organ retains essentially the same structure in the adult state, but in the group of Choanichthyes among the fishes and in the terrestrial vertebrates the structure of the olfactory organ is further complicated by the development of the internal nares, and in the mammals also by the naso-lacrimal duct.

The internal nares or **choanae** arise by a perforation of the nose sac cavity into the oral cavity. The actual perforation is preceded by the formation from a part of the ventral wall of the nose sac of an elongated tube stretching backward and down toward the oral cavity. This tube is the nasal canal, in which the epithelium lining the canal becomes thin and is thus fairly sharply segregated from the thicker epithelium giving rise to the sensory parts. With the elongation of the nasal canal the sensory part appears to be a dorsolateral outgrowth of this canal, although the sensory part is actually the older portion of the olfactory organ. The internal (posterior) end of the nasal canal fuses eventually with the epithelium lining the oral cavity and the intervening membrane becomes perforated as internal nares. The **Jacobson's organ,** where it is present, is another section of the olfactory sac retaining sensory function. It develops from a medioventral part of the sac.

The development of the naso-lacrimal duct will be discussed together with the development of the face (in section 12–2).

The Ear. The **auditory placode,** from which the internal ear is developed, initially shows considerable similarity to the olfactory placode, and is also converted by invagination into a saclike structure (Fig. 172), but in the early stages there already are important differences. Whereas in the amniotes the whole epidermal layer is involved in the formation of the auditory placode, and it later invaginates to form a sac which is, at least temporarily, open to the exterior, in the amphibians the auditory placode is formed by the thickening of the interior "sensory" layer of the epidermis, while the external, covering layer is not involved at all. As a result, when the placode invaginates, there is no opening or pit on the surface of the skin. In both cases, however, the opening of the sac becomes constricted and closed, so that eventually the rudiment of the ear takes the form of a completely closed vesicle—the **ear vesicle.** In the bony fishes the auditory organ is formed not by invagination but as a solid mass of cells on the inner surface of the epidermis, and is hollowed out secondarily.

The ear vesicle is the rudiment of the most essential part of the internal ear—the **labyrinth.** When first formed, it is somewhat pear-shaped, the pointed end directed upward. This pointed end later gives rise to the endolymphatic duct. Soon the ear vesicle starts expanding, pushing away the surrounding loose mesenchyme. Parts of the wall of the vesicle become very thin; the epithelial cells become flat. These will be the membranous areas of the labyrinth. Other parts, in particular parts of the medioventral wall of the vesicle, remain thick or even become thicker; the cells in these areas become columnar and give rise to patches of sensory epithelium which form the maculae of the internal ear. Even before it is subdivided into membranous and sensory parts, the ear vesicle gives off on its median surface a group of cells which become the acoustic ganglion (ganglion of nerve VIII) (see Campenhout, 1935).

The expansion of the ear vesicle is unequal so that the vesicle becomes constricted in some places and bulges out in others. As a result the shape of the organ becomes increasingly complicated so that it eventually deserves its name—the labyrinth (Fig. 173). The sacculus is subdivided by a constriction from the utriculus. The utriculus becomes drawn into three mutually perpendicular folds—the rudiments of the semicircular canals. The sides of the folds eventually stick together and become perforated while parts of the original cavity along the edges of the folds remain open and become the semicircular canals, opening at both ends into the cavity of the utriculus. A hollow outgrowth of the sacculus forms the rudiment of the lagena in lower vertebrates, and in higher vertebrates this outgrowth becomes very elongated and coiled to give rise to the cochlea. As the ear vesicle changes its shape

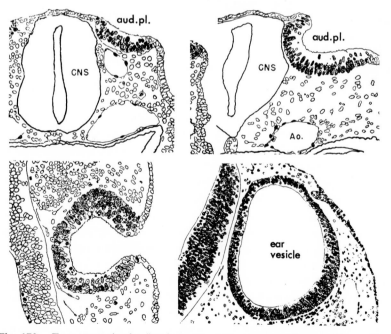

Fig. 172.   Four stages in the development of the inner ear from auditory placode —*aud. pl.*—to completed ear vesicle in a human embryo. *CNS,* Central nervous system; *Ao.,* dorsal aorta. (From Streeter, 1942, 1945.)

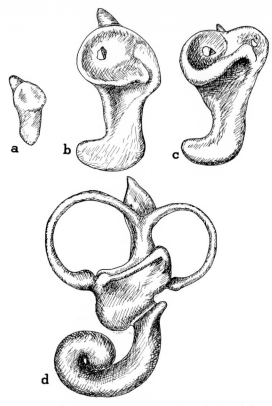

Fig. 173.   Development of the labyrinth from the ear vesicle in a human embryo.
(Modified from Streeter, 1906.)

and produces the various parts of the labyrinth, the sensory areas become
subdivided and further differentiated until each of the maculae have taken
up their final position in the fully developed labyrinth.

As the ear vesicle expands to produce the labyrinth, it becomes surrounded by mesenchyme cells which later give rise to cartilage and produce
the cartilaginous ear capsule which surrounds and protects the inner ear.
There is a direct causal relationship between the ear vesicle and the development of the ear capsule; if the ear vesicle is removed, the ear capsule does
not develop; if a foreign ear vesicle is transplanted in the tailbud stage, the
local mesenchyme cells may aggregate around it and produce an additional
cartilaginous capsule (Lewis, 1907). The mesenchyme that is used for the
ear capsule is of mesodermal origin, derived from the sclerotomes (see section 11–1). Mesenchyme of neural crest origin, as well as subcutaneous
mesenchyme, is apparently not capable of reacting to the induction by the
ear vesicle. As a result an ear vesicle transplanted heterotopically does not
always cause a good capsule to be developed around it. The most complete
capsules develop around ear vesicles transplanted to the immediate vicinity
of the normal ear, between the ear and the eye, where the grafted vesicle

can draw on the same supply of mesenchyme as the normal ear vesicle. However, the sclerotome mesenchyme of the trunk, that is, the mesenchyme giving rise to the cartilages of the vertebrate column and the ribs, reacts to the ear vesicle by forming large masses of cartilage, which may partially surround the grafted ear vesicle (Balinsky, 1925; Syngajewskaja, 1937).

The concentration of mesenchyme on the surface of the ear vesicle may be partly the result of its expansion, which would lead to the mesenchyme being compressed against its surface. However, there is no doubt that mesenchyme cells may travel considerable distances to reach the ear vesicle and to invest it with cartilage. In experiments of transplantation of the ear vesicle to the trunk region it can be seen that thick bars of cartilage grow out from the vertebral column to the ear vesicle, presumably indicating the pathway which had been followed by the mesenchyme. In the same experiments it can also be seen that the total amount of cartilage in the area is greatly increased, so that the ear vesicle must either stimulate the proliferation of skeletogenic mesenchyme, or increase the proportion of mesenchyme cells which become chondroblasts. The reverse occurs in the case of the removal of the ear vesicle: not only does the ear capsule not develop, but there is no superfluous cartilage in the area; in the absence of the ear vesicle the proliferation of the pro-cartilage cells is short of the normal, or else the cells that should have become cartilage cells differentiate along other paths.

The development of the ear gives us another example of a chain of inductions. The primary inductor—the roof of the archenteron, consisting of presumptive chordo-mesoderm—causes the development of the hindbrain. This as a secondary inductor stimulates the development of the ear vesicle (in conjunction with the direct action of the mesoderm on the presumptive ear ectoderm). The ear vesicle, as a tertiary inductor, causes the formation of the cartilaginous capsule.

One might have expected that the development of the middle ear would be related to the development of the inner ear, but this is not the case. The middle ear, consisting of the Eustachian tube, the ear ossicles (columella in the frog) in the cavity, and the tympanic membrane, develops normally after the removal of the ear vesicle. This may be because the middle ear is derived from the branchial apparatus, which is a very essential part of the vertebrate organization, deeply rooted in the basic mechanism of vertebrate development and thus not in need of stimulation from the inner ear. In this case the functional apparatus—the organ of hearing—is made up of two parts not causally connected in development, but linked together only through the medium of their definitive functioning (see Yntema in Willier, Weiss and Hamburger, 1955).

The Fin Fold.   Besides the organs developing from placodes, a number of structures are developed from the epidermis which in their early stages can be classed as "outgrowths," or more correctly as protrusions. These are: the unpaired fin fold, the external gills (in aquatic vertebrates), the "balancer" in the larvae of salamanders, and the paired limbs.

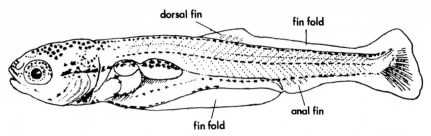

Fig. 174. Development of the unpaired fins from parts of the fin fold in a fish larva. (From Balinsky: Proc. Zool. Soc. London, *118*, 1948.)

The unpaired fin fold is a structure found in all fishes and in the larvae of amphibians. It is a vertical fold of skin, which starts in the posterior head region or the anterior trunk region, and stretches backward all along the back and dorsal side of the tail, bends over to the ventral side of the tail and can be traced forward, along the ventral side of the tail and the belly to the middle of the trunk. The fold consists of epithelium and connective tissue. In fishes, parts of the fold are invaded later by skeletogenous tissue which produces fin rays, thus transforming these parts into the unpaired fins of the adult (Fig. 174). Parts of the larval fin fold in between the unpaired fins of the adult disappear. In amphibians which metamorphose into terrestrial adults (all Anura and some of the Urodela) the fin fold disappears at metamorphosis, but it may persist in neotenic species or in some purely aquatic salamanders (Cryptobranchus).

The fin fold first appears as a longitudinal thickening of the epidermis, seen as a ridge on external inspection. The thickening increases by shifting upward and toward the midline of the adjacent strips of the epidermis. The cells moving in from the right and left flanks remain separated as two layers of epithelium, except at the crest of the ridge. There is, however, no hollow in between the two layers. The fold is hollowed out a bit later by the two epithelial layers separating in the middle, and then connective tissue cells of neural crest origin penetrate into the fold.

Although the neural crest cells enter into the formation of the fin fold in a later stage, they are actually responsible for the determination of the whole structure. If the neural crest cells are removed shortly after their formation, or the neural folds, from which they arise, are cut away, the fin fold is not developed in the region of the defect (Fig. 175, *A, B*). If a piece of the neural fold, or a mass of the neural crest cells, is transplanted under the epidermis in any part of the body, a fin fold develops at the site of the transplantation (Fig. 175, *C, D*) (Terni, 1934; Du Shane, 1935). The ability to induce the development of the fin fold is found in the trunk neural crest cells only. The neural crest cells of the head cannot induce a fin fold, although the epidermis of the head region is fully competent to react by developing a fin fold, if it is exposed to the action of the necessary inductor. The extent to which the fin fold reaches anteriorly is thus dependent on the extent to which the neural crest possesses the ability to induce the fold (Terentiev, 1941).

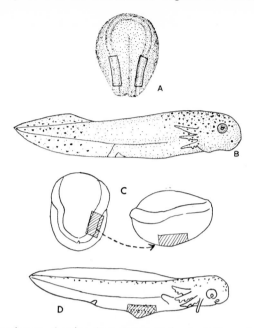

Fig. 175. Experiments showing dependence of the fin fold on the neural crest in salamander embryos. *A*, The neural folds removed on both sides in the neurula stage; *B*, result: no fin fold and no melanophores in the region of the operation. *C*, Transplantation of the neural fold onto the side of another embryo in the neurula stage (graft shaded); *D*, result: fin fold developed on the side. (*A* and *B* from Du Shane, 1935.)

The External Gills.   The external gills are hollow protrusions of epidermis, with connective tissue, blood vessels and muscle inside. They develop as hollow, outwardly directed pockets above the gill clefts and are found in some fishes (*Polypterus,* the lungfishes, *Misgurnus*) and in amphibian larvae. The original outpushing forms the shaft of the external gill; on this shaft develop secondary branches, which are formed at first as solid, outwardly directed thickenings of the epidermis of the shaft and are subsequently hollowed out (Fig. 176). Sometimes the shaft is so short that the gill appears to be a bunch of filaments which may branch in their own turn.

The pattern manifested in the development of the gills has been found in the Urodela to be dependent not on the epidermis, but on the inner layers of the embryo. The epidermis of the gill region may be lifted and replaced again after it has been rotated 90 or 180 degrees. If such an operation is carried out before the gill development begins, the external gills will appear in their normal position just as if nothing had happened (Harrison, 1921a). A piece of epidermis taken from the flank may be transplanted over the gill region to replace the local epidermis, and if the transplantation is carried out early enough (soon after the closure of the neural folds) the normal development of the external gills is not impeded. The mesoderm of the gill region (in the first type of experiment) may be included in the graft which is implanted in

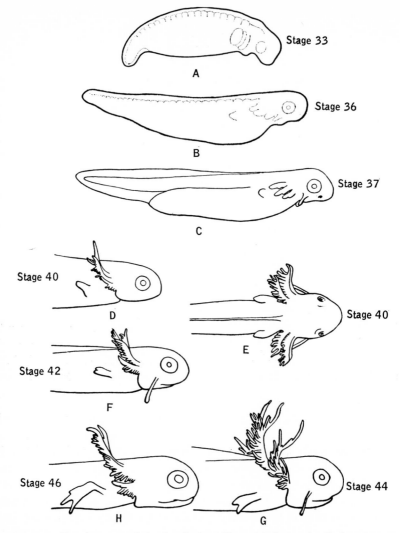

Fig. 176. Development of the forelimbs, gills and balancer in *Ambystoma punc-tatum*. The stages are shown after Harrison's normal table. (After Harrison, from Mangold, 1929.)

inverse orientation, and the gills will still develop in their normal position, al-though the development may not go as smoothly as when only the epidermis is involved. If, however, the endoderm of the gill region is rotated together with the other germ layers, the developing gills are dependent on the new position of the graft, even if it is disharmonious with the other parts of the embryo (Fig. 177). It is therefore the endoderm that determines the position of the developing external gills (Severinghaus, 1930).

The part played by the endoderm in the development of the external gills may also be tested by completely removing the endoderm of the gill region,

Fig. 177.   Transplantation of the gill rudiment in the salamander. *a,* Stage of operation, and area which was excised for transplantation. *b, c,* Gills in abnormal positions developed from grafts rotated 180 degrees during operation. (From Severinghaus, 1930.)

leaving the two other layers intact. The result is that the external gills do not develop at all (Mangold, 1936).

The Balancers.   The **balancers** are tentacle-like organs present in the larvae of many species of urodele amphibians (newts and salamanders). The organ is situated, one on each side, just behind the angle of the mouth underneath the eye. It is a slightly curved cylindrical process, consisting of epithelium and a connective tissue core. The connective tissue is especially dense just underneath the epithelium, where it forms a cylindrical supporting membrane. Proximally this membrane is attached to the quadrate. The epithelium on the tip of the balancer produces a mucous secretion and is therefore slightly adhesive. A newly hatched larva of a newt or salamander uses the balancers for support when resting on the ground, and to prevent the body from falling on one side in the stages when the forelimbs are not yet developed (Fig. 176, *F, G*). When the forelimbs become functional they take over the support of the body, and the balancers gradually degenerate (Fig. 176, *H*) (see Harrison, 1925a).

The balancers would not deserve our attention, if it were not that these simple organs lend themselves to some experiments of considerable interest. The epidermis of the balancer is normally derived from a part of the ectoderm lying just outside the neural fold in the vicinity of the eye rudiment. Other parts of the ectoderm, however, possess ability to develop into a balancer when stimulated by an inductor. Balancers have often been induced in experiments on the "primary organizer" and on the inducing ability of the archenteron roof. The development of balancers is an indication that the inductor possesses the regional specificity of an archencephalic inductor (see section 6–4).

The part of the organizer actually responsible for the induction of the balancer in normal development is probably the archenteron roof, but the adjacent portion of the neural plate, once it has been determined, also possesses the ability to induce a balancer (Mangold, 1931a). Further experiments on the development of balancers will be described in Chapter 13.

Tentacles are found in the vicinity of the mouth in many fishes, and also in the clawed toad, *Xenopus.* It is not known whether the development of these presents similarities to the development of the balancers of salamanders.

The early rudiments of paired limbs are similar to the gill rudiments in so far as they are outpushings of the epidermis which are filled with a mass of mesenchyme cells. In the case of the limbs, however, it is the differentiation of the mesenchyme into parts of the skeleton and muscle of the limb that deserves the greatest attention; the limbs will therefore be dealt with in conjunction with the organs derived from the mesoderm.

There is still one more structure of great importance which is derived from the epidermis; it is the mouth invagination (the stomodeum). Its formation and further development can be most conveniently dealt with together with the other parts of the alimentary canal, which are derived from the endoderm.

CHAPTER **11**

Development of the
Mesodermal Organs
in Vertebrates

11–1 THE FATE OF THE SOMITES AND THE ORIGIN
OF THE SOMATIC MUSCLES

The somites, when first formed, are masses of mesodermal cells with a
small cavity in the middle. The cells are arranged radially around the central
cavity. With further development, the shape of the somites changes, they
become extended in the dorsoventral direction, and flattened mediolaterally.
In the vertebrates with discoidal cleavage this leads to the elevation of the
dorsal parts of the embryo above the general level of the blastodisc. The flattening of the somite is accompanied by a change in the shape of its central

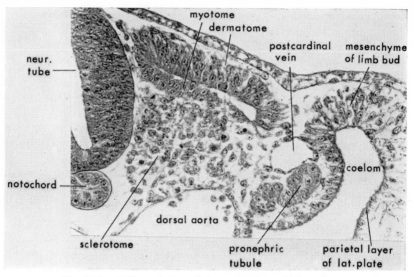

Fig. 178.   Differentiation of the somite in a chick embryo.

cavity (the myocoele); instead of being spherical, the cavity becomes a narrow vertical slit. An inner wall and an outer wall become clearly distinguishable, corresponding to the parietal and visceral layers of the lateral plates.
The inner wall of the somite becomes very much thicker than the outer wall.
The fate of the inner and the outer wall of the somites is completely different. The outer wall contributes to the formation of the connective tissue
layer of the skin, and is therefore called the **dermatome.** The inner wall produces skeletogenous tissue and the voluntary striated muscles of the body.
The skeletogenous tissue develops from the lower edge of the inner wall of
the somite, and this part of the somite is therefore called the **sclerotome**
(Fig. 178). The sclerotome breaks up into a mass of mesenchyme cells. The
cells migrate into the spaces surrounding the notochord and the spinal cord,
envelop these organs and later differentiate into cartilage, thus forming the
bodies and the neural arches of the vertebrae. The hemal arches in the tail
region and the ribs are of the same origin.

The dorsal part of the inner wall of the somite is the source of somatic
muscle in the vertebrate's body. It is therefore called the **myotome.** The cells
of the myotome rearrange themselves so that they become elongated in a
longitudinal direction. These longitudinally elongated cells differentiate subsequently into the striated muscle fibers. Originally each myotome becomes
a muscle segment, separated from the one anterior and the one posterior to
it by a connective tissue layer, the vertical myocomma. At one time in its
development the whole somatic musculature of the vertebrate consists of
such segments arranged in linear order.

In the lower vertebrates the myotome is the largest part of the somite, the
sclerotomes being small and rather inconspicuous. In the amniota, however,
the sclerotomes are much larger. Only the upper edge of the inner wall of the

somite adjoining the dermatome becomes the myotome. The size of this part increases rapidly; the myotomes grow downward to assume the same position, lateral to the neural tube and the notochord, that they occupy in the amphibians right from the start.

Developing as they do from the somites the myotomes are originally dorsal in position. In the course of the subsequent development, the muscle segments spread downward in the space between the skin on the outside and the somatic layer of the lateral plate on the inside, until the muscle segments on the right and the left side meet ventrally. It is in this stage, with minor alterations, that the somatic muscles persist in the fishes and in the aquatic larvae of amphibians. The segmentation of the lateral and ventral muscles, as well as of the dorsal muscles, is directly derived from the segmentation of the mesodermal mantle in somites.

In terrestrial vertebrates the primitive segmentation of the somatic muscles is more or less obliterated in connection with a change of locomotion. The segmented lateral bands of muscle are adapted to locomotion by lateral inflections of the body and the tail. The locomotion by means of two pairs of legs requires a completely different organization of the muscle system; as a consequence only traces of the original muscle segmentation can be discovered in the terrestrial vertebrates.

In *Amphioxus* muscle segments continue anteriorly almost to the tip of the snout. With the development of the brain and the skull in vertebrates the head region is exempted from the process of propelling the body by lateral inflections. The somatic muscles become superfluous in the head region. Nevertheless, mesodermal somites are formed in the head region of the embryo. The muscle segments derived from the somites lying posterior to the ear (the postotic somites) may persist and be linked up with the muscle segments of the neck region. The somites lying anterior to the ear vesicle are always very transitory structures. A part of the cells of these somites, however, differentiates as somatic muscle. The muscles derived from this source do not serve for locomotion, but are the six pairs of oculomotor muscles (the four rectus muscles and the two oblique muscles on each side).

11–2 THE AXIAL SKELETON: VERTEBRAL COLUMN AND SKULL

In most vertebrates the axial skeleton passes through three phases in its development. In the first phase the supporting system of the body is represented by the notochord. In the second phase cartilage develops partly in direct connection with the notochord, partly independently of the latter. This condition is preserved in the adult state of contemporary cyclostomes and elasmobranch fishes (we are not concerned with the question of whether this condition is primitive or secondary). In the remainder of the fishes and in tetrapods the cartilaginous skeleton is later replaced or supplemented by the bony skeleton. The bony skeleton and its relationship to the cartilaginous skeleton are amply dealt with in courses on comparative anatomy of verte-

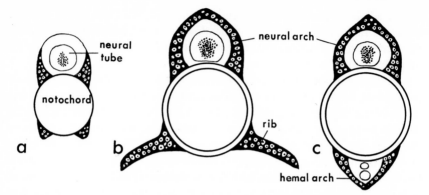

Fig. 179. Diagram of the dorsal and ventral pairs of arcualia (*a*) giving rise to the neural arch and the ribs in the trunk (*b*) and to the neural arch and the hemal arch in the tail (*c*).

brates and may, therefore, be entirely left out of consideration. We will consider here only the following aspects in the development of the skeleton:

1. The origin of cells giving rise to the cartilaginous skeleton.
2. The arrangement of the parts of the early cartilaginous skeleton in relation to the other organ rudiments of the embryo.
3. The dependence of the formation of the skeletal parts on the adjoining structures.

The material for the cartilaginous axial skeleton in the body and tail of vertebrates is derived, as was mentioned on page 286, from the sclerotomes. The sclerotomes, being parts of the somites, are segmental in origin, but once they become transformed into mesenchyme the segmental arrangement is largely lost, and the mesenchyme spreads out as a continuous sheath along the notochord, enveloping the spinal cord above and the caudal artery and vein below the notochord in the caudal region. In this continuous mesenchymal sheath nodules of cartilage, known as the **arcualia,** appear later in close apposition to the external surface of the notochord. Typically the arcualia appear in double pairs; one pair is formed dorsolaterally to the notochord and another pair ventrolaterally. The dorsolateral arcualia grow out dorsally alongside the spinal cord, and unite above the spinal cord to form the **neural arch.** The ventrolateral cartilages in the caudal region grow downward and unite underneath the caudal vein to form the hemal arch. In the cervical and thoracic region they give rise to lateral outgrowths—the rudiments of the ribs. At the same time the proximal parts of all four cartilages spread out around the surface of the notochord, contributing in varying degrees to the formation of the body of the vertebra (Fig. 179).

The cartilaginous neural and hemal arches do not bear a simple relationship to the somites (or sclerotomes); in many groups of vertebrates there are two series of cartilages (two pairs of dorsal elements and two pairs of ventral elements) developed in each mesodermal segment (*cf.* in elasmobranchs, in part also in amniotes). In other cases, although there is only one set of cartilages formed, these are situated intersegmentally at the junction of two myotomes (the myotomes retain the original segmentation of the somites). The way in which the original cartilage rudiments cooperate in the formation of the definitive vertebrae will not concern us here.

The fact that the cartilaginous (and later bony) vertebral column replaces the notochord functionally suggests that the location and the arrangement of the vertebral cartilages should be dependent on the notochord. In the absence of the notochord the cartilaginous axial skeleton is very irregular (see Fig. 153). Cartilage, however, is not completely absent in those sections of the body which do not have the notochord. The notochord is thus not indispensable for the formation of axial cartilages. The irregularity of the latter could be a secondary effect of the extirpation of the notochord, seeing that other systems, such as the spinal cord and segmented muscles, are greatly distorted as a result of the failure of the embryo to stretch normally.

In some experiments (Holtzer and Detwiler, 1953) in which the spinal cord instead of the notochord was removed from salamander embryos, it was found that the axial cartilages were either completely absent or reduced to insignificant vestiges. This could not be the result of damage to the sclerotomes since in other experiments very extensive defects of the somites including the sclerotomal region were completely restored at the expense of the remaining fragments of the somites. The results of extirpation experiments were corroborated by the transplantation of pieces of spinal cord into an incision on the lateral surface of the somites of another embryo (Fig. 180, a). It was found that complete and well developed neural arches were formed in association with the grafted spinal cord (Fig. 180, b).

The segmentation of the neural arches has not so far been considered. In normal development there is a definite relationship between the segmentation of the longitudinal dorsolateral muscles, the segmentation of the spinal nerves and ganglia and the segmentation of the vertebral skeletal elements, in particular that of the neural arches. There is a pair of spinal nerves with spinal ganglia corresponding to each muscle segment, the ganglia being situated opposite the median surface of the muscle segment. The neural arches alternate with the spinal ganglia and are thus situated intersegmentally with respect to the myotomes, so that each neural arch (and subsequently each vertebra) is connected to two consecutive muscle segments. The blood vessels also bear a relationship to this common pattern: an intersegmental artery arising from the dorsal aorta along each myocomma, close to the vertebral column (Fig. 181).

Experimental evidence is available to show that this whole system of metamerically arranged parts is originally dependent on the segmentation of the muscle rudiments (Detwiler, 1934). In a salamander embryo at the tailbud stage the anterior somites are somewhat broader than the posterior ones, so that if a block of several somites in the brachial region is removed

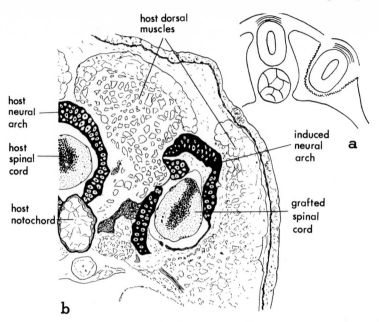

Fig. 180. Transplantation of a section of spinal cord into the dorsolateral meso-derm of a salamander embryo. *a,* Diagram of operation; *b,* result: the graft surrounded by an induced vertebra. (From Holtzer and Detwiler, 1953.)

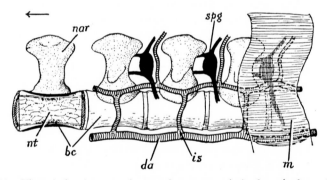

Fig. 181. The relation between the muscle segments (*m*), the spinal ganglia (*spg*), the intersegmental arteries (*is*) and the vertebrae in a larva of *Ambystoma*. *bc,* Bony cylinder of centrum of vertebra; *da,* dorsal aorta; *nar,* neural arch; *nt,* notochord. (From Goodrich, 1930.)

and replaced by a block of somites from the posterior trunk region, more somites may be fitted into the wound than had been cut out. Thus somites 7 to 12 could be substituted for somites 3 to 5 (Fig. 182). As a result the number of muscle segments on the operated side was increased in compari-son with the normal. It was found that the number of spinal nerves and spinal ganglia was increased also, but not necessarily in strict correspond-ence to the number of muscle segments, one nerve sometimes supplying more than one muscle segment. On the other hand the number of neural

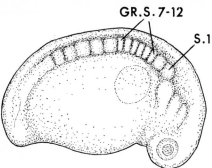

Fig. 182. *Ambystoma punctatum* embryo with somites 7–12 (Gr.S. 7–12) transplanted in place of somites 3–5. (After Detwiler, 1934.)

arches was found to be strictly in accord with the number of spinal ganglia, a bar of cartilage always appearing between two adjacent ganglia (Fig. 183). Since the ganglia are formed earlier than the cartilages, there can be no doubt that the neural arches are dependent on the ganglia and not the other way around.

The whole chain of reactions would then appear to be as follows. The neural crest cells which are produced along the whole length of the neural tube become aggregated opposite the median surfaces of the somites (or myotomes) and these aggregations become the rudiments of the spinal ganglia. Next the cells of the skeletogenic mesenchyme, produced by the sclerotomes, spread out over the notochord and neural tube, but they are apparently repulsed by the spinal ganglia and thus instead of forming a continuous sheet of cartilage enclosing the spinal cord, they give rise to a series of disconnected elements (the dorsal arcualia) alternating in posi-

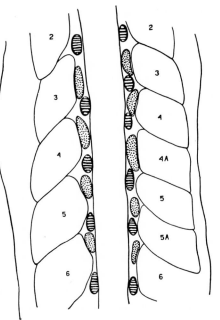

Fig. 183. Result of increasing the number of somites (on right side). The number of spinal ganglia (shaded) and neural arches (stippled) increased, as compared with control side (left). (From Detwiler, 1934.)

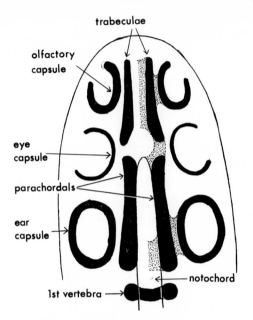

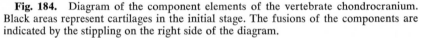

Fig. 184. Diagram of the component elements of the vertebrate chondrocranium. Black areas represent cartilages in the initial stage. The fusions of the components are indicated by the stippling on the right side of the diagram.

tion with the spinal ganglia. This is obviously not the complete picture, as it does not account for the cases in which two pairs of cartilages appear between each consecutive pair of spinal ganglia. It is thus likely that some structures other than the spinal ganglia, the neural tube and the notochord take part in determining the position of early cartilaginous rudiments.

The neural portion of the cranium in vertebrates in part bears the same relationship to the notochord and to the neural tube as does the axial skeleton in the posterior parts of the body. But there is little, if any, trace of a segmentation in the development of the cranium, and important parts of the cranium are quite peculiar in this respect.

In the lower vertebrates at the earliest stages of its development the cartilaginous cranium consists of several independent rudiments. These are: (1) The trabeculae; (2) the parachordals; (3) the capsules of the sense organs—the nose, the eye and the ear (Fig. 184).

The **trabeculae** (or **trabeculae cranii**) are a pair of elongated cartilages which appear in the most anterior part of the head, in front of the hypophysis. The trabeculae lie ventral and ventrolateral to the diencephalon and telencephalon, and their upper edges are wedged in between the brain on the inside and the rudiments of the nose and the eye on the outside. It has been established both by observation and experiment that the mesenchyme from which the trabeculae are developed comes from two different sources (*cf.* p. 271). The anterior part of the trabecula is formed of neural crest cells migrating forward and downward anterior to the eye cup. The posterior part of the trabecula is of mesodermal origin and is derived from the prechordal plate mesenchyme.

The **parachordals,** or parachordal cartilages, are derived from the mesenchyme produced by the sclerotomes of the somites in the head region. This mesenchyme spreads out on both sides of the notochord and eventually chondrifies in the form of two longitudinal rods, situated alongside the notochord and ending anteriorly at the same level as the notochord, that is, just posteriorly to the infundibulum and the rudiment of the hypophysis. The parachordals are very similar in origin and position to the rudiments of the cartilaginous vertebral column, but they lack the segmentation of the latter (possibly because the cranial ganglia, owing to the greater breadth of the neural tube in the head region, lie much further laterally and away from the region in which the development of the parachordals is taking place).

The cartilaginous capsules of the sense organs—the nose, the eye and the ear—develop from skeletogenic mesenchyme accumulating around the surface of the epithelial parts of these organs. The source of the mesenchyme may not be the same in all three cases. It is fairly certain that the ear capsule is formed by mesenchyme derived from the same sclerotomes as give rise to the parachordals. In fact the ear capsules are formed in close proximity and even in continuity with the parachordals, and strands of mesenchyme have been observed leading from the rudiments of the parachordals to the site in which the cartilaginous ear capsules start to develop (Filatoff, 1916).

It has been claimed (O. Schmalhausen, 1939) that cartilage of the nose capsules is derived from the epithelial nose rudiment itself. It would seem that this statement needs further corroboration. On the other hand, we have seen that the nasal placodes develop in very close proximity to the anterior transverse neural fold. The neural fold being the source of neural crest cells, it would not be very astonishing if these cells could produce the nasal cartilages as well as the epithelial parts of the olfactory organ.

In most vertebrates the eyeball is surrounded first by a connective tissue capsule, the sclera, which becomes cartilaginous in later stages. The source of the cells forming the sclera has not yet been established.

The capsules of all three sense organs are dependent in their development on the epithelial parts of the organs. This has been clearly shown in the case of the ear capsule (*cf.* section 10–4, The Ear) and is very probably true in the case of the olfactory capsule and the sclera of the eye, as these capsules fail to be formed if the epithelial parts of these organs are removed.

The parachordals, though spatially intimately associated with the notochord, show a high degree of independence from the latter organ. In experiments in which the notochord rudiment was removed (see p. 244), the development of the base of the skull does not seem to be affected to any extent. In particular, there is no foreshortening of the posterior part of the head, in sharp contrast to the shortening and stunting of the trunk and tail region in the absence of the notochord.

The further development of the cartilaginous skull (chondrocranium) is characterized by the enlargement and fusion of the initially formed cartilages. The right and left trabeculae fuse across the midline underneath the forebrain, and their posterior ends fuse with the tips of the parachordals. The parachordals envelop the notochord, particularly its dorsal side, and so give rise to the **basal plate** of the skull. At the point where the infundibulum and the hypophysis are situated, the cartilages leave a ventral opening (the hypophyseal fenestra) which persists for a long time and is closed only much later by cartilage or bone on the ventral side. The posterior ends of the parachordals grow upward and eventually fuse above the medulla, thus enclosing the foramen magnum of the skull. The nose capsule and the otic (auditory) capsule become firmly joined to the trabeculae and the parachordals respectively, thus contributing to the formation of the lateral walls of the chondrocranium. Gradually the lateral edges of the cartilaginous skull grow upward, and in the more primitive vertebrates (cyclostomes, many fishes and amphibians) form a roof over the dorsal surface of the brain. In teleost fishes and in all amniotes, however, the cartilaginous skull remains incomplete on the dorsal surface, and the cranial roof is formed by bones at a later stage. In higher vertebrates, in particular mammals, the initial stages in the development of the cartilaginous skull may be speeded up in such a way that the trabeculae and parachordals are fused right from the start—a condition which is achieved in lower vertebrates secondarily.

11-3 DEVELOPMENT OF THE PAIRED LIMBS

The paired limbs of vertebrates are very complex organs, built up of components derived from several different sources—from the lateral plate mesoderm, the epidermis and the somites, to name only the main components. Nerves and blood vessels are of course also indispensable components of differentiated limbs.

The first trace of the development of limbs may be found in the lateral plate mesoderm; the somatic layer of the lateral plate becomes thickened just underneath its upper edge. The cells of this thickening soon lose their epithelial connections and are transformed into a mass of mesenchyme without the somatic layer having lost its continuity. It is therefore a case of migration of mesenchyme cells from an epithelial layer, rather than that of the breaking up of epithelium into mesenchyme. The mesenchyme accumulates between the remaining lateral plate epithelium and the epidermis and soon becomes firmly attached to the inner surface of the epithelium (Fig. 185). The thickening of the lateral plate mesoderm and the subsequent formation of a mass of mesenchyme under the epithelium may coincide rather closely with the position of the two pairs of limbs, that is, they may appear in two disconnected regions—just behind the branchial region, and just in front of the anus. This is the case in the amphibians. In other vertebrates, however, the thickening and the mesenchyme accumulation may spread far beyond the actual region of limb development. In fishes the early limb rudiments are more elongated anteroposteriorly in the earlier stages than in the later stages of development. In the amniotes the thickenings and the mesenchyme gatherings are continuous throughout the whole length of the body in the form of horizontal ridges—the **Wolffian ridges** (Fig. 155, Stage 18). However, the most anterior and the most posterior parts of the ridge are thicker than the intermediate part, and it is only these anterior and posterior parts that develop progressively, giving rise to the forelimbs and hindlimbs. The intermediate part of the Wolffian ridge later disappears.

The epidermis over the mesenchyme mass becomes slightly thickened and bulges outward. This happens over the Wolffian ridge as well, but in the intermediate parts of the ridge the epithelial thickening disappears together with the gathering of mesenchyme. In the regions where the fore- and hindlimbs are to develop, the protrusion, consisting of a thickened epithelial covering and of an internal mass of densely packed mesenchyme, increases and becomes the **limb-bud.**

Of the two components contributing to the formation of the limb-bud, the mesoderm of the lateral plate and the epidermis, it is the former that starts the sequence of events leading to the formation of the limb. The presumptive limb mesoderm is determined as such at an early stage shortly after the closure of the neural tube. Pieces of lateral plate may be cut out in this stage and transplanted under the epidermis on the flank or on the

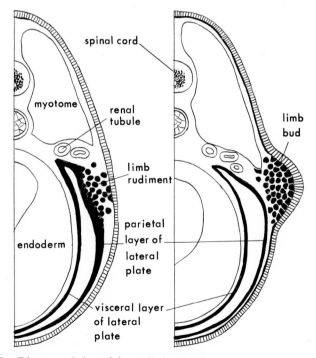

Fig. 185.   Diagram of the origin of limb mesoderm in an amphibian embryo.

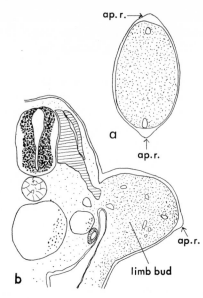

Fig. 186. Advanced limb-bud with apical ridge (*ap.r.*), in a chick embryo (*a*) (from Saunders, 1948) and cross-section of the forelimb-bud in a rat (*b*) (from Milaire, 1956).

head. The local epidermis will then become the epithelial component of the limb-bud, and a limb will develop heterotopically. The presumptive epidermis of the limb in the same stages, that is, before a limb-bud has been formed, does not possess any special properties, and if transplanted alone will not give rise to a new limb. Epidermis from any part of the body is able to cooperate with the presumptive limb mesoderm in forming a limb-bud. This can be shown by removing the epidermis in the limb region, and then covering the wound with a flap of epidermis taken from any part of the body (Harrison, 1918; Balinsky, 1931).

The epidermis is, however, by no means a passive component in limb development. This is especially clearly shown by some peculiarities of limb development in higher vertebrates. In the amniotes the limb-bud becomes slightly flattened at an early stage, and an epidermal thickening develops along the edge of the flattened bud. The thickening is in the form of a very sharply defined ridge, and sometimes (in reptiles) even takes the form of a solid fold of the epidermis. In cross-section the ridge looks like a nipple (Fig. 186). It is referred to as the ectodermal apical ridge.

The cells of the ridge differ from the ordinary epidermal cells not only in their arrangement, but also in their physiological properties. It was found that they contain more ribonucleic acid and more glycogen, and they differ very conspicuously from surrounding epidermal cells in their high content of the enzyme alkaline phosphatase (Fig. 187) (Milaire, 1956). All these biochemical properties may be taken as indications of active metabolism.

The apical ectodermal ridge is indispensable for the normal outgrowth of the limb rudiment. If the ectoderm of the apical ridge of a wingbud of a three-day chick embryo is removed without causing damage to the underlying mesoderm (Fig. 188) the distal parts of the wing fail to be formed,

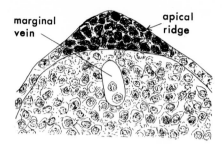

Fig. 187.   Part of cross-section of the limb-bud in a rat stained for alkaline phosphatase. The apical ridge shows intense positive reaction. (From Milaire, 1956).

although the proximal parts develop quite normally (Saunders, 1948). If the ectoderm covering the limb-bud in a chick embryo is removed and replaced by epidermis from another part of the body, the latter is found to be incapable of developing an apical ridge. The result is that the development of the whole of the distal part of the limb is suppressed. The girdle may, however, develop normally, and a short piece of cartilage representing the proximal part of the humerus or femur may also be formed. This is in some contradiction to the conditions found in amphibian embryos where flank or head ectoderm, as was indicated, may participate in the development of a limb, but then there are no apical ridges on amphibian limb-buds.

The investigation of the role of the epidermal apical ridge in the development of limbs in birds was extended further after it had been discovered that the mesoderm and ectoderm of a limb-bud may be separated very

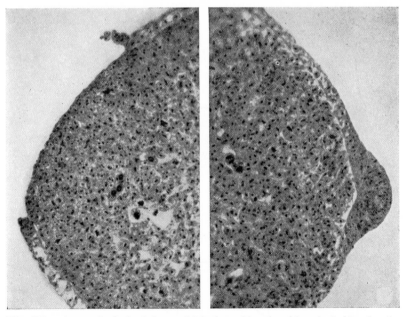

Fig. 188.   Apex of wingbud from which the epidermis with apical ridge has been removed (left), and control normal wingbud with apical ridge (right). (From Saunders, 1948.)

neatly by chemical instead of mechanical means (Zwilling, 1955). Treating a limb-bud with a trypsin solution causes the epidermis to separate from the mesodermal core of the bud. The mesoderm after this treatment is not fully viable, but the epidermis is quite healthy and may be used to cover the mesodermal part from which the epidermis is removed by immersion in a solution of Versene (the latter treatment destroys the epidermis, which comes off in flakes, but leaves the mesoderm in a very good condition).

After the mesoderm of a limb-bud is covered by epidermis of a different origin, the two stick firmly together, and such a composite limb-bud may be implanted onto the flank of a third embryo and there allowed to grow into a limb (Fig. 189). In this way it was possible to combine the mesoderm of a legbud with the epidermis of a wingbud and *vice versa*. The structure of the developing limb was in every case determined by the origin of the mesodermal component; it was a wing if the mesoderm was taken from a wingbud, and a leg if the mesoderm was that of a posterior limb rudiment. The origin of the ectoderm did not affect the nature of the developing limb. This experiment stresses the leading part played by the mesoderm in limb development.

The next experiment emphasizes the importance of the epidermis. In a "wingless' mutation in fowls the forelimb-bud appears approximately at the same time as the normal limb-bud, but fails to grow and produce a limb, though parts of the limb girdle may be present. It was noticed that the epidermis covering the wingbud in the affected embryos does not have an apical ridge. The experiment was therefore undertaken of combining

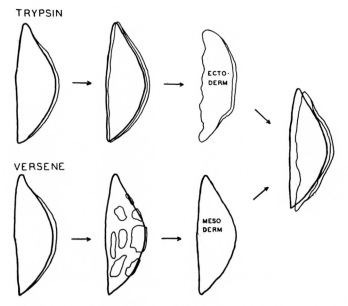

Fig. 189.   Separation and recombination of the limb-bud mesoderm and epidermis in the chick embryo. (From Zwilling, 1956.)

the mesoderm of a normal wingbud with the epidermis of an embryo of the wingless strain. The result was as had been expected; in the absence of an apical ridge the bud did not grow and no distal parts of the wing were formed (Zwilling, 1956).

The competence for limb development can be shown to be present all along the flank of the embryo between the forelimb and hindlimb region even if it does not manifest itself in normal development. In amphibia it is possible to induce a supernumerary limb by introducing an inductor into the region between the forelimb and the hindlimb rudiments (Fig. 190). Originally the ear vesicle was used as a limb inductor (Balinsky, 1925), but later it was found that other organ rudiments such as the hypophysis and the olfactory sac can also induce a limb if transplanted into the side of a young embryo. Under the influence of the induction, the local mesodermal cells accumulate as a compact mass under the epidermis, the epidermis is also made to react, and an additional limb develops. Depending on whether the induced limb lies nearer to the normal forelimb or hindlimb region, it may resemble a forelimb or a hindlimb in its structure (Balinsky, 1933).

Experiments on limb induction suggest that in normal development, as well, there must be some factor determining which part of the mesoderm competent for limb development actually produces a limb rudiment. Limb-buds are often induced together with other structures when "spino-caudal" inductors are introduced into embryos in the gastrula stage (p. 170). By transplanting the presumptive somite mesoderm together with the noto-chord into the lateral plate region of an embryo in the neurula stage, it is

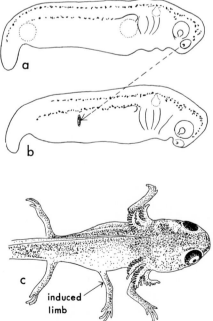

Fig. 190.   Induction of a supernumerary limb in *Triturus taeniatus* by means of a grafted nose rudiment. *a, b,* Diagrams of the operation. The position of the normal rudiments of the fore- and hindlimb shown in *a. c,* Larva with induced limb.

induced
limb

possible to induce the lateral plate mesoderm to develop into kidney (normally produced by the stalks of the somite). In these experiments it has often been observed that additional limb-buds developed together with the kidney tubules (Yamada, 1937). It follows that the determination of the limb mesoderm occurs in conjunction with the determination of other parts of the mesodermal mantle. The kidney in itself cannot, however, induce a limb.

Differentiation in the Limbs.   After the limb-bud has grown so far that its length exceeds its breadth, the differentiation of the subordinate parts of the limb sets in. We have already noted the slight flattening of the limb-bud. Now the distal portion of the bud becomes flattened even more, and at the same time it becomes distinctly broader than the proximal part of the limb rudiment. The flattened and broadened distal part is the hand (or foot) plate. The edge of the plate is initially circular, but soon it becomes pentagonal, the projecting points indicating the rudiments of the digits. While the tips of the digit rudiments keep growing out further, the intervening sections are retarded in their growth, so that the digits become separated by distinct incisions. The five digits appear simultaneously in all

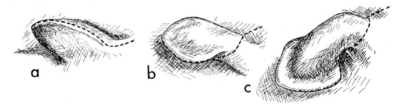

Fig. 191. Rotation of the developing forelimb in the lizard. (After Braus, from Hertwig, 1906.)

amniotes with pentadactyl limbs, but in amphibians, especially in urodeles, the first two digits appear earlier and digits 3, 4, 5, are formed one after another on the posterior edge of the limb. Where less than five digits are present in the adult limb, or more than five in cases of hyperdactyly, this condition is reflected in the structure of the hand (foot) plate.

In the early limb rudiments, the future flexor surface is ventral and the future extensor surface is dorsal, but as the limb elongates, a rotation takes place so that the flexor surface is turned posteriorly, and eventually it may even face in a posterodorsal direction. The preaxial edge of the limb, which is originally anterior, is then turned downward (Fig. 191). With the elongation of the limb it becomes bent at the elbow joint (or knee joint). A less pronounced flexion develops at the base of the carpus (or tarsus). The three main sections of the limbs thus become recognizable externally.

Concurrently with changes in the external appearance of the limbs, differentiation occurs in the interior of the limb rudiments. The mesenchyme cells which are closely packed in a young limb-bud become segregated into areas in which the mesenchyme is lying more loosely and into other areas in which the mesenchyme cells are crowded. The latter are the rudiments of the skeletal parts of the limb. The concentrated masses of mesenchyme in due course become converted into procartilage, and then, by further deposition of intercellular matrix, into cartilage. Whereas in the initial stage of mesenchyme concentration large sections of the limb skeleton are represented by a common mass of mesenchyme, in the procartilage stage individual elements of the skeleton are laid down as separate units; these may fuse together later.

The differentation of the limb skeleton generally proceeds in a proximo-distal direction though some deviations from this order are of fairly general occurrence. In amphibians the first skeletal part to become recognizable is the **stylopodium** (humerus or femur). Parts of the **zeugopodium** (radius and ulna in the forelimb, tibia and fibula in the hindlimb) are laid down next, and the autopodium differentiates considerably later. The girdle rudiments appear after the stylopodium, but earlier than the autopodium. In higher vertebrates the girdle tends to be developed simultaneously with the proximal elements of the limb. In the autopodium the proximodistal sequence is upset by the larger skeletal elements, the metacarpals and meta-tarsals differentiating more rapidly than the smaller elements, namely, the carpals and the tarsals. In the digits, however, the proximal phalanges are laid down earlier than the distal ones (see Sewertzoff, 1931).

The blood vessels appear in the limb-bud at an early stage, and although the pattern of the arteries, veins and capillaries is too variable to deserve much attention here, one blood vessel may be mentioned. It is situated along the edge of the hand (or foot) plate, just underneath the ectodermal apical ridge, and possibly it is responsible for supplying nutriment to this rapidly growing area of the limb (Fig. 192).

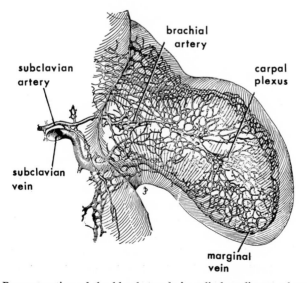

Fig. 192.   Reconstruction of the blood vessels in a limb rudiment of a pig embryo.
(From Woollard: Carnegie Contrib. Embryol., 22.)

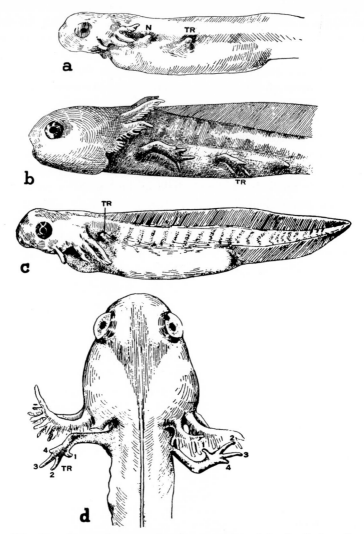

Fig. 193. Experiments demonstrating the properties of the forelimb rudiment of the salamander *Ambystoma punctatum* in the tailbud stage. *N*, Normal limb; *TR*, transplanted limb. *a,* Transplantation of a right limb rudiment to the left side, with inversion of the dorsoventral axis. *b,* Same animal later. *c,* Left limb transplanted orthotopically with inverted anteroposterior and dorsoventral axes. *d,* Same animal later; a.p. axis determined and retained in new position. (From Harrison, 1921b.)

The pattern according to which the various constituent parts (bones, muscles, blood vessels, nerves) are arranged in each limb is asymmetrical —proximal and distal ends, dorsal and ventral surfaces, anterior and posterior sides of the limb being different from one another. The first indication of this asymmetrical pattern can be noted when the tip of the limb-bud grows out in a slanting posterior direction, instead of growing straight out away from the side of the body (Fig. 176).

Special experiments have been carried out to find out how the asymmetry of the limb, and thus the basic pattern of its differentiation, is determined. Experiments were originally carried out on the embryos of the salamander, *Ambystoma punctatum*. Forelimb rudiments were transplanted at different stages after the end of neurulation in such a way that either one limb rudiment axis or two or all three were inverted (disharmonious) with respect to the host's body (Harrison, 1921b). The inversion of the proximodistal axis could only be carried out with the mesodermal part of the limb rudiment, but this is not of importance, as it is the mesoderm that is the carrier of the limb determination (Harrison, 1925b; Swett, 1937). The transplantations were done either **orthotopically,** that is, in place of a normal limb rudiment, or **heterotopically** on the flank. The experiments showed that the three axes of the limb rudiment were not determined simultaneously. In the earliest rudiments immediately after neurulation the anteroposterior axis is already fixed; limb rudiments, transplanted disharmoniously, with this axis inverted, had the tip of the limb-bud growing forward instead of backward, and the limbs later had their postaxial (ulnar) side placed anteriorly (Fig. 193). At the same time the inversion of the dorsoventral axis of the transplanted limb rudiment could still right itself; that is, the original upper part of the rudiment developed into the ventral (palmar) surface of the limb and the original lower part of the rudiment developed into the dorsal surface of the limb. This shows that the pattern of the limb differentiation was not yet fixed in respect to its dorsoventral axis and that this pattern could be imposed upon the limb rudiment by the host, that is, by the parts which were in connection with the limb rudiment in its new position. In a later stage of development when the tail rudiment begins to elongate, the dorsoventral axis of the limb was also found to be determined; if the limbs were transplanted in an inverted position, with the dorsoventral axis of the limb rudiment the inverse of the dorsoventral axis of the host, the limb developed with an abnormal orientation, its plantar surface facing upward. The surroundings could no longer change the pattern of limb differentiation in respect to the dorsoventral axis, as well as in respect to the anteroposterior axis (Fig. 194).

Fig. 194. Heterotopic transplantation of the limb rudiment in the late tailbud stage, with inverted dorsoventral axis. The palmar surface of the transplanted limb (partially reduplicated) facing upward. (From Swett, 1927.)

At the same time that the dorsoventral axis of the limb rudiment is determined, the proximodistal axis may still be inverted without impairing the normal development of the limb. Only in a later stage, when the limb-bud begins to be visible from the outside, does the proximodistal axis show signs of being determined, and if the limb mesoderm is transplanted with the axis inverted, the limb shows abnormalities in its development. (A limb cannot actually grow inward into the body because it would then be no longer in contact with the epidermis, and this contact is indispensable for the differentiation of parts of the limb.)

In cases when the transplanted limb rudiment grows out in a disharmonious orientation to the host, it is often observed that a kind of regulation occurs by means of the formation of a second limb-bud, whose orientation and differentiation is harmonious with the host's body. The appearance of the second limb-bud is the result of a sort of splitting of the original rudiment and is due to an influence of the host on the graft. This influence is not strong enough to invert the axial structure of the transplanted rudiment as a whole, but it is sufficiently strong to divert a part of the cells of the transplanted rudiment and to cause them to take on the axial structure of the host. If the original disharmonious limb were now to degenerate (as sometimes happens), the host would be in possession of a normal set of limbs.

The splitting of a limb rudiment (or any other organ rudiment) to produce two similar rudiments is known as **reduplication.** That reduplication is possible shows that the cells of the rudiment are not determined each to fulfill a definite part in the developing organ, even though the axial pattern of the rudiment as a whole is determined. From this it is further inferred that the determination of the axial pattern is not a matter of determining what each cell or group of cells has to do in the process of development of the organ, but it is a matter of polarity, of a heteropolar structure of the rudiment as a whole. This immediately recalls the polarity of the egg in the early stages of development, and some similarity between the rudiment of an organ and the early egg as a whole is indeed shown. Both possess the ability to develop a number of subordinate parts, each within its own scope of action (the whole animal in the case of the egg, one organ in the case of an organ rudiment), and without these subordinate parts being represented by discrete particles in the initial system.

Just as the egg can be split mechanically into parts and each part will develop into a miniature whole, so an organ rudiment, in a suitable stage of development, may be split mechanically into halves, and each half will develop into a complete organ. We have already mentioned similar results in the case of the eye rudiment. The same applies to the limb rudiment. The limb rudiment up to the stage of the limb-bud formation may be cut in two and each half transplanted separately, or the two halves left in place and kept apart by inserting a piece of extraneous tissue between them. Each will develop into a whole limb (Fig. 195) (Swett, 1926). The two limbs resulting from such a splitting may later grow to the normal size. Splitting of limbs may sometimes occur accidentally in young amphibian larvae developing in nature, or it may possibly be caused by the pressure of folds of the amnion in higher vertebrates, and abnormalities will be the result: limbs that are completely or partially reduplicated. If the splitting of the rudiment is due to a mechanical cause rather than being spontaneous, the two halves retain the same axial structure (polarity), thus being replicas of each other.

The limb girdles normally develop in intimate connection with the limbs themselves. The mesodermal material for the girdle is derived from the peripheral parts of the mesenchyme mass, the central part of which becomes the limb-bud. However, in their determination the girdles are partially

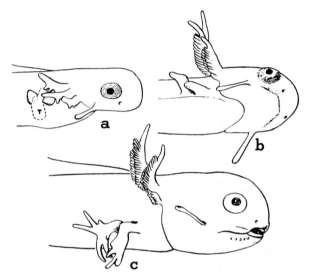

Fig. 195.   Splitting of the forelimb rudiment, with a strip of extraneous tissue (T) inserted across it (a) to produce two limbs (b, c). (From Swett, 1926.)

independent of the limbs. The limb girdles may develop independently of the limbs, if the development of the limb itself is suppressed in some way or other, as by the removal of the limb-bud (Detwiler, 1918). In limb induction experiments limb girdles sometimes are found where the limb itself fails to develop. The development of the girdle is in no way dependent on the existence of an interaction with the epidermis of the skin, as is that of the limb itself.

In birds the limb girdle may also develop fairly well even if the distal part of the limb is absent as a result of a failing cooperation between mesoderm and ectoderm, as in cases where the epidermis over the limb-bud lacks an apical ridge.

If the girdle (shoulder girdle or pelvic girdle) develops in the absence of the limb, it may be normal in its peripheral parts, but the fossa for the articulation with the humerus or femur does not develop in the absence of at least the proximal part of the stylopodium (humerus or femur) (Balinsky, 1931). The thickening of the girdle in the region of the articulation is also lacking in the absence of a limb. In so far as these parts are concerned, the amphibian girdles are dependent on the limbs. In bony fishes, in which the cartilaginous part of the shoulder girdle is reduced to a small skeletal element in the region of articulation, the bony **cleithrum** can develop independently of the pectoral fin (Balinsky, unpublished).

The mesenchyme of the limb-bud does not contain all the mesodermal cells used for the development of a limb; after the limb-bud is formed it receives an additional supply of mesodermal cells from the lower edges of the myotomes. These cells produce the muscles of the limb or at least a part of these. In fishes it is possible to observe **muscle buds** being formed at the lower edges of the myotomes. These muscle buds are protrusions of the cell masses of the myotomes. The muscle buds push downward and outward until they enter the limb-buds. In the limb-bud itself, the muscle buds derived from different myotomes fuse into a common mass of cells (myoblasts) from which the muscles of the limb subsequently develop. In the higher vertebrates, beginning with amphibians, muscle buds are not found although the migration of individual cells from the myotomes into the limb rudiments may take place.

In the fishes, if the lateral plate mesoderm of the limb region is transplanted to near the midventral line on the abdomen, it gives rise to fins which are devoid of muscles (Lopashov, 1950), thus proving that the lateral plate mesoderm is not capable of producing the limb muscles. Muscles are developed, however, if a part of the somite is included in the graft. In amphibians on the other hand, when the rudiment of the limb, consisting of lateral plate mesoderm with or without epidermis, is transplanted heterotopically, it will develop into a complete limb with muscles. This, however, does not definitely exclude the participation of the myotome material, since under experimental conditions some regulation could have taken place, just as half the rudiment may produce a whole limb in cases of reduplication.

The nerve supply to the developing limbs will be dealt with in the section on the differentiation of the nervous system (16–2).

11–4 DEVELOPMENT OF THE URINARY SYSTEM

The excretory organs in vertebrates are essentially aggregates of uriniferous tubules, connected originally at their proximal ends with the coelomic cavity by ciliated funnels (the **nephrostomes**), and communicating to the exterior by a system of ducts which, in the lower vertebrates, open into the cloaca. Both the tubules and the ducts are of mesodermal origin and they develop from the stalks of the somites (the **nephrotomes**). In the most primitive vertebrates such as the cyclostomes, and also in the Gymnophiona, one uriniferous tubule develops from the nephrotome in each mesodermal segment.

The nephrotome, prior to the development of a uriniferous tubule, is a strand of cells connecting the somites to the lateral plate mesoderm. The cells become separated into the parietal and visceral layer, and the cavity, which we may call the **nephrocoele,** is for a while continuous with both the myocoele and the definitive coelom between the two sheets of the lateral plate mesoderm. The connection of the nephrocoele with the cavity of the somite is soon obliterated, but the connection with the cavity of the lateral plate persists in the more primitive type of vertebrate excretory organs and becomes the nephrostome. The dorsolateral wall of the nephro-tome becomes drawn out into a hollow tube which is a prolongation of its cavity. The tube, which is in open connection with the coelomic cavity, becomes the uriniferous or renal tubule. The distal (outward) ends of the most anterior tubules (that is, the tubules formed in the anterior part of the trunk) soon turn backward and then fuse, thus giving rise to the common excretory duct, known as the **pronephric duct.**

It is an essential feature in vertebrates that the renal tubules are associ-ated with bunches of fine blood vessels (the **glomeruli**), through the endo-thelial walls of which the blood plasma containing excretory products is filtered into the uriniferous tubules or into the coelom in the immediate vicinity of the nephrostomes, so that the nitrogenous waste products may be carried through the tubes and the excretory ducts and removed from the body. Either the mass of blood vessels is invaginated into the wall of the renal tubule, which enlarges to contain the glomerulus and becomes **Bow-man's capsule,** or else the blood vessels form a bulge on the wall of the coelom. In this case the structure is referred to as the **external glomerulus,** or, if glomeruli of several segments are joined together, as the **glomus.** It is believed that the segment of the coelomic cavity into which the glomus projects is itself derived from an expansion of the nephrocoele (or several nephrocoeles) (Fig. 196).

The basic pattern of development of the excretory organs as described above becomes modified in various degrees. As with many other organ systems in vertebrates the development of the excretory tubules progresses in a craniocaudal direction, the tubules differentiating earlier in anterior segments than in posterior ones. The tubules which develop and begin func-

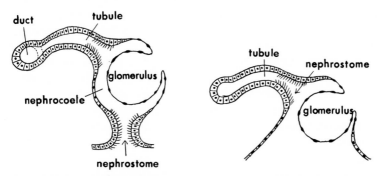

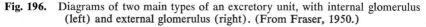

Fig. 196.   Diagrams of two main types of an excretory unit, with internal glomerulus (left) and external glomerulus (right). (From Fraser, 1950.)

tioning earlier, i.e., those of the anterior part of the trunk, tend to show more primitive conditions in their organization while those developing later are modified to a greater degree. In the most primitive vertebrates, the cyclostomes and some fishes, the differences between the anterior and posterior parts of the excretory system are only gradual, but in the higher vertebrates we may distinguish more or less clearly three sections of this system: the **pronephros,** the **mesonephros** and the **metanephros.**

The most anterior nephrotomes give rise to the pronephros, the mesonephros develops from those of the midtrunk region, and the metanephros is derived from the nephrotomes of the posterior part of the trunk. The resulting three types of kidneys can thus be distinguished as anterior, middle and posterior with respect to their place of origin. The pronephros is the first to develop, the mesonephros second, and the metanephros, if present, appears latest of all.

The three types of kidneys differ also in the way in which they are produced by the stalks of the somites, in the presence or absence of the nephrostomes, in the position of the glomerulus and in the origin and connections of the excretory ducts. In the pronephros the nephrostomes are well developed and the glomeruli tend to be replaced by a common glomus (see above), though in some primitive forms such as the Gymnophiona there may be glomeruli in each tubule of the pronephros. In the mesonephros a separate glomerulus is, as a rule, intercalated in the course of each uriniferous tubule. Nephrostomes are present at first, but may disappear later. In the metanephros the nephrostomes are not formed at all and the uriniferous tubules begin with the glomeruli (Bowman's capsules).

Further differences in the development of the tubules and ducts can best be described separately in relation to each section of the excretory system. Since all three types of kidneys are nothing but local differentiations of one system stretching throughout the body of a vertebrate, morphological features do not necessarily change abruptly from one section to another and transitional conditions may sometimes be found (see Fraser, 1950).

The Pronephros.   The description of the origin of a uriniferous tubule given on page 307 follows the observations made on the embryos of Gymnophiona (Brauer, 1902) and may be taken as representing the typical development of the pronephric tubules in the most archaic groups of vertebrates. The strictly segmental origin of the tubules is an essential feature of this development. In amphibians the pronephric tubules are formed from a common mesodermal thickening appearing beneath the third and fourth (in salamanders) or second, third and fourth somites (in frogs). Nevertheless, the number of the pronephric tubules corresponds to the number of segments participating in the development of the pronephros.

The rudiment of the pronephros can be traced back in amphibians to the neurula stage. By means of vital staining the presumptive material of the pronephros was found to lie in the mesodermal mantle just outside of the edge of the neural plate, posterior to the middle of the embryo. This cor-

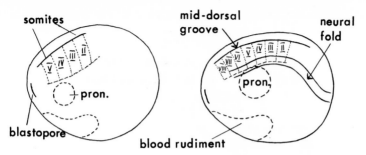

Fig. 197.   Position of the presumptive pronephros in the very early (left) and middle neurula (right). (After Yamada, 1937.)

responds, in the salamanders, to the third and fourth mesodermal segments (Fig. 197) (Yamada, 1937; Muchmore, 1951). In the neurula stage mesoderm of this region is capable of self-differentiation when transplanted heterotopically (Fales, 1935). At the same time, however, other parts of the mesodermal mantle also possess the ability to develop into renal tubules, as for instance when an inductor (the notochord) is transplanted into the lateral plate region of the embryo. The presumptive somites, when isolated from the notochord and cultivated *in vitro* (surrounded by a coat of skin epidermis for protection), may develop into renal tubules even though their normal destiny would be a different one (the development of muscle). After the end of neurulation (closure of the neural tube) the competence for the development of the pronephric tubules is restricted to the presumptive material of the pronephros. From this stage onward the pronephric rudiment cannot be replaced, and the removal of the rudiment leads to the absence of the pronephros.

The sequence of events during the transformation of the nephrotomes

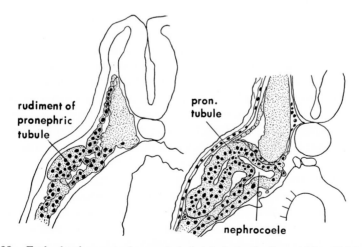

Fig. 198.   Early development of a pronephric tubule in the frog. (After Field, from Brachet, 1935.)

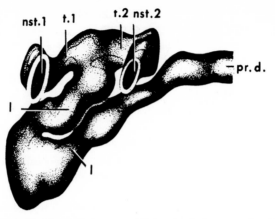

Fig. 199. Reconstruction of the pronephric canals in a newt embryo (median view). *nst. 1, nst. 2,* First and second nephrostomes; *t. 1, t. 2,* first and second pronephric tubules; *l,* loop formed by joint part of tubule; *pr. d.,* pronephric duct. (After Mangold, from Spemann, 1938.)

into the pronephric tubules, in frogs, salamanders and higher tetrapods, is not so clear as in the lower vertebrates. It is carried out by means of a rearrangement of cells, that is, by morphogenetic movements. The end result is, however, the same: several pairs of pronephric tubules are formed; they open by means of the nephrostomes into the coelomic cavity (Fig. 198) and fuse distally to form the pronephric duct (Fig. 199).

The formation of the pronephric duct, by fusion of the distal ends of the pronephric tubules, is a very important phase in the development of the excretory system, as this duct not only serves the pronephros, but is also instrumental in providing pathways for the outflow of urine from the mesonephros and metanephros, and, in the males, for the passage of spermatozoa.

In the salamanders it has been shown that the pronephric duct develops right from the start from a more caudal part of the mesoderm than the pronephric tubules; while the two pronephric canals develop from the mesoderm lying under the third and fourth somites, the rudiment of the pronephric duct lies under the fifth, sixth and seventh somites. If the rudiments of the pronephric tubules are removed, or if the embryo is bisected transversely between the levels of the fourth and fifth somites, the pronephric duct can still develop, thus showing that the pronephric tubules and the pronephric duct are determined independently of each other (Holtfreter, 1943c).

Once they have been formed, both the pronephric tubules and the pronephric duct elongate very considerably. The pronephric tubules as a result of their elongation are thrown into numerous loops, and eventually form a more or less spherical body, consisting of tangled and intertwined tubules. The pronephric duct, on the other hand, remains straight, and while it elongates, its posterior free end pushes itself backward along the lower ends of the somites. This backward movement of the tip of the pronephric duct ends when the duct reaches the cloaca and fuses with its wall, while the lumen of the duct opens into the cavity of the cloaca.

The backward elongation of the pronephric duct may be interrupted in various ways, as by making a deep incision across its path. If the wound remains gaping the pronephric duct cannot spread beyond the wound, and the duct is not continued into the posterior part of the body (Waddington, 1938; Holtfreter, 1943c). If, however, the wound is too small or is covered to a sufficient extent, the tip of the elongating duct may find its way around the wound and penetrate into the posterior part of the body. Behind the wound the duct returns to its level under the lower edges of the somites and eventually reaches the cloaca (Fig. 200, *a, b*). This experiment shows that the elongation of the pronephric duct is largely independent of the surrounding parts, although the latter do seem to direct the duct by furnishing a suitable path for its movement. The same is borne out with the utmost clarity by some further experiments (Holtfreter, 1943c), in which embryos in the closed neural tube stage were bisected transversely, one half reversed, and the two parts caused to heal together with opposite dorsoventral orientation (see Fig. 200, *c, d*). When the pronephric ducts reached the level of the operation in their backward elongation, the tip of the duct changed its direction, and struck a new path across the lateral body wall, until it reached the somites of the inverted posterior part of the embryo. It then elongated along the edge of the somites and reached the cloaca as usual.

The development of the pronephric duct shows a great similarity to the development of the lateral line (section 10–4). In both cases the migration of the posterior end of the organ rudiment along the length of the body is due to intrinsic tendencies of the cells of the rudiment, but the organs and

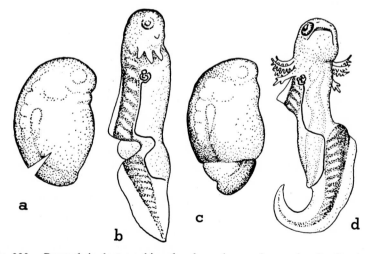

Fig. 200.   Pronephric duct reaching the cloaca by an abnormal path after transection of the embryo (*a, b*) and after rotation of the posterior part of the embryo through 180 degrees (*c, d*). (From Holtfreter, 1943c.)

tissues with which the rudiments come in contact influence the path taken by the migrating cells (see also section 9–3).

The Mesonephros. The mesonephric tubules are derived from the nephrotomes as in the case of the pronephros, but their interrelation is by no means so simple. Only in the more archaic groups of vertebrates (selachians, Gymnophiona) are the renal tubules formed directly from the nephrotomes (Fig. 201). In most amphibians and in all higher vertebrates the mesodermal cells of the nephrotomes dissolve into a mass of mesenchyme stretching on each side of the body along the dorsal edge of the lateral plates. This elongated mass of mesenchyme is known as the **nephro-**

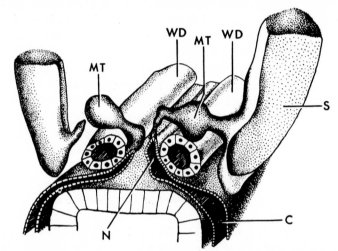

Fig. 201. Relation of the mesonephric tubules to other mesodermal parts in the embryo of a sturgeon. The rudiment of the mesonephric tubule (*MT*) is shown as still connected to the somite (*S*) on one side, and just separated on the other. *C*, coelom; *N*, nephrostome; *WD*, Wolffian duct. (After Maschkowzeff, from Fraser, 1950.)

genic cord or nephrogenic tissue. The mesonephric tubules are developed from the nephrogenic tissue by a secondary aggregation of the mesodermal cells.

The aggregating cells begin by forming epithelial vesicles which stretch and elongate to become tubes. The number of the mesonephric tubules does not correspond to the number of segments, several tubules being developed in the region of each segment. Even if at first only one tube is formed in each segment it soon gives rise by budding to secondary and tertiary tubes, and so on. If present at first the nephrostomes may disappear later. In the frogs the nephrostomes lose their connection with the rest of the nephric tubule but open secondarily into the veins (connecting the veins to the coelom).

The nephrogenic tissue is not at once determined for differentiation as mesonephros. If the presumptive material of the mesonephros of a newt embryo in the neurula stage is transplanted into the pronephric area of a tailbud stage embryo, the grafted tissue gives rise to excretory tubules of the pronephric type. On the other hand, the same material taken from a tailbud stage embryo on grafting will produce mesonephric tubules even if it comes to lie at the site where a pronephros should have developed (Machemer, 1929).

The mesonephric tubules do not produce a duct of their own. As the tubules are formed, their distal, free ends join up with the pronephric duct, which thus becomes the duct of the mesonephros as well and is then called the **Wolffian duct.**

The development of the mesonephros has also been found to be dependent on the pronephric duct in another way. The nephrogenous tissue develops into the mesonephric tubules only if stimulated to this by the pronephric duct. In the preceding section experiments have been mentioned in which the penetration of the pronephric duct into the posterior half of the body was prevented by placing an obstacle (in the form of a gaping wound) in its path. When the operation is successful, that is, if the duct does not reach the region where the mesonephros normally develops, the nephrogenous tissue fails to form the renal tubules, or forms only poorly developed tubules (Waddington, 1938; O'Connor, 1939). Apparently some stimulus (induction) from the pronephric duct is necessary for the normal development of the mesonephric tubules from the nephrogenous tissue.

The Metanephros.   The metanephros develops from the posterior part

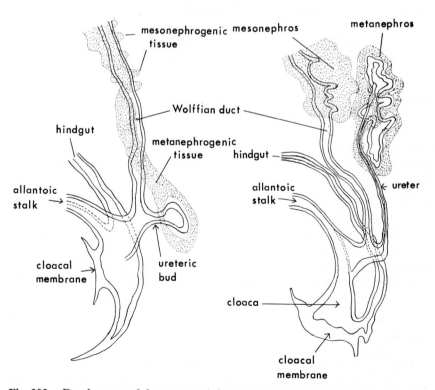

Fig. 202.   Development of the ureter and the metanephros in a rabbit embryo. (After Schreiner, from Brachet, 1935.)

of the nephrogenic cord, the part adjacent to the cloaca. There is no trace of a relation of the tubules to the nephrotomes, although the nephrogenic tissue is primarily derived from the latter. As in the case of the meso-nephros, the metanephros does not develop a duct of its own, but uses the pronephric duct as a means of removing the urine that it excretes. How-ever, the connection between metanephros and the pronephric duct is established not directly, but by means of a special outgrowth or branch of the duct. Before the metanephros starts differentiating, a bud is formed on the pronephric duct a small distance in front of the point where the duct joins the cloaca. The bud elongates in the direction of the posterior part of the nephrogenic cord and eventually penetrates into the mass of nephrogenic tissue. The duct formed in this way becomes the **ureter** and the bud from which it develops is the **ureteric bud.** Having reached the nephrogenic tissue, the end of the ureter begins to branch, the branches later becoming the collecting tubules of the kidney (Fig. 202). The point at which the branching begins expands to form the renal pelvis. The nephrogenic mesenchyme accumulates around the tips of the collecting tubules and in due course differentiates into the renal tubules with their glomeruli. No trace of nephrostomes can be discovered in the metanephros at any stage.

11–5   DEVELOPMENT OF THE HEART

The heart in vertebrates develops from the mesoderm forming the ventral edges of the lateral plates in the pharyngeal region of the body. It will be most convenient to describe first the development of the heart in animals such as the amphibians and to consider later the development of the heart in higher vertebrates.

The Heart in Lower Vertebrates.   During gastrulation in the amphibians the sheet of mesoderm advances forward from the blastopore, penetrating between the ectoderm and the endoderm. The rate of movement of the mesoderm is greatest in the dorsal region of the embryo, intermediate laterally and least ventrally. In the neurula stage the dorsal and dorsolateral parts of the mesodermal mantle have reached the head region of the embryo; ventrally there remains an approximately triangular area in which there is no mesoderm intervening between ectoderm and endoderm. This triangle, roughly corresponding to the oral and pharyngeal region of the embryo, has its apex posteriorly and the broad base anteriorly. The posterior part of the area is the site of development of the heart, and it is later filled in by the mesoderm participating in the formation of the heart. The most anterior part of the area remains free of mesoderm, and here the mouth breaks through after a fusion between the endoderm and the ecto-derm has taken place.

The presumptive material of the heart is found in the edges of the meso-dermal mantle bordering the mesoderm-free area on the right and on the left (Fig. 203). The presumptive heart mesoderm may be excised in the neurula stage and cultivated in a saline solution (Goerttler, 1928). To prevent the mesoderm from disintegrating, it is isolated together with a flap of ectoderm; the ectoderm then closes into a vesicle with the mesoderm inside. The mesodermal cells under these conditions differentiate into muscle tissue, and this begins to pulsate rhythmically—an unequivocal indication that the developed muscle tissue is cardiac muscle, as only this type of muscle is capable of autonomous rhythmical contraction. The morphological structure (shape) of the heart, however, does not develop if the presumptive heart mesoderm is taken from the embryo in the neurula stage. In these early stages the development of the heart is in some way dependent on the endoderm. By cutting through the ectoderm and the mesodermal mantle and removing the whole of the endoderm it is possible to produce newt embryos which consist of ectoderm and mesoderm only (Mangold, 1936; Balinsky, 1939). The embryos survive the operation quite satisfactorily, but their term of life is limited because they do not possess the food supply normally contained in the yolky cells of the endoderm. The endodermless embryos show various defects in the ectodermal and mesodermal organs due to the lack of inductive influences emanating from the endoderm. One of such defects is the complete absence of the heart, although the mesodermal layer, from which the heart rudiment is derived, may remain intact. (See also pp. 282 and 325.)

After the end of the neurulation the free edges of the mesodermal mantle gradually converge toward the middle of the mesoderm-free area and become thickened in the heart region, foreshadowing the formation of the heart rudiment. At this stage it can be noticed that between the free edges of the mesodermal mantle converging from the right and the left there lie a number of loose cells, similar to mesenchyme cells in their structure. These cells are derived from the ventral edge of the mesodermal mantle. They are the rudiment of the **endocardium,** that is, the endothelial lining of the cavity of the heart. The endocardial cells soon accumulate in the midline

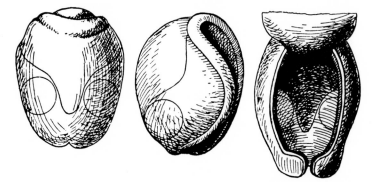

Fig. 203.   Position of the heart rudiments (circles) in the newt embryo in the neurula stage. (After Goerttler, from Huxley and de Beer, 1934.)

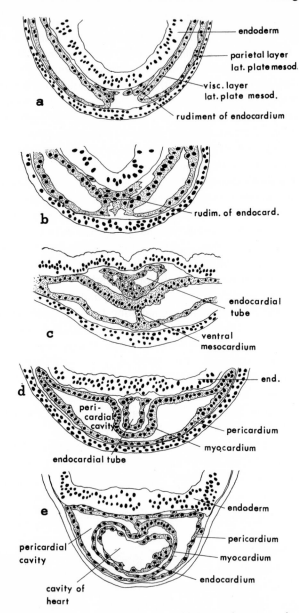

Fig. 204. Development of the heart in amphibian embryos. *a, d, e, Triturus; b, Salamandra; c, Rana.* (From Mollier, in Hertwig, 1906.)

as a longitudinal strand and eventually become arranged in the form of a thin-walled tube. The lumen of the tube is the cavity of the heart (Fig. 204). The endocardial tube bifurcates at both ends; at the anterior end its two prolongations are the ventral aortae, and at the posterior end it receives the two vitelline veins—the first venous blood vessels to reach the heart. All these vessels are at first similar to the endocardial tube in that they consist only of a thin layer of endothelium, produced by the mesenchyme cells joining together.

While the endocardial tube is being formed, the edges of the mesodermal mantle, which at this stage may also be called the edges of the lateral plate mesoderm, close in along the midline and fuse with each other. The fusion first occurs under the endocardial tube, between it and the ectoderm. Very soon, however, the visceral layer of mesoderm envelops the endocardial tube on the dorsal side as well. By fusion of the mesodermal layers of the right and left side, epithelial partitions are formed above and below the endocardial tube. In analogy to the dorsal and ventral mesentery these are called the dorsal and ventral mesocardium. The ventral mesocardium is of a very ephemeral existence, it is very soon perforated, and the coelomic cavities of the right and left side become continuous underneath the endocardial tube. The dorsal mesocardium persists longer, but is also dissolved at a later stage. The coelomic cavities expand in the heart region to form the pericardial cavity. The pericardial cavity is initially only a part of the general body cavity, the coelom. It becomes completely separated from the remainder of the coelom, largely owing to the failure of the coelom to develop in the region of the branchial pouches, which lie immediately dorsal to the heart rudiment. Posteriorly the connection of the pericardial cavity with the rest of the coelom becomes occluded by the developing liver, to which the posterior end of the endocardial tube becomes closely connected. A connective tissue wall developing in this position (at the anterior boundary of the liver) is the **septum transversum.**

From the above description it will be clear that the pericardial cavity is lined by lateral plate mesoderm. The parietal layer of this mesoderm persists as the epithelial wall of the pericardial cavity, or the pericardium proper. The visceral layer adheres to the endocardial tube. This layer differentiates as muscle tissue, and thus becomes the **myocardium** of the heart.

The heart is at first an almost straight tube and does not show a subdivision into its various chambers. Later the tube becomes inflected in a very characteristic way. Starting from behind, the tube first runs forward, then bends downward and to the right and eventually again to the left, upward and forward. The heart thus becomes coiled in the shape of an S. The degree of the twisting in higher vertebrates is greater than in lower ones, so that in the former the tip of the second inflection comes to lie well posterior to the tip of the first inflection (Fig. 205). The tubular heart rudiment becomes constricted in some places and dilated at others, and is thus subdivided into its four main parts. The sinus venosus lies posteriorly, the atrium develops at the tip of the first inflection of the heart, the descending part, from the first to the second inflection, becomes the ventricle, and the part going forward from the second inflection becomes the conus arteriosus. Before this subdivision is performed, however, the functioning of the heart starts; it begins to pulsate at a regular rhythm. The pulsations of the heart start very early in the development of the embryo, even before

Fig. 205. Twisting of the heart rudiment in a chick embryo (ventral view). *A,* Atrium; *V,* ventricle; *C,* conus arteriosus. (Modified from Patten, 1958).

the peripheral blood vessels are ready to receive the blood stream. This is important so that no delay may arise because of the heart's being unready to function when otherwise circulation might begin.

As the presumptive heart rudiment is being formed, its capacity for performing the further stages of development independently of the normal surroundings increases perceptibly. If the heart rudiment is excised after the end of neurulation and transplanted into an abnormal position or allowed to develop *in vitro,* enclosed in a vesicle of skin, the differentiation goes much further than in the experiments referred to above. Not only pulsating muscle tissue develops, but a cardiac tube is formed, and the tube becomes inflected just as does the heart in normal development (Stöhr, 1924). The rudiment of the heart in this stage, however, is by no means strictly determined in all its parts. The left or right half of the heart rudiment may be excised, and the remaining half then develops into a complete whole (Ekman, 1925; Copenhaver, 1926). What is more, half of the heart rudiment may be explanted or transplanted, and it still develops into a complete heart. The ability of a half of the heart rudiment to form a whole heart may be used to produce two hearts in the same embryo. For this an incision should be made lengthwise before the two halves of the heart rudiment unite in the middle. Inserting a piece of extraneous tissue, a somite for instance, or leaving the wound to gape may prevent the halves of the heart rudiment from coming together. Two complete pulsating hearts develop under these conditions (Ekman, 1925). The reverse can also be done: a complete heart rudiment may be superimposed on the intact presumptive heart mesoderm of a host embryo. The two rudiments then fuse into one whole and produce a normal heart (Ekman, 1925; Copenhaver, 1926).

The main points emerging from the above will now be summarized:

1. The heart develops in a very anterior position, in the pharyngeal region. The position of the heart in the thorax (as in an adult tetrapod) is thus secondary and is due to a displacement of the organ in later development.
2. The heart develops from a paired rudiment, uniting in the midline secondarily.

3. Each half of the heart rudiment is able to differentiate even before the two halves fuse.
4. Two heart rudiments may fuse into one organ of normal structure.
5. The functioning of the heart begins at an early stage of development.

The Heart in Higher Vertebrates.   With all the above in view, it is easy to understand the development of the heart in the vertebrates having yolky eggs and partial cleavage.

Because of the physiological requirements of the developing embryo (the necessity of establishing circulation) the heart in meroblastic vertebrates develops precociously, before the body of the embryo becomes separated from the yolk sac. The embryo in this stage is still lying flat on the surface of the yolk, and its lateral plate mesoderm is to be found toward the outer parts of the blastodisc. The lateral plate mesoderm is prevented from uniting on the ventral side of the embryo by the intervening yolk. As a result the two halves of the heart rudiment begin differentiating independently of one another. Two endocardial tubes are actually formed; each becomes invested by the myocardium and surrounded laterally by the pericardial cavity. When the body folds undercut the anterior end of the embryo and the foregut becomes separated anteriorly from the yolk sac, the right and left heart rudiments become able to meet in the midline under the pharynx. The two endocardial tubes come to lie alongside each other and soon fuse into one tube in the cardiac region, while in front and behind the heart the endothelial tubes remain separate, thus leading to a state described earlier for the lower vertebrates (Fig. 206). The visceral walls of the right and left pericardial cavities also meet and fuse above and below the endocardial tubes, forming a single pericardial cavity. The single endocardial tube is now completely surrounded by the myocardium. In view of the experimental results mentioned above, the fusion of two heart rudiments to form one single organ is not at all surprising.

Of all the organs of a vertebrate, the heart is the one which starts its definitive function earliest. It is also greatly dependent in its development on function. It is essential for the development of the heart that a blood stream should actually flow through it. It is true that a heart of a typical shape with recognizable parts will develop even in complete isolation, as in the explantation experiments (Stöhr, 1924), but such an isolated heart soon stops developing further. Also, a heart in its normal position is arrested in its development if in some way or other (interruption of the afferent blood vessels) it is deprived of circulation. The degree of the heart's development and growth appears to be dependent on the volume of blood passing through it or on the size of the animal which the heart supplies with blood. Heart rudiments have been transplanted reciprocally between large and small species of salamanders (Copenhaver, 1930, 1933) and it was found that the hearts of small species grew beyond their usual size in large hosts, and the hearts of large species were undersized in small hosts. In every combination the

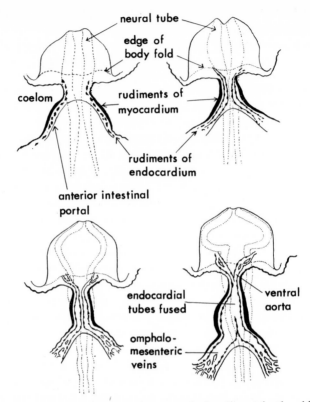

Fig. 206. Development of the heart from a paired rudiment in the chick embryo, viewed from the dorsal side (semidiagrammatic.) (After Patten, 1958.)

transplanted heart grew to approximately the same size as the host heart should have grown.

In the vertebrates developing a pulmonary circulation the heart becomes separated to a greater or lesser degree into a right half, carrying blood to the lungs, and a left half, receiving the blood from the lungs by way of pulmonary veins and sending it to the rest of the body. The first indication of this separation is found in the lungfishes (Dipnoi). In the amphibians the atrium becomes subdivided by a partition wall, arising between the points of entry into the atrium of the sinus venosus and the pulmonary vein. After the separation is completed, the right atrium receives blood from the sinus venosus and the left atrium from the lungs by way of the pulmonary vein. Both atria, however, pour out their blood into the ventricle which remains undivided.

The Mammalian Heart. The most elaborate transformations are found to occur in the development of the heart in mammals. The task of directing the blood into its various channels is complicated not only by the presence in the adult state of a double circulation (the systemic and the pulmonary circulation) but also by the necessity of providing for a rapid change from fetal to postnatal circulation. The subdivision of the heart into right and left halves is dependent on the cooperation of several structures, and carried out in stages. The atrium is separated into right and left halves first by a primary partition wall (septum primum), which is similar to that found in amphibians. As soon as this partition wall subdivides the cavity of the atrium, however, it is perforated by an opening allowing for intercommunication between the right and left atrial cavities. This is necessary so as not to deprive the left atrium of a blood flow (on which, as we have seen, the development of the heart is dependent in a high degree) during the intrauterine life, when the amount of blood passing through the lungs is very small. At a later stage a second partition wall (septum secundum) develops in the atrium just to the right of the first, but this remains incomplete, a large **foramen ovale** remaining open in its lower part (Fig. 207, a). The foramen ovale does not quite coincide in position with the opening in the primary partition wall; nevertheless blood may pass through both openings so long as the blood pressure in the right atrium is greater than that in the left atrium, as it actually is during fetal life. After birth, however, when the pulmonary circulation is fully established, the pressure in the left atrium becomes higher than that in the right atrium. The primary partition wall becomes pressed against the secondary wall, and as a result the previously existing passage becomes closed (see Fig. 207, b). Thus the atria become finally separated. Eventually the primary and secondary partition walls of the atrium fuse together.

The atrium originally communicates with the ventricle by way of the atrio-ventricular opening, which is a narrowed part of the heart tube. The separation of this single opening into two is performed by means of connective tissue outgrowths developing on the dorsal and ventral edge of the openings. The outgrowths, known as the **endocardial cushions,** fuse in the middle and are also joined by the lower edge of the primary interatrial partition

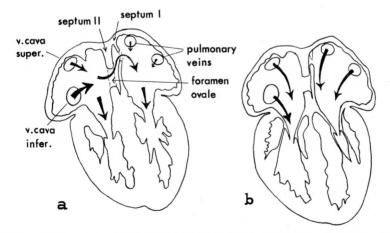

Fig. 207.   Section of mammalian heart (diagrammatic), showing the change-over from fetal to postnatal circulation. Arrows show the direction and amount of blood flow.

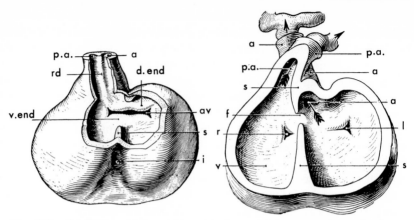

Fig. 208. Subdivision of the atrio-ventricular canal in mammals. *a,* Aorta; *a.v.,* atrio-ventricular foramen; *d.end,* dorsal endocardial cushion; *f,* interventricular foramen; *i,* indentation of ventricular myocardium; *l,* left atrio-ventricular foramen; *p.a.,* pulmonary artery; *r,* right atrio-ventricular foramen; *rd,* ridge separating the conus arteriosus in two channels; *s,* parts of interventricular septum; *v.end,* ventral endocardial cushion; *v,* cavity of right ventricle. (After Kollmann, from Prentiss and Arey, 1917.)

wall; as a result two canals are left, one leading from the right atrium into the right ventricle and the other leading from the left atrium into the left ventricle. The endocardial cushions also participate in the formation of the atrioventricular valves (Fig. 208).

The ventricle becomes partially separated into left and right halves by an indentation of the myocardium at its tip (at the apex of the posteroventral inflection of the heart tube). The developing partition wall is therefore largely muscular, but is incomplete. It is supplemented by a membranous (connective tissue) part arising in conjunction with a wall subdividing the conus arteriosus into two channels: one to serve for systemic circulation, and the other for pulmonary circulation. The latter partition wall cuts in backward between the origin of the fourth and sixth pairs of aortic arches in such a way that a ventral channel leads into the arches of the third (carotid) and fourth (systemic) pairs, while the dorsal channel leads into the sixth (pulmonary) pair of arches. The partition wall in the conus arteriosus is spirally twisted so that the ventral channel becomes connected to the left ventricle and the dorsal channel to the right ventricle.

It will be seen that the blood entering the right atrium from the sinus venosus, including the blood from the placental circulation (via the umbilical veins, pp. 207 and 218), is directed into the pulmonary arch (except for the volume of blood passing from the right into the left atrium via the foramen ovale). The pulmonary blood vessels, during the fetal life, cannot take all this blood, neither is there any necessity to have the blood pass through the lungs, as they are as yet not serving as respiratory organs. The greater part of the blood passing through the right half of the heart is therefore returned to the systemic circulation by way of the arterial duct (ductus arteriosus) which connects the pulmonary arch to the dorsal aorta. After parturition the arterial duct becomes closed by a violent contraction of its muscular wall, and all the blood from the right part of the heart is forced into the blood vessels of the lungs. The double circulation thus becomes finally established.

It is a further peculiarity of the mammalian heart that the opening between the sinus venosus and the right atrium broadens out to such an extent that the sinus venosus becomes incorporated into the wall of the right atrium. As a result the main veins (anterior and posterior venae cavae, the coronary vein), which originally entered the sinus venosus, now have separate openings into the right atrium. Similarly the proximal part of the pulmonary vein is incorporated into the wall of the left atrium, so that there are two, or sometimes even four pulmonary veins entering the left atrium independently of each other.

11–6 DEVELOPMENT OF THE BLOOD VESSELS

In the adult vertebrate, blood vessels large and small permeate almost all parts of the body. Supernumerary parts of the body, whether they are grafts or results of induction, become vascularized, and the blood vessels, which in this case are additional to the ones produced in normal development, link up with host arteries and veins and are included in the host's blood circulation. It would appear that blood vessels are attracted to penetrate any parts which are in need of a blood vessel supply. Furthermore it appears that a suitable situation for blood vessels is duly reserved in the pattern of organization of any organ. It must be noted, however, that in vertebrates the blood vessels are invariably situated in spaces occupied by connective tissue or its derivatives. Where capillaries seemingly penetrate into epithelium, this is actually tantamount to the invasion of channels or other spaces between epithelial cells by connective tissue from which the endothelial cells of the blood vessels are derived.

The rudiments of the blood vessels are laid down as aggregations of mesenchyme cells. The cells participating in the formation of blood vessels are called **angioblasts.** They are probably always of mesodermal origin. We have seen that in the vascular area of amniotes the walls of the blood vessels are developed from the blood islands in conjunction with the development of the first blood cells (section 7–1). Similar, though more concentrated, blood islands occur in vertebrates with holoblastic cleavage (in particular in amphibians) on the ventral side of the abdomen. In the amniote body proper and in all areas except for the ventral body wall in the anamniotes, the blood vessels develop independently of blood corpuscles. The aggregations of angioblasts become arranged in the form of a flat epithelium surrounding a cavity. The epithelium is the endothelium of the blood vessel; the outer layers of the blood vessel walls are differentiated much later. A student of the anatomy of the adult vertebrates is accustomed to find the blood vessels in the form of tubes of a constant diameter over relatively long stretches. The first blood vessels laid down in the embryo are only rarely in the form of straight

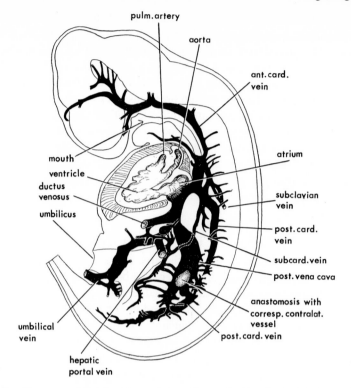

pulm.artery

aorta

ant. card.
vein

mouth

ventricle

ductus
venosus

umbilicus

atrium

subclavian
vein

post. card.
vein

subcard. vein

post. vena cava

anastomosis with
corresp. contralat.
vessel

umbilical
vein

post. card. vein

hepatic
portal vein

Fig. 209.   Venous system of a 12 mm. pig embryo, showing irregular shape of blood
vessels, especially in the posterior part of the body. (After Patten, 1958.)

tubes. Over large areas the blood vessels are initially laid down in the form
of a network (Figs. 130, 192). The further development of individual canals
in the network depends on the amount and the direction of blood flow. Those
channels which happen to come in the line of the greatest blood flow become
increased in diameter, develop the connective tissue and the muscular layers
and become arteries or veins. The channels that receive less blood remain in
the form of capillaries, or degenerate and disappear completely. Another
form in which early blood vessels appear is as sinuses—extensive irregular
spaces surrounded by endothelium. Large sections of some of the larger veins
appear in this form (Fig. 209). Some of these sinuses later acquire a more
regular tubular form, others remain as such in the adult animal; for instance,
the postcardinal "vein" becomes a tubular blood vessel in urodeles, but
remains essentially in the form of a sinus in the dogfish.

Once the network of endothelial tubes—the rudiments of the blood vessels—has been established, new blood vessels continue to be formed by
sprouting and outgrowth of those already present. Lateral branches may
form on already existing capillaries, or a capillary may become interrupted
and each free end grow out in a new direction. Two outgrowths from dif-

ferent blood vessels may contact each other and fuse, thus establishing a new channel for circulation (Fig. 210).

As a result of the extreme plasticity of the blood vessel system the eventual arrangement of arteries, veins and capillaries in any part of the body is largely dependent on the amount and direction of blood flow in the part in question. In their first formation the embryonic blood vessels also appear in conjunction with other rudiments, suggesting a dependent mode of differentiation, though the dependence is not of a functional nature since the blood vessels develop prior to the establishment of circulation. If the heart rudiment in an amphibian embryo is removed before circulation starts, the main blood vessels continue to develop for some time, until the embryo dies. The exact position of the blood vessel rudiments is probably determined by a process similar to an induction, though this has not been shown conclusively.

Certain locations in the embryo offer preferential conditions for the development of blood vessels. These locations are: (1) between the visceral mesoderm and the endoderm (see Fig. 131); (2) around the kidneys, especially the pronephros and the mesonephros. Close networks of capillaries are always found in these two locations and they are the site of development of some of the major blood vessels. The heart obviously belongs to the first group. We have seen that the determination of the heart occurs at an early stage of development, and that as a consequence it shows a high degree of autonomy in its further differentiation. It is noteworthy, nevertheless, that the heart does not develop in embryos from which the endoderm has been removed (Mangold, 1936; Balinsky, 1939) (see p. 315). The blood vessels which form the (paired) continuation of the heart tube anteriorly and posteriorly also develop in conjunction with the endodermal parts. The two poste-

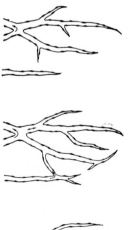

Fig. 210. Outgrowth of new capillaries, establishing new channels of circulation. (After Clark, from Arey, 1947.)

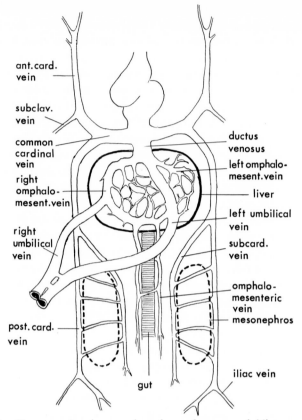

ant.card.
vein

subclav.
vein

common
cardinal
vein

right
omphalo-
mesent.vein

right
umbilical
vein

post.card.
vein

ductus
venosus

left omphalo-
mesent.vein

liver

left umbilical
vein

subcard.
vein

omphalo-
mesenteric
vein

mesonephros

iliac vein

gut

Fig. 211.   Venous system in an early embryo of a mammal (diagrammatic).

rior vessels are the vitelline veins, which collect the blood from the network
on the surface of the gut and, in amniotes, from the network on the yolk sac
(via the omphalo-mesenteric veins, which are a posterior prolongation of the
vitelline veins). The anterior prolongations of the heart tube are the ventral
aortae, which become connected along the partition walls between the endodermal branchial pouches (see section 12–3) with the dorsal aortae, a pair
of blood vessels developing on the dorsal side of the endodermal gut.

The aortic arches develop in a craniocaudal sequence, and the first to
appear is the mandibular arch which ascends between the edge of the mouth
and the first (spiracular) branchial pouch. The second aortic arch develops
between the first spiracular and the second (first true branchial) pouch, and
subsequent aortic arches develop between the more posterior branchial
pouches. Six aortic arches are laid down in the embryos of all vertebrates
except for some very archaic forms (cyclostomes and some selachians)
whose number of gill clefts is larger than five. The dorsal aortae become
extended forward into the head and backward throughout the length of the
trunk and to the base of the tail. In the trunk region the two dorsal aortae
later fuse into one unpaired dorsal aorta, while in the branchial and head

region the two aortae remain separate. In the trunk, the dorsal aorta gives off at the level of each myocomma an intersegmental artery. Larger arteries which do not show a very regular arrangement serve to convey blood to the viscera.

The blood vessels developing in conjunction with the excretory organs are the cardinal veins. The first rudiments of these appear, in amphibia, in the form of a venous sinus around the pronephros. Very soon prolongations of this sinus are found anteriorly and posteriorly. The posterior prolongation is the postcardinal vein, which develops along the groove between the somites and lateral plates dorsolaterally to the nephrotomes and later becomes closely associated with the mesonephric kidney. The anterior prolongation is the anterior cardinal vein. This runs forward on the same level as the post-cardinal vein, just above the branchial pouches into the head. The anterior cardinal vein and postcardinal vein join at the level of the anterior edge of the pronephros forming the common cardinal vein (= duct of Cuvier), which runs inward to join the vitelline veins where they enter the heart (Fig. 211). At a later stage a second pair of veins develops in tetrapods on the median side of the mesonephric kidney—the **subcardinal veins.**

The above basic pattern of the main blood vessels is laid down with great regularity in embryos of all classes of vertebrates, but it becomes greatly modified as development goes on, especially in the higher vertebrates.

Of the six pairs of aortic arches, the anterior two pairs become degenerate to a greater or lesser extent, but even in the higher vertebrates remnants of the ventral portions of these arches take part in the formation of the external

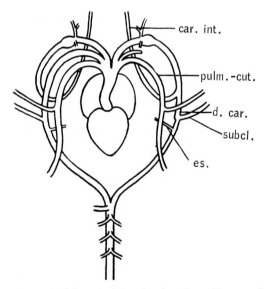

Fig. 212.   Arrangement of the arterial arches in a frog, *Xenopus laevis,* which had the fourth left aortic arch destroyed in the tadpole stage (ventral view). The blood flows to the dorsal aorta through the third (carotid) arch and the persisting carotid duct. (From Millard, 1945.)

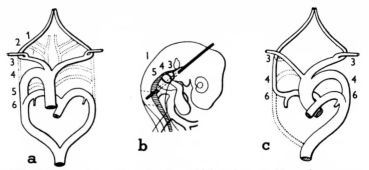

Fig. 213. Ligating the aortic arches in a chick embryo. *a,* Normal arrangement of aortic arches in later stages of development (ventral view). *b,* Thread in position to ligate the aortic arches on the right side. *c,* Suppression of the fourth right aortic arch and retention of the fourth left arch. (After Stephan, from Waddington, 1952.)

carotid, and some of the blood vessels in the ventral region of the head. In the fishes and in the larval amphibians a number of aortic arches from the third onward supply blood to the capillary network of the gills. In the terrestrial vertebrates (including amphibians after metamorphosis) the third pair of aortic arches becomes the carotid arch. The anterior parts of the dorsal aortae now serve to forward the blood from the carotid arch to the dorsal part of the head and to the brain. This blood vessel becomes the internal carotid artery. The fourth pair of aortic arches becomes the main channel for the blood flow from the heart to the dorsal aorta and hence to the body. These are the systemic pair of arches (in amphibians and reptiles). The fifth pair of aortic arches degenerates in all tetrapods except in urodeles, where they carry part of the blood to the dorsal aorta. The sixth pair of aortic arches gives off blood vessels to the lungs (and to the skin in amphibians). It is thus the pulmonary arch. The connection of the pulmonary arch to the dorsal aorta becomes obliterated. In birds and mammals a further simplification of the aortic arch system is introduced by the degeneration of one of the two systemic arches (of the fourth pair); in the birds only the right arch is left, and in the mammals only the left one.

The reduction of the aortic arches is obviously dependent on the blood flow through these vessels, in the same way as the blood flow has been found to mould the development of vessels from the original capillary networks. The following experiments prove this. In frog tadpoles all four (third to sixth) aortic arches persist up to metamorphosis and only then the transformations occur which lead to the adult condition. In the tadpole of the clawed toad, *Xenopus laevis,* the systemic arch on one side was destroyed shortly before metamorphosis. The result was that the carotid arch of the same side took over the work of the systemic arch; the connection of the carotid arch to the dorsal aorta, which is interrupted in normal development, persisted, and the blood from the heart flowed to the dorsal aorta by way of the carotid arch (Fig. 212) (Millard, 1945). Even more interesting, perhaps, is an experiment performed on the chick embryo: the normally persisting right systemic arch was ligatured. The effect was that the left systemic arch now took over the blood flow from the heart to the dorsal aorta and became permanent instead of degenerating (Fig. 213).

The vitelline veins on their way to the heart pass through the region in which the liver is developed at a later stage. When the rudiment of the liver is formed (see section 12–4) it envelops the vitelline veins. These break up into a system of capillaries (or sinuses) associated with the liver lobules and thus give rise to the hepatic portal system. The veins in front of the liver and the heart become hepatic veins, and the parts of the veins caudal to the liver give rise to the hepatic portal vein. The two vitelline veins join together in amniotes, and the common trunk is known as the **venous duct** (ductus venosus). It also receives later the umbilical veins, which return blood from the allantois and the placenta.

The blood vessels of the cardinal system undergo extremely complicated transformations in the course of ontogenetic development and of evolution in vertebrates. The anterior cardinal veins are least changed; they become the internal jugular veins, and the common cardinal veins become the anterior venae cavae. The postcardinal and subcardinal veins after a series of transformations, which will not be described here, give rise, in air breathing vertebrates, to the inferior vena cava, which uses the venous duct as its path of entry into the heart.

11–7   DEVELOPMENT OF THE GONADS

The gonads in vertebrates develop from the upper edge of the visceral layer of lateral plate mesoderm in the posterior half of the body. The first rudiment of the gonad appears as a longitudinal strip of mesodermal epithelium lining the body cavity immediately lateral to the dorsal mesentery. This thickening is called the **germinal ridge.** Initially the ridge is produced by the mesodermal cells assuming a high columnar shape, but soon the

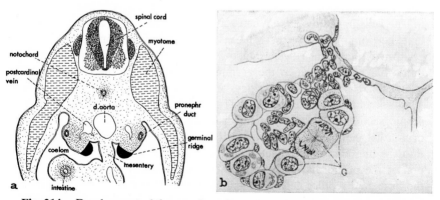

Fig. 214.   Development of the gonads. *a,* Transverse section of a mouse embryo to show the position of the germinal ridges and their relation to the mesonephric rudiment and the dorsal mesentery. *b,* Gonad rudiment of a frog, showing distinction between the large primary gonocytes and the much smaller somatic cells. (After Bouin, from Brachet, 1935.)

cells become arranged into a compact mass several cells thick, and the ridge protrudes into the coelomic cavity (Fig. 214, a). Still later, the connection between this cell mass and the peritoneal wall becomes constricted laterally, and the gonad remains suspended from the peritoneal wall only by a double layer of peritoneum—the **mesorchium** or **mesovarium.**

The genital ridges are situated in the same region of the body as the rudiment of the mesonephros and in close spatial association with the latter, the genital ridge being medial to the mesonephros. Although in the later development of vertebrates the genital apparatus becomes associated with the excretory system, the early stages of the two organ systems are completely independent of each other.

What has been said so far about the formation of the gonad rudiment refers only to its gross morphology. A more detailed microscopic examination of the germinal ridge discloses that the ridge consists of two types of cells. The majority of cells are very similar to the other cells of the peritoneal epithelium, even though they may become columnar in the initial stages of the development of the ridge. The cells of the second type have a very different appearance. They are much larger than the ordinary mesodermal cells, they contain large, more or less vesicular nuclei and their cytoplasm differs from that of the surrounding cells in its staining properties (amphophil instead of acidophil in birds) and (in amphibians) in its higher yolk content. The shape of the cells is nearly spherical and they do not participate in the epithelial arrangement of the typical mesodermal cells, but appear to be interspersed between the cells of the mesodermal epithelium. Cells of this peculiar type are known as the **primordial germ cells** (Fig. 214, b).

The Origin of the Primordial Germ Cells. According to views held by many embryologists only the primordial germ cells are destined to give rise to the gametes (eggs and spermatozoa), while the ordinary mesodermal cells of the genital ridges differentiate into the somatic cells of the adult gonad: the Sertoli cells in the testes, the superficial epithelium and the follicle cells in the ovaries. Whatever their eventual fate, it appears certain that the primordial germ cells have a different origin from the rest of the cells in the germinal ridges, that they first appear in parts of the embryo other than those in which the germinal ridges lie, and that the primordial germ cells reach the germinal ridges after a more or less extensive migration.

In amphibians the primordial germ cells may be traced back in microscopic sections to a position just above the dorsal mesentery. Here the primordial germ cells lie surrounded by mesenchyme before they migrate into the mesodermal epithelium of the germinal ridges. The median position of the primordial germ cells is probably already a secondary one. In the birds and reptiles cells identical in their cytological properties with the primordial germ cells have been found in early developmental stages— before the beginning of segmentation of the mesoderm—in the endodermal layer of the extraembryonic part of the blastoderm. Similarity of appearance cannot, however, be considered as proof that the cells found in the extraembryonic endoderm are the same cells as appear later in the germinal ridges.

To test whether the primordial germ cells are derived from the cells in the extraembryonic endoderm, the latter may be destroyed prior to their migration. In the chick the cells in question occupy a crescentic area in front of the head end of the embryo. This area can be destroyed in various ways, as for instance by cauterizing it with a hot needle or irradiating it with radium emanation (Dantschakoff, 1941). As the crescentic area lies well beyond the embryo proper, the development of the latter is unimpaired. At the time when the germinal ridges should have been formed, no primordial germ cells are to be found in the treated embryo, and the germinal ridges themselves are absent or very poorly differentiated, although the mesoderm from which they develop had not been tampered with.

Two conclusions are drawn from these experiments:

1. That the peculiar cells of the extraembryonic endoderm are actually the primordial germ cells, which are later incorporated into the gonad.
2. That the germinal ridge does not develop in the absence of the primordial germ cells, which stimulate the peritoneal epithelium to develop into the somatic part of the gonad, while they themselves are the source of the generative cells in the gonad.

These conclusions are further supported by the following experiment: if the primordial germ cells are exposed to very weak irradiation with X-rays, while still in the extraembryonic endoderm, they are damaged to a greater or lesser extent, but not killed at once. Some of the damaged cells succeed in reaching the site of gonad formation and are to be found there for a short time. In this case the genital ridges are formed, thus showing that even damaged primordial germ cells may exert an inducing influence on presumptive gonad mesoderm. After a short time all the primordial germ cells perish, and there remains a gonad consisting of mesodermal cells only. The gonad may grow and develop further without showing any signs of producing generative cells (Dantschakoff, 1941). Such a gonad, devoid of generative elements, may be called a sterile gonad.

As to the means by which the primordial germ cells travel from the extraembryonic endoderm to their later position in the germinal ridges, it is supposed that they do so largely through the blood vessel system. The cells have been observed to leave the endodermal epithelium, to move slowly into the space between endoderm and mesoderm, and then to penetrate into the blood vessels of the area vasculosa. The primordial germ cells may actually be seen in the blood stream. Although most of them eventually reach the germinal ridges, some go astray and are found in various parts of the body, where they eventually degenerate. The dependence between the mesodermal part of the germ ridges and the primordial germ cells is thus a reciprocal one: the germinal ridges do not develop without a stimulation from the primordial germ cells, and the latter cannot persist if they fail to become embedded in the suitably differentiated mesoderm of the germ ridges.

Further experimental work on the origin of primordial germ cells has been carried out in the amphibia. Two suggestions have been made as to the origin of the cells. Some embryologists consider the primordial germ cells to be derived from the endoderm as in birds; others hold that the cells originate in the lateral plate mesoderm. The experimental evidence is rather in favor of the second view.

In the neurula stage the lateral plate mesoderm of the trunk region is found near the blastopore and even around its ventral and lateral edges, as can be proved by localized vital staining. This area was removed in experiments to test the origin of the primordial germ cells (Nieuwkoop, 1946). As a result no primordial germ cells could be found later in the operated embryos. It is interesting to note that the peritoneal epithelium in the operated embryos obviously developed at the expense of the more dorsal and cranial parts of the mesodermal mantle. Contrary to what was found in the chick, the germinal ridges were also present in operated embryos, though they were of a smaller size than normal. The germinal ridges were, however, sterile. The result is in keeping with the suggestion that the primordial germ cells are derived from the ventrolateral mesoderm, and that they have a different source of origin from the germinal ridges. Still more convincing are results of some heteroplastic transplantations.

It has been stated earlier (section 11–5) that in salamander embryos it is possible to remove the whole of the endoderm in the neurula stage. The endoderm from another embryo may then be inserted into the empty ecto-mesodermal shell. If the operation is carried out heteroplastically, and if the two species are chosen in such a way as to allow the cells of each to be distinguished by some peculiarity, then the derivation of the primordial germ cells, or any other cells, from the endoderm or the mesoderm respectively can be determined with certainty. The experiments were carried out with the purpose of investigating the origin of the primordial germ cells using *Triturus cristatus* as one of the components, and *Triturus alpestris* or *Ambystoma mexicanum* as the other. All cells of the latter two species contain in their cytoplasm considerable amounts of pigment granules, while in the cells of *Triturus cristatus* (except for the melanophores) pigment granules are either completely absent or very few in number. The operation is a rather difficult one, but a sufficient number of operated embryos survived (Nieuwkoop, 1946). The result was a very clear-cut one: all the primordial germ cells found in the operated embryos had the specific properties (pigmentation) of the species to which the ectoderm and mesoderm belonged and not of the species to which the endoderm belonged. The primordial germ cells were, in this experiment, embedded in germinal ridges consisting of cells of the same species.

Now the source of the primordial germ cells, the ventrolateral mesoderm, can also be transplanted heteroplastically. Experiments, using the same species for the heteroplastic combination, have been performed on embryos in the late gastrula stage. In this stage the lateral plate mesoderm is to be found around the ventrolateral lips of the blastopore (Nieuwkoop, 1946). Depending upon little variations in the position of incisions in both embryos the final position of the heteroplastic graft varied to a certain extent. Sometimes the graft came to lie dorsally in the somite region or further cranially than was expected. In such cases no primordial germ cells were differentiated from the graft. In other cases, however, the graft occupied a position in the region of the genital ridge or immediately beneath it. In such cases a variable number of primordial germ cells were derived from the graft, while others were produced by the host. The proportion of graft and host primordial germ cells was roughly the same as the proportion of host and graft tissue in the lateral plate of the operated embryo. Some of the somatic cells of the germinal ridge were also developed from the graft, depending on the position in which it came to lie. In one case the graft occupied exactly the position of the germinal ridge, and practically all the somatic cells of the ridge were derived from the grafted tissue (*Triturus alpestris*). The primordial germ cells, which originate in the more ventral part of the lateral plate, showed the typical features of host cells (*Triturus cristatus*). The result is as good a proof as could be wished that the germinal ridge and the primordial germ cells have a different origin, and that the two become united secondarily.

Experiments have also been adduced in support of the endodermal origin of primordial germ cells in the amphibians. Bounoure (1939) irradiated the vegetal pole of newly fertilized frog's eggs with ultraviolet light and found that the number of primordial germ cells in later stages was greatly reduced. Monroy (1939) removed parts of endoderm in frog neurulae and found that the embryos were completely sterile if the ventral endoderm was removed, whereas the removal of the dorsal endoderm did not affect the development of the primordial germ cells. The result suggests that the primordial germ cells are derived from the ventral endoderm. Also Nieuwkoop (1946) observed the absence of primordial germ cells in embryos in which the whole of the endoderm was removed in the neurula stage. In no experiments, on the other hand, could it be shown that cells recognizable in any specific and indubitable way travel from the endoderm into the mesoderm of the germ ridges. The above experiments are therefore open to an alternative explanation. It is suggested that the endoderm is not the source of cells that become primordial germ cells but that it exercises an influence (induction) which is necessary for the development of the primordial germ cells (Nieuwkoop, 1946, 1950). Thus in the absence of the endoderm or of its inductive action, the primordial germ cells would not develop even if the cells normally undergoing this development are present.

The conclusion seems inevitable that the primordial germ cells are derived from different germinal layers in the amphibians and in the birds.

In many invertebrates, however (*Ascaris, Sagitta,* some insects), the primordial germ cells are known to be distinguishable in very early cleavage stages even before the germinal layers become segregated. The primordial germ cells can thus be considered as not belonging to any of the germ layers. This is in contrast to the somatic parts of the gonads and the auxiliary parts of the sexual organs (ducts, etc.) which are mesodermal.

As to the mechanism by which the primordial germ cells in the amphibians reach their destination, it is ameboid movement of the cells themselves. There is no reason to suppose that blood vessels play any part in their transport.

No experimental investigations have been carried out on the origin of the primordial germ cells in human embryos, but careful observations show that these are first found in the endodermal epithelium of the yolk sac in the vicinity of the allantoic stalk, and that from there the germ cells migrate into the adjoining mesenchyme and eventually take up their position in the germinal ridges (Witschi, 1948).

The Role of the Primordial Germ Cells. The exclusive role of the primordial germ cells as the sole source of generative cells has been challenged on various grounds. It has been claimed that after starting the development of the gonads the primordial germ cells degenerate and are replaced by secondary germ cells derived from the mesodermal epithelium of the germ ridge. Alternatively it is claimed that the external epithelium of the ovary (known as the germinal epithelium) even in adult life is capable of producing oocytes which further differentiate into ova (Allen, 1923). The difficulty in arriving at a definite decision on this point lies in the fact that in later stages of the development of the gonads the primordial germ cells lose their peculiar appearance, and it becomes increasingly difficult to distinguish them from the other cells of the gonad. Even if it is true that the germinal epithelium of the ovary may produce new ova during adult life, it is by no means clear that the cells differentiating as oocytes are not derived originally from the primordial germ cells. Certain experimental evidence can be adduced in favor of dualism existing in the cellular composition of the gonads, that is, of the presence, throughout life, of separate somatic and generative cells. The generative cells of the gonads show a peculiar sensitivity to X-rays, and a weak irradiation with these rays may cause a complete destruction of all generative cells in the gonad without impairing the somatic cells; the testis of a mammal after treatment remains intact, the seminiferous tubules can be seen in microscopical sections, but the spermatogonia and cells in other stages of spermatogenesis are absent.

On the other hand, it has been claimed (Parkes, Fielding and Brambell, 1927) that ovaries in mammals can regenerate after being removed by operation. The regeneration would proceed from the peritoneal epithelium, and the oocytes found in the regenerated ovary would be derived from the same source. This observation, if correct, would invalidate the idea that the primordial germ cells are the sole source of generative cells. It still remains true, however, that the so-called primordial germ cells are indispensable for the initial development of the gonads in the embryo.

In the foregoing, no distinction has been made between the development of the ovaries and the testes. In the early stages of organogenesis with which we are concerned here, there is practically no difference in the development of the gonads of the two sexes. Only with the onset of histological differentiation do the testes and the ovaries become increasingly different from each other. The later stages of the development of the gonads, as well as the development of the ducts and accessory organs of the genital apparatus, cannot be reviewed here.

CHAPTER 12

Development of the

Endodermal Organs

in Vertebrates

12–1 THE RELATION BETWEEN THE ARCHENTERON AND
THE DEFINITIVE ALIMENTARY CANAL

The relation between the archenteron and the definitive alimentary canal
is very different in vertebrates with complete cleavage from that in those with
incomplete (meroblastic) cleavage.

The Relation in Lower Vertebrates. The archenteron in *Amphioxus*
is originally lined with presumptive endoderm, presumptive mesoderm and
presumptive notochord cells. When the notochord and the mesoderm are
segregated from the endoderm, the latter closes the gap on its dorsal side
(see section 5–3) and the resulting cavity becomes the cavity of the
alimentary canal.

336

In holoblastic vertebrates—cyclostomes, ganoid fishes, lungfishes and amphibians—the presumptive notochord and mesoderm also participate in the lining of the archenteron, forming its roof. This roof is segregated from the endoderm and the endodermal cavity which is closed by the fusion of the free edges of the endodermal layer.

The resulting endodermal cavity consists of two unequal portions. The anterior portion is dilated and lined by a relatively thin endodermal epithelium. This part is usually referred to as the foregut. The posterior portion is called the midgut. The cavity of the midgut is narrower. Dorsally it is lined by a rather thin epithelium, but the ventral wall consists of a mass of large cells containing abundant yolk; the wall here is therefore very thick.

The fate of the various parts of the endodermal lining of the fore- and midgut in amphibians has been elucidated by the method of local vital staining (in newts by Balinsky, 1947; in frogs by Nakamura and Tahara, 1953). To stain the inner surface of the gut, cuts were made through the body wall (in some experiments through the neural plate) and pieces of agar soaked in vital stain were applied to the endoderm from the inside. After the stain was taken up by the endodermal cells the agar was removed; the wound healed easily, and the embryos continued to develop apparently quite normally.

In the neurula stage the ventral wall of the foregut becomes flattened out and later even folded slightly upward. The cavity is thus subdivided into two pocketlike recesses. The larger anterior one, lying immediately underneath the brain, gives rise to the cavities of the mouth and of the branchial region. The posterior pocket, bordering on the mass of the yolk-laden cells of the midgut, is known as the liver diverticulum, and in frogs it gives rise to the rudiments of the liver and the pancreas, but in salamanders it also participates in the formation of the stomach and the duodenum (Balinsky, 1947). It is rather remarkable that in frogs the liver diverticulum is extended downward and backward until the endoderm is perforated, and the cavity of the gut opens into the space between the endoderm and the mesoderm on the ventral side of the embryo. The significance of this perforation is not known.

The further development of the gut will be described mainly as it occurs in the frogs. Some difference between the processes in the frog embryos and in the newt embryos will be indicated where it seems desirable.

By vital staining it has been shown that the foregut endoderm gives rise to the following parts of the alimentary canal: part of the lining of the mouth, the pharynx, the esophagus, the stomach and the anterior half of the duodenum, as well as the lungs, the liver and part of the pancreas. The midgut contains the presumptive material for the posterior half of the duodenum, part of the pancreas, the intestine and the cloaca. The transformation of the foregut into the anterior part of the alimentary canal involves a very marked stretching of the whole endodermal rudiment with concurrent narrowing of the gut and constriction of its walls. This can best

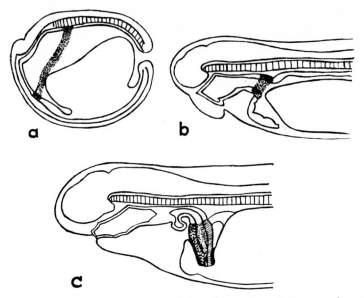

Fig. 215. Development of the stomach in a frog embryo (after experiments with vital staining). *a,* Position of the presumptive gastric endoderm (stippled) in the neurula stage. *b,* Constriction of the alimentary canal in the stomach region. *c,* Elongation of the stomach rudiment. (From Nakamura and Tahara, 1953.)

be shown by considering the development of the stomach in the frog embryo. In the neurula stage the presumptive material of the stomach is arranged in the form of a narrow ring (Fig. 215, *a*), the ventral part of which lies just in front of the liver diverticulum; the dorsal part occupies the roof of the gut just at the mouth of the midgut, and the right and left parts obliquely cross the lateral walls of the foregut. Figure 215, *b* shows how this ring contracts and at the same time broadens until eventually the rudiment of the stomach becomes more or less barrel-shaped, with a fairly narrow cavity.

The part of the gut in front of the stomach rudiment also elongates, but instead of becoming tubular it is flattened dorsoventrally and expanded sideways. Most of this cavity becomes the pharynx. The floor of the pharynx is raised partially as a result of the increase of the heart rudiment, which develops just underneath. The lateral edges of the pharynx are drawn outward even more and form the gill pouches, which will be dealt with in section 12–3. The epithelium lining the pharynx becomes rather thin, in contrast to the lining of the rest of the alimentary canal.

The extremely constricted part between the pharynx and the stomach becomes the rudiment of the esophagus, which is very short in the embryo. Posteriorly the stomach communicates with a section of the gut in which, for a while, the internal cavity remains rather broad where it extends downward into the liver diverticulum and backward into the midgut. The anterior wall of the liver diverticulum later gives rise to the liver parenchyme, while part of the cavity of the diverticulum becomes the cavity of the gall bladder.

The original communication of the liver diverticulum with the rest of the gut cavity is narrowed and becomes the bile duct. (Further details of the liver development will be given in section 12–4.) The midgut is greatly extended concurrently with the elongation of the embryo, which is rapid at this time (see section 9–4), and the cavity of the midgut becomes very narrow. The yolk contained in the cells of the midgut (as well as in the

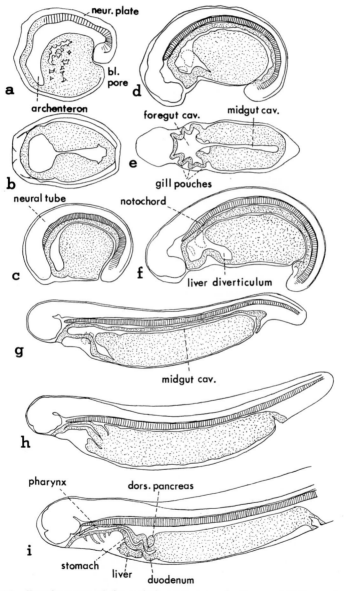

Fig. 216.   Development of the endodermal organs in the newt *Triturus taeniatus,* from the neurula stage to swimming larva. *b, e,* Frontal sections; all others are median sections. (From Balinsky, 1947.)

anterior parts of the alimentary canal) becomes gradually used up, and the cells are arranged in the form of a columnar epithelium lining the duodenum and the intestine.

The posterior end of the midgut becomes extended into the tail rudiment as a postanal gut (Fig. 159, *b*). The postanal gut has only an ephemeral existence and is soon broken up and disappears. The terminal part of the persisting midgut also gives rise in amphibians to the urinary bladder, which develops as a ventral evagination of the gut in late stages of larval life.

It will be noted that the cavity of the gut throughout its length becomes, in the frogs, the cavity of the alimentary canal. This is not so in the urodeles; in the newts the cavity of the midgut becomes occluded by the yolky endodermal cells. Some of these cells later degenerate and break down completely. For a time there is no cavity in the midgut, and only later a new

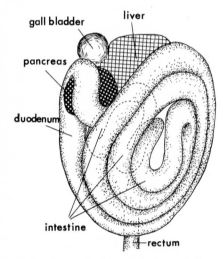

Fig. 217. Coiling of the intestine in a tadpole of the frog *Xenopus laevis*.

cavity develops by expansion of the cavity of the "liver diverticulum." The latter, in urodeles, is not only related to the development of the liver, but is actually incorporated into the main channel of the alimentary canal. The posterior end of the "liver diverticulum" at a later stage penetrates into the mass of the yolk-laden endoderm of the midgut. The central cells of the midgut become resorbed and the remaining cells, mainly those which were situated nearer to the surface of the midgut rudiment, align themselves into the epithelium of the intestine (Fig. 216).

As the main portions of the alimentary canal begin to take shape, the canal as a whole becomes twisted in a characteristic way. At an early stage the stomach has already assumed a slanting position (see Fig. 215). Subsequently the posterior end of the stomach is shifted to the left, while the adjoining part of the duodenum comes to lie transversely, going from left to right. The distal part of the duodenum is then bent in such a way that it reaches to the anterior end of the intestine, which is more or less in a dorsal position. The alimentary canal thus performs a complete spiral revolution, which may be referred to as the gastro-intestinal loop, and which occurs with greater or lesser modifications in all vertebrates. Where the loop is in its lowest position, it leaves a space or saddle on the dorsal side, and this space is taken up by the rudiment of the pancreas (see also section 12–4). The posterior part of the alimentary canal also becomes twisted into folds and loops as the result of the elongation of the alimentary canal which exceeds the elongation of the body. The folding and twisting of the duodenum and intestine is especially prominent in the tadpoles of the frogs which, in connection with their herbivorous diet, have a very long intestine (Fig. 217). The pattern of the twisting varies somewhat in different frogs and need not be followed here. In urodeles and also in fishes, the intestine does not elongate to the same degree and its folding may be very limited though the gastro-duodenal loop is always present.

The Relation in Higher Vertebrates.   In vertebrates having a meroblastic type of cleavage the development of the alimentary canal presents very different problems, and the processes leading to the formation of the definitive cavity of the alimentary canal are quite peculiar. In the following we will describe this process as it occurs in birds and mammals. It must be noted in the first instance that in birds the archenteric cavity is often lacking altogether, and if it is present, as a canal leading forward from Hensen's node, it is very small and its walls are not endodermal. The starting point for the development of the alimentary canal proper is a sheet of endoderm lying flat under the ectodermal and mesodermal parts of the embryonic region of the blastodisc. The sheet of endoderm lies flat on the yolk of the yolk sac or forms the roof of the mammalian yolk sac, which is a space

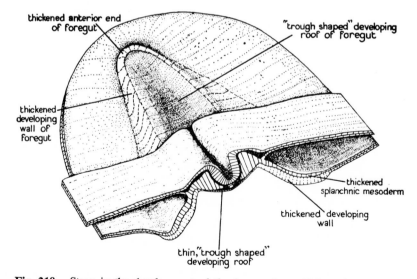

Fig. 218.   Stage in the development of the foregut in a chick embryo. Ectoderm with neural folds and the parietal mesoderm are shown as being cut away at the anterior end of the embryo. (From Bellairs, 1953.)

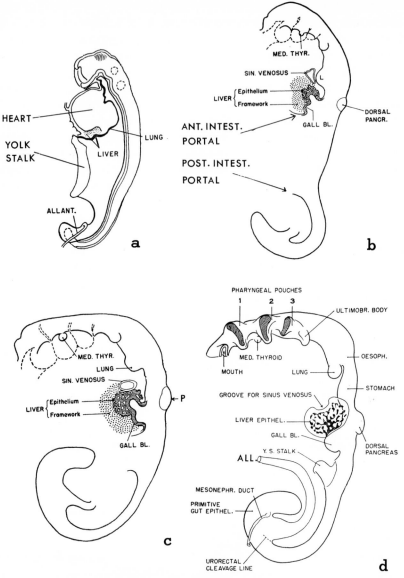

Fig. 219. Development of the endodermal organs in a human embryo (semidiagrammatic). (From Streeter, 1942.)

filled with fluid. In both cases the cavity of the alimentary canal is separated from the cavity of the yolk sac by a process of infolding.

During this infolding the median strip of the endoderm lying immediately under the notochord and somites remains in this position, while the immediately adjoining strips on the right and left become inflected downward, and the crests of the folds converge toward the middle and eventually fuse (Fig. 218). The inner surfaces of the folds contribute to the formation of the floor of the gut, while the outer surfaces of the folds are continuous with the endodermal lining of the yolk sac. Concurrently with the movement of the endodermal layer in a transverse plane a complicated shifting has been found to occur longitudinally; the median strip which forms the dorsal wall of the gut slides forward, while in the folds closing the gut laterally and ventrally endodermal material moves obliquely backward. During their movement downward the lateral strips of endoderm are accompanied by the visceral layer of the lateral plate mesoderm, which closely adheres to the endoderm throughout this whole series of formative movements. It is probable that the dynamic force of the movement is due to the endoderm and mesoderm jointly (Bellairs, 1953).

As a result of the downward movement of the visceral layer of the lateral mesoderm, the coelomic cavity becomes considerably expanded locally. The coelom is reduced again, however, when the body folds undercut the embryo, as explained on page 203. The alimentary canal becomes separated from the yolk sac cavity first of all at the anterior end of the embryo. This part corresponds to the foregut of the amphibian embryo. Somewhat later, the posterior part of the endodermal groove closes into a canal, and this becomes the posterior part of the alimentary system corresponding to the midgut in the amphibians. Accordingly we will refer to it as the midgut. Between the foregut and the midgut there remains a gap where the endoderm does not close to form a canal, or even a groove, and where it is in open communication with the yolk sac (Fig. 219).

The edges of the folds separating the foregut from the yolk sac where they meet in the middle form a ridge, known as the **anterior intestinal portal.** A similar edge at the anterior end of the midgut is the **posterior intestinal portal.** The gap between the two is very large at first, but its relative size diminishes with the growth of the embryo, and eventually it is reduced to the opening of the **yolk stalk,** connecting the cavity of the gut to the cavity of the yolk sac.

From the beginning the foregut is much broader than the midgut, and is flattened in cross-section. As in the amphibians, the foregut gives rise to the endodermal lining of all anterior parts of the alimentary canal including most of the duodenum, the liver and pancreas developing just in front of the anterior intestinal portal. The oral and especially the pharyngeal parts of the foregut remain expanded in a transverse direction, and the pharynx becomes drawn out to form the branchial pouches, but the posterior part of the foregut, corresponding to the esophageal, gastric and duodenal parts of the alimentary canal, eventually becomes round in cross-section. The esophagus in higher vertebrates is soon greatly elongated, in connection with the development of the neck. The midgut is from the start narrower than the foregut and soon becomes round in cross-section too.

Although the foregut in higher vertebrates is quite different in shape from the foregut of amphibians, it develops the gastro-duodenal loop in much the same way. The anterior (cardinal) end of the stomach is displaced to the left side, the posterior (pyloric) end is turned downward and toward the middle, while the anterior part of the duodenum assumes a transverse direction, thus bringing about the familiar position of the stomach in adult mammals and birds. The intestine becomes convoluted in a pattern that varies greatly not only as between different classes, but even in one class, as in various mammals. A peculiarity occurring in mammals is that sections of the intestine adjoining the yolk stalk, both anteriorly and posteriorly, sink down into the umbilical cord and lie for a time practically outside the body of the embryo proper as a sort of umbilical hernia. The convolutions of the intestine begin forming inside the umbilical cord, but well before birth the definitive intestine is withdrawn into the body and only the yolk stalk remains in the cord (this occurs in the human embryo during the third month of pregnancy).

The posterior end of the midgut gives rise to the cloaca and also to a postanal gut, which disappears later. The ventral wall of the cloaca produces the allantoic diverticulum (the endodermal part of the allantois), as has been indicated in section 7–1. In higher mammals (in particular in man), however, the allantoic diverticulum is formed very early as an outgrowth of the yolk sac at the posterior end of the embryo, even before the embryo becomes subdivided into the embryonic body and the extraembryonic parts. Later the allantoic diverticulum is incorporated into the ventral floor of the midgut and assumes the same position as it has in lower amniotes.

At the time of their formation both the foregut and the midgut are blind diverticula, without openings to the exterior at the front and hind end of the embryo. The formation of the mouth opening at the anterior end of the foregut occurs in much the same way in both holoblastic and meroblastic vertebrates. It will be dealt with summarily in the next section (12–2). The development of the anal or cloacal opening is, however, different. In vertebrates with holoblastic cleavage the blastopore, or part of it, persists, as a rule, as the anal (cloacal) opening. As in higher vertebrates there is no patent blastopore leading into an endodermal archenteron, the cloacal opening has to develop by a perforation of the body wall at the posterior end of the midgut. The point at which this perforation occurs is discernible as early as the primitive streak stage and lies at the posterior end of the streak (Fig. 147, a). When the primitive streak shrinks in the late gastrulation stages, it leaves in front the three germinal layers: the ectoderm, the mesoderm and the endoderm, lying one above the other. At the posterior end, however, the separation of germinal layers does not occur; the ectodermal and endodermal layers do not become separated by the intervening mesoderm and remain in close contact with one another. The resulting double-layered plate is the **cloacal membrane.** When the midgut becomes separated from the yolk sac by folds, the cloacal membrane is incorporated into the wall of the gut. The ectodermal side of the cloacal membrane is originally dorsal, but this position is inverted by the development of the tailbud, occurring just anterior to the cloacal membrane (Fig. 147, b). The tail rudiment protrudes backward, and the midgut above the cloacal membrane develops a diverticulum entering the tail rudiment— the postanal gut. As a result the cloacal membrane now comes to lie at the root of the tail with the ectodermal side facing downward. The part of the midgut adjoining the cloacal membrane is somewhat dilated and becomes the rudiment of the cloaca into which open the Wolffian ducts. The ectoderm is slightly depressed in the region of the cloacal membrane, forming the **proctodeum.** The cloacal membrane separates the cavity of the cloaca from the cavity of the proctodeum till late in embryonic development, but it is eventually ruptured and thus a free passage from the alimentary canal to the exterior is allowed.

In metatherian and eutherian mammals the cloaca becomes subdivided by a partition into a dorsal part, serving for the passage of feces, and a ventral part, communicating with the genital and urinary ducts and the urinary bladder. This separation is carried out well before the rupture of the cloacal membrane.

12–2 DEVELOPMENT OF THE MOUTH

In all vertebrates the mouth opening appears rather late in the embryonic life after all the primary organ rudiments have already been formed. The mouth opening breaks through where the anterior end of the endodermal part of the alimentary canal touches the ectodermal epidermis beneath the front end of the neural tube. The ectodermal epidermis here sinks in to form a pocketlike depression, the ectodermal mouth invagination or the **stomodeum.**

The ectodermal epithelium and the endodermal epithelium fuse here and become a **pharyngeal membrane** (Fig. 147, b). The membrane becomes very thin, and eventually disappears. Forthwith, the boundary between the ectodermal and endodermal epithelia becomes indistinguishable, and it is not an easy matter to determine later which part of the oral cavity is derived from the stomodeum and which from the endodermal gut. There are, however, indications that the stomodeal epithelium reaches (in mammals) to about the middle of the tongue ventrally and to the beginning of the pharynx on the dorsal side.

The determination of the stomodeum begins at the end of gastrulation, and the development of the ectodermal mouth invagination is induced by the endoderm when the anterior end of the archenteron comes in contact with the ectoderm of the presumptive mouth region (Ströer, 1933). No mesoderm ever penetrates here between the ectoderm and endoderm, and the endoderm remains in contact with the ectoderm throughout the subsequent stages until the rupture of the pharyngeal membrane. After the endoderm and ectoderm come into contact in the oral region, the oral ectoderm acquires, to some extent, an ability to differentiate as stomodeum. The anterior part of the archenteron may now be removed, and

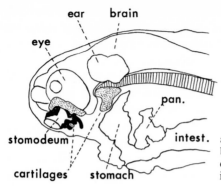

Fig. 220. Development of the mouth and stomodeal invagination in a newt embryo after removal of the anterior endoderm in the neurula stage. (From Balinsky, 1939.)

the remaining ectoderm produces a mouth invagination, although there are no endodermal oral and branchial parts behind it. The mouth invaginations thus developed are small and their shape is not normal. If, however, the operation is performed a bit later, in the neurula stage, the stomodeum may develop in an almost completely normal way, with the sole reservation that it does not lead into the endodermal alimentary canal (Fig. 220). As the oral ectoderm acquires the ability for self-differentiation, all the rest of the epidermis loses the competence for development into stomodeum. If the region where the mouth is to develop is covered with a flap of epidermis taken from a different part of the body, the development of the stomodeum is suppressed, in spite of the presence of oral endoderm. The presumptive epidermis of the stomodeum can also be transplanted in the neurula stage, and if the site of the transplantation is in the immediate vicinity of the normal mouth, an additional stomodeal invagination is formed by the graft, and this invagination may break through into the endodermal cavity of the host (Balinsky, 1948).

Besides giving rise to a part of the oral epithelium, the stomodeal invagination furnishes the cells which become the rudiment of the anterior lobe of the **hypophysis.** This rudiment is formed as a solid bud, or a small pocket **(Rathke's pocket),** on the dorsal side of the stomodeal invagination, just in front of the pharyngeal membrane. The rudiment pushes backward through the connective tissue and eventually comes to rest underneath the diencephalon. The original connection with the stomodeum becomes interrupted, while the floor of the diencephalon furnishes the posterior lobe of the hypophysis.

One of the most characteristic differentiations of the oral cavity is the **teeth.** The rudiments of the teeth consist of an epithelial cap (the **enamel organ** which secretes the enamel), and a connective tissue **papilla** which produces the dentine (Fig. 221). The enamel organs of the teeth may develop from both the ectodermal and the endodermal epithelium. This can be proved by experiments similar to the ones quoted above, in which either the stomodeum or the endodermal oral epithelium is prevented from developing. The mouth invaginations, developing in the absence of the anterior end of the archenteron and the parts derived therefrom, often have well differentiated teeth. When the endodermal oral cavity develops without a stomodeum being formed, or when the two are not in contact, teeth can be found developing in the endodermal epithelium.

In cyprinid fishes one or two rows of teeth develop on the inner surface of the last branchial arch (the pharyngeal teeth). There can be no doubt that the enamel organs in these teeth are of endodermal origin.

In the lower vertebrates, the fishes and amphibians, the connective tissue papillae of the teeth project from the inside into the stratified epithelium lining the mouth, and the enamel organs of the teeth are developed from the malpighian layer of this epithelium. In mammals, however, the epithelium sinks down at the edges of the jaws into the connective tissue in the form of ridges. The rudiments of the individual teeth are formed at the

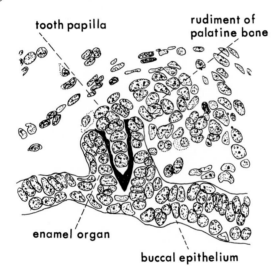

Fig. 221.   Rudiment of tooth in a salamander embryo. (After de Beer, 1947.)

edge of the dental ridge, at the expense of a connective tissue papilla and of the layer of innermost cells of the dental ridge which adjoin the papilla and form the enamel organ. The teeth are thus formed and begin to grow deep in the tissue of the jaw, and they erupt to the surface only when they have almost reached their full development.

While the cells of the enamel organ may be ectodermal or endodermal, the connective tissue cells of the dental papillae are derived from the neural crest. The neural crest is also the source of skeletogenous cells for the development of the mandibular arch: the quadrate and the Meckel's cartilage (see section 10–3). The development of the mouth as a whole depends therefore on the harmonious cooperation of cells coming from different primary rudiments—the foregut, the epidermis and the neural crest.

The development of the mandibular arch is, in part at least, directly dependent on the influence exercised by the ectodermal mouth invagina-

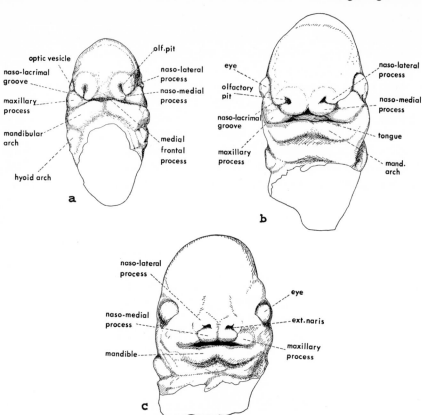

Fig. 222. Development of the face of a pig embryo. (After Patten, 1944.)

tion; if the ectodermal mouth invagination fails to be formed the ventral part of the mandibular arch is not formed. An additional Meckel's cartilage can develop in connection with a mouth invagination resulting from the transplantation of the presumptive stomodeal ectoderm. The quadrate, on the other hand, may develop even in the complete absence of any sort of oral cavity; it is thus not dependent on the latter (Balinsky, 1948).

The development of structures surrounding the edges of the mouth contributes very much to the formation of the face in man and of corresponding parts in higher mammals. The structures in question are a number of swellings, consisting of actively growing mesenchyme and covered by ectoderm, which are formed around the stomodeal invagination (see Fig. 222). At the dorsal edge of the stomodeum medially there is a slight swelling —the **medial frontal process.** Lateral to the frontal process on each side there develops a U-shaped swelling encircling the nasal pit. The free ends of the U are directed downward. The inner branch lies just alongside the frontal process and impinges on the edge of the mouth as the **medial nasal process.** The outer branch, lying lateral to the nasal pit, is the **lateral nasal process,** and it does not quite reach to the edge of the mouth. Around the angle of the mouth on each side another U-shaped swelling develops with the upper branch ending on the edge of the mouth as the **maxillary process.** The lower branch spreads along the lower edge of the mouth as the **mandibular process.**

The two medial nasal processes grow downward and toward the midline until they fuse and exclude the frontal process from participating in the formation of the edge of the mouth. The maxillary processes grow forward and eventually fuse with the lateral edges of the medial nasal processes, thus completing the upper edge of the mouth. The opening of the nasal pit remains just above the line of fusion of the maxillary and medial nasal processes. The upper edge of the maxillary process also fuses with the lateral nasal process. The furrow lying between the maxillary process and the lateral nasal process, as can be seen in Fig. 222, leads from the angle of the eye to the nasal pit. The infolded epidermis lining this furrow gives rise to a ridge of epithelial cells, which later becomes hollowed out and establishes a communication between the space underneath the eyelids and the nasal cavity. This is the **nasolacrimal duct.**

The lower edge of the mouth acquires its final shape after the median fusion of the two mandibular processes.

The nasal, maxillary and mandibular processes, as has been stated above, are essentially proliferating masses of mesenchyme covered externally by ectodermal epithelium. The mesenchyme later ossifies and gives rise to some of the most important parts of the facial skeleton. The mesenchyme in the lower portions of the medial nasal processes ossifies as the pair of premaxillary bones. The maxillary bones are produced by the ossification of the mesenchyme of the maxillary processes. The mesenchyme of the mandibular processes gives rise to the mandibular bones. The upper parts of the medial nasal processes together with the medial frontal process develop into the back of the nose.

12–3   DEVELOPMENT OF THE BRANCHIAL REGION

The development of the branchial region is no less dependent on the cooperation of parts of different origin than is the mouth region. The leading part in this development belongs to the endoderm. The endodermal cavity in this region, from its beginning, is distended in a transverse direction. In the amphibians it is derived from the inflated part of the foregut. In the stage immediately following the closure of the neural tube, the lateral walls of the pharyngeal cavity bulge out and produce a series of outwardly directed pockets on each side. These pockets are the **gill pouches.** The gill pouches are developed one after another beginning with the first pair, that is, the one lying just posterior to the mandibular arch. As the endodermal gill pouches reach the epidermis, having pushed aside the intervening mesoderm, the epidermis becomes folded inward, to meet the gill pouches. A series of branchial grooves is thus developed on the surface of the embryo,

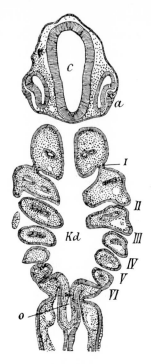

Fig. 223. Frontal section of an embryo of a skate, *Raja*, showing branchial pouches and clefts, which are indicated by Roman numerals. *a*, Eye; *c*, cavity of brain; *Kd*, cavity of pharynx; *o*, oesophagus. (After Maurer, in Hertwig, 1906.)

each groove corresponding to an endodermal pouch (Fig. 223). The outer wall of the endodermal pouch and the inner wall of the epidermal groove fuse into a **closing plate,** similar to the pharyngeal membrane. A **branchial cleft** is formed when the closing plate becomes perforated, so that an open communication is established between the pharyngeal cavity and the outer medium. In aquatic vertebrates gill filaments are then developed on the walls of the gill clefts; these are the **internal gills.**

The development does not always reach the final stage—the formation of the gill filaments serving for respiration. Even in aquatic vertebrates some of the gill pouches may not develop into gill clefts: thus in the bony fishes and the amphibians the first pair of gill pouches (the one lying behind the mandibular arch) does not reach the epidermis, and remains a blind diverticulum of the pharyngeal wall. In the amniotes the gills never function as respiratory organs, but the gill pouches and the branchial grooves nevertheless develop (Figs. 155, 156). The further development is then arrested; the closing plates remain unperforated, or if they do become perforated, no gill filaments are developed in the clefts. The whole system is later reduced to a greater or lesser extent, and its remnants are used up for the development of parts having nothing to do with the respiratory function; thus the first pair of branchial pouches becomes the eustachian tube in the terrestrial vertebrates. The third, fourth and fifth gill pouches give rise to a series of glands of internal secretion: the **thymus,** the **para-**

thyroids and the **ultimobranchial** bodies. All these organ rudiments develop from masses of endodermal cells that become detached from the walls of the branchial pouches and are then shifted downward and backward until they come to lie in the neck or in the anterior trunk region. The importance of these glands for the well-being of the animal gives a possible explanation for the persistence of the branchial pouches in the embryos of terrestrial animals although they had long since lost their original functional significance.

A further important gland of internal secretion associated with the pharynx in its development is the **thyroid gland.** The thyroid gland develops in vertebrates as a ventral pocket in the floor of the pharynx. Subsequently the pocket becomes closed and separated from the pharyngeal wall. The thyroid rudiment is then displaced in a caudal direction, until, in terrestrial vertebrates, it comes to lie ventral to the trachea.

It has already been stated that the endodermal gill pouches are the initiators of all the developments in the branchial region. The epidermal branchial grooves are induced by the endodermal pouches when they touch the epidermis; without the endodermal gill pouches the epidermal grooves do not develop. Neither do they develop if the endodermal pouch, though present, does not reach the epidermis (as in the case of the first pouch in amphibians and bony fishes).

The relations between the endoderm and the external gills have been discussed already (section 10–4). For the development of the external gills in the urodeles it is also necessary that the endodermal gill pouch reach the epidermis, otherwise the gills fail to appear. The external gills in the urodeles are later supplemented by internal gills—gill filaments developing on the branchial arches. Both external and internal gills function simultaneously, until both are reduced during metamorphosis. In the anurans the external gills function only temporarily, during a short period after the hatching of the larvae. Soon after the little tadpoles begin to swim, a fold of skin appears anterior to the external gills—the **opercular fold.** The opercular fold spreads backward over the gill region, covering both the external gills and the gill slits. Both are thus included in a branchial cavity. The posterior edge of the opercular fold becomes attached to the skin behind the branchial region, so that only a narrow opening—the **branchial aperture**—leads from the branchial cavity to the exterior. At the same time the external gills are reduced in size and the internal gills develop on the branchial arches beneath the external gills and function throughout the whole period of larval development.

The visceral skeleton is an important integral part of the branchial region. As has been stated above, the visceral skeleton develops from cells of the neural crest. In their downward migration the neural crest cells are split by the gill clefts into several languettes moving in between the adjacent gill pouches. Later the branchial arches are formed by the chondrification of the neural crest mesenchyme in about the same position as the streams of migrating cells were to be found. The dependence of the visceral arch development on the branchial clefts is, however, a more intimate one than would follow from the above. In the absence of the gill pouches the neural crest mesenchyme does not chondrify, and no branchial arches are formed (Balinsky, 1948). If the number of branchial pouches is reduced (after an operation in which part of the endoderm of the branchial region has been removed), the number of visceral arches is similarly reduced; one skeletal arch is developed on each side of the remaining branchial clefts, thus the number of arches is one more than the number of clefts present. In normal development the visceral arch that is developed in front of the first gill pouch is the hyoid arch; the rest are the branchial arches proper.

For the development of the visceral arches it is not necessary for the endodermal gill pouches to establish a connection with the epidermis and for the gill cleft to break through; the presence of the endodermal pouches, even if they are represented by blind pockets, is sufficient to induce the development of the skeletal arches.

We can now review the dependence of the development of the various parts of the visceral skeleton on the adjoining parts of the alimentary canal.

1. The upper part of the mandibular arch (the quadrate and the region of articulation with the lower jaw) develops independently of parts of the alimentary canal.
2. The lower part of the mandibular arch is dependent on the ectodermal mouth invagination.
3. The hyoid arch and the branchial arches are dependent on the endodermal gill pouches.

A very peculiar feature in the development of the amphibian visceral skeleton is presented by the second basibranchial. This skeletal element is developed from mesodermal mesenchyme, not from neural crest mesenchyme, and it is also independent of the endodermal gill pouches; it is formed even if the whole of the endoderm of the branchial region has been removed.

The visceral arches, which are dependent in their development on the epithelial parts of the alimentary canal, appear to exercise some influence on the development of the teeth. The tooth rudiments, consisting of the ectodermal or endodermal enamel organ and the papilla, derived from neural crest mesenchyme, are formed in connection with certain skeletal elements: Meckel's cartilage, the rudiments of the vomer and palatine bone, and later of the premaxilla, maxilla and dental bone.

With the reduction of the branchial pouches in adult terrestrial vertebrates the visceral skeleton becomes modified. The lower end of the hyoid arch persists as the body of the hyoid bone; the lower part of the first branchial arch becomes utilized in the formation of the horns of the hyoid bone. Parts of the subsequent branchial cartilages contribute to the formation of the thyroid cartilage and cartilages surrounding the trachea.

12–4   DEVELOPMENT OF THE ACCESSORY ORGANS OF THE ALIMENTARY CANAL: LUNGS, LIVER, PANCREAS

The Lungs.   The lungs develop from a rudiment which is a pocketlike evagination of the endodermal epithelium on the ventral side of the alimentary canal, just posterior to the branchial region. The pocket at first projects straight downward. At its tip it bifurcates, and the two branches grow out to the sides and backward. The unpaired medial part of the rudiment becomes the trachea; the two branches give rise to the two bronchi and to the lungs themselves.

In the lower vertebrates the lungs are developed as saclike expansions at the ends of the bronchi, the walls of which become folded to various degrees. In the warm-blooded vertebrates, birds and mammals, in which the lung respiration attains the highest efficiency, the greater degree of differentiation of the air spaces in the lungs already becomes manifest in the earlier stages of lung development. In mammals the distal ends of the bronchi as they grow out become branched in a more or less dichotomous fashion, the branches representing the secondary, tertiary, etc., bronchi, and the bronchioles. The alveoli are eventually developed on the terminal branches of this system (Fig. 224).

The unpaired part, the rudiment of the trachea, may elongate greatly by subsequent growth. Although the first visible rudiment of the lungs is ventral and unpaired, there is good reason to believe that the lungs are derived from originally paired and lateral rudiments. By local vital staining in amphibians it has been ascertained that the presumptive epithelium of the lungs lies, in the neurula stage, in the lateral walls of the foregut, just posterior to the presumptive endoderm of the gill pouches. The presumptive lung endoderm later shifts downward toward the midline (Balinsky, 1947).

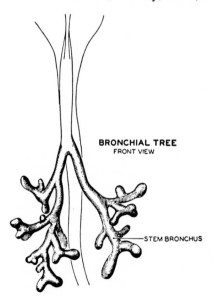

BRONCHIAL TREE
FRONT VIEW

STEM BRONCHUS

Fig. 224.   Branching of the lung rudiment in a 35-day-old human embryo. (From Streeter, 1948.)

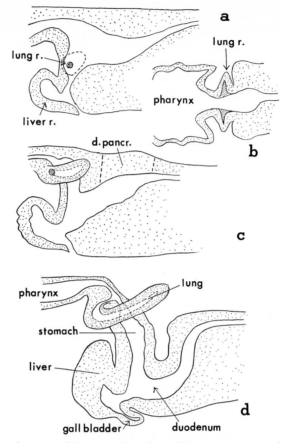

Fig. 225.   Development of the lungs in a frog, *Xenopus laevis* (semidiagrammatic).
a, c, d, Projection on the median plane; *b,* frontal section.

In the clawed toad *Xenopus* (and possibly also in other frogs), the lung
rudiments first become noticeable as two separate lateral pockets in a
stage when the parts of the alimentary canal are not yet clearly separated
from one another. When the gastric part of the gut becomes inflected downward, the lung rudiments are found just behind and ventral to the crest of
the transverse fold separating the pharyngeal section of the gut from the
esophagus. Following this the part of the gut cavity connected to the
lateral lung rudiments protrudes forward and becomes a distinct pocketlike
evagination (Fig. 225). This cavity is the rudiment of the trachea (which
is very short in frogs). The cavity of the trachea, which is continuous with
the cavities of the lung rudiments, becomes temporarily separated from the
esophagus and later opens into the pharyngeal cavity. This new opening
is the glottis (it does not coincide with the mouth of the original invagination which gave rise to the trachea) (Nieuwkoop and Faber, 1956).

The lateral and independent origin of the lung rudiments in amphibians
makes it probable that in the early history of the terrestrial vertebrates the

lungs developed from the last pair of gill pouches, which failed to break through to the exterior and became adapted to the retention of air gulped in through the mouth; thus they became an organ in which the oxygen could diffuse into the blood vessels supplying the organ.

The swim bladder of fishes is similarly a pocket, growing out from the endodermal wall of the alimentary canal posterior to the branchial region. In many fishes the cavity of the swim bladder remains permanently in communication with the esophagus, and air can be taken into the swim bladder through this canal. It is fairly obvious that the unpaired lung of the lungfishes (Dipnoi) is the same organ as a swim bladder but adapted to respiration. Whether the rudiment of the swim bladder may be compared to a gill pouch has not been investigated.

The Liver.   The liver in all vertebrates develops from the endodermal epithelium on the ventral side of the duodenum.

In amphibians the site of liver development is the anterior wall of the liver diverticulum, referred to on p. 337. At the stage when the main parts of the alimentary canal begin to take shape, the anterior wall of the liver diverticulum bulges forward, so that the slitlike cavity of the diverticulum enlarges locally. This pocketlike enlargement of the gut cavity is the **primary hepatic cavity** (Fig. 226). At the time when this occurs, in frogs, the liver diverticulum is still in communication with the space between the endoderm and the mesoderm. Next the front wall of the hepatic rudiment becomes thrown into folds, and the folds occlude most of the primary hepatic cavity leaving only the most posterior part open. At the same time the original communication between the primary hepatic cavity and the duodenum is constricted, and becomes gradually transformed into the **bile duct.** Simultaneously the opening into the submesodermal space becomes

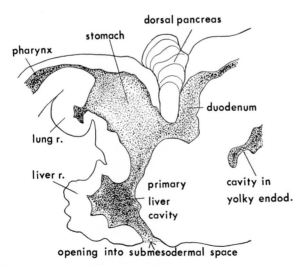

Fig. 226.   Reconstruction of the lung, liver and pancreas rudiments of the frog *Xenopus laevis* (median section).

closed by a sheet of cuboidal endodermal epithelium, and the adjoining posterior remnant of the primary hepatic cavity becomes the cavity of the gall bladder. The epithelial folds of the anterior wall of the liver rudiment soon break up into strands of cells, which for a short time may appear as tubules, with their lumen opening into the remainder of the primary hepatic cavity and thus also communicating with the rudiment of the bile duct and the gall bladder. The strands of liver cells become interwoven with blood vessels and sinuses produced by ramifications of the vitelline veins (see p. 329).

In the embryos of amniotes the structure of the gut in early stages is very different from that of amphibians owing to the absence of the yolk in the endoderm, and the liver rudiment accordingly has also a very different appearance, but it develops in a corresponding position, namely in the ventral wall of the gut posterior to the section which gives rise to the stomach. The first visible rudiment of the liver can be found as a pocket-like evagination on the anterior intestinal portal (see p. 343) (Fig. 219, *a*), at a stage when the opening from the yolk sac into the definitive gut is still quite wide. As the floor of the gut continues to close in an antero-posterior direction, the liver rudiment is later found some distance in front of the anterior intestinal portal, well within the foregut. The endodermal cells then begin migrating forward from the original pocketlike evagination, in the form of solid strands or cords of cells (Fig. 219, *b, c*). The strands of liver cells form a meshwork and enclose the blood vessels (the vitelline veins and their ramifications) lying in the region posterior to the heart. From an interaction of the cords of liver cells and the blood vessels there emerges eventually the complicated structure of the adult liver. A fine lumen may be seen sometimes running along the length of the strands of epithelial cells, so that they may appear as thin tubules.

The gall bladder is formed as a secondary hollow outgrowth at the posterior edge of the original hepatic rudiment.

The liver increases in size very rapidly, and soon becomes a large and massive organ, although the part of the duodenal wall from which it develops is relatively a very small one, and, even of that, a large proportion is used up for the formation of the gall bladder and cystic duct.

The Pancreas. The pancreas develops from two rudiments—a ventral rudiment (or two ventral rudiments) and a dorsal one. The ventral rudiment develops from a part of the ventral wall of the duodenum just posterior to and in close association with the rudiment of the liver. It may appear as a pocketlike evagination and very soon becomes subdivided distally into a system of proliferating epithelial tubules. The original evagination becomes the duct of the ventral pancreas.

The dorsal pancreatic rudiment in amniotes is also a pocketlike evagina-tion of the duodenum, but it appears on its dorsal side slightly in front of the liver rudiment.

In frogs the dorsal pancreas develops from a section of the roof of the midgut, posterior to the rudiment of the stomach. A part of the roof becomes cut out by transverse crevices reaching from the cavity of the gut right into the submesodermal space above the gut. At the same time that part of the gut wall which is destined to become the pancreas makes an abortive attempt to form a pocketlike evagination, but eventually the dorsal pancreatic rudiment in amphibians, as well as the ventral rudiment, is a solid mass which becomes transformed into a system of alveoli and ducts secondarily by rearrangement of cells.

The dorsal and the ventral pancreatic rudiments may remain completely independent throughout life, as in the dogfish, but in amphibians and amniotes the two rudiments approach each other and fuse completely. The system of pancreatic ducts becomes reorganized in later life and is very variable in different vertebrates.

12–5   DETERMINATION OF THE ENDODERMAL ORGANS

Experimental investigation of the endoderm in early stages of development shows that the endodermal organ rudiments, like those derived from other germinal layers, are not determined right from the start, but that the endodermal cells destined to participate in the formation of the various organ rudiments are no more determined for their respective fates than are the cells of the other germinal layers. In the earlier stages of development their fate is a function of the position that each cell or group of cells occupies in the embryo as a whole. This can be proved by isolating parts of the presumptive endoderm and cultivating them apart from the rest of the embryo or by transplanting them into an abnormal position.

In an extensive series of experiments pieces of presumptive endoderm of young gastrulae were cultivated in the "Holtfreter solution." Various tissues were observed to differentiate from such isolated pieces; some conformed to the normal destiny of the isolated parts, and some did not. The range of differentiations included not only oro-branchial epithelium, stomach epithelium, liver, pancreas and intestine, but also notochord and muscle, which should not have developed from the presumptive endoderm if it were to keep to its prospective significance (Holtfreter, 1938a, 1938b). Thus the fate of the endoderm is not laid down finally in the early gastrula stage. Much greater deviations from the prospective significance of the various endodermal parts could be observed when these parts were placed in surroundings which, unlike the saline solution, could actively influence the differentiation of these parts.

In the early neurula stage it is possible to separate the whole of the endoderm of a newt embryo from the ectoderm and mesoderm. The endoderm is removed as a whole through a slit on the ventral side of the embryo, leaving the ectoderm and mesoderm as an empty shell. The isolated endoderm can then be inserted again into the ecto-mesodermal shell of the same embryo, or of another embryo of the same species, or even into the ecto-mesodermal shell of an embryo of another species; the endoderm of the small *Triturus taeniatus* has been successfully implanted into the ecto-mesodermal shell of the larger *Triturus alpestris* (Mangold, 1949). The implantation may be carried out so that the orientation of the endoderm is in harmony with the orientation of the ecto-mesoderm, or the endoderm may be implanted in an inverted position. In the first case a completely normal larva has been observed to develop. A normal embryo also develops if the endoderm is implanted with its dorsoventral orientation reversed. This shows that the determination of the dorsal and ventral parts in the endoderm is not fixed in the endoderm itself, but is imposed on the endoderm by the surrounding ecto-mesoderm. Here we may recall that in the experiments with the transplantation of the dorsal lip of the blastopore (primary organizer) the endo-

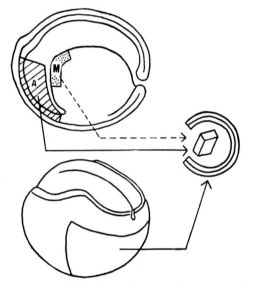

Fig. 227. Testing the determination of parts of the neurula endoderm in the newt *Triturus pyrrhogaster* (diagram of operation). *A*, Anterior endoderm; *M*, middle endoderm. (From Okada, 1955.)

derm is often observed to develop a secondary lumen of the midgut, just underneath the notochord developed from the transplanted organizer. This secondary lumen is of course part of the dorsal differentiation of the endoderm. However, if the anteroposterior axis of the endoderm was inverted with respect to the axis of the ecto-mesodermal shell, the development was highly abnormal, thus showing that the differentiation of the endoderm along the anteroposterior axis cannot be dominated by the ecto-mesoderm.

A similar result emerged in experiments in which small pieces of endoderm taken from a late gastrula or early neurula stage were implanted in various positions into another embryo. When the pieces of endoderm were taken from embryos in the gastrula stage, the grafts were often smoothly incorporated into the endoderm of the host. The use of heteroplastic transplantation made it possible to distinguish the grafted cells from the host cells (differences in cell size in grafts between *Triturus taeniatus* and *Ambystoma mexicanum*), and thus to make sure that the graft was not destroyed, but had fitted into the construction of local tissues. Thus presumptive oro-branchial endoderm was found to be able to develop into intestinal epitehlium and *vice versa*. Stomach epithelium was developed from endoderm having a different prospective significance. Occasionally, however, grafts differentiated out of harmony with their surroundings, and the later the stage of the embryo from which the graft was taken, the oftener this occurred. After the end of the neurulation the grafts differentiated, in the main, according to their prospective significance (Balinsky, 1948).

In another experiment different parts of the neurula endoderm were transplanted into parts of the ecto-mesodermal shell, either from the anterior part of the neurula or from the posterior part (Fig. 227) (Okada, T. S., 1955a, 1955b). The endoderm taken for this experiment was either a part of the foregut endoderm, mainly destined to become pharynx, or endoderm from the midgut, normally differentiating as stomach and intestine. It was found that midgut endoderm grafted into the anterior ecto-mesoderm produced pharynx (in addition to other parts). Foregut endoderm surrounded by posterior ecto-mesoderm was in part differentiated as intestine. In both cases endoderm produced parts which were not in accord with the prospective significance of the endodermal cells, and it seems plausible that these differentiations were induced by the adjoining mesoderm. Again we find that the endoderm, as well as the ectoderm, is dependent on the mesoderm in its differentiation.

In all experiments with explantation and transplantation of endoderm it may be observed that liver tissue very rarely develops outside its normal position. In isolation experiments liver differentiated only as a rare exception, and then only atypically. In transplantation experiments liver often differentiated if endodermal grafts came to lie in the liver region, but even presumptive liver endoderm did not differentiate as liver heterotopically unless taken in rather late stages of development, that is, toward the end of neurulation. Even then only small fragments of liver tissue could be seen to develop heterotopically, although other endodermal tissues such as stomach epithelium, oro-branchial epithelium, pancreas and intestinal epithelium developed heterotopically just as well as in their normal position, if a sufficiently old graft were taken for the transplantation. All this indicates that quite special conditions are necessary for the development of the liver rudiment, and that these special conditions are found only in the region where the liver develops normally. This is another instance (besides the dorsoventral differentiation of the endoderm) where the development of the endodermal rudiments is definitely shown to be dependent on the influence of the surroundings.

There is as yet very little information concerning the earliest determination of endodermal organs in higher vertebrates. In birds some information has been derived from experiments in which parts of the chick blastoderm were grafted to the chorio-allantoic membrane of another chick embryo. Vari-

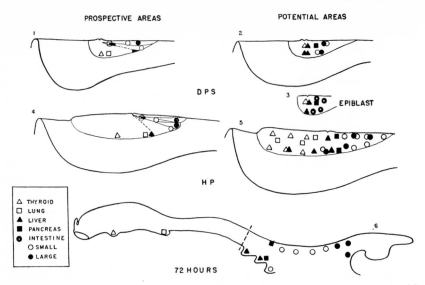

Fig 228. Diagrams showing the development of prospective as compared with potential areas of the blastoderm: *1*, Prospective areas for thyroid, lung, liver, pancreas (dorsal) and intestine in the definitive streak stage. The blastoderm is represented as folded longitudinally so that median structures appear in their future dorsal position. Data from marking experiments. *2*, Potential areas for the above organs in the definitive primitive streak stage: from results of differentiation in chorio-allantoic grafts. All three germ layers included in transplants. *3*, Potential areas as in Diagram 2 when hypoblast was not included in the grafts. *4*, Prospective areas as in Diagram 1, for the late head-process stage. *5*, Potential areas in the head-process stage. *6*, Diagram of the nearly closed gut tube of the three-day chick. At this stage most organ rudiments are visible, and potencies correspond with prospective significance. (From Rudnick, 1952.)

ous endodermal organs were seen to differentiate from the grafts, namely, pharyngeal epithelium, thyroid, lung, liver, large and small intestine. However, these tissues developed without a very definite relationship to the origin of the grafts, if the grafts were taken from early blastoderms up to the head process stage. In other words the endodermal differentiations in grafts were by no means strictly in accord with the prospective significance of the areas transplanted. Figure 228 shows the relation between the prospective significance of parts of the blastoderm, in so far as the endodermal organs are concerned, and the differentiation obtained by explanting these same parts. The lack of correspondence is obvious. The interpretation of these experiments is complicated even more by the fact that the endoderm (or hypoblast), taken alone, differentiates extremely poorly. Information on the differentiation of endodermal organs is obtained by grafting sections of the blastoderm containing all the germinal layers.

It has been concluded (Rudnick and Rawles, 1937; Rudnick, 1952) that in itself the endoderm has a very low power of differentiation, and that it needs interaction with mesoderm for its development, at least until the completion of gastrulation. It was noted that liver and thyroid usually differentiate in explants which also show the presence of the heart (Willier and Rawles, 1931), and that intestine is accompanied by mesoderm forming coelomic spaces (Rudnick and Rawles, 1937). This would be broadly in correspondence with what has been found in the amphibian embryo. The role of the upper layers of the blastoderm may possibly go even beyond that. It has been observed that the epiblast alone, without the hypoblast, when cultivated on the chorio-allantois will produce various endodermal tissues, such as thyroid, liver, pancreas and intestine, almost in the same way as a whole blastoderm consisting of both epiblast and hypoblast. This may mean that the actual formative cells of the endodermal organs are derived from the epiblast in conformity with what has been said about migration of cells from the primitive streak into the hypoblast (p. 153) (Rudnick, 1952).

Only when the endodermal gut becomes separated from the yolk sac during the second day of incubation does the differentiation of endodermal explants correspond to their prospective significance.

Experiments have been reported concerning explantation of parts of early mammalian embryos to the chick chorio-allantois (Nicholas and Rudnick, 1933), to omentum or under the connective tissue capsule of the kidney of adult rats (Waterman, 1936). Only grafts from fairly advanced embryos (ten-day embryos of rats, eight- to nine-day embryos of rabbits) produced endodermal tissue (in particular, liver and pancreas). The earliest embryos whose endodermal tissues showed signs of differentiations in grafts were rabbit embryos already in the eight-somite stage. A regular and clear differentiation of pancreas and liver was found to occur in still older, nine-day, embryos, in which the foregut had already been developed. The visible rudiments of liver and pancreas appeared one day later.

CHAPTER 13

The Genetic Control

of Organogenesis

It has been shown in Chapter 8 that gastrulation signifies a turning point in regard to the genetic control of development. Before gastrulation, development seems to be dominated by cytoplasmic factors, and the nuclear genes can exercise control only in so far as they determine the structure of the cytoplasm of the egg. From the beginning of gastrulation the nuclear genes can interfere in the development more directly, and from this stage therefore the paternal genes, brought into the egg by the spermatozoon, may make themselves felt. Organogenesis must therefore also be under the control of nuclear genes.

At present it is not possible to show exactly what an individual gene does when its action influences the process of development in animals. That the activity of individual genes is indispensable for any developmental process is inferred from disturbances of development following a change in the genic composition of the organism due to a mutation. In a mutant one or more genes become different from the "normal" allele, or the arrangement of the genes in the chromosomes is changed (in translocations), and this change comes to our notice by producing a deviation from the normal development of the organism. Quite often the deviation is of the nature of a developmental arrest, and from this it may be inferred that the gene or genes in their normal, unmutated state are somehow involved in producing the normal course of development. If therefore we find that in a mutant some process leading to formation of organ rudiments is disturbed, we conclude that this process is under the control of the normal allele of the mutated gene, or is dependent on the normal arrangement of genes in the chromosomes.

The changes caused by mutant genes in the developing organism so far as organogenesis is concerned may be, rather arbitrarily, grouped under three headings: defective development (decrease in the size or the number of organs), excessive development (increase in the size or the number of organs) and qualitative change in the organ rudiments (substitution of one rudiment in place of another).

13–1   GENES AS THE CAUSE OF DEFECTIVE DEVELOPMENT

The genes causing defective development may be illustrated by the following examples.

A mutant line is known in guinea pigs which shows various degrees of abnormalities of the head. The abnormalities are of the nature of cyclopic defects (see section 6–4); paired organs of the head tend to approximate each other on the ventral side and fuse into unpaired organs. In animals defective to a greater degree the more anterior parts of the head disappear altogether, and even the whole of the head may be absent, while the organs of the trunk region are more or less normal (Fig. 229) (Wright and Wagner, 1934). These defects are so similar to cyclopic defects which can be produced in amphibians by removing the anterior part of the archenteron that it is hardly possible to doubt that in the mutant guinea pigs the origin of the abnormalities is a similar one; the abnormal genetic constitution in some way inhibits the action of the primary organizer (the chordo-mesoderm and the endoderm of the head region). The external ear is also involved in abnormal guinea pigs, and this is a direct indication that the branchial region of the foregut developed abnormally, as the external ear is connected in its development with the first branchial pouch, and its abnormal position could only arise if the arrangement of the branchial pouches was defective right from the start. A similar mutation is known in mice (Little and Bagg, 1924), and it probably occurs in other animals as well.

An example of genetic control of organogenesis in a later stage is presented by Danforth's short tail mutant in the mouse (symbol Sd). Externally the mice carrying this gene differ from the normals by having a shortened tail or by the complete absence of the tail. Another feature which interests us

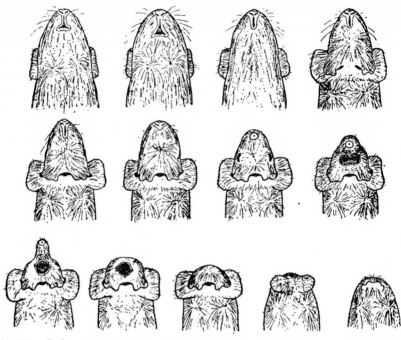

Fig. 229. Series of cyclopic defects of increasing severity in guinea pigs of the mutant strain studied by Wright. (From Needham, J., 1942.)

here is the reduction in size or complete absence of one or both kidneys found in these mice. As has been explained previously (section 11–4) the metanephros in mammals develops from two separate rudiments: the metanephrogenic tissue and the ureter, which sprouts from the Wolffian duct. In Danforth's short-tailed mice the ureter buds off from the Wolffian duct as usual, but it tends to remain shorter and sometimes does not reach the metanephrogenic tissue at all. If the ureter does not reach the metanephrogenic tissue, the kidney does not develop, as the induction from the ureter is necessary for the kidney tubules to be differentiated. If the ureter reaches the metanephrogenic tissue, then at least a small kidney develops. The size of the developing kidney depends on the degree of branching taking place at the end of the ureter, as only those parts of the metanephrogenic tissue become differentiated into renal tubules which lie in the immediate vicinity of the tip of the ureter or its branches (the latter becoming the collecting tubules of the differentiated kidney). Thus the cause of the kidney defect is the arrest of development of the ureter as a result of the changed genetic composition of the affected animals. The arrest in the development of the ureter prevents the establishment of the spatial relationship between inductor and reacting system (the ureter and the metanephrogenic tissue), which is necessary for the induction of the kidney to take place (Gluecksohn-Schoenheimer, 1943, 1945).

13–2   GENES AS THE CAUSE OF EXCESSIVE DEVELOPMENT

The second group of genes, the genes causing excessive development, may be represented by the genes causing the increase in the number of digits on the forelimb or hindlimb. Quite a number of such genes are found in different animals. The development of this condition has been studied recently in a mutation of the mouse called "luxate" (symbol lx) (Carter, 1954). In these mice additional toes, one or two, appear on the preaxial side of the foot, that is, on the inner side of the hallux. The anomaly can be traced back to the limb-bud stage, when the hindlimb-bud is excessively broad on its anterior edge. When, in the next stage, mesenchyme condensations appear indicating the rudiments of the digits, the number of these condensations is greater than normal. As the size of each digit rudiment corresponds to the size in normal limbs, it seems plausible that the excessive number of digit rudiments is the result of the excessive amount of material provided for the digit development in the abnormally broad limb-bud.

There are several mutations in mice which produce a partial or complete duplication of the whole body. The duplication may be posterior, involving the tail, sacral and hindlimb region (Danforth's posterior reduplication, Danforth, 1930), or more generalized ("Kinky" homozygotes, symbol Ki, Gluecksohn-Schoenheimer, 1949). The duplications could not have been produced later than the time of primary organ formation, possibly even during the gastrulation stage. It is not known what is the mechanism by which duplication is achieved, but in this connection it is sufficient to know that the genetic constitution of the embryo may influence the morphogenetic processes involved in the formation of primary organ rudiments.

13–3   GENES AS THE CAUSE OF QUALITATIVE CHANGES IN ORGAN RUDIMENTS

The third type of change wrought by a change in genetic composition is the substitution for one organ rudiment of a rudiment of a different kind. No genes of this group are known in vertebrates, but some are found in insects, in Drosophila in particular. Many organs of adult insects having a complete metamorphosis develop from special rudiments—the **imaginal discs.** The imaginal discs are found in the larvae long before metamorphosis, but they remain quiescent as groups of cells under the epidermis and do not participate in the current functions of the larval body. Shortly before metamorphosis the imaginal discs begin to grow and each develops into some organ or other, the organ becoming functional in the adult insect. So eyes, legs, antennae, and other parts of the adult fly develop from special imaginal discs. Each imaginal disc is determined for the development of a specific organ; the imaginal discs of the larva may be transplanted, and they then develop into heterotopic organs. Mutations are known, however, which change the specific properties of certain imaginal discs. In the mutation "Aristopedia" (symbol sa) the imaginal discs of the antennae are changed in such a way that the developing

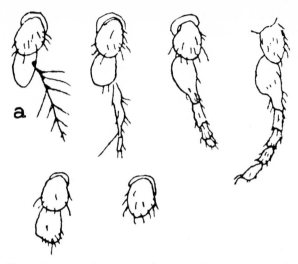

Fig. 230. Normal antenna in *Drosophila* (*a*) and antennae transformed by the aristopedia gene into tarsus-like organs. (After Timofeeff-Ressovsky, from Waddington, 1939b.)

organ is a small leg, with a pair of claws at the end (Fig. 230) (Balkaschina, 1929). Similarly a leg may develop in place of part of the compound eye in the mutation "ophthalmopedia" (Gordon, 1936) or of part of the wing—"podaptera" mutation (Goldschmidt, 1945). The change of the antennal or wing imaginal disc into the imaginal disc of a leg must have been effected by the change of genetic constitution (the presence of a mutated gene instead of the normal allele of this gene).

Genes of a similar action may have played some part in the evolution of vertebrates. In some of the extinct amphibians and contemporary reptiles an unpaired pineal eye is found instead of the pineal organ, supposed to be a gland of internal secretion. The transformation of one organ into another may have been due to the action of a gene similar to the "aristopedia" gene in *Drosophila*. A similar type of change may have been involved in the transformation of the posterior pair of branchial pouch rudiments into lung rudiments, as has been indicated (section 12–4). In this connection we may consider differences arising in the development of related animals, even though the genetic background of such differences is, as yet, unknown. In section 10–4 it was indicated that the balancer of salamander larvae develops as a result of an induction, the stimulus being given off probably by the roof of the archenteron, and the reacting system being the epidermis in the neurula stage. The balancers are found in most newts and salamanders, but a few species do not possess them. Experiments have been performed to determine the cause of this difference. If, in the neurula stage, the epidermis of a species possessing a balancer (viz., *Triturus taeniatus*) is transplanted to the side of the head of a species which does not have balancers (viz., *Ambystoma mexicanum*), the transplanted epidermis will develop a balancer in exactly the same position in which it is usually found, that is, underneath the eye near the angle of the mouth. A reverse transplantation, that is, transplantation of epidermis from *Ambystoma mexicanum* to the site of balancer development in *Triturus taeniatus,* results in the absence of a balancer on the operated side of the head (Mangold, 1931a; Rotmann, 1935). These results show that the absence of balancers in *Ambystoma mexicanum* is due to the failure on the part of the epidermis to react to the stimulus of the inductor, or in other words that the epidermis lacks an appropriate competence. The inducing stimulus, on the other hand, is present both in species possessing balancers, and in species not having these organs. As the inability of *Ambystoma mexicanum* to develop balancers is genetically fixed, it is to be concluded that the hereditary factors (genes or their combinations) responsible for this particular peculiarity of *A. mexicanum* directly affect the competence of the ectoderm, while the inducing systems remain unchanged.

Another example illustrating the same principle is found in the development of the mouth in anurans and urodeles. The larvae of urodeles (salamanders and newts) have typical teeth consisting of the pulp and the layers of dentine and enamel. The teeth are situated inside the mouth and attached to the jaws and the bones of the palate. The tadpoles of the frogs and toads have no true teeth; instead the edges of the jaws are covered by horny sheaths, and rows of small horny teeth and epidermal papillae are developed on an oral disc surrounding the mouth. It has been shown in section 12–2 that the ectodermal mouth parts develop under the influence of an induction from the oral endoderm. The inductor, however, determines only the position of the ectodermal oral invagination, but not its specific peculiarities. It is possible to transplant ectoderm from an early frog embryo to a salamander embryo in such a way that the grafted ectoderm covers the mouth region. As a result the graft is induced to develop a mouth. This mouth is, however, in every respect the mouth of a frog, with horny jaws and rows of horny teeth and oral papillae (Fig. 231) (Spemann and Schotté, 1932). It is thus evident that the influence of the mouth inductor is similar in the salamanders and the frogs, or at least sufficiently similar for frog ectoderm to be able to react to a salamander's inductor. What is different is the competence of the ectoderm; to the same stimulus the ectoderm of different animals reacts in its peculiar way. It is the competence or nature of reaction of the ectoderm that is affected by the hereditary factors responsible for the differences in the development of the mouth in urodeles and anurans.

It is also possible, however, for the hereditary factors to modify an inducing system without changing the competence of the reacting tissues. It has been proved that the dorsal fin fold in amphibians is induced by the neural crest (section 10–4, the fin fold). In most salamander larvae the fin fold reaches anteriorly almost to the occipital region, but in *Eurycea bislineata* it is present only on the tail, the trunk being devoid of a fin fold. This peculiarity is not a result of the inability of the trunk epidermis to develop a fin fold. If the trunk epidermis of *Eurycea bislineata* is transplanted to the back of

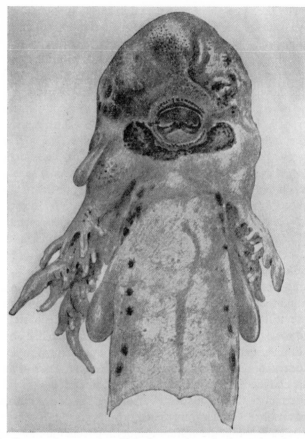

Fig. 231. Larva of a newt in which the ectoderm in the oral region was replaced by ectoderm of a frog embryo. The grafted ectoderm developed horny jaws and teeth and the adhesive organ found in frog tadpoles. (After Schotté, from Spemann, 1938.)

an embryo of *Ambystoma maculatum* (a species having a fin fold in the trunk region), the transplanted epidermis will participate in the formation of a fin fold (Bytinski-Salz, 1936). The peculiarity of the fin fold of *Eurycea* is thus due to a failure of the trunk neural crest to act as inductor.

We have seen that the position of the external gills found in amphibian larvae, and the very fact of their development, is dependent on the endoderm, while both the epidermis which covers the gills and the mesenchyme which forms the connective tissue and the blood vessels of the gills react to the inducing influence of the endoderm. Now the number of pairs of external gills in tadpoles of frogs and toads varies in different species from one to three. In the South African toad *Bufo carens* there are three pairs of external gills, while in another species, *Bufo regularis,* only two pairs are formed. If the presumptive epidermis of *Bufo regularis* is transplanted to an embryo of *Bufo carens,* and if it lies in the branchial region, it will form gills in response to an induction from the host endoderm. It was found that three gills were

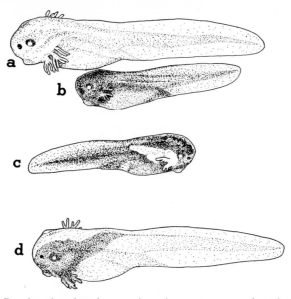

Fig. 232.   Results of exchanging ectoderm between two species of toads differing in the number of external gills. *a*, Normal tadpole of *Bufo carens; b*, normal tadpole of *Bufo regularis; c*, tadpole of *B. regularis* with grafted ectoderm of *B. carens; d*, tadpole of *B. carens* with grafted ectoderm of *B. regularis*.

formed on one side, the number typical for the host and not for the reacting epidermis (Fig. 232). In the reciprocal transplantation, that is, transplantation of the presumptive epidermis from *Bufo carens* to *Bufo regularis*, two external gills develop, again the number typical for the host. It is evident that the epidermis can form any number of external gills, and the number that actually develops is determined not by the reacting system (grafted epidermis) but by the inductor—the host endoderm (Balinsky, unpublished). The ability of a part of the embryo to act as inductor may thus be affected by the specific genetic constitution of an animal.

PART V

Differentiation
and Growth

PREFUNCTIONAL AND FUNCTIONAL PERIODS OF DEVELOPMENT

With the formation of all or of most organ rudiments of the embryo, the main features of the organization of the animal have already been indicated. What may be called the morphological plan of the animal is established; by morphological plan is meant here the kind and number of the organs, their relative position and the general features of structure of each organ. However, organ rudiments at this stage are not capable of performing their specific functions, on which depends the ability of the animal to lead an independent existence. The cells of the organ rudiments lack the peculiar structures that are necessary for specific functions; the organ rudiments are usually too small, and the animal as a whole is likewise far from the adult size. All the developmental processes dealt with so far may be grouped together as the **prefunctional** stages of development. Now a new phase of development sets in, which brings the animal to its **functional** state. The main processes involved are **growth** and **differentiation.** Some new organs may appear in late stages of development, especially in animals passing through a larval stage, and minor morphological adjustments may occur in the organs formed earlier, but the processes of growth and differentiation are predominant. Growth and differentiation may proceed concurrently, or growth may precede differentiation, so that the cells first proliferate and then differentiate, growth ceasing with the onset of differentiation. The functioning of the organs usually begins before growth has come to a standstill and very often before the final stage of differentiation has been achieved—the initial stages of differentiation making the cells already capable of function. Once differentiation has started, and often even somewhat earlier, the parts of the embryo in question are as a rule irrevocably determined for their future fate. This determination concerns, however, only the general type of organ or tissue which is to be developed; the details are "polished" in the process of development, and sometimes the functioning of the organs plays a part in the final adjustments in the various organs. For convenience's sake differentiation will be discussed before growth.

NUTRITIONAL REQUIREMENTS FOR DIFFERENTIATION AND GROWTH

As the development of the embryo progresses its nutritional requirements increase; this is due to quite a number of factors, all working in the same direction. First, in the course of time the original supply of reserve materials stored in the egg, as yolk, glycogen, lipid granules and the like, becomes exhausted, and new supplies from without become necessary if development is to continue. (If new supplies are not furnished, as for instance when larvae emerged from the egg are kept under conditions of starvation, large amounts of proteins may be used up, which should otherwise be utilized in a different way; this, however, is an abnormal condition—see Løvtrup and Werdinius 1957.) In meroblastic eggs, even with maximal separation of active cytoplasm and yolk, restricted amounts of food supplies in the form of glycogen and small yolk granules may be present in the cytoplasm and are used in the early stages of development, but later the embryo must draw for its food requirements on the main yolk depot, which is extraembryonic (yolk sac). The second factor increasing the nutritional requirements of the embryo is the rise in respiration with development, which has been dealt with in section 5–9. Lastly, a third and most important factor comes in at the beginning of growth, which is practically absent during cleavage, very restricted during gastrulation, but becomes predominant in the later stages of development. Growth which may be considered as equivalent to synthesis of new cytoplasm (for a more accurate definition see Chapter 15) can proceed only if large amounts of materials are made available to the embryo in some way or other.

What are the materials which have to be supplied? It is not possible to deal here with the mechanisms of nutrition and growth in general; this is a subject which falls into the field of general physiology. Some facts relating to the nutritional requirements of the early embryo will, however, be relevant.

The nutritional requirements of the embryonic tissues cannot be very well studied in those forms in which reserve materials are stored inside the cells, as it is not easy to determine what particular substances are being utilized and in what order and in what form (some data, however, have been given in sections 2–7 and 5–9). A much more favorable object is an embryo developing after meroblastic cleavage; such an embryo can be separated from the yolk and cultivated *in vitro*. Experiments of keeping blastodiscs of birds on plasma clots have been referred to repeatedly in the preceding text. The question naturally suggests itself whether the blood plasma + embryo extract, used in the above experiments, may not be replaced by a synthetic medium. A complete synthetic medium was accordingly worked out (Spratt, 1948) (see Table 11).

Table 11. Composition of "Complete" Synthetic Medium for Cultivation in vitro of Chick Blastoderms. Components Shown in Milligrams per 100 cc. of the Solution

Salts:		Vitamins:	
NaCl	700.0	thiamine	0.01
KCl	37.5	pyridoxine	0.05
Ca(NO$_3$)$_2 \cdot$ H$_2$O	21.0	niacin	0.05
MgSO$_4$	27.5	riboflavin	0.01
Fe(NO$_3$)$_3 \cdot$ 9H$_2$O	0.14	inositol	0.05
Na$_2$HPO$_4 \cdot$ 12H$_2$O	24.5	biotin	0.01
KH$_2$SO$_4$	2.6	Ca-pantothenate	0.01
NaHCO$_3$	55.0	choline-HCl	0.50
Carbohydrates:		carotene	0.01
Glucose	850.0	vitamin A	0.01
Agar	250.0	ascorbic acid	0.05
Amino acids:		cysteine HCl	0.10
dl-lysine-HCl	15.6		
dl-methionine	13.0		
dl-threonine	13.0		
dl-valine	13.0		
l-arginine-HCl	7.8		
l-histidine-HCl	2.6		
dl-isoleucine	10.4		
dl-phenyl-alanine	5.0		
l-leucine	15.6		
dl-tryptophane	4.0		

The agar was added to provide a solid substrate for the embryo, and in some cultures was omitted without deleterious effect.

On this "complete" medium the chick embryo, explanted in the primitive streak or head process stage, will develop just as well as on a blood plasma + embryo extract medium. Not all of the components of the synthetic medium are of equal importance, as could be ascertained by omitting some of

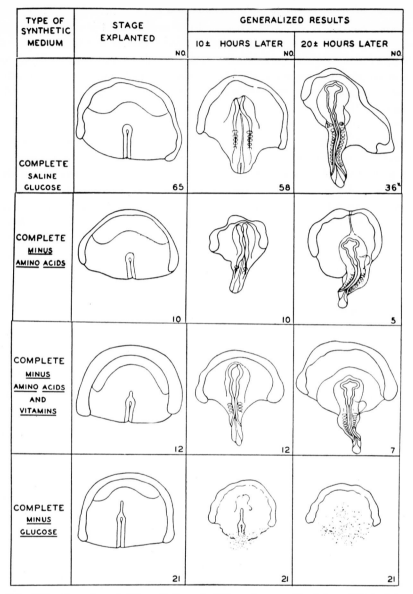

TYPE OF SYNTHETIC MEDIUM	STAGE EXPLANTED	GENERALIZED RESULTS	
	NO.	10± HOURS LATER — NO.	20± HOURS LATER — NO.
COMPLETE SALINE GLUCOSE	65	58	36
COMPLETE MINUS AMINO ACIDS	10	10	5
COMPLETE MINUS AMINO ACIDS AND VITAMINS	12	12	7
COMPLETE MINUS GLUCOSE	21	21	21

Fig. 233. Growth and differentiation of chick embryos cultivated on different synthetic media. Figures in the lower right hand corner of each square indicate number of cases observed. (From Spratt, N. T., Jr., 1948.)

them in special experiments. Deleting glucose from the medium, with all the other components present, immediately stops all developmental processes and leads to a rapid disintegration of the embryo (see Fig. 233). Obviously, of all the substances included in the complete medium, glucose provides the only source of energy for the embryo, and it cannot be dispensed with. The exclusion of all the amino acids, on the other hand, permitted development to proceed but there was practically no growth of the embryo. Some morpho-genetic processes were also retarded, such as the development of the ear vesicles and of the posterior end of the embryo (the latter is perhaps mainly due to the absence of growth). The withdrawal of the vitamins from the medium, in addition to the absence of amino acids, did not further affect development; growth was stopped but differentiation proceeded (see Fig. 233).

From these experiments it can be seen that morphogenesis and differentiation may be carried out by means of reorganization of the substances already present in the embryo, provided that an energy source is available to carry out this reorganization, but that the intake of additional building materials is required for growth. Incidentally the experiments also show that morphogenesis (the elaboration of structure) may proceed without growth; we have already reached the same conclusion before when analyzing the role of "morphogenetic movements" in early development.

CHAPTER 14

Differentiation

14–1 HISTOLOGICAL DIFFERENTIATION

Histological differentiation is the process as the result of which the parts of the organism acquire the ability to perform their special functions. In the case of multicellular animals the parts in question are the cells and groups of cells. The special functions of cells in the above definition are distinguished from the basic functions of life which are common to all living cells. Every cell is capable of performing the processes of metabolism (respiration, synthesis, and so on), possesses to a certain degree the ability for ameboid movement, shows irritability and is able to react to external stimuli. These functions are found in both undifferentiated and differentiated cells. Differentiated cells, however, are able to perform special functions, or to perform them in such a way as other cells cannot. Thus the nerve cells are capable of conducting nervous impulses to great distances and at a high speed. The liver cells secrete bile (besides their other functions). The melanophores produce granules of pigment in their cytoplasm. These are the special functions of nerve cells, hepatic cells and melanophores.

Invagination and epiboly (expansion) are special functions of the cells of the archenteron and of the gastrula ectoderm respectively, and are based on special properties of the cells of the presumptive archenteron and the presumptive ectoderm not present in other cells, for instance in the cells of the early blastula. Thus the cells of the gastrula are also differentiated in a way. The differentiations of cells in the early embryo have been dealt with in the previous chapters. In this chapter we will be concerned with the differentiation leading to the acquisition by the tissues of their definitive functions, that is, the functions which they are to perform in the fully developed organism. The term **histological differentiation** is restricted to this latter type of developmental process.

The ability to perform special functions is dependent on the existence of specific mechanisms in the differentiated cells. These mechanisms are sometimes visible in the form of organoids of the cell, such as the myofibrils of muscle cells, cilia of epithelial cells of the trachea, long processes of nerve cells. These tangible morphological properties of the cells are also called **differentiations,** a practice that is legitimate as they are actually the visible expression of the process of differentiation. In other tissues the differentiation becomes visible not so much as a change in the structure of the cell itself, but as the result of the production by the cells of intercellular structures, such as fibers in the connective tissues, matrix of cartilage and bone, cuticle on the outer surface of the skin in invertebrate animals. These extracellular parts are also called differentiations in the same sense as the organoids of the cells. The special function of a cell may be the secretion of some substance that does not remain in the tissue as a permanent part, but is removed or dissolved in the surrounding medium. In this case granules of secretion may sometimes be seen in the histological preparations as a morphological expression of the function of the cell.

The functional mechanisms of histologically differentiated cells are cytoplasmic. The building up of these mechanisms therefore causes a shift in the relative volume of the nucleus and the cytoplasm. As has been stated before, the nuclei of the cells are their conservative part; they do not change essentially with the onset of development (section 4–5). It is believed that even in the case of fully differentiated cells the nuclei in all the various tissues of the animal's body have the same chromosomes and genes. In differentiated cells the mass of cytoplasm increases while the nuclei do not increase, or do not increase in the same proportion. The ratio—mass of cytoplasm to mass of nucleus—increases with differentiation. This ratio gives a rough quantitative estimate of the degree of differentiation. The estimate may be made even more exact if chemical substances and not morphological parts are taken into consideration. The basic substances of the nucleus are the chromosomes, with deoxyribonucleic acid as their essential component. The cytoplasmic structures are built up of proteins. Furthermore, the enzymes, which are the essential part of the functional mechanism of a cell, are of a protein nature. It is thus possible to substitute the amount of deoxyribonucleic acid for the basic structure of the cell and the amount of protein for its changing functional mechanisms. Direct measurements show that the relation protein/deoxyribonucleic acid changes with differentiation in agreement with expectation (Davidson and Leslie, 1950).

In the developing chick embryo the amounts of protein nitrogen and of deoxyribonucleic acid phosphorus were determined every day between the eighth and the nineteenth days of incubation. The organs were brain, heart, liver and leg muscle. In all these tissues the ratio of nitrogen to phosphorus increased steadily throughout the period of observation. The highest increase was in the brain, the organ which can rightly be called the most highly differentiated of all the organs of a higher vertebrate. The next highest increase of nitrogen was found in the leg muscle—a tissue which is rapidly differentiating in the chick embryo. The liver showed a rather slow increase of nitrogen, which is understandable, as the embryonic liver does not attain full functional differentiation. Furthermore it remains a blood-forming organ in the embryo and contains therefore masses of proliferating, undifferentiated cells. The lowest increase of nitrogen was found in the heart. This can be accounted for by the fact that the heart starts functioning very early. By the eighth day it has already achieved a degree of histological differentiation which changes but little during the further twelve days of incubation.

14–2   THE CHEMICAL BASIS OF DIFFERENTIATION

Whatever the type of differentiation of the cell, it is doubtless based on the chemical constitution of the cell. In every case where the function of the cells consists in the elaboration of some substances, whether in the form of structural elements or in the form of secretions, there must be a specific enzyme or specific combination of enzymes responsible for the reaction. As the substances produced by cells are very diverse, a correspondingly varied assortment of enzymes may be postulated as being present in different tissue cells. It may therefore be said that "differentiation is the production of unique enzymatic patterns" (Spiegelman, 1948; see also Boell, 1948, 1955). In the present state of the development of science it is not possible to ascertain for every type of differentiated cell what is the chain of chemical transformations which are going on, and what are the enzymes which control these reactions.

If the differentiated tissue is distinguished by the presence of a specific chemical substance, it may be questioned at what time this specific substance first appears and when the final concentration of the substance is reached. In a few cases it has been possible to answer the above question by direct measurements.

Hemoglobin.   The red blood cells perform their function of oxygen transport by means of the hemoglobin accumulated in the cytoplasm. The hemoglobin is elaborated in the descendants of stem cells, which by active proliferation increase the number of cells later destined to differentiate into erythrocytes. By comparing the mass of cells in every stage of differentiation with the mass of the stem cells it is possible to reconstruct a sort of growth curve, showing the increase of the volume of the group of cells produced by one stem cell. Against this growth curve the amount of hemoglobin per single cell was determined spectroscopically, by measuring the absorption of light by the hemoglobin (Thorell, 1947). It was found that the stem cells contain no hemoglobin. There are only traces of hemoglobin in the cells while growth is going on. After the volume of the cells becomes stabilized, that is, after the cells have ceased to grow and proliferate, the amount of hemoglobin rapidly increases until the full amount is produced, which is

found to be 28.10^{-6} microgram per cell (microgram $= \dfrac{1}{1000}$ milligram).

The final stage in the differentiation of a mammalian erythrocyte, the pycnotic degeneration and extrusion of the nucleus occurs after the full amount of hemoglobin is reached.

Myosin.   The specific substance of striated muscle is the protein **myosin.** In the adult muscle, myosin amounts up to 10 to 12 per cent of the fresh weight or nearly 50 per cent of the dry weight.

The amount of myosin in the protoplasm of the cells undergoing differentiation into striated muscle fibers has been estimated and compared with the morphological differentiation of the muscle (Nicholas, 1950). In the rat embryo the differentiation of striated muscle proceeds in the following way:

12th-13th day after fertilization: the myotomes expand into the lateral and ventral body wall.

14th day: the syncytial muscle fibers have been formed; on the same day the muscles become functional: they start contracting.

17th day: transverse striation of the myofibrils becomes visible.

The changes in the chemical constitution of the muscle tissue during embryonic and postembryonic development are shown in Table 12.

Table 12.   Changes of Chemical Composition of Muscle Tissue during Development in the Rat. (The figures have been recalculated from the data of Nicholas and his collaborators; see Nicholas, 1950, Herrmann and Nicholas, 1948, 1949)

Stage	Dry weight in % of fresh weight	Deoxyribo- nucleic acid in % of fresh weight	Myosin in % of fresh weight
13th day after fertilization	5.5	0.57	0.1
At term (21 days after fertilization)	9.0	0.34	1.0
At weaning	14.0	0.22	4.5
Adult rat	20.0	0.08	10.0

Several conclusions can be drawn from the above data:

1. In the prefunctional stage the amount of myosin in the presumptive muscle cells is very small, and this amount increases greatly in the period when the functioning of the muscle begins (between the thirteenth and the twenty-first day after fertilization).

2. The differentiation of the muscle tissue is by no means accomplished when the tissue begins to function. This is shown by the tenfold increase of myosin in the postembryonic stages of development. This increase is due mainly to the elaboration of additional quantities of myofibrils in the muscle fiber. As a result the strength of the fiber increases greatly. As measured by the breaking load of the fibers, their strength increases more than 400 fold between the seventeenth day after fertilization and the adult stage.

The specific physical organization of the myofibrils can be studied by testing the differentiating muscle for birefringence. Positive birefringence was found to be present in muscle tissue on the fourteenth day after fertilization. The birefringence is a definite proof that the myosin molecules are arranged in long chains. This arrangement thus precedes the appearance of contractility, while the transverse striation of the fibrils appears after the muscles have started to contract.

The data on the development of striated muscle bring out again the change in proportions between the nuclear apparatus and the functional mechanism of differentiating cells. While the amount of myosin (the functional substance) increases, the amount of the deoxyribonucleic acid decreases as compared with the other substances of the cells (fibers).

There is another lesson to be learned from the development of muscle tissue. It concerns the relationship between organ rudiment formation and histological differentiation. The organ rudiments—the myotomes in this case—are formed in an early stage of embryonic development, about the seventh or eighth day after fertilization. The cellular materials for the development of the muscles take up their final position in the lateral and ventral body wall during the twelfth to thirteenth day of development. But the elaboration of myosin and its arrangement in long chains (as shown by positive birefringence) follows only two days later.

Enzymes.   The development of enzymic mechanisms may be illustrated by the appearance of the proteolytic enzymes pepsin and trypsin in the digestive tract of a salamander, *Ambystoma punctatum* (Dorris, 1935). The gastric glands which secrete pepsin are developed in the walls of the stomach after all the parts of the alimentary canal are already clearly recognizable (see section 12–1). In stage 40 the walls of the stomach consist of columnar epithelium, with a large number of yolk platelets in the cells (see Fig. 234, *a*). At stage 41 the deeper lying cells of the epithelium become clumped together, each clump representing the rudiment of one gastric gland (Fig. 234, *b*). In stage 42 the cells of the gland rudiments are clearly arranged in a spherical layer, and a cavity appears in the middle— the lumen of the gland (Fig. 234, *c*). In stage 43 the lumen of the gland

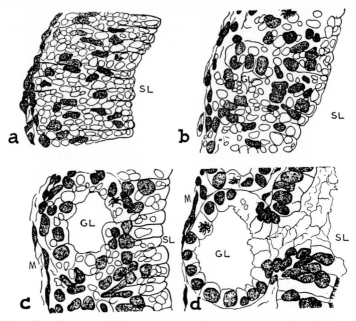

Fig. 234. Development of the gastric glands in the stomach of *Ambystoma punc-tatum* embryos, stages 40(*a*), 41(*b*), 42(*c*) and 43(*d*). *GL.* Cavity of gastric gland; *M,* visceral mesoderm lining alimentary canal; *SL,* stomach lumen. (From Dorris, 1935.)

is lined by a smooth cuboidal epithelium and it communicates by means of a narrow duct with the cavity of the stomach (Fig. 234, *d*). The yolk granules in the cells have disappeared by this time.

Pepsin first appears in the gastric glands between stages 42 and 43. Prior to and in stage 42 no trace of pepsin can be found in the stomach, and at no stage either earlier or later is pepsin present in other parts of the body. We see that the glands are already clearly distinguishable morphologically (in stage 42) before they can perform their specific function—the production of pepsin. The morphological differentiation of the glands is, however, not quite completed by that stage, as there is some further progress till stage 43 when the glands attain their final structure. The presence of trypsin in the pancreas can first be discovered in stage 43, very slightly later than pepsin. The pancreatic acini first become distinguishable in parts of the organ (especially in the dorsal pancreas) at stage 41, they are more distinct in stage 42 and well differentiated in the dorsal pancreas in stage 43. As in the gastric glands, the structure of the secretory parts is laid down first, and the specific function (production of trypsin) sets in after that. It may be added that the mouth in salamander larvae breaks through at about stage 42 and feeding begins normally in stage 44.

We shall now consider an enzyme which is more widely distributed in the body of the animal, namely alkaline phosphatase, the enzyme which causes hydrolytic splitting of monoesters of the phosphoric acid in an alkaline medium (Moog, 1946) and which may also function as a phospho-transferase, that is, it transfers the phosphate radical from one molecule to another. In early mammalian embryos the enzyme is present in only small quantities and in a diffuse state, except that it seems to be always present in the nuclei of the cells (Danielli, 1953). At the time of the onset of differentiation it appears in large quantities, but in only a few tissues. It is found to be concentrated in the subcutaneous tissue of the embryo in cells concerned with the development of the subcutaneous connective tissue layer. A little later alkaline phosphatase is found in cartilages and in the hair papillae (Hardy, 1952). In all these three sites the enzyme is supposed to be connected somehow with the elaboration of fibrous proteins—collagen fibers in the connective tissue and cartilage matrix, fibers of keratin in developing hairs. In later stages of development the enzyme is very abundant in the periosteum of bone and in the matrix of bone. In the latter position (where the enzyme is extracellular) the alkaline phosphatase splits off the phosphoric acid from glucose phosphates in the form of calcium phosphate, which impregnates the bone matrix. Further sites of alkaline phosphatase concentrations are in the brush border of proximal convoluted tubes of the kidney and in the cells of the intestinal mucosa. In both of these sites the alkaline phosphatase is concerned with the transfer of glucose from the lumen (of the renal tubule and the intestinal lumen respectively) into the internal medium of the body. In the adult the quantities of alkaline phosphatase in the kidney, the intestinal mucosa and the bone surpass by far the quantities found in other tissues, as may be seen from Table 13.

It is worth while tracing the timing of the appearance of the alkaline phosphatase in the hair papillae. The rudiments of hairs in the mouse

Table 13. Distribution of Alkaline Phosphatase in Different Tissues (After Greenberg, from Spiegelman, 1948)

Tissue	Enzymatic Activity of Alkaline Phosphatase (in arbitrary units)
Liver	4
Hyperplastic breast	9
Lymph nodes	8
Bone marrow	23
Spleen	17
Kidney	1072
Skeletal muscle	2
Cardiac muscle	12
Skin	5
Lung	36
Intestinal mucosa	2789
Gastric mucosa	17
Thymus	3
Pancreas	1
Brain	12
Bone	420

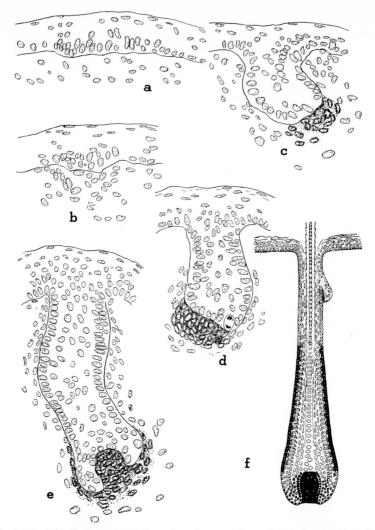

Fig. 235. Distribution of alkaline phosphatase (shown by dark stippling) in developing hair follicles (*a, b, c, d, e*) and in a fully differentiated hair follicle of a mouse. (*f* From Hardy, 1952.)

embryo first appear 14 days after fertilization in the form of epithelial thickenings. There is no trace of the connective tissue papilla in this stage. In 15-day-old mouse embryos mesenchyme cells accumulate under the epithelial thickenings, thus forming the rudiments of the future papillae. The differentiated papilla, as has been previously indicated, contains large amounts of alkaline phosphatase, which is connected with the function of the papilla as the organ supplying the materials for the formation of the hair itself, which largely consists of fibrous keratin (see Birbeck and Mercer, 1957).

In the rudiment of the papilla, when it is first detectable, there is no alkaline phosphatase. Only in the more advanced hair rudiments of a 15-day-old embryo can small amounts of alkaline phosphatase be demonstrated. In the hair papillae of 16-day-old embryos large quantities of the enzyme may already be found (Fig. 235) (Balinsky, 1950).

The formation of the rudiment of the hair papilla thus precedes the appearance of the specific substance (enzyme) which is a part of the functional mechanism of the differentiated organ.

14–3   RESULTS OF IMMUNOLOGICAL INVESTIGATIONS

In cases where the specific substances of differentiated tissues cannot be determined chemically, they can still be traced by the use of immunological methods. A suitable experimental animal, usually a guinea pig or rabbit, is immunized against the tissue that is being studied. For this purpose the tissue, crushed into a brei or in the form of an extract, is injected into the animal that is to be immunized. The injected animal develops **antibodies** against the protein of the tested tissue in its blood plasma. The proteins which are used for immunization are called **antigens.** If the antibodies are again brought into contact with the same antigens, a reaction of a high degree of specificity will take place—the antibodies reacting only with the same antigen or with very closely related substances (and then to a weaker degree). The type of reaction depends on the conditions in which antibody and antigen are caused to meet. If a tissue extract containing antigens is mixed with the blood plasma of the immunized animal, a reaction of precipitation takes place: the antigen is agglutinated by the antibodies and forms a precipitate. Another possible way of testing the presence of the antigen is to inject the blood plasma of the immunized animal containing antibodies into the animal in which the presence of the same antigen is suspected or, in the case of an embryo, to place it in the immunized blood plasma or add the plasma to the medium in which the embryo or its parts are cultivated. In these circumstances the antibodies may attack the antigen contained in the cells of the host and damage the cells or cause some anomaly in their behavior (for review see Cooper, 1948; Tyler in Willier, Weiss and Hamburger, 1955).

The following experiment may serve as an example of the results obtained by the precipitation method (Burke, Sullivan, Petersen and Weed, 1944). Experimental animals were immunized by injecting tissues of the adult chick. In this way anti-organ sera were prepared against brain, lens, kidney, bone marrow, erythrocytes, ovaries and testes. The antisera were then tested against tissue extracts from organs of the chick embryo at different stages of development. A reaction between the two proves that the specific tissue substances (antigens) of the adult animal are already present in the organs of the embryo. The following table shows the earliest stage (expressed in days of incubation) at which the specific adult tissue antigens could be discovered:

Brain	11 days
Lens	7 days
Kidney	9 days
Erythrocytes	4 days
Ovaries and testes	11 days

In every case the antigen was found to be present when the tissues were well on their way to histological differentiation. Some "cross reactions" were also found, that is, reactions of an antiserum with tissue other than the one used for immunizing the donor of the serum. This means that the substances of the various organs could not be completely told apart.

The technique used in the above experiment appears not to have been up to the highest standard of perfection possible by means of the precipitation method, as was shown by subsequent work (ten Cate and Doorenmaalen, 1950). The latter work, however, was restricted to the development of the specific substance of the lens. Rabbits were immunized by injecting extract of chick and frog lens respectively.

Tables 14 and 15 show the results obtained.

Table 14.   Adult Lens Antigen in Chick Embryos

Age of Embryo (Hours)	Number of Experiments	Number of Lenses	Result of Reaction
60	3	40	+ + ++
58	1	60	+
54	3	50, 30, 30	± ± −
51	1	50	−
48	1	50	−

Table 15.   Adult Lens Antigen in Frog Embryos

Stage of Development (Shumway, 1940)	Antigen	Result of Reaction
25	lens + eye cup	+
23	lens + eye cup	+
23	rest of embryo	−
21-22	lens + eye cup	+
21-22	ventral epidermis	−
19-20	lens + eye cup	+
19	lens + eye cup	+
19	lens + eye cup	+
19	lens + eye cup	−
17-18	presumptive lens ectoderm	−

In the chick the earliest unambiguously positive reaction was with lens tissue of a 58-hour-old chick; in the frog, with the lens of an embryo in stage 19.

The first positive reactions were obtained in these experiments with lenses in a very early stage of differentiation. The lens of a 58- to 60-hour-old chick embryo is in the form of a vesicle which is still connected with the epidermis, with the inner wall not yet thickened by the initial stages of fiber differentiation. The lens rudiment of a frog embryo in stage 19 is still less developed; it has not yet even formed a vesicle, but only a thickening lying in the pupil of the eye cup. It is thus shown that the organ-specific substance may be present as soon as the organ rudiment is formed and before it shows any visible traces of histological differentiation.

The second way of testing the presence of an antigen in the embryo is to bring the antiserum into contact with the embryo and to check on the development of the organ rudiments in these conditions. An experiment of this type was performed some time ago by Guyer and Smith (1918), who immunized fowls by injecting into their blood crushed lenses from adult rabbits. The antiserum was then introduced into the blood of pregnant rabbit does. A considerable number of the offspring showed defects of the lens (opacity) as well as of other parts of the eye. Although this early investigation caused some criticism, the essential part of the result, that is, the injury to developing organs by an antiserum, has also been observed in experiments carried out with modern techniques. In the experiments of Burke and collaborators (1944) adult lens antiserum injected into the incubated chick egg caused cytolysis in the developing lens.

In another study of this type chicken embryos were cultivated outside the egg in a medium of agar and albumen, and antisera against organs of the adult chicken were added to the culture medium (Ebert, 1950). Adult brain, heart and spleen were used as antigens. Large amounts of antiserum added to the culture medium produced only a generalized lethal effect. If sufficiently high dilutions of the antiserum were used, the effect showed a high degree of specificity, though not an absolute specificity. Brain antiserum caused defects in the development of the forebrain only. Antisera for heart and spleen produced slightly more generalized effects; cytolysis occurred in the somites and lateral plates, thus involving most of the mesoderm (no mesoderm defects were produced by the brain antiserum). The most interesting result concerns the heart; embryos treated with heart antiserum failed to develop a pulsating heart. Thus the earliest heart rudiment already possesses the antigen (or antigens) present in the adult heart and as a result may be destroyed or arrested in its development by the adult heart antiserum.

The result of the latter experiment is quite in keeping with the findings on the lens rudiment with the precipitation method. As to a comparison between immunological and purely chemical investigations, it will be noted that the immunological investigations have permitted the tracing of organ-specific substances to earlier stages of development than the chemical analysis.

14–4 CONDITIONS FOR DIFFERENTIATION

The Problem of Reversibility of Differentiation. Differentiation may be reversible to a certain extent. The morphological and physiological peculiarities of tissues require for their maintenance the environment which surrounds them in the normal organism. Except for manifestly non-living parts, such as the chitin cuticle in insects or the hairs in mammals, all other animal structures may become changed, or may even dissolve and disintegrate if the normal conditions in the organism are changed. A disintegration of the animal's morphological organization is even more to be expected if the integrity of its body is interfered with, as in the case of wounding or in experiments with explantation of parts of the animal's organs and tissues. The cultivation of small portions of tissues in a clot of blood plasma or any other suitable medium (tissue culture) is an especially effective method of investigating the extent to which the histological differentiation of tissues may be reversed. In tissue cultures intercellular structures (fibers of connective tissue, matrix of the bone and cartilage) become destroyed, the normal arrangement of cells in the tissues becomes dissolved, and the specific organoids of cells may also disappear (myofibrils, cilia). Cells derived from different tissues may acquire a very similar appearance, an appearance not unlike that of cells which have not yet undergone differentiation. These phenomena may be conveniently called **dedifferentiation.**

We have seen that in differentiation there is an increase of proteins in the cells in relation to nucleic acids. The reverse is true in the case of dedifferentiation. Dedifferentiation involves a breakdown of functional mechanisms of cells (composed of protein), and it can be expected that the amount of protein in the cells would diminish as compared with the amount of nuclear material. This is actually the case. Determinations of protein nitrogen and deoxyribonucleic phosphorus were carried out on a culture of fibroblasts taken from a chicken's heart and grown *in vitro*. After six days of cultivation the amount of protein nitrogen per unit of deoxyribonucleic acid phosphorus was diminished by one-half (Davidson and Leslie, 1950).

The dedifferentiation of cells under conditions of tissue culture, or the retention of a lower state of differentiation, if embryonic cells are being cultivated, is the result of the change of environment in which the cells are being kept. The separation of a part of the body from other parts and the damage to the tissues caused by cutting them may contribute toward these changed conditions (the latter factor will be dealt with in the discussion on regeneration, Chapter 19). However, the main factor which causes the cells to lose their differentiation and to start growing and proliferating instead is the medium which surrounds them in tissue culture, and which is different from the medium (the fluid bathing the cells) in the intact organism. The standard medium for tissue cultures consists of blood plasma, embryo extract and some modification of the Ringer saline solution. The salts are necessary for the upkeep of ionic balance between the cells and the surroundings. The blood plasma contains fibrin which clots and forms the solid substrate on which the cells can spread out, and which also becomes slowly dissolved and supplies some of the necessary nutrients for the cells. The embryo extract is added as a growth-stimulating agent. Without embryo extract, on blood plasma and saline alone, the cells grow only very slowly, if at all. With embryo extract added the cells start growing and proliferating rapidly, and dedifferentiation occurs, as described above.

The embryo extract naturally contains some of the substances surrounding the cells in the early embryo. The behavior of cells in tissue cultures may thus be attributed to the fact that the medium bathing the cells tends to keep them in a condition characteristic of the embryo in the stage from which the extract had been taken. In a seven-day-old embryo, often used for the preparation of embryo extract, growth is rapid (see section 15–3) and there is as yet not much histological differentiation. If this interpretation were essentially correct, it should be possible to produce progressive differentiation of cells by exposing them to media containing extracts from consecutively older embryos. This has been done in the following experiment (Gaillard, 1942). Osteoblasts derived from a 16-day-old chick embryo were kept in cultures in two series. The tissues were grown in flasks (Carrel flasks) on the surface of a blood plasma clot and suffused with embryo extract, which was changed every two days. In one series the embryo extract was always the same, prepared from seven-day-old chick embryos. In the other series older embryos were taken to prepare the extract at each subsequent change, namely: 10-day, 12-day, 15-day, 18-day embryos, then the extract from the heart of a newly hatched chick and lastly blood serum of an adult hen. The changing extracts were to imitate the changes in the tissue fluids with progressing development. The experiment gave results according to expectation; the first culture grew and proliferated without any differentiation, while bone developed in the second culture.

In spite of the apparent simplification of the cells which have undergone dedifferentiation, they do not revert to the state of embryonic cells. Evidence from numerous experiments, which cannot be considered here in detail, proves that dedifferentiated cells retain their histological specificity and do not acquire new competences. If cultivated alone, renal tissue may become disintegrated and its component cells grow out in the form of a disorganized sheet or layer, but the cells remain kidney cells and, given suitable conditions, they again arrange themselves into the shape of tubules (see also p. 396).

Similarly cartilage may be dedifferentiated and the cartilage cells grow as a disorderly mass, scarcely distinguishable from a mass of connective tissue cells. But if the culture is kept under conditions which do not favor rapid growth, as when the culture medium is poor in growth-promoting substances (small amounts of embryo extract) or if it is not changed often enough to a fresh medium, a new differentiation becomes possible. When this happens the former cartilage cells again secrete cartilage matrix, thus showing that they retained their functional specificity in spite of morphological simplification.

In tissue cultures cells are dissociated from one another as a result of the breakdown of the bonds which bind them together in a normal tissue. The dissociation is not complete, especially in the case of epithelial tissues, which remain joined in a sheet even when cultivated *in vitro,* nor is it quite under the control of the experimenter. Therefore the attempt was made to separate tissue cells by more direct methods so as to test how far individual cells can retain the peculiarities that they had acquired previously in conjunction with other cells. By grinding tissues, especially embryonic tissues, in a specially prepared small glass mortar they may be disaggregated and a sufficient number of individual cells remain alive and may be used for further study (Weiss and Andres, 1952). A more delicate method is to treat tissues with a weak solution of trypsin in a calcium- and magnesium-free saline. This treatment causes the cells to separate from each other. A suspension can be obtained in this way consisting almost exclusively of completely separated individual cells which appear to be quite healthy (see Fig. 236) (Moscona, 1952). The suspension may then be put into a medium in which the cells can reaggregate and, under favorable conditions, resume differentiation. In some experiments the medium was that used for tissue culture with reduced embryo extract to facilitate differentiation (see above). In other experiments the cell suspension was injected into the veins

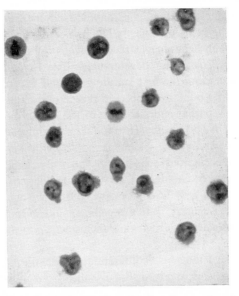

Fig. 236. Suspension of completely dissociated cells, produced by treating chick embryo tissues with trypsin. (Courtesy of Dr. A. A. Moscona.)

of chicken embryos, and the cells became disseminated through the vascular route. Individual cells or small clusters of cells then settled at various sites in the body of the embryo or on the chorio-allantois and were either incorporated in the tissues of the host or gave rise to small local growths —**teratomas** (Andres, 1953).

The main result of these experiments is that the cells, which had passed through a condition of complete disaggregation, were able to resume specific tissue differentiation, whether on a plasma clot or on the chorio-allantois of a living embryo. Masses of brain tissue, muscle, cartilage, bone, nephric tubules, glandular tissue or epidermis, usually in the form of cysts with clearly differentiated feather germs, were observed in various experimental series. The assortment of tissues appearing in any experiment depended on the origin of the cell suspension, the stage of the embryo from which the cells were derived, and the part of the embryo taken. The results available so far are compatible with the assumption that every type of cell differentiates after disaggregation in conformity with its previous differentiation. So if cell suspensions were prepared from whole embryos, the teratomas contained a very wide variety of tissues including nervous tissue, musculature and glands. If only limb-buds are used for preparing the suspension, the teratomas contained epidermis, cartilage, bone and mesenchyme but no nervous tissue, muscle or glands (Andres, 1953).

It is remarkable that structures developing from cell suspensions do not present a chaotic assemblage of different types of differentiation. Rather, they produce parts resembling organ rudiments of a normal embryo, with the various tissues segregated from each other and each tissue arranged into a recognizable morphological unit: nerve cells form brain vesicles with a central lumen; cartilage cells are sometimes arranged into elongated rods with perichondrial ossification; epidermis cells are arranged in layers with a clear distinction between proximal and distal surfaces; feather germs may show a very high degree of internal organization (Fig. 237). As it is unlikely that each unit of tissue is always derived from one single cell, it follows that the cells sort themselves out in some way—that cells of any one kind join together and group themselves anew in an order similar to what they had before they had been separated. This is the same process that we have found in relation to cells at the time of gastrulation and neurulation (section 9–3) and is a further evidence of the "affinities" between particular kinds of cells.

In the dedifferentiated state the kidney cells do not excrete urine, and the cartilage cells do not produce chondrin. Their functional mechanisms are inactivated or perhaps even broken down. A renewal of histological differentiation would then require a rebuilding of the functional mechanism of the tissue cells (of their "unique enzymatic pattern"). The way the cells do this is quite an important problem, especially if it is realized that the cells in the dedifferentiated or disaggregated state may go through a series of mitotic divisions and nevertheless preserve their original specific prop-

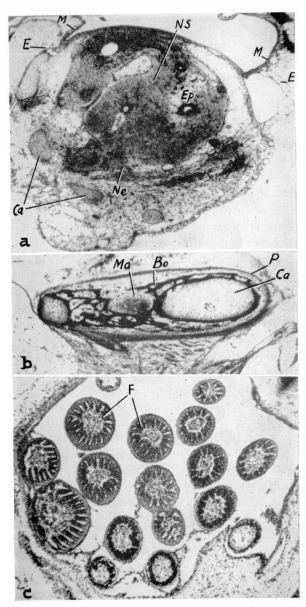

Fig. 237. Teratomas developed from completely dissociated cells of a chick embryo and injected into the blood vessels of an intact chick embryo. *a,* Teratoma consists mainly of a large mass of brain tissue (*NS*) surrounded by some muscle tissue and cartilages (*Ca*). *M,* Visceral mesoderm of the host; *E,* endodermal epithelium of the host; *Ne,* nerves; *Ep,* ependyma. *b,* Teratoma consists of an elongated piece of skeleton with cartilage (*Ca*) and bone (*Bo*), surrounded by a layer of perichondrium (*P*) and also of masses of muscle and adipose tissue. *Ma,* Bone marrow. *c,* Teratoma is a large epidermal cyst with feather germs (*F*) seen in cross-section. (From Andres, 1953.)

erties. We have reason to believe that the specificity of each kind of cell rests on their chemical composition, on their "unique enzymatic pattern" or, in the last instance, on the peculiar arrangement of amino acids in the protein molecules and on the specific spatial configuration of the molecules. This peculiar sequence of amino acids and the specific configurations of protein molecules must emerge every time new molecules are produced in the cell, be it in cell divisions, in cell growth or in cell differentiation. It has been argued that such a highly specific arrangement of particles can conceivably arise only if there exists in the cell a model for each type of complex molecule. As the types of molecules are more or less the same in successive generations of animals, it is natural to seek for the models in the hereditary substance of the cell: in the chromosomes and the genes. This is the starting point for the theory according to which the chromosomes or their component parts, the genes, represent a set of moulds or templates on which are modeled all the complex proteins of the nucleus and the cytoplasm.

Now it is not so easy to ascribe to the nucleus the persistence of different histological types of cells in one and the same species. The experimental evidence is that the nuclei in cells developing into different tissues still preserve the full complement of hereditary factors necessary for differentiation into any kind of cell inside the specific range. It seems logical therefore to assume that the persistence of histological specificity in cells growing and proliferating in the dedifferentiated state resides in the ability of the cytoplasm to ensure the perpetuation of the same kinds of protein molecules. This could be achieved if there existed in the cytoplasm some kind of copy of the nuclear moulds or templates, but only of the special types which are characteristic of a particular tissue. These templates could originally be formed in the nucleus and find their way into the cytoplasm and reproduce there. Self-reproducing copies of nuclear genes persisting in the cytoplasm have been called **plasmagenes** (see Wright, 1941, 1945; Darlington, 1944; Lindegren and Lindegren, 1946; Spiegelman, 1948; Weiss, 1953; R. E. Spratt, 1954). With the electron microscope one can see in the cytoplasm of cells minute granules containing ribonucleic acid. (Palade and Siekevitz, 1956.) It seems that these granules, called **ribosomes,** may actually perform some of the functions attributed to plasmagenes.

Control of Differentiation by Chemical Substances. Since the actual differentiation of tissue cells is dependent on conditions in their environment, it is possible, as we have seen (p. 389), to direct their development along definite pathways by exposing them to appropriate treatments. Some further experiments along these lines will now be described, first of all in relation to the control of differentiation by such chemical substances as vitamins and hormones.

Stratified epithelium in vertebrates appears to be a suitable object for experiments of this kind. Under normal conditions stratified epithelium takes on various forms, both in different groups of vertebrates and in different parts of the same animal. In terrestrial vertebrates stratified epithelium is squamous and cornified on its surface, but in fishes the epidermis, though stratified, is not squamous and contains mucus-secreting cells. The degree of cornification in the former varies, being very strong on the surface of the body, but weak in that lining the oral cavity, the pharynx and the esophagus. In the vagina the stratified epithelium undergoes cycles of cornification accompanying the menstrual cycle. The epidermis gives rise to a number of glands, among them the mammary glands, in which the epithelium becomes simple columnar, but under pathological conditions it may revert to the cornified type, as in some kinds of cancer (Pullinger, 1949). Lastly, in the endodermal part of the alimentary canal of mammals the esophagus is lined with stratified epithelium while the posterior parts, starting with the stomach, are lined with columnar mucus-secreting epithelium.

The transformation of stratified epithelium from the cornified to the non-cornified type may be achieved by purely chemical methods under conditions of explantation *in vitro*. If small pieces of vaginal wall of juvenile female mice are cultivated in the standard media for tissue cultures, the epithelium remains without any traces of cornification. If, however, the female sex hormone $3,17\beta$ estradiol is added to the culture medium, the epithelium becomes squamous and cornified on its surface. Other preparations of estrogens have a similar effect (Hardy, 1953).

The opposite transformation may be achieved by treatment of the cells with vitamin A. If the skin from young seven- to eight-day incubated chicken embryos is grown in the ordinary tissue-culture medium (blood plasm + embryo extract), it develops a squamous cornified layer on the surface of the epithelium. An addition of vitamin A to the medium causes a complete transformation of the epithelium, which now becomes a cuboid or columnar mucus-secreting one. It is remarkable that the cells do not need to be continuously in a vitamin A enriched medium to be converted into the non-keratinized type: a short treatment with the vitamin suffices to switch the development from the one channel into the other. To enable the vitamin to reach each epithelial cell in a short time, the skin of a chick embryo was dissociated into single cells by trypsin treatment (as described on p. 390), and these were then immersed into a 0.06 per cent solution of vitamin A for 15, 30 or 60 minutes. After this the cells were put onto a plasma clot and cultivated for several days. The cells reaggregated and in the best cases formed cysts or vesicles, with the distal surface of the epithelium turned inward and the outer surface surrounded by connective tissue. Typical stratified squamous epithelium developed in controls not exposed to vitamin A (see Fig. 238, *a*). As to the vitamin A treated preparations, it was found that even a 15 minute sojourn in the vitamin solution was sufficient to transform the cells into the non-keratinizing type, while additional periodic washings (30 minutes every two days) furthered the development of the epithelium into a typical columnar epithelium with goblet cells (Fig. 238, *b*) (Weiss and James, 1955).

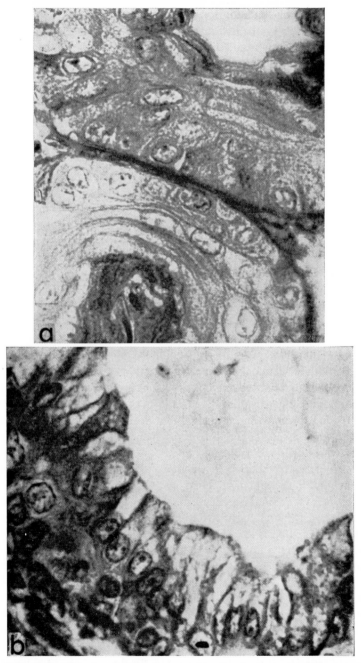

Fig. 238.   Influence of a short treatment with vitamin A on the differentiation of dissociated and reaggregated epidermal cells. *a,* Control culture, differentiating as typical stratified squamous epithelium. *b,* Treated culture, developing into columnar epithelium with goblet cells. (From Weiss and James, 1955.)

A special type of condition for the differentiation of some tissues is the interaction between epithelium and mesenchyme. The organized growth of many epithelial structures and the very preservation of the epithelial arrangement of cells is dependent on the presence of connective tissue on one (the proximal) surface of the epithelium. An example of this is presented by the embryonic epidermis of the early amphibian embryo (see p. 169). When isolated without mesoderm or mesenchyme of any kind, embryonic epidermis soon loses its epithelial arrangement; the cells acquire a reticulate arrangement and eventually degenerate and die off. In the presence of mesodermal mesenchyme the epithelial arrangement is preserved; the ectoderm remains healthy and differentiates as normal skin epidermis.

When epithelial tissues are grown in tissue culture they tend to grow as sheets of cells, which, though preserving contact with each other, do not follow the arrangement that was present in the original tissue or that should have arisen in the course of development, if the part taken for cultivation was the early rudiment of some epithelial structure. For example, if epithelial cells of renal tubules are cultivated alone, they form a disorganized sheet or layer spreading out on the surface of the plasma clot. If, however, some connective tissue cells are added to the culture, the spreading out of the sheet of renal cells is arrested, and they become reconstituted into tubules, reminiscent of the normal tubules of the kidney (Drew, 1923).

The dependence of epithelia on the connective tissue can be demonstrated very clearly when rudiments of glands are cultivated *in vitro* with or without connective tissue. The submandibular glands of a mouse embryo appear on the thirteenth day of gestation in the form of a pair of solid, club-shaped buds, growing from the buccal epithelium down into the connective tissue layer, one on each side of the tongue. Soon the epithelial bud becomes surrounded with dense mesenchyme forming a "capsule." On the fourteenth day the tip of the epithelial bud becomes indented; this is the beginning of the branching of the gland rudiment and the two primary branches grow out, each bearing a knoblike thickening at its end. The end knobs divide repeatedly and eventually give rise to the secreting acini of the gland, while the more proximal parts likewise become split and form the ramified system of ducts. While this outgrowth and branching of the epithelial parts goes on, the capsule mesenchyme surrounds the branches and penetrates between them, forming the connective tissue of the gland. The ducts and acini become hollowed out at a later stage.

This development can be observed on whole gland rudiments cultivated *in vitro* in a plasma clot (Fig. 239) (Borghese, 1950). By treatment of the early thirteenth day rudiment of the submandibular gland with a 3 per cent solution of trypsin for three to five minutes in the absence of calcium and magnesium ions, the cohesion of the mesenchyme with the epithelium can be destroyed and the two components of the rudiment may be separated from each other. On placing them in a culture medium, both components are found to be fully viable but their differentiation is no longer normal:

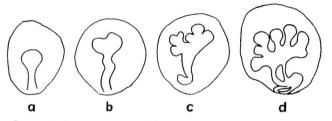

Fig. 239. Camera lucida drawings of the capsule and the epithelial part of a mouse's submandibular gland growing *in vitro*. (From Borghese, 1950.)

the capsule connective tissue produces a typical culture of mesenchyme, with individual cells crawling out radially from the initial piece of tissue; the epithelial bud loses shape and becomes transformed into a sheet of growing cells. This behavior, however, does not show that the cells of either the capsule or the epithelial part have changed in their essential nature: if the two components are set in the culture medium near to one another, the mesenchyme comes to surround the epithelial rudiment, whereupon the latter starts sprouting and branching in a very nearly normal fashion. The interaction between the epithelium and mesenchyme has restored the system to its normal state (Fig. 240) (Grobstein, 1953a, 1953b).

The mutual influence of the epithelium and mesenchyme of the submandibular gland rudiment may be considered to be a special case of induction, and, as has been done with respect to inductions occurring in earlier stages, it may be questioned whether the influence of one component on the other is specific or not. To answer this question, isolated epithelium

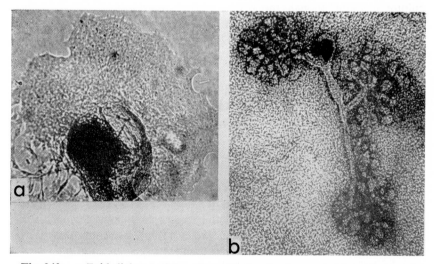

Fig. 240. *a,* Epithelial part of the submandibular gland separated from the capsular mesenchyme and grown *in vitro*. *b,* Epithelial part of the submandibular gland separated from the capsular mesenchyme as in *a* but then reunited with the capsular mesenchyme in the culture. (*a,* From Grobstein, 1953a; *b,* from Grobstein, 1953b.)

of the submandibular gland was cultivated with mesenchyme of a different origins: mesenchyme from the rudiment of maxilla, from the somites, from the lateral plate, or from the lung rudiment. With all foreign mesenchyme the epithelial rudiment of the submandibular gland failed to sprout and ramify; the sheetlike spreading out of the epithelial cells was arrested, however, and the rudiment developed eventually into an epithelial cyst. When the epithelial rudiment was implanted into a culture of capsular mesenchyme, which had been previously killed off by heat, the rudiment behaved very much the same as when it was surrounded by foreign mesenchyme; the flattening out and spreading of the epithelium was suppressed, but no growth or ramification took place. Apparently the stimulus necessary for the normal development of the duct and acini system of the submandibular gland can only be given off by the capsular mesenchyme of this organ and only in the living state (Grobstein, 1953b).

In another case of tissue interaction the inducing stimulus has been found to be less specific. We have seen (p. 364) that the development of renal tubules of the metanephros is dependent on the presence of the growing and ramifying ureter bud. Under the influence of the latter the loose mesenchyme of the metanephric rudiment (the metanephrogenic mesenchyme) becomes partially converted into epithelium and forms convoluted tubes—the renal tubules—which in normal development link up with the terminal ramifications of the ureteric ducts. The interaction of the ureteric bud and the metanephrogenic mesenchyme can be observed *in vitro* in the same way as the interaction of epithelium and mesenchyme of the submandibular gland, that is, after separating the epithelial and the mesenchymal components by trypsin treatment and reuniting them in the culture medium, though the results of the reaction here are of course essentially different. The experiment referred to previously (p. 396) differs from the ones described now in the stage at which the renal tissue was taken for cultivation: in the previous experiment the epithelium was derived from already differentiated renal tubules; in the present experiment it is the initial formation of the renal tubules that is under consideration.

By combining metanephrogenic mesenchyme taken from an 11-day-old mouse with other tissues it was found that the ureteric duct is not the only part that causes the mesenchyme to be converted into renal tubules. The epithelial part of the submandibular gland cultivated together with metanephrogenic mesenchyme produces the same effect, although there was no reciprocal action; the gland epithelium remained unbranched. The epithelium of the submandibular gland can thus serve as an "abnormal inductor" of renal tubules, just as adult liver can serve as an abnormal inductor of a neural plate. Furthermore, the spinal cord, especially its ventral half, proved to be a very efficient inductor of renal tubules, when placed in a culture of metanephrogenic mesenchyme (Grobstein, 1955).

The renal tubules in the above experiments always developed in the immediate vicinity of the inducing tissue. An attempt was therefore made to test whether immediate contact is necessary for the induction to take place. For this purpose the inducing and reacting tissues were separated by thin membranes of various degrees of porosity. Cellulose ester membrane filters were used, varying in thickness from 20 μ to 150 μ, and with pores approximately 0.8 μ, 0.4 μ and 0.1 μ in diameter. The experiment consisted essentially in arranging for two tissue cultures to grow one on each side of the membrane filter. The inducing culture (spinal cord) was grown on one side, and the reacting tissue (the metanephrogenic mesenchyme) on the other side of the membrane. It was soon discovered that the inductive influence could easily pass through the coarser filters (with pores 0.8 μ and 0.4 μ in diameter) of up to 60 μ in thickness, but not if the filter was 80 μ thick (or consisted of four or more layers each 20 μ thick) (see Fig. 241). With finer filters (pores *ca.* 0.1 μ in diameter) the induction became weaker, and the influence could cross only a thin membrane not exceeding 30 μ in thickness. A still finer filter, such as a cellophane membrane 20 μ thick, effectively stopped the inducing influence.

Positive results have also been obtained when the submandibular gland components were cultivated on the opposite sides of a membrane filter, the epithelial part on one side and the capsular mesenchyme on the other. The epithelial bud produced numerous ramified outgrowths, thus showing that the inducing principle could penetrate through the filter (Grobstein, 1953c).

The electron microscope was applied to see what was going on in the pores of the filter membrane separating the inductor and the reacting tissue. In ultrathin sections made perpendicular to the separating membrane, it could be seen that in the case of coarser membranes the pores contained cytoplasmic outgrowths of cells, coming both from the neural tissue and from the mesenchyme. The possibility was therefore not excluded that the processes of the two kinds of cells met somewhere inside the membrane and thus established a direct contact between the inducing and reacting cells. These processes were, however, more scarce when filters with 0.4 μ pores were used, and in the filters with 0.1 μ pores there was practically no penetration of the filter by cytoplasmic processes of the cells, except for a few small impocketings which did not go more than 1 to 2 μ into the substance of the filter (Grobstein and Dalton, 1957). The results of these experiments are very important: they show that the interaction of cells responsible for the normal differentiation of tissues can cross a narrow gap between the cells. The inducing principle is therefore a diffusible substance. Since this substance does not penetrate through cellophane, it must be a macromolecular substance, most probably a protein or a nucleoprotein. The similarity of these results to those obtained with the neural inductor in early amphibian development (see p. 184) is very obvious and very suggestive. Much remains to be done. We still have to learn why the inducing substance does not spread over greater distances—whether this

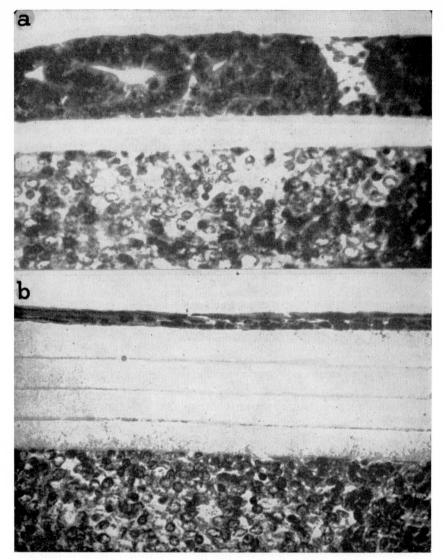

Fig. 241. Induction of kidney tubule differentiation through cellulose ester membrane filters. Below the filters is the inductor, spinal cord tissue; above the filters is the reacting nephrogenic mesenchyme. *a,* Successful induction through one layer of filter (20 μ thick). *b,* Four layers of filter preclude induction. (From Grobstein, 1957.)

is the result of extremely small quantities of the substance produced, of its instability or of some other unknown factor. The chemical nature of the inducing substance has to be elucidated. The way for further research is now open and new discoveries will probably not be long awaited.

CHAPTER **15**

Growth

Growth is the increase in size of an organism or of its parts due to synthesis of protoplasm or of **apoplasmatic** substances. Protoplasm in this definition includes both the cytoplasm and the nucleus of cells. Apoplasmatic substances are the substances produced by cells and forming a constituent part of the tissues of the organism, such as the fibers of the connective tissue, the matrix of bone and cartilage, and so on, as opposed to substances produced by the cells and subsequently removed from the organism, such as the secretions of digestive and skin glands, or substances stored as food, such as fat droplets in cells of the adipose tissue. Imbibition of water or the taking of food into the alimentary canal before the food is digested and incorporated into the tissues of the animal, although it may increase the weight of the animal, does not constitute growth.

Growth is the result of the preponderance of the anabolic (synthetic) over the catabolic (destructive) processes in the organism. If synthesis and decomposition go on at the same rate, there is no increase in the bulk of the organism—no growth. In certain conditions decomposition dominates over synthesis, as for instance in prolonged inanition, when synthetic processes are impossible because of lack of food supply, while catabolic processes (oxidations, etc.) continue to satisfy the current requirements for energy. After the internal food reserves (fat in the adipose tissue) have been exhausted, the energy is produced at the expense of the proteins of the protoplasm, and the result is a decrease in the mass of living matter—which may be called **degrowth** (Needham, J., 1942).

In ordinary life growth is often identified with increase in height. From a biological view point, however, growth is an increase of mass of living substance; only the weight can therefore be considered as an index of growth. The increase in linear dimensions (height, length) may accompany increase of mass, but the connection is by no means a simple one, and this must always be borne in mind, if for any reason the height or length of a growing animal is measured instead of its weight.

By weighing a growing animal at regular intervals and plotting the weight against time on a diagram we get a **growth curve,** showing the increase of the mass of the animal with time. The shape of the growth curve very often resembles the letter S: there is an initial part, where the curve rises very gradually, then a middle part, where the curve rises steeply, and the last part, where the rise of the curve is again slowed down and it asymptotically approaches a horizontal line signifying the limit of growth in each particular case. A curve of this shape is known as the **sigmoid curve** (Fig. 242).

If the increments of growth for equal time intervals are measured, the increase in different periods of life of the same individual may be estimated. The increase, taken as a difference between the final and initial size (weight) of an animal for any period of time irrespective of other factors, is called the **absolute increase.** The sigmoid curve is actually a graphic representation of the absolute increase, and it shows that the absolute increase per unit of time is, in this case, greatest in the middle part of the growth cycle. The absolute increase is, however, not a correct indication of the rate of growth and cannot well serve for comparing the growth at different periods of life and for comparing the growth of different organisms. It is obvious that if a small animal and a big animal show the same absolute increase in a given time, their rate of growth is not the same: the small animal has to grow at a greater rate to have the same absolute increase as the big one. The rate of growth is measured as the relative increase—the increase related to the initial mass of growing substance. The exact definition of the rate of growth is then

$$v = \frac{dW}{dt} \cdot \frac{1}{W} \qquad (1)$$

where v is the rate of growth and W is the size (weight) of the animal at any given time, t.

Growth thus defined is, in its essence, an increase in geometrical progres-

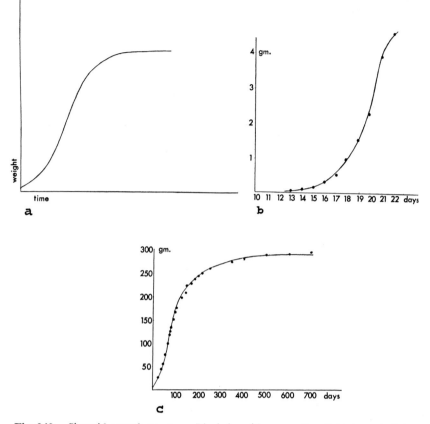

Fig. 242. Sigmoid growth curves. *a,* Ideal sigmoid curve, after Robertson. *b,* Intra-uterine growth of the white rat. *c,* Postnatal growth of the white rat. The points show the actual measurements. (After Donaldson, Dunn and Watson, from Fauré-Fremiet, 1925.)

sion, the increase being proportional to the initial quantity of growing substance.

Increase in geometrical progression is called exponential; growth is thus an **exponential process.** The general formula for exponential growth, show-ing the increase of the growing mass with time, is

$$W = e^{vt} \qquad\qquad (2)$$

where W as before is the weight of the animal at any given time t, v is the observed rate of growth, and e is the base of natural logarithms (equal to 2.71828 . . . , accurate to the fifth decimal place). The above formula is correct if the initial quantity of growing matter is infinitely small and very nearly accurate if the initial size is negligible. If the initial size of the growing animal cannot be ignored, the formula must be slightly changed by the introduction of a constant, as follows:

$$W = be^{vt} \qquad\qquad (3)$$

where b is a constant and is equal to the initial size of the growing organism.

The rate of growth may be calculated from the same data as those on which the sigmoid curve of growth is based.

15–1 GROWTH ON THE CELLULAR AND INTRACELLULAR LEVEL

The essentially exponential nature of the growth process can best be verified by considering growth on the cellular level. In a mass of growing cells, other conditions being equal, the increment of growth will be greater, the greater the number of growing cells. If each cell produces, by mitosis, two cells after a certain period of time, then the total number of cells will be doubled, and the same will occur after another similar interval, provided that all the time neither the cells nor the conditions for their existence change. Such conditions are very nearly fulfilled in the case of cells growing in tissue cultures with sufficiently frequent transfers of cells to new portions of nutritive medium. Cell divisions under these circumstances have been observed in some experiments to occur at intervals of from 10 to 23 hours. The exponential growth is thus based on the ability of the cells to duplicate themselves.

A similar case is the growth of unicellular organisms (bacteria, protozoa and unicellular plants) in constant and favorable conditions. The principle of exponential growth applies in the same way to the increase of populations of multicellular animals reproducing in optimal conditions, provided that the food supply is infinite, and that the animals can disperse themselves in space to avoid overpopulation. Theoretically the growth of living matter (in this case proportionate to the number of animals) may continue indefinitely, always having an exponential nature. Practically, of course, such an indefinite increase can never occur since the conditions for it cannot be provided.

Matters become more complicated if growth is considered on the intracellular level. Growth being the result of the preponderance of anabolic over catabolic processes, the rate of growth would depend on the factors controlling each of these. The synthetic processes in the cell are supposedly due to the presence of enzymes catalyzing in succession each step of the building up of the molecules of which the protoplasm of the cell consists. The degradation of the substances in the cells through oxidation and fermentation is similarly due to a number of enzymes, each controlling one step in the decomposition of each complex organic compound. The over-all rate of the synthetic process in the cell may therefore be dependent on any one of the enzymes participating in the chain of synthetic reactions, and whichever is in the shortest supply may obstruct the chain of reactions, or put a ceiling on the amount of substances that can be synthesized in any given time. The same applies to the catabolic processes. Furthermore, the rate of action of an enzyme may depend, besides on the quantity of the enzyme, on the presence of coenzymes or activators. Any one of these can then also limit the amount of turnover in a chain of anabolic or catabolic reactions.

The rate of growth on the intracellular level thus is not determined by the initial mass of growing cytoplasm, but is dependent on the performance of an enzyme, or set of enzymes, that set the limit to the amount of complex organic compounds that may be synthesized. Exponential growth of the protoplasm as a whole can take place only if the crucial enzyme, or set of enzymes, increases proportionately to the total amount of protoplasm of a cell. That this is so, is by no means self-evident. On the contrary, there is evidence that the growth of the cell between two divisions does not proceed at a uniform rate. This has been found by direct measurements of the growth of unicellular animals.

The growth of infusorians with elongated bodies was studied by measuring their length at regular intervals between two divisions (Schmalhausen and Syngajewskaja, 1925), and in a heliozoan, *Actinophrys*, which has a spherical shape, the growth was estimated by measuring the diameter of the cell (Syngajewskaja, E., 1935). Recently the remarkable achievement has been performed of weighing individual cells of *Amoeba* throughout its life cycle (Prescott, 1957). The results of all these measurements conform very well with one another, and show that in these unicellular organisms growth is most rapid after a cell division, and slows down later (see Fig. 243).

On the other hand, it has been indicated (McIlwain, 1946) that the most important enzymes controlling the growth of cells may be present in very minute quantities, perhaps one molecule, or a very few molecules, of the enzyme per cell. Under these circumstances the amount of the enzyme cannot increase gradually, parallel to the increase of the whole protoplasmic body; the increase of the enzyme can only be discontinuous, one unit (molecule) at a time. If it is one of the crucial enzymes, then the synthesis should also of necessity proceed in disconnected periods and with different rates. All discrepancies in the composition of the cellular body must be ironed out somehow before the cell divides into two daughter cells; otherwise the daughter cells would have different properties from the mother cell at the beginning of its individual existence, which is contrary to what is observed with indefinitely reproducing cells in tissue cultures, or protozoans under favorable conditions. It follows that, in cells that reproduce and grow at a constant velocity, it is possible to study the increase of the mass of living matter without inquiring into the intracellular mechanisms that ensure that the daughter cells have the same constitution as the mother cells. The mechanism of intracellular growth is, however, of the greatest importance for the understanding of cases in which the properties of the cells do change, as time goes on, and this is the case with the cells of the developing embryo.

If all the enzymes, with their coenzymes and activators, taking part in the metabolism of the cell could be measured, it might be possible to determine in every case which of the enzymes is the limiting factor in the growth of

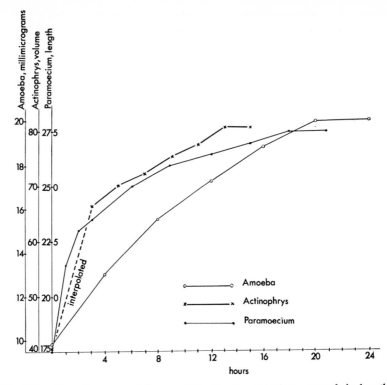

Fig. 243. Growth curves of three protozoans. *Paramoecium*—growth in length, in arbitrary units (divisions of ocular micrometer) (after Schmalhausen and Synga-jewskaja, 1925). *Actinophrys*—growth in volume, recalculated from original measure-ments of diameter (after Syngajewskaja, 1935). *Amoeba*—growth in weight, in milli-micrograms (after Prescott, 1957). In *Actinophrys* growth could not be measured during division, which takes about three hours to complete, and is interpolated for that period.

any particular cell. This cannot be done at the present stage of our knowl-edge. It is possible, however, to determine the presence in the cells of certain substances which, though they are not enzymes, are supposed to be con-nected in some way with the growth of cells. The first of such substances is the nucleus.

The nucleus is itself a growing part of the cell, and it is possible to esti-mate the changes in the amount of nucleic substance in the cells in different stages of cell growth. The substance, the amount of which can actually be measured, is deoxyribonucleic acid, the essential component of the chromo-somes. The amount of this acid per cell may be estimated either chemically or spectroscopically (by ultraviolet light absorption due to the presence of nucleic acid), and it was found that in any species of animal all somatic cells having a diploid set of chromosomes possess the same amount of deoxyribonucleic acid. Table 16 shows the amounts of deoxyribonucleic acid per cell found in different vertebrates.

Table 16.   Amounts of Deoxyribonucleic Acid in Single Nuclei in Various Animals (in mg. x 10^{-9}) (Vendrely and Vendrely, 1949; Mirsky and Ris, 1949)

Bull	6.4
Pig	5.1
Guinea pig	5.9
Dog	5.3
Man	6.0
Rabbit	5.3
Horse	5.8
Sheep	5.7
Mouse	5.0
Duck	2.2
Fowl	2.4
Turtle	5.1
Toad	7.3
Frog	15.7
Carp	3.3
Trout	5.8

The average quantities of deoxyribonucleic acid per cell found in different somatic tissues of the same animal are the same. The figures apply to the nuclei in the resting stage.

During mitosis the amount of deoxyribonucleic acid is doubled, as the daughter cells show the same amount of the substance generation after generation. Direct (spectroscopic) measurements in individual cells in different stages of mitosis showed that the reduplication of the deoxyribonucleic acid content takes place in the very early prophase stage; during the metaphase the acid is split into two halves, and in each daughter cell the amount of nucleic acid remains constant until the prophase of the next mitosis (Swift, 1950). As the cells grow during the resting stage (interphase), the conclusion can be formed that the ratio between the amount of deoxyribonucleic acid in the cell and the amount of cytoplasm is highest during mitosis and diminishes gradually during the interphase. This should be compared with the observation mentioned above that the growth of a cell (a protozoan) is most rapid just after division and slows down later. This is in good agreement with the observation that the nucleus is in some way necessary for the growth of the cell as a whole.

Another substance which is in some way connected with growth is ribonucleic acid. Most of the ribonucleic acid of a cell is contained in the cytoplasm, though smaller amounts are also found in the nucleolus. In the cytoplasm the ribonucleic acid is normally bound to the structural elements of the cytoplasm—the microsomes. The amount of ribonucleic acid per cell is by no means constant; the ratio of ribonucleic acid to the deoxyribonucleic acid varies in different kinds of cells from 0.2 in the thymus to as much as

8.1 in the pancreas (Davidson, 1947). The observation has been made (Brachet, 1941) that the ribonucleic acid content is highest in those cells which are especially active in synthesizing proteins. Such are some of the secretory cells (pancreas), but also the actively growing cells. Cells that cease growing lose their ribonucleic acid. This has been directly demonstrated in the development of some tissues, in which growth and reproduction of mother cells ("stem cells") precedes the specific histological differentiation of the cells produced in this way. The differentiated cells lose the ability to grow and divide, as in the case of red blood corpuscles or of the permanently growing teeth of rodents, where the dentine is perpetually produced at the expense of a group of reproducing cells at the root of the tooth. In the stem cells, from which the red blood corpuscles are eventually produced, the content of ribonucleic acid is 5 per cent. As the differentiation progresses the percentage of ribonucleic acid drops almost to zero (Thorell, 1947).

On the other hand, if cells are stimulated to more intensified growth and reproduction, the percentage of ribonucleic acid goes up. This happens when differentiated cells are explanted *in vitro* and allowed to grow as a tissue culture. Measurements of the ribonucleic acid have been made on explants of chick heart tissue, and the amount of ribonucleic acid was compared with the amount of deoxyribonucleic acid, which as we know remains constant per cell. The measurements showed that at explantation the amount of ribonucleic acid was about double that of deoxyribonucleic acid, but it rose to about five times the amount of deoxyribonucleic acid after four days of cultivation *in vitro* (Davidson and Leslie, 1950).

15–2   GROWTH ON THE ORGANISMIC LEVEL

The relation between growth of cells and growth of whole organisms is not the same in every case. It will be convenient to distinguish three types of such relationships.

Type 1.   The volume of the animal's body increases owing to the growth of the individual cells, without an increase in their number (**auxetic growth** after Needham, J., 1942). This rather rare case is found in nematodes, in rotifers, and Larvacea among the tunicates. In the development of the nematodes the cell divisions stop in the early stages of organogenesis. The number of cells in the fully grown nematode is thus the same as in a young one just emerged from the egg (see Hyman, 1951). The number of cells in each rudiment may be definitely fixed in this type of development. Thus the whole excretory system of a nematode consists of only three cells. In the rotifer *Hydatina senta* the total number of somatic cells of the body could be estimated and was found to be 959 (Martini, 1912). The gonads, however, are an exception among the other organ rudiments; the mitoses in the sex glands are not restricted to the early embryonic period, and the gametocytes continue to proliferate in the adult animals.

While the number of the cells is restricted owing to the cessation of mitoses, the growth of the individual cells continues, and the growth of the animal is proportional to the increase in size of its constituent cells.

Type 2.   The animal's growth may be a result of the increase in number of its constituent cells (**multiplicative growth** after Huxley, 1932). The increase in number is brought about by mitotic division of all the cells, while the average size of the cell remains the same or nearly so. This type is found quite commonly in the growth of embryos; it is especially characteristic of the prenatal growth of the higher vertebrates. If the size of the cells remained strictly constant and no other processes were involved, growth of the animal's body would be directly proportional to the number of its constituent cells. Actually the size of the cells does not remain constant; as the embryo develops and its tissues become differentiated, the cells, or many of them, increase in size. This increase in size of the individual cells is, however, very limited and can account for only a very small proportion of the over-all increase of the body. As the animal develops, the mechanism of its growth becomes modified by the differentiation of its tissues, so that a further type of growth should be recognized.

Type 3.   In this type, the growth of the animal is based to a greater or lesser degree on the activity of special cells, retaining their ability to divide mitotically, while other cells have lost this ability more or less completely and so cannot proliferate any more (**accretionary growth,** Huxley, 1932). These latter cells are the differentiated cells of the body, performing various physiological functions necessary for the maintenance of the animal's life. The former may be called the reserve cells, for they provide a supply of new cells, capable of reinforcing and replacing, in case of necessity, the functioning differentiated cells. The multiplying cells can also be called undifferentiated cells, inasmuch as they lack the specific morphological and physiological properties of functioning cells. However, the competence of the proliferating cells may be already curtailed to a large extent, so that they can differentiate only into one type of functioning cell, or only into a limited number of types.

The epidermis of terrestrial vertebrates may serve as an example of this type of growth. In the outer layers of the epidermis the cells do not divide and do not grow. Their cytoplasm becomes keratinized, thus forming a protective layer on the surface of the skin. The fully keratinized cells are no more vital, and are perpetually being peeled off. The Malpighian or generative layer of the epidermis consists of cells which are not keratinized, and which possess the capacity to proliferate; later the outermost cells of the layer replace the keratinized cells, undergoing keratinization in their own turn.

The cells of the bone tissue (the osteocytes) do not grow and do not proliferate. The growth of the bones is dependent on the activity of the cells of the periosteum, which are capable of proliferation and can become osteocytes, while they secrete additional quantities of intercellular bone matrix.

In a number of organs centers are found, regions or layers which alone contain proliferating cells. In the vertebrate eye the proliferating cells are found in a ring in the region of the ciliary body. From there new cells are added onto both the retina and the iris. In the intestine of vertebrates proliferating cells are found at the bottom of the intestinal glands, and the same is true in other organs.

15–3 GROWTH CURVES AND THEIR INTERPRETATION

If the growth of an animal were a strictly exponential process, the growth curve would have the shape of a hyperbola, the curve produced by an exponential increase of a quantity. The curve would begin at a point near zero of the coordinate system and rise with an ever-increasing steepness.

A prerequisite for the growth being truly exponential is that the rate of growth should be constant. This in turn is possible only if neither the animal itself nor the environmental conditions change as growth proceeds. Such conditions apply to the growth of populations in perfectly favorable conditions, as has been mentioned previously. Purely exponential growth has been found in rod-shaped bacteria (Schmalhausen and Bordzilowskaja, 1930). In round bacteria—cocci—growth becomes retarded as the organism increases in size, presumably because the relative decrease of the surface in relation to the mass of protoplasm places the coccus in a worse position with regard to exchange of substances with the surrounding medium.

In multicellular organisms exponential growth is never found in a pure form. The nearest approach to it is the growth of larvae of insects having a complete metamorphosis. In caterpillars of moths growth starts at about the same rate after each moult; the absolute increase, therefore, is most rapid toward the end of larval development, as would be expected with exponential growth. However, the growth curve in this case is not a smooth hyperbola, as growth is very much retarded before and during moulting. The curve is thus broken up into a series of spurts and level portions (Lewitt, 1932).

In most animals the rate of growth does not remain the same, but diminishes quite regularly from the beginning to the end of the cycle of growth. The resulting curve, as has already been mentioned, in many cases is in the shape of a letter S—the sigmoid curve. The absolute increase is small at the beginning, becomes greatest in the middle of the period of growth and diminishes again later, until growth ceases altogether. If, however, the rate of growth is calculated for the various parts of the growth cycle it is found that the rate of growth is highest at the beginning and declines throughout the period of growth. The low absolute increase at the beginning is due to the fact that only a small mass of living substance is growing. In the middle part of the cycle the mass of growing substance has increased, and the absolute increase is, therefore, greater even though the rate of growth is not so high as at the beginning. In the last part of the growth period the growing mass is even greater than at the middle of the cycle, but the rate of growth has sunk so low that the absolute increase slows down and comes to a standstill. The growth curve approaches asymptotically to a horizontal line, which is the limit of growth, or in other words the maximal size that can be attained by the animal whose growth is being considered.

The limit to growth is very distinct in such animals as birds and mammals, but not in some of the lower vertebrates such as reptiles and fishes. In these animals growth can go on indefinitely, although with ever retarded rate. As a result, some individual fishes and reptiles may attain an extraordinarily large size if they escape accidental death which puts an end to the growth of other members of the same species. The growth curve in these animals approaches the shape of a parabola: it starts a little above the zero point (the initial mass of growing substance can never be 0) and gradually rises. The concavity of the curve is greatest near the starting point, and the curve flattens out as growth proceeds, approaching asymptotically a straight line inclined at an angle to the axes of the system of coordinates. This type of growth has been called **parabolic growth** (Fig. 244) (Schmalhausen, I. I., 1927, 1930a).

It is interesting to note that the sigmoid curve is also found in the growth of populations. Animal populations under ideally favorable environmental conditions should increase exponentially, but under actual conditions a shortage of food and space soon causes a decline in the rate of reproduction, with the result that the growth curve of the population becomes sigmoid. In the end the population may stop increasing further, having achieved the limit which is permitted by the environment given in each case.

The progressive decline in the rate of growth has attracted much attention, and several theories have been proposed to account for this decline. The

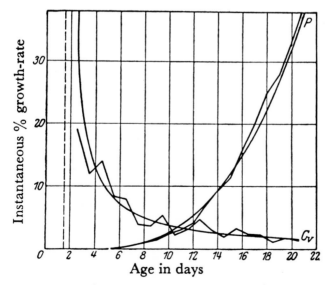

Fig. 244.   Increase in weight (*P*) and true rate of growth (*Gv*) in parabolic growth (growth of the chick embryo). The curves based on actual measurements and the smoothed theoretical curves are shown side by side. (After Schmalhausen, from Needham, J., 1931.)

theories fall into two main groups, depending on whether the sigmoid curve or the parabolic curve had been considered as better representing the nature of the growth in metazoa.

The first interpretation, mainly propounded by Robertson (1908) and more recently followed by L. von Bertalanffy (1948, 1949), starts with the assumption that the organisms possess an intrinsic potential growth rate. This potential growth rate is curtailed by obstacles arising in the process of growth, and setting a limit to it. As the organism approaches the limit of growth the growth rate diminishes, and eventually growth stops altogether. These relationships have been given the following mathematical form:

$$\frac{dW}{dt} = bW\frac{L-W}{L} \qquad (4)$$

where W is the size of the organism at any given time, t, b is the potential growth rate and L is the maximum size (limit of growth or asymptote of the growth curve). The equation (4) is known as the **logistic equation.** The sigmoid curve is also sometimes referred to as the logistic curve.

The logistic equation represents growth as an essentially exponential process, the increase $\left(\dfrac{dW}{dt}\right)$ being proportional to the growing mass (W).

On the other hand, the potential growth rate can actually be observed only in the initial stages of growth, while W is very small in relation to L (the actual size being very small as compared with the final size). At this stage the factor $\dfrac{L-W}{L}$ is near unity. As W increases, this factor becomes less than unity, and when W = L, it becomes 0. The increase then also becomes zero.

The logistic equation has been applied both to the growth of individuals and to the growth of populations. It appears, however, that the equation is much more fitted to the latter case. Several objections may be brought against the application of the equation to the growth of individuals.

1. The nature of the limiting factor has not been made clear. Bertalanffy suggests that the decline in the rate of growth is due to a discrepancy between the increase of surfaces and volumes. (The surfaces increase as the square of linear dimensions, the volumes as the cube of linear dimensions.) The ultimate growing unit, however, is the cell, and the sizes of cells as a rule do not change so as to account for the retardation of growth. The change in the relative size of any other surfaces in the animal's body would not explain the universal nature of the slowing down of growth.

2. Animals having no upper limit of size (having a parabolic growth curve) still show a retardation of growth. The logistic equation is not applicable in this case.

3. Holometabolic insects do not show a retardation of growth, although their growth is limited. The existence of a growth limit thus does not cause in itself a slowing down of growth during the earlier periods of development.

The second way of interpreting the growth curve makes use of the fact that the rate of growth (specific rate of growth) is inversely proportional to the time that elapses since the beginning of growth:

$$v = \frac{k}{t} \qquad (5)$$

$$\text{or } v \cdot t = k \qquad (6)$$

where v is the growth rate, t is time and k is a constant which is different for each organism.

The time since the beginning of growth is an indicator of the age of the animal. Minot (1891, 1908) was the first to make this observation and he accordingly stated that the decrease of the rate of growth is due to increasing age of the animal. The decline of growth is thus one of the many expressions of **ageing.** Minot also showed the connection between ageing and the differentiation of cells in the developing animal. The more the tissues of the animal become differentiated the less they are able to grow. This aspect of Minot's theory has been further elaborated by Schmalhausen.

According to Schmalhausen (1930b), the differentiation of parts of the developing animal and not age as such is the cause of retardation of growth. Inasmuch as the degree of differentiation of some cells of the animal's body may be preserved at a certain level, their rate of growth may remain constant, and exponential growth would take place in such groups of cells or tissues. This would account for the type of growth found in caterpillars: the caterpillar of a later instar is not more differentiated than one of the first instar so far as its larval organs are concerned. The organ rudiments of the adult moth (the imaginal discs) do not change very much until the pupal stage, and anyway they form only an insignificant part of the bulk of a caterpillar and do not essentially influence the growth curve. Given equal levels of differentiation at the start of every instar the growth rate remains constant, and this has actually been observed.

Growth may also be exponential in the growth centers of a vertebrate's body (Malpighian layer of the epidermis, ciliary body of the eye), but the proportion of proliferating cells to those which have partially or completely lost their ability to proliferate may diminish with age, and this becomes noticeable in the decrease of the over-all rate of growth. Besides, the cells of the growth centers are already partially differentiated and do not proliferate as rapidly as the cells of an early embryo. Lastly, growth centers may disappear altogether. The growth of the long bones of the limbs in vertebrates is dependent on the presence of an actively proliferating layer of cells between the diaphysis of the bone and its epiphyses. At about the time of sexual maturity the proliferating layer disappears, and the bony cores of the epiphyses are firmly joined onto the diaphysis. The result is that the growth in length stops.

The residual growth centers are, however, an exception rather than the rule among the tissues of a higher animal. Most of the cells become differentiated, which is of course essential for the functioning of the various organs, and this leads to a retardation of growth.

There appears to be an inverse proportion between the degree of differentiation and the rate of growth: the more highly an organ or tissue is differentiated, the slower it grows. The most highly differentiated tissue of the vertebrate's body is the tissue of the brain and spinal cord. The nerve cells in the adult state completely lose the ability to proliferate, and in the brain and spinal cord there is no residue of undifferentiated cells functioning as a growth center. The result is that the brain and spinal cord, throughout life, grow more slowly than other parts of the body. The sense organs have a degree of differentiation that is scarcely lower than in the brain tissue. These also grow very slowly. Organs like the heart and the kidneys grow at a higher rate than the brain and sense organs. The heart and the kidney, however, are rather highly differentiated organs, with no growth centers consisting of undifferentiated cells to provide for proliferation, and their growth is consequently not very fast. Higher rates of growth are found in the muscle and skeletal system. Although bone tissue has no growth of its own, and the striated muscle possesses a highly specialized differentiation, in both cases there are residues of undifferentiated cells providing for rapid growth. There is the periosteum in the case of bone (perichondrium for cartilage, in addition to a certain degree of interstitial growth). New elements of muscle tissue—new cells or muscle fibers—are not normally produced after the earlier part of the embryonic period has passed, but the individual muscle fibers, or muscle cells in the case of the smooth muscle tissue, have the ability to increase in size to a very great extent and thus produce a rapid growth of the muscular tissue.

The highest rates of growth are found in parts of the intestinal tract and in the skin of vertebrates. The skin possesses a permanent growth center in its Malpighian layer. There are groups of growing and proliferating cells in the gut (at the bottom of the intestinal glands and crypts); furthermore, the epithelial cells of the intestinal tract do not have a complicated morphological differentiation. The connective tissue is also capable of rapid growth; while the products of differentiation in this tissue are the various intercellular fibers, the cells themselves (fibroblasts and histiocytes) do not become highly differentiated, and even in the adult animal they can at any time be mobilized to proliferate and produce new connective tissue fibers. Such a mobilization of the connective tissue cells takes place in wound healing.

In any given organ the rate of growth is highest before histological differentiation has set in, or at least when it has not progressed very far. Even in the case of very highly differentiated organs, such as the brain, the early rudiment of the organ may possess and actually does possess a very high growth rate, as is shown in Table 17.

Table 17.   Growth of Organs of the Chick Embryo from the Fourth Day of Incubation to Hatching (after Schmalhausen, I. I., 1927)

	Initial Weight in Milligrams	Final Weight in Milligrams	Weight Increased by a Factor of	Growth Constant $k = v \cdot t$
Brain	10	1,020	× 102	2.10
Lens	0.08	8.8	× 110	2.10
Forelimb	0.75	540	× 720	2.93
Metanephros	1.4 (on ninth day)	130	× 93	3.58
Whole embryo	53	41,000	× 773	

The inverse relationship between growth and differentiation can also be shown in experimental conditions. In the intact organism the tissue cells differentiate and their growth becomes restricted. When bits of tissues are cultivated *in vitro,* the tissues become dedifferentiated, and at the same time they regain the ability to proliferate at a rapid rate. It is possible, however, to retard the growth of a tissue culture by supplying it with a medium poor in growth-promoting substances or by keeping it longer on the same plasma clot. Under these circumstances the cells may show signs of renewed differentiation; intercellular fibers (or cartilage matrix) are produced in the culture, depending on the nature of the cultivated cells. Differentiation goes hand in hand with reduced proliferation.

Dedifferentiation and a concomitant increase of the proliferation rate of cells also occurs during regeneration of lost parts, as will be described in Chapter 19.

The contention that progressive differentiation is the actual cause of the decline in the growth rate of animals with time is thus very well substantiated. On the other hand, this does not quite explain the existence of a definite growth limit found in many animals, such as birds, mammals and also insects. The nature of the limitation of growth may perhaps be different in the arthropods and in the vertebrates. In the arthropods (insects) the limit to growth is quite obviously connected with the mechanism of metamorphosis and will be discussed in Chapter 18. In the higher vertebrates a regulation of growth by means of hormones exists. The hypophysis is known to produce a growth-stimulating hormone. Animals (dogs) treated with this hormone may be caused to grow in excess of the usual size (Evans and collaborators, 1933). In humans dwarfism is attributed to an insufficient production of the growth hormone by the hypophysis. The thyroid hormone is also necessary for normal growth, and the removal of the thyroid in a young animal causes the growth to be below normal. The cessation of growth in mammals (and probably also in birds) seems to be the result of an interplay of perhaps numerous hormones controlling growth. The details of this mechanism are not yet completely understood.

15–4 PROPORTIONAL AND DISPROPORTIONAL GROWTH OF ORGANS

The growth of different organs and of parts of the same animal (embryo) very seldom goes on at the same rate. As a rule some grow faster and some grow slower. The result is that the proportions of the animal change with growth. In the case of vertebrates the central nervous system and the sense organs are distinguished by their particularly low rate of growth, and the size of these organs diminishes relatively throughout the whole embryonic and postembryonic development. In the early embryo (of a chick for instance) the head is at first quite as large as the rest of the body, and a large portion of the head is made up by the relatively enormous eyes. At the time of hatching the head is already much smaller than the body, and in the adult fowl the head is relatively quite small.

To compare the growth of different organs their rates of growth may be estimated. However, the rate of growth changes with age, so that it is not a very convenient quantity for purposes of comparison. The growth constant k (see preceding section) may be calculated for each organ as well as for the whole animal, and the constants of different organs may be conveniently compared.

A different method of analyzing the unequal growth of organs has been proposed by J. Huxley (1932). Huxley found that if two parts of an animal grow at different rates, their sizes at any given moment stand in a very simple relationship to each other. The relationship is expressed by the following formula, known as the formula of allometric growth:

$$y = bx^k \qquad (7)$$

where y is the size of one of the organs, x is the size of the other organ, b is a constant, and k is known as the **growth ratio,** because it shows the relation of the rates of growth of the two parts which are being compared. If the growth ratio equals 1, the two organs grow proportionally or **isometrically.** The constant b in this case shows the proportion of one organ

to the other: $b = \dfrac{y}{x}$ If the growth ratio is $\neq 1$ the growth is disproportional

or **allometric,** the relative size of the two organs changing as the growth proceeds. If the growth ratio is > 1 the organ y grows at a quicker pace than the organ x. The organ y is called **positively allometric** to the organ x. If the growth ratio is < 1 the organ y grows at a slower pace than the organ x; it is **negatively allometric** to x.

The formula for allometric growth can be given in a logarithmic form:

$$\log y = \log b + k \log x \qquad (8)$$

In other words, the logarithms of the sizes of two organs growing at different rates are proportional to each other. This form of the equation is especially instructive as it allows of a very simple graphic representation of

allometric growth (Fig. 245). If the logarithms of the sizes of one organ
are plotted against the logarithms of the sizes of the other organ, the points
lie on a straight line crossing the ordinate axis at a distance from the zero
point $=$ log b and ascending with an inclination a, determined by the
equation:

$$\tan \angle a = \frac{\log y - \log b}{\log x} = k \qquad (9)$$

The latter formula discloses the true significance of the exponent k, and also
shows how this exponent can be determined in practical work.

A great advantage of the formula for allometric growth is that it elimi-
nates time, which enters into all the other growth equations as the independent variable. As a result the formula can be applied for the analysis
of growth of parts in groups of specimens whose timing (age) is not known,
as when a series of animals of different sizes is collected in nature and
not bred and reared under permanent observation. The formula of allometric growth is also applicable to linear dimensions as well as to volume
or weight, and is in this respect more pliable than the other equations used
for analysis of growth.

The organs to be compared, x and y, may be chosen at will. A special
case of the application of the formula is when the size of an organ is compared with the size of the body less the size of the organ, the growth of

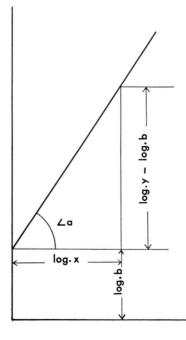

Fig. 245.   Graphic representation of allometric growth (in logarithmic
transformation).

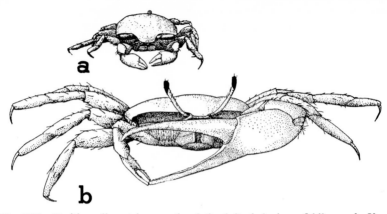

Fig. 246. Positive allometric growth of the left chela in a fiddler crab *Uca. a,* Young male crab with both chelae of equal size. *b,* Adult male with disproportionately large left chela. (From Morgan, 1927.)

which is being investigated. It can then be learned whether an organ grows in excess of the other parts (the remainder of the body) or not.

It has been found that organs attaining especially large proportions in certain animals often owe their increased size to positive allometric growth. In the fiddler crab, *Uca pugnax,* one of the chelae in the males attains an extraordinarily large size, up to 38 per cent of the weight of the body as a whole. The large chelae are positively allometric in respect of the rest of the body, with a growth ratio of 1.62. This means, by the way, that in a young male crab the chelae are very nearly equal in size, and that the relative size of the larger increases as the crabs grow (Fig. 246).

In most mammals the facial region of the head, including the jaws, grows at a greater rate than the cranial part of the head. The length of each of these parts of the head has been measured and the growth compared during the postembryonic life of sheep dogs. In a puppy the length of the cranium is almost double the length of the facial region (42 mm. and 22 mm.). In the adult dog the cranial region is only slightly longer than the facial region (120 mm. and 112 mm.). The facial region is thus positively allometric in respect to the cranial region. The growth ratio is in this case 1.49. The lower rate of growth of the cranial region is mainly due to the rate of growth of the brain, which is enclosed by the skull and which is noted for its low rate of growth; it is negatively allometric in respect to the body as a whole.

The formula of allometric growth can be used not only for the comparison of the growth of morphologically definable parts of the animal's body, but also for the increase of various chemical components of the body. In differentiating tissues the amount of protein increases in respect to the amount of nuclear chromatin, as has been indicated in section 14–1. This can be represented as a case of allometric growth; in Table 18 the quantities

**Table 18.   Comparison of Increase of Protein Nitrogen and Deoxyribo-
nucleic Phosphorus in Different Organs of the Chick Embryo
(after Davidson and Leslie, 1950)**

Organ	Incubation Period in Days	Growth Ratio of Protein Nitrogen to Deoxyribonucleic Phosphorus
Brain	8–13	2.55
Brain	15–19	1.05
Heart	10–19	1.05
Liver	8–19	1.10
Muscle (leg)	12–19	1.40

that have been compared are the deoxyribonucleic acid phosphorus and the protein nitrogen in several organ rudiments of the chick embryo.

The high rate of increase of the protein nitrogen (large growth ratio) corresponds, according to what was said before, to a rapid rate of differentiation. The high figures for the brain and the leg muscle can therefore be easily understood. The heart shows little increase of differentiation because in the 10-day chick the heart is already fully functional, that is, already differentiated. The liver is an example of an organ which becomes differentiated late, and even there the level of morphological differentiation is in this case rather low.

CHAPTER **16**

Correlations

Once an organ rudiment has been formed and starts growing and differentiating, it becomes independent of its surroundings to a certain extent. The rudiment of an organ, an eye for instance, may be transplanted to another region of the body, and it will continue its development in spite of the abnormal surroundings, and it may also influence its surroundings in such a way as to make its own development more complete. In the case of a transplanted eye rudiment this may mean that a lens is induced to develop from the local epidermis. The embryo as a whole after the end of neurulation seems to have become a complex of different organ rudiments, each of which is capable of developing on its own. This is sometimes expressed by a statement that the embryo has entered into a "mosaic" phase of development (Huxley and de Beer, 1934). As a mosaic is built up of many different colored stones, so the embryo is built up of a number of different organ rudiments.

Actually the independence of the various organ rudiments is only a very relative one, and the parts of the embryo continue influencing one another's development. The influence of different parts of the embryo on each other's development is known as **correlation.** Strictly speaking, the dependent development of organ rudiments in the earliest stages of development may also be termed correlation, but the means of correlation in the earliest stages of development and in the later stages are different, and the results of the correlation are also different.

In the earliest stages of development the interdependence of parts of the embryo is by means of (1) a gradient, or (2) embryonic induction.

The result of the interaction is the segregation of the embryo into the various organ rudiments.

In the later stages of development the result of the interaction lies in modifying or stimulating the development of existing organ rudiments. We shall distinguish the following types of correlations:

(1) Growth correlations.
(2) Correlations resulting from the activity of the nervous system.
(3) Correlation through the mechanical functions of the organism: correlations in the skeletal-muscle system; correlations in the vascular system.
(4) Correlations by means of hormones circulating in the blood.

We shall discuss some examples of these correlations.

16–1   GROWTH CORRELATIONS

It can be shown in some favorable cases that the growth of one part of the embryo is dependent on the growth of other adjoining parts or on the growth of the embryo as a whole. Such a dependence can be called a growth correlation. A beautiful example of a growth correlation has been investigated in the case of the growth of the eye cup and the lens (Harrison, 1929). The investigation was made on two species of American salamanders—*Ambystoma punctatum* and *Ambystoma tigrinum*. These two species differ in the rate of growth and in the eventual size of the animal. *A. punctatum* grows more slowly and attains a smaller size. *A. tigrinum* grows more rapidly and is considerably larger in the adult state. Parts of the eye were transplanted heteroplastically between the two species. In some experiments the eye cup was transplanted from *A. punctatum* into *A. tigrinum* in place of the excised host eye rudiment. The lens in these experiments developed from the local *A. tigrinum* epidermis. In other experiments only the lens-forming epidermis was transplanted from the *A. punctatum* embryo over the intact eye cup of *A. tigrinum*. Reciprocal transplantations were also performed; that is, the eye cup or the lens-forming epidermis was transplanted from *A. tigrinum* embryos onto *A. punctatum* embryos, while the other part of the eye consisted of host tissue.

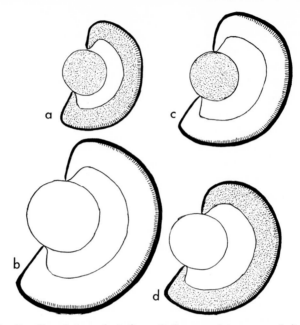

Fig. 247. Results of transplantation of the complete eye and its parts from *Ambystoma tigrinum* to *Ambystoma punctatum*. *a,* Normal eye of host, *A. punctatum*. *b,* Complete eye of *A. tigrinum* grafted to *A. punctatum*. *c,* Eye cup only of *A. tigrinum* transplanted; lens from *A. punctatum* cells. *d,* Lens-forming epidermis of *A. tigrinum* transplanted; eye cup from *A. punctatum*. (Drawn to scale after Harrison, 1929.)

It was found that in every case when an *A. punctatum* graft was placed into the quickly growing *A. tigrinum* embryo, the growth of the graft was accelerated as compared with the normal *A. punctatum* organ. As a result, the transplanted eye cup or the transplanted lens grew to a larger size than normal *A. punctatum* organs of the same age. They did not attain, however, the size of the corresponding host organs. On the other hand, they retarded the growth of the host parts with which they were associated; so the host lens remained relatively small if it developed in association with a grafted *A. punctatum* eye cup. Also the host eye cup remained smaller than normal if it was supplied with an *A. punctatum* lens. The result of this correlation in the growth of the eye cup and the lens was that the eye with parts of different origin was very nearly harmonious in respect to the proportion between the size of the eye and the size of the lens.

A correlation of exactly the same nature was observed when eye cup or lens was grafted from *A. tigrinum* to *A. punctatum*. Here the grafted parts were retarded in their growth, but the host part, whether it was the eye or the lens, grew in excess of the normal rate and became larger than the parts not involved in the operation and thus met the grafted components of the eye halfway. Again the resulting composite eyes had the eye cup and the lens in approximately harmonious size proportions (Fig. 247).

16–2 CORRELATIONS IN THE DEVELOPMENT
OF THE NERVOUS SYSTEM

Here the most striking feature is the dependence in development of the central nervous system (brain and spinal cord) and the peripheral nervous system (the nerves and ganglia) on the peripheral organs.

It is common knowledge that the nerves and nerve centers show a certain correspondence in the degree of their development with the organs they supply. Thus the spinal nerves supplying the limbs in terrestrial animals are stronger than the other spinal nerves, the corresponding ganglia are larger, and the spinal cord itself has swellings in the cervical and the lumbar region, from which the fore- and hindlimbs are innervated.

It can be shown experimentally that, at least in part, this correspondence in the degree of development of the nervous system and the peripheral organ is due to a direct correlation between the two.

If the forelimb rudiment of a salamander embryo is removed and the limb fails to develop, the nerves of the brachial plexus remain smaller (thinner) than they would have been if the limb was there. Also the spinal ganglia III to V are smaller. The number of cells in each ganglion may be reduced to 50 per cent as compared with the normal or the unoperated side of the same animal. If an additional limb rudiment is transplanted, the local spinal nerves supply the nerves to the transplanted limbs, and then these nerves increase in thickness and the corresponding ganglia increase in size. The increase in the number of cells may be up to 40 per cent (Detwiler, 1926a). In similar experiments performed on the chick embryo it could be shown that the sensory and motor parts of the spinal cord were similarly reduced in the absence of a limb and increased in cases of periph-

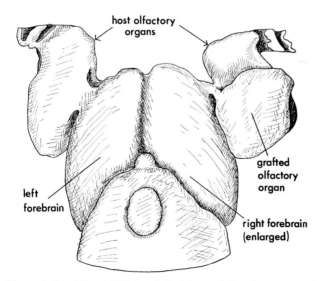

Fig. 248. Hyperplasia of the right lobe of the telencephalon due to transplantation of an additional olfactory organ in *Ambystoma*. (From Burr, 1930.)

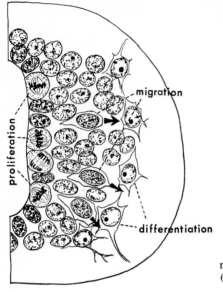

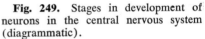

Fig. 249. Stages in development of neurons in the central nervous system (diagrammatic).

eral overloading (transplantation of an additional limb-bud) (Hamburger, 1934; Bueker, 1947). Reduction occurs both in the gray matter (the cells) and the white matter (the fibers) of the spinal cord.

It is of great importance that in the brain an increase in peripheral organs also causes excessive development of the parts connected with such organs. The forebrain is dependent in its development on the olfactory organs, as has been shown by removing the olfactory rudiments and by transplanting additional ones. In the absence of an olfactory organ the corresponding lobe of the forebrain remains underdeveloped (Burr, 1916). If an additional olfactory organ is transplanted into the region beside the normal one, the fibers of the olfactory nerve grow from the transplanted organ into the adjoining lobe of the forebrain, and this causes an increase in the size of the latter (Fig. 248) (Burr, 1930). A similar dependence exists between the eye and the roof of the midbrain in amphibians and between the eye and the optic tectum in fishes (Dürken, 1911; Harrison, 1929; White, 1948).

The mechanism by which the size of the nerve center (a spinal ganglion or a part of the brain or spinal cord) is changed by a change of the periphery is fairly complicated. The volume of any nerve center is dependent on several processes. These are (see Fig. 249):

1. *Proliferation.* The cell divisions in the central nervous system are restricted to the stratum immediately adjoining the lumen of the neural tube. The amount of proliferation, which can be roughly judged from the number of the mitoses in any region, differs from one part of the neural tube to another, and it has been claimed that the formation of outgrowths or swellings on the surface of the neural tube, which have been dealt with in section 10–1, is always preceded by an increase in the number of

dividing cells near the lumen in the area where the swelling is to appear (Frank, 1925).

2. *Migration.* Some of the dividing cells remain near the lumen of the neural tube and so preserve the continuity of the center of proliferation (*cf.* section 15–2), while others migrate outward and settle down on the outer border of the mass of cells and there start differentiating into the functional neurons of the brain and spinal cord. In this way a layer of young neurons is formed just underneath what will become the white matter of the brain or spinal cord (see Fig. 249). This layer is often referred to as the **mantle.** In addition to this primary migration, nerve cells may carry out more complicated migrations up or down the length of the neural tube, forming concentrations of nerve cells known to anatomists as the various nuclei of the brain and spinal cord. One very special type of migration is the one occurring only in the cerebellum and cerebral hemispheres in higher vertebrates, as the result of which nerve cells pass beyond the layer of nerve fibers, the white matter of the central nervous system which originally surrounds the mass of cells (the gray matter), and accumulate at the surface, giving rise to the cortex of the cerebellum and the cerebral hemispheres.

3. *Cell differentiation.* The cells constituting the early neural tube and also the spinal and other ganglia initially appear to be all alike. Later, however, some of the cells become neurons of various types, while others become neuroglia cells and never acquire nervous functions. Lastly, quite a considerable number of cells in the various parts of the central nervous system degenerate during the embryonic period (during the fifth to sixth day of incubation in the chick, after Hamburger and Levi-Montalcini, 1949), thus enhancing the differences in cell numbers in the various areas. At present there is no means of finding out in advance what will be the fate of any given cell, but the experiments referred to above show that the fates of individual cells are probably not firmly fixed in early stages.

When the periphery connected to a nerve center (ganglion or center inside the brain or spinal cord) is increased, the increase of nerve cells in the center may be brought about by:

(a) Increased proliferation—this has been found in some cases, but seems to be of minor importance.

(b) Increase in the number of cells which differentiate as neurones.

(c) Decrease in the number of degenerating cells.

All three of these factors may act simultaneously or become prominent in changing proportions in various cases. The effect of a decrease in the periphery seems to be mainly to cause a degeneration of large numbers of nerve cells, sometimes as much as 90 per cent in restricted areas having no other peripheral connections. One of the most remarkable facts in this dependence of the nerve centers on the periphery is that cells may apparently be affected although they have no direct connections with peripheral organs. Nerve cells may degenerate even though the organs to which they should have been related are removed before nervous connection between the central nervous system and the periphery is established. In the case of overloading, cells which do not normally send their processes into the area concerned are drawn into supplying the additional periphery. (See further Piatt, 1948; Weiss in Willier, Weiss and Hamburger, 1955; Hamburger, 1956.)

Not only the volume (thickness) of the nerves is dependent on the organs which they supply, but the paths which the nerves take, and thus the whole configuration of the peripheral nervous system is determined by the periphery.

The mode of development of the nerves was a subject of controversy as long as embryology relied on purely descriptive methods. The conflicting theories were those of His and Hensen. According to His the nerves consist of processes growing out from the nerve cells and eventually reaching their organs of destination, or making contacts with the processes of other nerve cells in the central nervous system. According to Hensen, the nerve fibers develop from intercellular bridges connecting all the cells of the multicellular animal from the earliest stages of development. The controversy was solved by Harrison (1908), when he tried cultivating in a plasma clot pieces of neural tube taken from frog embryos. He could observe directly the formation of outgrowths from the nerve cells and could see the free ends of the processes push forward through the medium. Harrison's findings were corroborated by numerous investigators, and it was also shown that the processes of nerve cells can establish new connections *in vitro,* if nervous tissue is cultivated together with a different type of tissue (muscle tissue for example).

The problem now arises as to what directs the nerve fibers in their outgrowth from the central nervous system or from the cranial and spinal ganglia to the peripheral organ. The answer is given by the results of embryonic transplantations. If the forelimb rudiment of an amphibian embryo, prior to the outgrowth of the nerves, is cut out and transplanted to a position very near the original one, the brachial nerves will deviate from their normal paths and will be deflected in the direction of the transplanted limb. If the distance of the transplanted limb from the original position is not too great, the brachial nerves will penetrate into the limb and ramify in it just as if the limb was in its normal position. The limb in this case becomes fully functional and moves in coordination with the other limb. The same may happen if an additional limb is transplanted into the immediate vicinity of the host limb (Fig. 250). The brachial nerves will develop branches running out to the additional limb and will supply it (Detwiler, 1920, 1926b). If the normal path of the nerves is blocked by some obstacle, the outgrowing nerves may avoid the obstacle, go around it and still reach their normal destination. This has been observed when a piece of mica was inserted into a frog embryo between the spinal cord and the region where the hindlimb rudiments were to develop. The nerves formed loops around the mica plate and still reached the hindlimb rudiments (Fig. 251) (Hamburger, 1929). However, if the limb rudiment is placed further away from the normal limb site, or if the obstacle between the spinal cord and the limb rudiment is too great, the nerves fail to be attracted to the limb. If the limb rudiment is placed on the side of the embryo, too far for the normal forelimb or hindlimb nerves to reach it, it still will attract the local spinal nerves. These nerves will grow into the limb, but they cannot provide for the normal functioning of the limb: the limb cannot move.

It appears that only the areas of the spinal cord from which the nerves of the brachial and lumbar plexuses originate possess the properties necessary for controlling the function of limbs. These properties, as we have seen previously (section 10–1), are established at some time between the neurula and the tailbud stage. The centers controlling the movements of the fore- and hindlimbs are interchangeable, however: when forelimb buds were transplanted in place of hindlimb rudiments, they acquired normal mobility. Limbs transplanted to the head may be supplied by fibers of the cranial nerves and then they can be seen to move synchronously with the respiratory movements of the jaws and gills. The movements of the limbs are, however, rather of the nature of twitchings and differ from the coordinated movements of normal limbs (Detwiler, 1930).

The last experiment has already shown that the attraction of the nerves

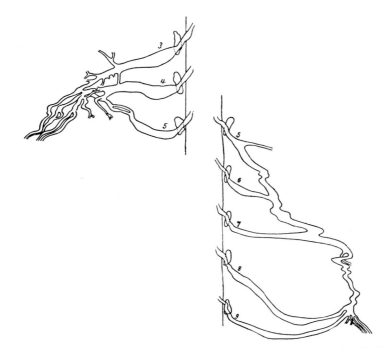

Fig. 250. Nerve supply to a transplanted limb. Only the nerves supplying the limbs are drawn. The numbers indicate the spinal ganglia. (After Detwiler, from Mangold 1928.)

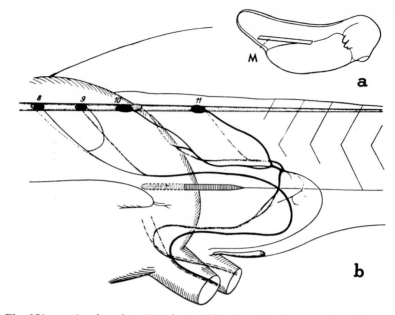

Fig. 251. *a,* A mica plate (*M*) inserted in the path of outgrowth of the spinal nerves to the hindlimbs in a frog embryo. *b.* The nerves have grown around the obstacle and have reached the hindlimbs. (From Hamburger, 1929.)

by the peripheral organs may be unspecific to a certain extent, the nerves growing out to organs other than the ones they normally supply. This is further borne out by the following experiment. An eye was transplanted into the side of a salamander embryo after the forelimb rudiment was removed. The brachial nerves were deflected from their normal path and grew out toward the transplanted eye. Having approached the eye, however, the brachial nerves failed to penetrate into the eye and establish an actual connection with it, but stopped with free ends in the tissue surrounding the eye (Detwiler and van Dyke, 1934). Two aspects of the nerve supply to organs must thus be recognized: one aspect is the outgrowth of nerves toward the organ, and the other aspect is the actual establishment of a connection between the nerve and the organ. The attraction of the out-growing nerves to peripheral organs seems to be unspecific to a very high degree; possibly any growing mass of tissue will attract a nerve that is sufficiently near to it. The connections between the nerve and the end organ can be made only if the two correspond to each other, at least in a general way.

The nature of the attraction of nerves to peripheral organs has been investigated in special experiments. It was shown that the nerves are not directed by any chemical substances diffusing from the peripheral organs, neither can their outgrowth be controlled by electrical currents or magnetic fields. It is thus not a case of chemotaxis or galvanotaxis. What actually directs the outgrowing nerve fiber is the ultramicroscopic structure of the colloidal intercellular matrix, through which the tip of the nerve fiber is moving. It is thus the same factor as that which directs the movements of mesenchyme cells. The nerve follows pathways consisting of bundles of oriented molecules or colloidal micellae (Fig. 252). The peripheral organs can influence the direction of nerve outgrowth by altering the submicroscopic structure of the intercellular matrix surrounding them. This they do probably by withdrawing water from their surroundings and causing a shrinkage which orients the particles of the matrix along lines radiating from the intensively growing organ. This radial arrangement of particles of the matrix naturally causes the nerves to converge toward the center around which the matrix is so polarized (Weiss, 1934).

A counterpart to the dependence of the nervous system on the peripheral organs is the influence which the nerves exercise on the organs which they supply. This influence does not concern the initial stages in the development of organs but sometimes is very important for their subsequent differentiation. The muscles are originally formed before the nerves are developed, and the differentiation of muscle tissue may proceed for some time in the complete absence of nerve supply, as for instance when the whole of the neural plate is removed in an early stage of development. The histological differentiation may proceed so far that the muscles become functional, that is, they may show contractions, spontaneous or as reactions to direct stimulation. If the innervation of the muscle does not occur, however, the muscle fibers undergo a fatty degeneration and are eventually resorbed (Ham-

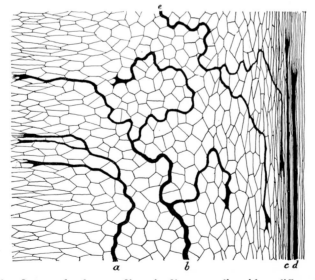

Fig. 252. Outgrowth of nerve fibers in fibrous media with a different degree of ultrastructural organization (random arrangement in center turning into prevailing horizontal orientation on left and strict vertical orientation on right). (From Weiss, in Willier, Weiss, Hamburger, 1955.)

burger, 1929). Thus some sort of **trophic** influence of the nerves is necessary for the persistence of muscle tissue.

Some sense organs depend for their persistence on a continued influence of the nerve endings supplying them. The gustatory organs in fishes and in man degenerate if the nerve which supplies them (the glossopharyngeal nerve) is interrupted. If a regeneration of the nerves takes place, the regenerating fibers on reaching the epithelium cause the epithelial cells to differentiate as gustatory buds in place of the ones that had previously degenerated (Detwiler, 1926b).

16–3 CORRELATIONS THROUGH MECHANICAL FUNCTION: THE SKELETON-MUSCLE SYSTEM

It is a well known fact that the skeletal muscles are stimulated to an increased growth by exercise, and that they undergo a reduction if they are deprived of function or if they are not exercised sufficiently. The whole training of athletes and sportsmen is based on this principle. It is also common knowledge that muscles degenerate in certain pathological conditions, for instance if the nerves or the nerve centers which control them are destroyed (paralysis). Also if a joint is **ankylosed** and loses mobility, the muscles which would normally move the bones in this particular joint tend to degenerate. In both the latter cases the reduction of the muscles is a secondary effect of their lack of function.

It is not commonly known, however, that the function of the skeletal muscles is also reflected in the structure of the skeletal parts to which the muscles are attached. At the points where muscles or their tendons are joined to the bone, the bone forms thickenings or outgrowths in the form of tubercles, trochanters or cristae.

There is experimental evidence that these outgrowths may not be produced by the bone independently of the muscles. In the mammalian skull there are certain ridges which serve for the attachment of the muscles moving the lower jaw; the ridge of the interparietal bone serves for the insertion of the masseter muscle, and the temporal muscles are attached to the temporal crests. In an experiment performed on newborn puppies the temporal muscle was removed on one side of the head (Anthony, 1913; see review in Brash, 1929). Nine months later it was found that the temporal crest on the side of the operation was not developed. The cranial wall on this side was also thinner than normal.

To reduce the activity of the chewing muscles the teeth rudiments were excised in young mammals (dogs, rabbits), so that the animals, as they grew up, could not bite or chew their food properly. As a result it was found that the whole shape of the skull was changed. In an operated dog the skull was lower and slightly more elongated; it thus acquired a certain similarity to the skulls of anteaters, who do not have teeth and do not chew their food (Anthony, 1913). If the teeth are removed on one side only, the skull develops a distinct asymmetry; on the operated side it is much weaker than on the normal side (see Brash, 1929).

In general, the bones even in adult animals show a dependence on the stresses (pressure) to which they are subjected (see Jansen, 1920).

These stresses arise from the action of muscles, on the one side, and from the weight of the body which the bones have to support, on the other side. If the functioning of a limb is reduced for any reason, such as failure of the limb bones to be joined together after a fracture, ankylosis of a joint due to any pathological process, or absence of nerve supply, resorption of the bone starts in the skeleton of such a limb. Both the thickness of the compact bone and the amount of cancellous bone may be reduced to a very marked extent (see Harris, 1933). On the other hand new bone develops if the stresses to which the skeletal parts are exposed become modified, as when bones are united in an abnormal position after an injury. The over-all thickness of a bone may be increased if it has to bear an increased weight.

There is no doubt that the general shape of the bones and their relative sizes may initially develop independently of function. Parts of the limb rudiments of chicken embryos have been cultivated on the chorio-allantois (Murray, 1926), or on a plasma clot *in vitro* (Fell, 1928), and the various parts of the limb skeleton developed very nearly normally, in spite of the absence of movement or supporting function of the limbs. In another experiment (Hauschka, 1951) mouse embryos, already possessing limb-buds, were cut into tiny fragments and these were then injected into the peritoneal cavity of adult mice. The fragments of the embryo became attached to the peritoneum and developed into various organs, among which were recognizable parts of the skeleton: ribs, vertebrae and parts of the limb skeleton.

The shape of the skeletal parts and their relative sizes in the above experiments must have been determined by hereditary factors independent of the functions which the skeleton performs in the intact animal. In normal development the initial formation of the skeleton takes place in the pre-functional stage. With progress of development, however, the further growth and even the persistence of bone appear to become increasingly dependent on the functional stresses. What is inherited is thus not only the general pattern of the skeleton as laid down in the earlier stages of ontogenesis, but also the ability of the bone tissue to react to increased or diminished functional stress by an additional proliferation or resorption as the case may be.

Another example of correlations exercised through mechanical function is found in the dependence of the structure of the circulatory system (heart and blood vessels) on the blood flow. This dependence has been dealt with in sections 11–5 and 11–6. See in particular pages 319 and 328.

16–4   CORRELATION BY MEANS OF HORMONES

In this connection the term "hormone" is used in the strict sense, that is, a chemical substance secreted into the blood, circulating in the blood stream and eventually influencing the activities of the cells, tissues and organs or some of them, to which the substance is conducted in this way. Some chemical substances which mediate influences of parts of the animal on one another (such as the substance inducing the neural plate) will not be called hormones. The control of various functions of the organism by means of hormones, especially in vertebrate animals, has been in the past the subject of innumerable investigations, and a special science, **endocrinology,** deals with this field of biology. Predominantly this field of biology deals with processes which should be classed as physiological, that is, processes which are involved in maintenance and cyclical change in the course of the life of an individual, rather than in processes of progressive development. The processes of development, however, are also under the control of hormones. Naturally these are not the earliest developmental processes, as in the initial stages of development the blood vascular system does not exist, and therefore hormones cannot be instrumental in correlating the parts of the embryo. It is not possible here to give even a brief review of the hormones having a morphogenetic action; it will only be indicated what morphogenetic processes are under hormonic control.

1. **Growth.** One of the hormones produced by the hypophysis has a specific stimulating effect on growth. The treatment of normal animals with the growth hormone causes gigantism (Evans and collaborators, 1933) and the abnormal excessive growth in humans suffering from acromegalia is attributed to an excessive production of the hormone due to tumors of the hypophysis.

The deficiency of the thyroid hormone causes, in mammals, a retardation of growth.

2. **Metamorphosis.** The metamorphosis of amphibian larvae is under the control of the thyroid hormone. The function of the thyroid gland is in its turn under the control of the hypophysis. These phenomena will be dealt with in the section on metamorphosis.

3. **Secondary Sex Characters.** The primary sex characters distinguishing a male from a female animal are the sex glands—the testis and the ovary, respectively. By secondary sex characters are meant the distinctions between the sexes other than the presence of the sex glands. These distinctions fall into two groups: the organs which are essentially necessary for reproduction, such as the genital ducts, the copulatory organs, the uterus in the female of viviparous animals, and organs or characters playing an indirect part in reproduction, such as the spurs and comb in the cock, the beard in man, distinctions in coloration, stature, the shape of feathers, and so on.

In vertebrates all secondary sex characters appear to be under the control of the sex hormones produced by the sex glands: by a male hormone produced by the testis and a female hormone produced by the ovary. This is most easily demonstrated in the case of the second group of secondary sex characters. Coloration, shape of feathers or stature becomes changed in castrated animals or in animals to which, after castration, a gland of the opposite sex has been grafted. The characters of the first group, those taking an immediate part in reproduction, are not so readily influenced by a change in the hormonal milieu because the critical phases in their development are passed in the earlier part of the life of the individual, during the intrauterine or embryonic period. The hormones must therefore be administered during that period if they are to have effect. Hormones have been injected into the eggs of birds and into the amniotic cavity of mammalian embryos before birth, with the result that the accessory parts of the sex organs (the ducts, etc.) were changed in accordance with the hormone used. This group of secondary sex characters may thus also be dependent on the sex hormones (Dantschakoff, 1941).

4. **Pigmentation.** Pigmentation in vertebrates is, in some cases, dependent on the normal functioning of the endocrine glands. The removal or the spontaneous atrophy of the hypophysis causes in amphibians a permanent contraction of the melanophores and thus a paler coloration. In birds a state of hyperthyroidism causes newly developing feathers to be devoid of pigment, old feathers retaining their pigmentation (see Willier, 1952). The examples quoted here are of course on the borderline of the field of purely physiological changes controlled by hormones.

CHAPTER **17**

Genetic Control

of Growth and

Differentiation

Growth and differentiation proceed under the constant control of genetic factors. Most of the known genes affect either the growth or the differentiation of already formed organ rudiments, or both. Only a minority of known genes can be shown to affect the earlier stages of development—cleavage, gastrulation and organogenesis. It would therefore be both impossible and out of place to give a survey of the genes affecting the later periods of development. It is important, however, to give an indication of the way in which the activities of genes link up with the developmental processes described in the preceding sections. A few examples, picked out from the abundance of known facts, will serve to indicate the connection between hereditary factors and differentiation and growth. For a fuller consideration of the gene action the student is referred to the books on physiological genetics (Goldschmidt, 1938; Waddington, 1939b; Wagner and Mitchell, 1955).

17–1 EFFECTS OF GENES ON GROWTH

Differences in the sizes of animals are hereditary, and this means of course that the process of growth is controlled by the genotype. There may be, however, very different ways of such control.

Table 19. Comparison of Numbers of Cells in Early Embryos of a Large and a Small Breed of Rabbit

Hours After Fertilization	Average Number of Blastomeres	
	Flemish Giant	Polish Breed
32½	4.41	4.06
40	9.04	8.29
41	11.64	8.62
48	21.75	14.00

In one well investigated case (Table 19) the genetic constitution acts directly on the ability of cells to grow and proliferate. The case is not one of single gene action, but concerns the differences between the growth of two breeds of rabbits, probably due to a number of genes. In the Flemish giant breed the average adult weight is 4750 gm.; in the Polish breed the average weight of the adult is only 1680 gm. In both breeds the eggs are of the same size, but in the cleavage stage a difference already appears: the Flemish giant embryos cleave at a quicker rate than the Polish breed, so that at any given time there are more cells in the embryo of the Flemish giant (Castle and Gregory, 1929; Gregory and Castle, 1931). The blastocyst of the Flemish giant at the age of six days is 47.8 microns in diameter, that of the Polish race 40.5 microns in diameter. At the age of 12 days the embryo of the large race is 23.1 mm. long, the embryo of the smaller race is 18.1 mm. long.

The rate of cleavage is of course not to be confused with the rate of growth. The size of the embryo after 12 days of development is, however, due to growth, occurring as a result of an intake of food stuffs from without (through the placenta). It is thus evident that the early stages of growth in the two breeds already proceed at different rates. The difference in the rate of growth persists throughout the later days of gestation and during the postembryonic development. The unequal rate of cleavage serves to support the conclusion that the different rates of growth are due to intrinsic properties of the cells, and are not the secondary effect of some other distinctions.

The dwarf mutation in mice (pituitary dwarf, symbol *dw*) presents an example of quite a different nature. The character is dependent on a single recessive gene. The homozygotic mice, which alone show the character, are born indistinguishable from normal individuals. At the end of the first week of postembryonic life, however, the dwarf individuals begin to show a slightly retarded growth as compared with the controls. In the third, the dwarfs become conspicuously smaller than normals, and after weaning they increase in weight only slightly. The adult dwarfs are only one third to one fourth of the weight of normal mice. The retardation of growth in this case has been traced to an abnormality of the hypophysis, which is reduced in size and fails to produce the **growth hormone** (see section 16–4). This has been proved experimentally by transplanting pieces of fresh rat hypophysis subcutaneously into dwarf mice. The result was that the treated mice resumed growth and reached the size of normal individuals (Smith and Macdowell, 1930). In this case therefore the gene does not directly affect the intrinsic growth rate or the cells' ability to proliferate, but the primary effect of the gene is to modify in a certain way the differentiation of one of the organs (the anterior lobe of the hypophysis). The arrest of growth is then the visible expression of the deficiency of a growth-stimulating substance normally produced by the hypophysis.

The genetic constitution may affect the growth of different parts in unequal proportion. The mutant gene short ear (symbol *se*) in the mouse will serve to illustrate such a case. The pinnae of the newborn mice are the same size in normals and mutants. Subsequently the growth of the pinna in the mutant is retarded, as shown by the following figures (after Keeler, from Gruneberg, 1952):

	Length of the Pinna	
Age	*Normal*	*Short Ear*
14 days after birth	7.1	6.0
28 days after birth	11.6	7.6
Percentage increase	63%	27%

The most affected part of the pinna is its distal part, while the proximal part is hardly changed at all. The defect has been traced to an abnormality in the development of the cartilage supporting the pinna. The cartilage in the mutant mice is laid down as a collection of mesenchyme but this does not differentiate beyond the precartilage stage. A true perichondrium is not formed, and the rudiment eventually dissolves altogether. Without the cartilage to support it, the pinna of the ear does not grow out. In short-eared mice, the ear cartilage is not the only one that fails to differentiate; parts of the cartilaginous skeleton of the ribs, sternum and vertebral column (in the tail region) do so as well. As a result multiple defects are present.

There are two points to note in the case of the short-eared mice:

1. The growth of the ear is affected not directly, but through the derangement of differentiation. This is in agreement with the concept (see

section 15–3) that rates of growth are dependent on the differentiation of the organs and tissues.

2. A single primary abnormality (in this case a failure of part of the precartilage to differentiate into cartilage) may cause deviations in the structure of many parts of the animal. This is what in genetics is known as the **pleiotropic effect** of a gene.

17–2   THE PLEIOTROPIC EFFECTS OF GENES

Whereas the term pleiotropic effect refers to the multiplicity of the manifestations of a gene, from the point of view of embryology it is important in this connection to distinguish between the primary effect and the secondary effects of a gene. If there are several manifestations of a gene these different manifestations are often reducible to one primary cause. This principle will now be illustrated by examples concerned with histological differentiation rather than with growth.

The gray lethal mutant in the mouse (symbol *gl*) is produced by a single recessive gene. The heterozygotes are normal. The homozygotes, as the name of the mutant implies, are not viable and die at the age of three or four weeks after birth. The gene causes multiple defects of the whole skeleton, arrest of the growth of bones, failure of the teeth to erupt. All the defects may be traced back to the incapacity for secondary bone resorption. Normally the bones containing an internal cavity (including the long bones of the limbs) grow in such a way that layers of bone are deposited on the surface, while inside the bone is destroyed (resorbed) by osteoclasts. The solid bones also undergo a secondary resorption of part of their matter during the process of growth. In the gray lethal this resorption does not occur, although osteoclasts are present. The interior of the long bones of the limbs

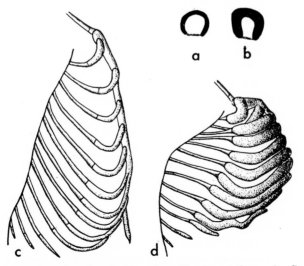

Fig. 253.   Cross-section of trachea (*a, b*) and profile view of thorax (*c, d*) of normal rats (*a, c*) and rats with "emphysemic" gene. (From Gruneberg, 1938.)

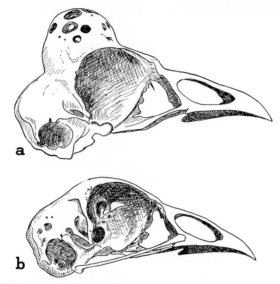

Fig. 254.   Skull of a Houdan fowl with a dome-shaped reticulated protuberance in the frontal region (*a*), and skull of a normal fowl (*b*). (Modified from Schmalhausen, I. I., 1949.)

remains filled with spongy bone, instead of developing cavities filled with bone marrow. The shape of the bones is abnormal, as well as the shape of the skull (Gruneberg, 1936). The teeth are prevented from erupting, because in normal development the alveoli containing the tooth rudiments have to be partially resorbed at the edges, so as to let the teeth grow out and cut through the gums. The resorption of the bone is, however, impossible and the teeth remain encapsulated in the bony matter of the jaws.

The normal bone resorption is dependent on the activity of the parathyroid glands. It is supposed therefore that the primary effect of the gene may consist either in preventing the parathyroid glands from producing their hormone or in some sort of inactivation of the parathyroid hormone.

Another gene causing widespread defects by interfering with a process of differentiation is the "emphysemic" gene in the rat (Gruneberg, 1938). The primary effect of the gene is to cause an abnormal thickening of most of the cartilages, especially those of the ribs, the trachea and the nasal region (Fig. 253). The thickened ribs immobilize the thorax and so interfere with normal breathing. Thickened cartilages of the nose and trachea create further difficulties in respiration. The result is an emphysema of the lungs. As a further tertiary effect, the right ventricle of the heart is hypertrophied to compensate for the insufficient aeration of the lungs. Abnormal individuals eventually die from one of several causes: inability to suckle due to the blockage of the nostrils, slow suffocation due to defective respiration, heart failure, or hemorrhages from lung capillaries due to increased blood pressure.

Most of the characters caused by genes referred to so far are of the nature of defects or developmental arrest. We shall now consider a case in which an apparently new structure is produced. The structure in question is the peculiar crest of feathers found on the head of the Houdan fowl (Schmal-hausen, 1949). The crest or tuft of feathers takes the place of the fleshy crest found in the ordinary fowl. The abnormality is by no means restricted to the integument: the skull is abnormal, the roof of the skull bulges upward and is perforated by numerous openings, the bulge of the skull encloses a bloated brain. The primary effect of the gene appears to be the increased secretion of the cerebrospinal fluid by the cells of the forebrain. The fluid is found in vacuoles lying under the ependyma, and the anterior ventricles of the brain become swollen. The cerebrospinal fluid apparently penetrates through the thin roof of the third ventricle and causes a dissolution of the skeletogenous tissue destined to differentiate into the bones of the cranial roof (parts of the frontal bones). Only the lateroventral parts of the frontal bones develop, and the skull remains open dorsally. The skeletogenous mesenchyme from the disorganized bone rudiments concentrates under the skin epidermis and gives rise to the rudiments of feathers (the feathery crest). Finally, during postembryonic development, islands of bone develop underneath the crest. They fuse into a network of bony trabeculae and so form the new roof of the skull, adapted to the shape of the bloated brain (Fig. 254).

17–3   THE COOPERATION OF GENES

The examples of multiple secondary effects produced by one primary action of a gene illustrate one principle of genetic control of development. Another principle involved is the cooperation of several genes to produce one final effect.

The coloration of insects is produced by several groups of pigments, and the actual color is usually dependent on the combination of more than one pigment, as well as upon the structure of the surfaces bearing the pigment and the position of the pigment granules in respect of the surface. Any of the various pigments may appear independently of the others, and genes are known which affect certain pigments or groups of pigments and do not affect the others. One group of dark pigments, known as the ommo-chromes, is found in the eyes of insects (the fruit fly, *Drosophila,* the meal moth, *Ephestia,* and a number of other moths, flies and wasps). Similar or identical pigments are also present in the internal organs of the same animals, as in the gonads or the brain. In the meal moth one pigment of this group, an **ommin,** gives a black color to the eyes of both the adult moth and the cater-pillar, while a related pigment, an **ommatin,** is found in the skin of the caterpillar. In *Drosophila* the brown pigment of the eye is an ommatin.

It has been found (for a review see Kühn, 1941, Ephrussi, 1942) that all the pigments of the ommochrome group are derived primarily from the amino acid tryptophan, which is processed by a series of steps, controlled by several genes. The same genes, that is, genes with an identical chemical activity, are found in different insects having the ommochrome pigments. All the genes of a series have to be present in an active state if the final pigmenta-tion is to be normal.

The first gene of the series is the vermilion (symbol v^+) gene of *Drosophila,* or the corresponding a^+ gene of *Ephestia.* If the gene mutates to a state v or a, the dark pigment of the eyes is not formed, no matter what the remaining genetic constitution of the animal. The color of the eyes in the mutants becomes light red, this color being produced by a different pigment, unrelated to the ommochromes and masked in normal insects by the presence of the darker ommochrome. It is now known that the normal genes v^+ and a^+ are instrumental in transforming tryptophan into the related substance, **kynurenin.** The defective coloration in mutants not possessing the normal v^+ or a^+ genes can be restored to normal if the necessary quantities of kynurenin are introduced from without. This can be done in several different ways:

1. By transplanting into the mutant larva or early pupa tissue from an animal bearing the normal gene v^+ (or a^+). The rudiment of the eye or the gonad has been used for such transplantation. The kynurenin is produced by the cells of the graft possessing the necessary normal gene. Kynurenin, being a diffusible substance, can spread into the tissues and cells of the host and there be used for the development of pigment, provided that the other mechanisms for this development are available.

2. By injecting into a mutant individual an extract from a normal individual. This method, by the way, has been used to prove the identity of the substances produced by the gene v^+ in *Drosophila* and the gene a^+ *Ephestia:* an extract from a normal *Drosophila* individual caused an a *Ephestia* mutant to develop black eyes, and an extract from a normal *Ephestia* caused a v mutant of *Drosophila* to develop wild type eye coloration.

3. By feeding the larvae (in the case of *Drosophila*) on cooked pupae of the insects possessing the normal gene. The kynurenin, contained in the boiled tissue and taken in with the food, can be used by the larvae for producing pigment.

4. By cultivating eye rudiments of mutant individuals in an extract from insects (whether of the same species or a different one) possessing the normal gene. The eye rudiment attained normal color *in vitro.*

5. By cultivating eye rudiments of mutant insects in a saline solution to which kynurenin had been added.

6. By adding kynurenin to the food of mutant *Drosophila* larvae.

7. By injecting kynurenin into v mutants of *Drosophila* or a mutants of *Ephestia.*

In all the variations of these experiments a substance, which could not be produced by the tissues of the animals and which is necessary for the further development of the pigment, was introduced in some way or other so that the broken chain of reactions could be restored, and the later processes of pigment development could then proceed in an apparently normal fashion. The experiments listed under 5, 6, 7 prove that the substance in question is kynurenin.

Kynurenin is a colorless substance, and further transformations are necessary before the actual pigment is formed. These transformations can be broken off at the next stage by the failure of the gene cinnabar (symbol cn^+) in *Drosophila*. An analogous gene must exist in other insects. The mutation of the gene cn^+ into cn does not prevent the formation of kynurenin, but kynurenin is not processed into the next link in the chain of substances leading to pigment formation. The ommochromes are thus not produced.

The failure of the cn^+ may be compensated in the same way as the failure of the v^+ gene by implanting into the body of mutants tissues of individuals bearing the normal gene. The action of the normal cn^+ gene has also been replaced by a chemical substance, which is sometimes found as an impurity with kynurenin. It has not been possible to separate this substance from kynurenin in a pure form. This prevents an exact analysis of the substance, but at the same time suggests that it is a very nearly related substance and probably a product of transformation of kynurenin.

The next gene which is known to participate in the development of ommochromes is the gene white (symbol w^+) found in *Drosophila* and in many other insects. The mutation of w^+ into w removes all eye pigments from the eye. The tissues of mutant w individuals may be used as a source of the substances restoring to normal the pigment development in v and cn individuals. This proves that the earlier stages of pigment development proceed normally in the white mutant individuals, but that the chain is broken at a later stage.

The pigmentation of the eye of a w mutant cannot be restored by the implantation of tissues of a normal w^+ animal, nor can this be done by introduction of extracts or chemical substances. The gene w^+ must actually be present in the cells of the tissue (the eye rudiment) which is to become pigmented. Under no known circumstances can the actual colored substance be elaborated unless the tissues in which the pigment is found normally have been present. It is concluded, therefore, that the link of the chain controlled by gene w^+ is intimately connected with the cellular structures, and that the substances involved cannot diffuse beyond the limits of the cells. It is possible that the final chemical modeling of the pigments occurs in the granules in which they are found in the fully differentiated cells.

There is hardly any doubt that the three genes mentioned above are not the only ones that are involved in the transformation of tryptophan into ommochrome pigment. Very many more genes would be needed to account for the whole chain of reactions. All these genes must thus cooperate to produce the normal differentiation of the eye. The picture becomes still more complicated if it is considered that the eye rudiment itself must be there for the pigments to be deposited. The development of the eye rudiment is dependent on other genes having nothing to do with the genes of coloration. Thus in *Drosophila* a mutation is known which prevents the development of the eyes (eyeless). In such a mutant no eye pigments are produced, although all the genes otherwise necessary for the elaboration of pigment in the eye may be present.

PART VI

Morphogenetic Processes in the Later Part of Ontogenesis

Taken as a whole, ontogenetic development appears to slow down with the increasing age of the individual animal. In the later periods of life, spectacular changes such as those occurring during cleavage, gastrulation and organ formation do not take place. When histological differentiation starts it gives rise to various tissues, but further differentiation serves mainly to support and only sometimes to increase the already established initial differences. Even growth, as has been stated, diminishes in rate with age. There are, however, special cases in which morphogenetic processes may be aroused again at a late stage of ontogenesis, so that the changes produced by these processes may equal in volume those observed in the earlier stages of the individual's development. These special cases will now be considered.

CHAPTER **18**

Metamorphosis

The first case, in which morphogenetic processes may be reactivated after development has almost reached a standstill, is observed in animals in which the embryo develops into a larva, and the larva is transformed into the adult by way of **metamorphosis.** Larval forms and an accompanying metamorphosis are found in most groups of the animal kingdom though by no means in all representatives of each group. The larvae usually have special names distinguishing them from the adult forms. The following is a survey of the occurrence of larval forms, with their names:

1. Porifera — amphiblastula
2. Coelenterata — planula
3. Platyhelminthes
   Turbellaria polycladida — Müller's larva
   Trematodes — miracidium, cercaria
   Cestodes — onchosphaera
4. Nemertinea — pilidium

445

5. Annelida	— trochophore
6. Mollusca	— trochophore and veliger
7. Crustacea	
Entomostraca	— nauplius
Malacostraca	— zoea
8. Insecta	— nymphs of insects with incomplete metamorphosis. Caterpillars, grubs, etc., of insects with complete metamorphosis.
9. Bryozoa ectoprocta	— the cyphonautes larva
10. Echinodermata	— pluteus, bipinnaria, auricularia, etc.
11. Enteropneusta	— the tornaria larva
12. Ascidiacea	— the tadpole larva
13. Cyclostomata	— ammocoete
14. Amphibia	— tadpoles of frogs, aquatic larvae of salamanders

The morphogenetic processes in all these groups differ both in the nature of transformation and in the mode of causation of the whole sequence, it is thus impossible to describe them in common terms. Two cases are chosen here for closer study: the metamorphosis in insects and the metamorphosis in amphibians.

18-1   CHANGES OF ORGANIZATION DURING METAMORPHOSIS IN AMPHIBIANS

In amphibians metamorphosis is associated in typical cases with a transition from an aquatic to a terrestrial mode of life. Superimposed on this change of environment is, in the anurans (frogs and toads), a change in feeding. The tadpoles of most frogs and toads feed on vegetable matter—particles of plants, living and decaying—which they scrape off submerged objects with the aid of the horny teeth surrounding their mouths. Some are detritus feeders, passing through their guts the mud and detritus collected from the bottom, and others, as the tadpole of the clawed toad, *Xenopus,* are plankton feeders. Adult frogs are carnivorous, living on insects, worms, and the like, but sometimes also on larger prey, such as smaller frogs and even little birds and rodents, which they catch, overpower and swallow. In the case of urodeles, there is no substantial change of diet, the larvae being as carnivorous as the adults, though naturally feeding on smaller animals (mainly crustaceans and worms). The changes in the organization of the animals during metamorphosis are in part progressive and in part regressive, and may be grouped into three categories:

1. The organs or structures necessary during larval life but redundant in the adults are reduced and may disappear completely.

2. Some organs develop and become functional only during and after metamorphosis.

3. A third group of structures while present and functional both before and after metamorphosis become changed so as to meet the requirements of the adult mode of life.

In *anurans* the differences between the mode of life of the larva and the adult are much more profound, and accordingly the changes at metamorphosis are more extensive than in the urodeles. We will consider them first.

The regressive processes occurring during the metamorphosis in frogs are the following: The long tail of the tadpole with the fin folds is resorbed and disappears without a trace. The gills are resorbed; the gill clefts are closed and the peribranchial cavities disappear. The horny teeth of the perioral disc are shed as well as the horny lining of the jaws. The shape of the mouth changes. The cloacal tube becomes shortened and reduced. Some blood vessels are reduced, including parts of the aortic arches as indicated in section 11–6.

The constructive processes involve first the progressive development of the limbs, which increase in size and differentiation. The forelimbs, which in the frogs develop under cover of the opercular membrane, break through to the exterior. The middle ear develops in connection with the first branchial pouch (the pouch situated between the mandibular and the hyoid arch). The tympanic membrane develops, supported by the circular tympanic cartilage. The eyes protrude on the dorsal surface of the head and develop eyelids. The tongue is developed from the floor of the mouth. The organs which function both in the larva and the adult, but change their differentiation during metamorphosis, are primarily the skin and the intestine. The skin of the tadpole is covered with a double-layered epidermis. During metamorphosis the number of layers of cells in the epidermis increases and the surface layers become cornified. Multicellular mucous and serous glands develop as pockets sinking from the surface into the subcutaneous connective tissue layer. The lateral line sense organs, present in the skin of tadpoles, disappear during metamorphosis in most frogs. The pigmentation of the skin is changed; new patterns and colors appear. The intestine, which is very long in tadpoles, as in most herbivorous animals, becomes greatly foreshortened, and most of the coils which it forms in the tadpole become straightened out. The metamorphosis is very rapid and takes only a few days.

In *urodeles* the changes at metamorphosis are far less striking. The tail is retained, only the fin fold disappears. The branchial apparatus is reduced; the external gills become resorbed and the gill clefts close. The visceral skeleton becomes greatly reduced. The head changes its shape, becoming more oval. The progressive changes are less conspicuous than in the metamorphosing frog tadpoles. They are restricted mainly to the changes in the structure of the skin and the eyes. The eyes bulge more on the dorsal surface of the head and develop lids. The skin becomes cornified and multicellular skin glands become differentiated. The pigmentation of the skin changes. The legs, contrary to those of tadpoles, suffer hardly any change at all, and the same may

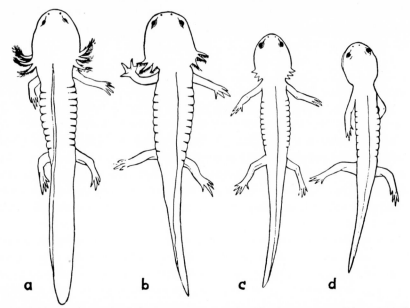

Fig. 255. Metamorphosis in the axolotl. *a,* Animal in the larval condition. *d,* Fully metamorphosed animal. *b, c,* Intermediate stages of metamorphosis. (From Roth, 1955.)

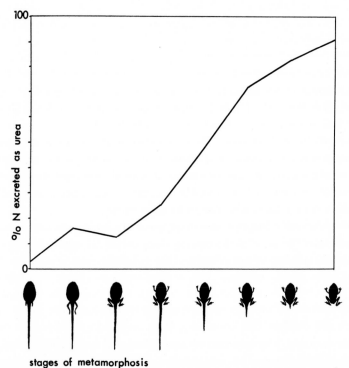

Fig. 256. Changes in the amounts of ammonia and urea excreted by tadpoles of the frog *Pyxicephalus delalandii* during metamorphosis. (After unpublished data by J. Balinsky.)

be said of the alimentary canal. The metamorphosis is on the whole more gradual, and may take up to several weeks (Fig. 255).

It is worth noting that the lungs do not undergo drastic changes during metamorphosis in both anurans and urodeles; they develop very gradually, and become fully functional in the larval state. Long before metamorphosis the larvae of frogs as well as of salamanders start coming up to the surface and gulping air into their lungs and thus supplementing their aquatic respiration. This may be of considerable importance where the larvae develop in stagnant and polluted waters, as is often the case.

Hand in hand with morphological changes during metamorphosis go physiological changes, of which we shall point out the following. In the frog tadpoles the endocrine function of the pancreas starts at metamorphosis, and this is connected with the increased role of the liver in the turnover of carbohydrates (glycogen) (Abeloos, 1956). A profound change takes place in the excretory mechanism. In the tadpole the end product of nitrogen metabolism is ammonia, which is easily disposed of (by diffusion) in an aquatic medium, but which in a terrestrial animal might accumulate and become dangerous because of its high toxicity. Metamorphosed frogs, however, excrete most of their nitrogen in the form of urea and only small amounts as ammonia. The change-over occurs in the late stages of metamorphosis, and is, of course, due to a changed function of the liver, which performs the synthesis of urea (Munro, 1939) (Fig. 256).

The reduction of the gills and the tail in tadpoles is effected by autolysis of the component tissues of these organs with active participation of ameboid macrophages which phagocytose the debris of the disintegrating cells. The same mechanism, though of a limited scope, is in action when the external gills and fin folds are reduced in urodele amphibians.

Since destructive processes play such a considerable role in metamorphosis, and since, in addition, food intake may be interrupted during the crucial part of the transformation, especially in tadpoles, the mass of the body at the end of metamorphosis is smaller than at the beginning ("degrowth"—see p. 402). The reduction of the body mass is not only due to loss of some parts (gills, tail), but the remaining parts, excepting the actively growing organs, also appear to shrink during metamorphosis; the head and trunk of metamorphosed amphibians are smaller than in larvae just before the commencement of metamorphosis.

18–2 CAUSATION OF METAMORPHOSIS IN AMPHIBIANS

The concurrent changes in so many parts of the animal's body during metamorphosis suggest the existence of some common cause for all the transformations. It has been found that this common cause is a hormone released in large quantities from the thyroid gland of the animals entering

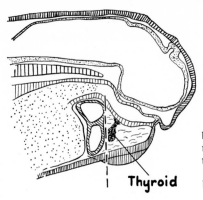

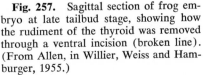

Fig. 257. Sagittal section of frog embryo at late tailbud stage, showing how the rudiment of the thyroid was removed through a ventral incision (broken line). (From Allen, in Willier, Weiss and Hamburger, 1955.)

the stage of metamorphosis. A first indication of this was obtained when Gudernatsch fed some frog tadpoles on dried and powdered sheep thyroid gland and observed that they metamorphosed precociously (1912). Feeding tadpoles with preparations of other glands did not have the same effect. This experiment made it very probable that tadpoles are capable of reacting to the thyroid hormone by metamorphosing.

The proof that the thyroid hormone is actually the cause of metamorphosis in normal development was further given by the following two experiments. The rudiment of the thyroid gland was removed in frog embryos in the tailbud stage (through an incision on the ventral side, Fig. 257). The operated tadpoles were fully viable and showed normal growth but failed to metamorphose (Allen, B. M., 1918), although they were kept alive almost a year after the control animals had become little froglets. The thyroidless tadpoles continued to grow and attained a much greater size than normally: total length up to 123 mm. instead of about 60 mm. as at the beginning of metamorphosis in normal tadpoles. It was thus proved that metamorphosis cannot set in without a stimulus emanating from the thyroid gland. The third and final experiment consisted in supplying the thyroidless tadpoles with thyroid hormone from without, either by feeding them on dried thyroid gland (Allen, B. M., 1918) or by immersing them in water containing soluble extracts from thyroid glands. The tadpoles treated in this way immediately proceeded to metamorphose, thus showing that their own thyroid glands are not necessary so long as they are supplied in some way or other with thyroid hormone (see Allen, B. M., 1938).

Similar experiments were carried out on urodeles. A very favorable animal for the experiments is the axolotl, *Ambystoma mexicanum,* which under ordinary conditions does not metamorphose at all, but may be induced to metamorphose by thyroid treatment (see Marx, 1935).

As already indicated, the active principle of the thyroid gland may be introduced into the animal's body in several different ways. Normally it emanates from the animal's own thyroid gland. The same effect may be produced by

(a) implanting bits of live thyroid gland;

(b) feeding the animals on thyroid gland;

(c) injecting them with preparations of thyroid gland;

(d) keeping them in water containing soluble extracts of thyroid glands.

The latter fact clearly shows that the active principle of the thyroid gland is a chemical substance—a hormone. A saline extract of fresh thyroid tissue contains protein, **thyroglobulin,** which retains the activity of the thyroid gland. An important characteristic of thyroglobulin is that it contains iodine, which, as we will see, is of great importance for the working of the thyroid hormone. Thyroglobulin has a molecular weight of about 675,000; its molecules are thus very large and it is unlikely that thyroglobulin can penetrate as such through cellular membranes, which would be necessary if it were to leave the thyroid gland and reach the cells eventually reacting to the thyroid treatment. It seems probable, therefore, that thyroglobulin is broken up into some compound or compounds with smaller molecules and it is these that actually reach the effective organs. By breaking down thyroglobulin under experimental conditions two iodine-containing amino acids may be obtained —di-iodotyrosine and thyroxine. Of the two, the thyroxine is far more potent, but it is by no means certain that thyroxine under natural conditions acts alone, and not in a combination with some other substances (joined with other amino acids into a polypeptide?). The structure of the thyroxine molecule is known, and it may be synthesized in the laboratory.

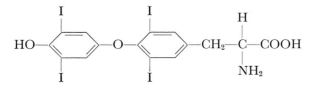

The molecule of thyroxine contains 4 atoms of iodine. It has been tested whether iodine alone can cause metamorphosis in amphibians, and positive results have been obtained either by keeping frog tadpoles or newt larvae in water containing the element iodine (in solutions as weak as 0.0000003833% !), by injecting animals with iodine or by implanting iodine crystals into the body cavity. The element iodine caused metamorphosis even in thyroidectomized axolotls, so it must have been acting directly and not through the increased production of hormone by the animal's own thyroid gland. On the other hand, it appears that the degree of activity of the iodine atoms may be greatly influenced by the type of protein (or amino acid) with which the iodine is bound. This can be shown clearly by comparing the activity of the two amino acids di-iodotyrosine and thyroxine. By placing tadpoles in solutions of each amino acid separately it was found that the same amount of iodine was 300 times more active when

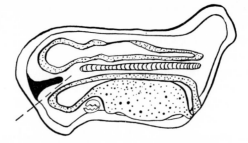

Fig. 258. Sagittal section of frog embryo at early tailbud stage, showing how the rudiment of the hypophysis (black) was removed through an incision indicated by the broken line. (From Allen, in Willier, Weiss and Hamburger, 1955.)

forming a part of the thyroxine molecule than when incorporated into the di-iodotyrosine molecule.

The thyroid gland is not the only one which is involved in the causation of metamorphosis in amphibians. It has been discovered that the hypophysis plays an important part as well. If, in frog tadpoles, the hypophysis is destroyed (Adler, 1914) or if the rudiment of the hypophysis in a late frog embryo is excised (Fig. 258) (see Allen, B. M., 1929) the tadpoles do not metamorphose, just as if their thyroid glands had been removed. The removal of the hypophysis rudiment may later be compensated by implanting pieces of hypophysis from metamorphosed or adult frogs, but only if the thyroid gland of the animal is intact. If the thyroid gland has been removed, no amount of hypophyseal tissue implanted can induce metamorphosis. From this it may be concluded that the hypophysis does not act upon the tissues directly, but only by way of stimulating the thyroid gland. In fact the thyroid gland of hypophysectomized animals remains underdeveloped, and does not accumulate the thyroid hormone (in the form of "colloid") in its follicles. Implantations of active hypophyses can also be used to stimulate metamorphosis in the axolotl.

The agent necessary for activating the thyroid gland is produced in the anterior (epidermal) lobe of the hypophysis and has been isolated in the form of a **thyrotrophic hormone.**

In the case of larval amphibians the hypophysis does not produce the thyrotrophic hormone until the time when metamorphosis normally occurs. This has been proved by taking hypophyses from tadpoles of various ages and transplanting them into tadpoles whose own hypophyses had been removed previously. Whereas hypophyses taken from tadpoles in stages of metamorphosis or from metamorphosed frogs compensated for the removal of the animal's own hypophysis and restored the tadpole's ability to metamorphose, hypophyses taken from younger stages were not effective.

The following pattern of amphibian metamorphosis emerges from these studies. The initial signal for metamorphosis is given by the anterior lobe of the hypophysis when it reaches a certain degree of differentiation and becomes capable of producing the thyrotrophic hormone. The thyrotrophic hormone activates the thyroid gland, which builds up and releases the thyroid hormone. The thyroid hormone (of which thyroxine is the most powerful component) affects the tissues directly, causing the degeneration and necrosis of some cells and stimulating the growth and differentiation of others.

18–3 TISSUE REACTIVITY IN AMPHIBIAN METAMORPHOSIS

One of the most intriguing aspects of amphibian metamorphosis is the diverse nature of reactions in different tissues to one and the same condition—the presence of the thyroid hormone. Whereas some of the tissues (the tail, the gills) become necrotic owing to the action of the hormone, others (the limbs) react to the hormone by increased growth and progressive differentiation. It can easily be shown that the character of the reaction is not due to the position of the parts in question, nor to an uneven distribution of the active principle, but solely to the nature of the reacting part. Parts of a tadpole's tail transplanted to the trunk undergo metamorphosis together with the host's tail and become absorbed. On the other hand, an eye transplanted to the tail of a tadpole before metamorphosis remains healthy while all the surrounding and underlying tissues undergo necrosis. As the tail shrinks, the eye is brought nearer to the trunk, and eventually becomes fused to the body in the sacral region after the whole of the tail has disappeared (Fig. 259) (Schwind, 1933). The same experiments, incidentally, prove that the stimulus of the thyroid gland is carried through the blood vascular system, as it is only by this route that it can reach any part independently of its actual position. The secretion of the thyroid gland is thus a hormone in the narrow sense of the term. What happens to a tissue under the influence of the thyroid hormone is determined by the reactive properties of the tissue itself, or by what we have earlier called its **competence** (see p. 169). The competence of tissues to react to the thyroid hormone is not directly dependent on their histological differentiation; in tadpoles, while the myotomes of the tail become resorbed during differentiation, the myotomes of the trunk are not so affected.

It has further been noticed that different parts of the body reacting to the thyroid hormone (whether by degeneration or progressive development) are not equally responsive to the dosage of the hormone. Very weak dosages applied to frog tadpoles cause an acceleration of the growth and differentiation of hindlimbs and a shortening of the intestine. Further processes may not be set into motion at all or follow only after a lengthy treatment. A higher dose of the thyroid hormone causes the breakthrough of the forelegs. An even greater dose of the hormone is necessary to cause the resorption of the tail. There is evidently some kind of threshold value for each part, which has to be attained before the reaction sets in. Different parts of the tail have different threshold values, the tip of the tail reacting more

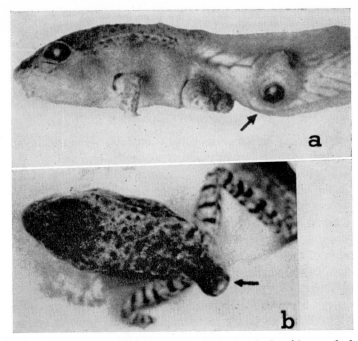

Fig. 259. *a*, Metamorphosing frog (*Rana sylvatica*) tadpole with a grafted eye in the tail. *b*, Metamorphosed froglet, the tail has disappeared, but the eye is retained in the sacral region. Eye indicated by arrow. (From Schwind, 1933.)

readily than the proximal parts. In general the degree of sensitivity to the thyroid hormone is reflected in the order in which the metamorphic changes proceed in normal development: the parts which have a low threshold (legs, reacting by growth) respond earlier than the parts having a high threshold value (tail, reacting by reduction) (Etkin in Willier, Weiss and Hamburger, 1955; Abeloos, 1956). When heavy dosages of thyroid hormone are supplied to young tadpoles, all processes start at once and the normal sequence of events becomes upset, the destructive processes being capable of proceeding faster than the constructive processes: the forelimbs break through before becoming differentiated, the tail becomes reduced before the legs are sufficiently developed to take over locomotion. The result is, of course, the death of the animal. In urodeles the bulging of the eyes seems to be the reaction which may be elicited by the weakest doses of the thyroid hormone. Next follow, more or less concurrently, the reduction of the fin fold and the shortening and disappearance of the external gills. The closure of the gill clefts and the transformation of the skin are the result of a maximal stimulation, and accompany complete metamorphosis in the course of normal development. The earlier stages of metamorphosis, including the shortening of the external gills, are partly reversible: if the thyroid treatment is stopped, the gills may elongate again to a certain extent.

18–4 PROCESSES OF INDUCTION DURING AMPHIBIAN METAMORPHOSIS

Although as a general rule the processes of metamorphosis are a direct reaction to the thyroid hormone which reaches each tissue, there are some notable exceptions to this rule. The skin covering the tail of a tadpole, although subject to necrosis in normal metamorphosis, remains healthy if transplanted to the body, provided that it is transplanted without the underlying muscles. Skin transplanted with underlying tail muscle will necrotize in any position on the body. It is thus evident that the direct action of the thyroid hormone is on the muscle tissue, and the overlying skin becomes involved in the process of resorption secondarily.

A more complicated case is presented by the development of the tympanic membrane in frogs. The middle ear, with its cavity connected by the Eustachian tube to the pharynx, is one of the structures that develop progressively during metamorphosis. The tympanic membrane first becomes differentiated toward the end of metamorphosis. It is supported in frogs by a cartilaginous ring, **the tympanic cartilage,** which develops as an outgrowth from the posterior edge of the quadrate cartilage. The skin which later participates in the formation of the tympanic membrane is originally no differ-

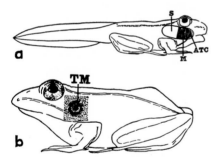

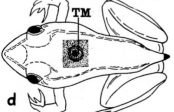

Fig. 260. Dependence in the development of the tympanic membrane on the tympanic cartilage. *a,* Stage of operation: skin (*S*) turned back to reveal tympanic cartilage (*ATC*). *b,* Normal tympanic membrane (*TM*) developed on control side. *c,* No tympanic membrane developed after tympanic cartilage had been removed (*EC,* scar from operation). *d,* Tympanic cartilage transplanted subcutaneously on the back has caused development of tympanic membrane from local skin. (From Helff, 1928.)

ent from the skin covering the rest of the body. During metamorphosis the connective tissue layer of the skin in the area of the tympanic membrane becomes reorganized, the original layer of fibers—the stratum compactum— is broken up with the participation of phagocytes, and a completely new, somewhat thinner fibrous layer is developed in its place. In the fully differentiated tympanic membrane the skin is less than half the thickness of ordinary skin, but much more compact, and it also differs in its pigmentation.

It has been found that the differentiation of the tympanic membrane is not due to a direct action of the thyroid hormone, but that it is induced by the tympanic cartilage (Helff, 1928). If the tympanic cartilage is removed before metamorphosis, the tympanic membrane does not develop. If the area in the otic region is covered by skin from the flank or from the back, the skin will develop a tympanic membrane. Lastly, if the tympanic cartilage is inserted under the skin on the flank or the back in tadpoles approaching metamorphosis, the local skin becomes differentiated as tympanic membrane (Fig. 260).

We can draw up a complicated chain of interactions that must take place before a tympanic membrane becomes differentiated. The first step is the formation of the rudiment of the hypophysis. The latter is developed in conjunction with the stomodeal invagination (see p. 346) and presumably is induced by the oral endoderm (p. 345). The hypophysis in due time secretes the thyrotrophic hormone and activates the thyroid gland. The thyroid gland releases the thyroid hormone, and this causes the posterior edge of the quadrate to become differentiated as the tympanic cartilage. Lastly the tympanic cartilage induces the skin to differentiate as tympanic membrane.

18–5 MOULTING AND ITS RELATION TO METAMORPHOSIS IN INSECTS

In any consideration of metamorphosis in insects it has to be taken into account that metamorphosis in these animals is a special form of moulting (the periodic shedding of the cuticle of the skin which necessarily accompanies growth, because the strongly sclerotized parts of the cuticle cannot stretch or cannot stretch sufficiently to accommodate the growing mass of the body). A large proportion of the external features of an insect is embodied in the sclerotized parts of the cuticle, such as the details of shape of parts of the body, the hairs and spines on the surface of the skin, the sculpture of the surface of the cuticle and to a certain extent the pigmentation. In the process of moulting these features are lost with the discarded cuticle. The external characters of the insect, in so far as they find their expression in cuticular structures, have to be produced anew at each moult, though on a larger scale. The new cuticle is secreted by the epidermis of the skin, and it is therefore this layer of cells which is directly responsible for the external features of the insect emerging from a moult, whether the moulted insect is an enlarged copy of the previous stage or whether it shows some new characters.

The moult in every case is quite a complicated process. In between two moults the cells of the epidermis are quiescent, they are more or less flat, and the epithelial layer may be rather thin. The epidermis cells adhere closely to the inner surface of the cuticle. Before each moult, however, the cells of the epidermis become activated, they detach themselves from the cuticle and enter a phase of rapid growth and proliferation. Numerous mitoses are observed. (Proliferation of epidermal cells is not observed, however, during the larval moult of cyclorraphe dipteran larvae, in which the cells of the larval epidermis do not divide between the stages of the egg and the pupa, and are eventually discarded and replaced by imaginal epidermis.)

The number of epidermal cells produced by mitosis may be in excess of what is necessary, and some of the cells at this stage undergo degeneration by pycnosis. In spite of the degeneration of some of the cells, the layer of epidermis becomes thicker and the remaining cells become arranged in a regular columnar epithelium. The surface of this epithelium foreshadows the shape of the animal emerging from the moult. In those parts of the body which are to be increased as the result of the moult the epidermis is thrown into folds which expand and straighten out after the insect has emerged from its old skin. The folding is especially great where new or greatly increased appendages (such as the wings) are to be developed.

The epidermal cells now produce on their surface a thin layer of hardening secretion which becomes the outermost layer of the new cuticle—the **epicuticle,** consisting of a substance of lipoprotein nature, **cuticulin.** A fluid produced mainly by special moulting glands is now poured into the space between the surface of the new epicuticle and the innermost surface of the old cuticle (Fig. 261). The fluid contains enzymes which digest the inner layers of the old cuticle until little more than the old epicuticle is left. The fluid with the substances digested from the old cuticle later becomes reabsorbed into the body of the insect. At the same time as the old cuticle is being digested, the epidermis produces further layers underneath the new epi-

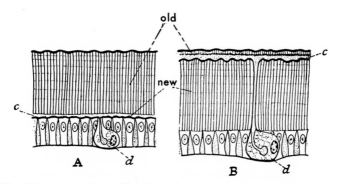

Fig. 261.   Two stages of moulting in an insect. *A,* Old cuticle separated from epidermis which has produced a new epicuticle (thick black line). *B,* Deep layers of old cuticle dissolved, and layers of new endocuticle produced beneath the new epicuticle. *d,* Moulting glands; *c,* moulting fluid. (From Wigglesworth, 1939.)

cuticle: the **exocuticle,** containing large quantities of cuticulin and also phenolic substances which are later oxidized to produce the dark pigment in the cuticle, and eventually the **endocuticle** consisting of protein and chitin, which is a nitrogenous polysaccharide.

When the old cuticle is reduced to a thin shell, it is ruptured at the back of the head and thorax, and the insect crawls out of its old coat. The new cuticle is by no means quite completed at this stage; after moulting the cuticle hardens and visible pigment is produced in the cuticle from colorless precursors (phenolic substances). Further layers of endocuticle are deposited by the epidermal cells on the inner surface of the cuticle for days and even weeks after the moult has taken place.

It will be seen that some elements of amphibian metamorphosis, namely, destructive processes (resorption of the old cuticle, necrosis of part of the epidermal cells) as well as constructive processes (rearrangement of epidermal cells, formation of new cuticle), are present in an ordinary moult in insects. It depends on the condition of the epidermal layer whether the new structures produced during the moult are similar to the old ones or different. In the first case the moult contributes to the growth of the animal; in the second it becomes a mechanism for progressive development. If the changes achieved after a moult are considerable, the result is metamorphosis. In the primarily wingless archaic insects, the Apterygota, the young insect emerging from the egg is essentially similar to the adult, differing only in size and in the immature state of the sexual organs. Moulting in these insects leads only to growth, and the advent of sexual maturity is not related in any way to moulting; in fact, moulting and growth continue even after the attainment of reproductive ability.

In all other insects, the Pterygota (winged or secondarily wingless forms), there is a distinct imaginal stage, which is attained after a specific imaginal moult, after which the insect does not moult any more. Except for secondarily wingless insects the imaginal stage differs from the larval stages by the presence of wings. The imago also differs from the larval stages by the full development of external genital organs. (The gonads, on the other hand, may become fully functional only some time after metamorphosis.) In the more primitive winged insects the wings appear gradually, the rudiments of the wings in the form of flat outgrowths of the second and third thoracic segment being visible in the later larval, or as they are often called, **nymphal** stages. These rudiments increase with every subsequent moult, but at the last imaginal moult there is an abrupt and very marked increase in the size of the wings, and after this moult the wings become functional (only in the mayflies the first winged stage, the **subimago,** moults again before it turns into an imago). The insects in which the rudiments of the wings develop on the surface of the body are called Exopterygota; these comprise the locusts, cockroaches, dragonflies, mayflies, bugs and other related groups. In the most advanced orders of insects, however, the wings develop internally, as folded appendages concealed during the larval stages in deep pockets (infoldings) of the epidermis (see Fig. 262). The epidermis covering these wing rudiments retains an embryonic character throughout larval life, and although the rudiments continue growing slowly their epidermis does not participate in the formation of the external cuticle of the larva and comes into action only when the larval stage is drawing to an end. Such rudiments, concealed under the surface of the body in the larval stage and reaching full differentiation in the imago, are called **imaginal discs.** The insects in which the wings develop internally as imaginal discs are called Endopterygota (caddis flies, beetles, butterflies, bees and wasps, mosquitoes and flies).

Although the development of the wings in the adult insect attracts the greatest attention, the other parts of the body also change at the time of metamorphosis from a larva or nymph to the adult (imago). Even where the larva or nymph leads the same mode of life as the adult and has a fairly similar general appearance, as in the locusts or bugs, many finer features of structure change. In the bug *Rhodnius prolixus,* for instance, which has been studied by Wigglesworth (1954), the fine structure of the cuticle and the pigmentation of the adult bug are very different from the last larval stage, so that even small areas of skin of the larva and the adult can be easily distinguished.

In the endopterygote insects the difference between the larvae and adults is much greater. Not only wings but also mouth parts, antennae and legs may be developed from imaginal discs, while the larval appendages become discarded as in the case of butterflies (see Fig. 263). In some parasitic wasps and in flies such as *Drosophila* and *Musca,* the whole larval epidermis is discarded and replaced by the imaginal epidermis derived from a series of imaginal discs (Fig. 264). Concurrently with the formation of appendages and other external parts, the internal organs are also reorganized. As the locomotion of the winged adult is so completely different from the locomotion of a crawling larva, the muscle system may have to be radically changed. During metamorphosis of higher insects the larval muscles become broken down and their

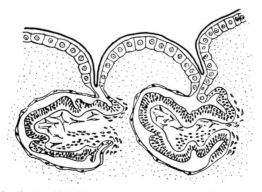

Fig. 262. Longitudinal section through the imaginal discs of the wings in full-grown larva of an ant, *Formica.* (After Perez, from Wigglesworth, 1954.)

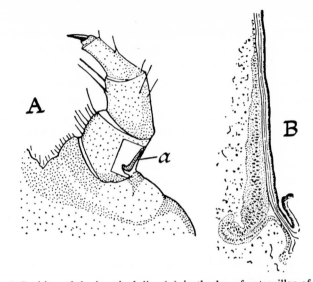

Fig. 263. *A,* Position of the imaginal disc (*a*) in the leg of caterpillar of a butterfly, *Vanessa. B,* Detail of the imaginal disc. (After Bodenstein, from Wigglesworth, 1939.)

remnants are consumed by phagocytes. The adult muscles, in particular the muscles operating the wings (the flight muscles), are then developed.

The eyes of the adult insects in the more advanced orders are very different from those of the larvae and are developed from special imaginal discs. Most of the cells of the alimentary canal of the larva may undergo resorption, and the alimentary canal of the imago is lined by a new epithelium produced at the expense of pockets of small reserve cells, which are found between the functioning cells of the larval intestine.

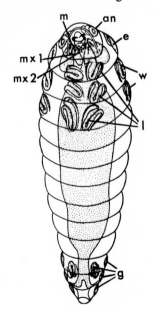

Fig. 264. Imaginal discs in the last stage larva of a parasitic wasp, *Encyrtus. an,* Antennae; *e,* compound eyes; *m,* mandibles; *g,* genital organs; *mx1, mx2,* first and second pairs of maxillae; *l,* legs; *w,* wings. (After Bugnion, from Kühn, 1955.)

While the moderate amount of transformations that occur in ectopterygote insects can be performed in one moult, the reorganization that is needed to produce the imago in most endopterygote insects is so profound that a resting stage—the pupa—is intercalated between the larval and the adult condition. In the pupa the pockets containing the imaginal discs—wings, limbs, antennae, etc.—are brought to the surface. Internally, the formation of adult parts is, however, not yet completed, and while the reorganization takes place in the pupal stage the insects do not take food and are very restricted in their movements, if they move at all.

When the reorganization is completed, another moult takes place, and the imago emerges from the pupa. Metamorphosis which includes a pupal stage is called complete metamorphosis, and the insects having this type of metamorphosis are called Holometabola; those not possessing a pupal stage and thus having an incomplete metamorphosis are Hemimetabola. The holometabolous insects are the Endopterygota; the terms, so far as systematics are concerned, are synonyms, though they stress different properties of the same group of insects.

18–6   CAUSATION OF MOULTING AND METAMORPHOSIS IN INSECTS

In an ordinary moult **(larval moult)** all parts of the body must participate in the process and carry it out at the same time, if the moulting is to be successful. This suggests a common cause to which all parts of the insect are subjected. The existence of a common cause is even more obvious in the case of metamorphosis in which the involvement of both external and internal organs may be more far reaching and radical. This common cause may be expected to be either external or internal. Cases are actually known in which under natural conditions an external factor is necessary to start a moult. In the blood-sucking bug *Rhodnius,* such a factor is the intake of food. The bugs of this species feed only once in the interval between two moults, taking up so much blood that their body weight may increase many fold. Moulting occurs regularly 12 to 15 days after a feed in the case of the first four larval stages. The same dependence of moulting on food intake holds true for the last, the fifth larval stage, only the interval is somewhat longer, namely, about 28 days, and the result is different: the moult transforms the larva into a winged imago. Another example in which an external factor is necessary to initiate a moult is the case of the pupa of the moth, *Platysamia cercopia.* After pupation the insect falls into a quiescent state with reduced rate of metabolism—the **diapause**—which continues throughout winter. It is essential that during this time the pupa be exposed to cold, otherwise the diapause is prolonged indefinitely. However, the diapause may be broken precociously if the pupa is treated with cold (3 to 5° C) for at least two weeks. The temporary cooling activates the vital processes in the pupa and on return to a warmer environment the pupa moults, and so the development is completed with the emergence of the imago.

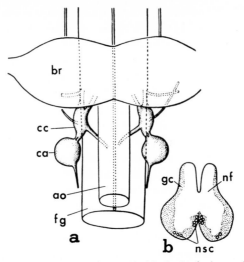

Fig. 265. The endocrine glands associated with the brain in moths. *a*, Dorsal view of the brain in a hawk moth pupa. *b*, Transverse section of the protocerebrum in a caterpillar of the meal moth. *ao*, Aorta; *br*, brain; *ca*, corpora allata; *cc*, corpora cardiaca; *fg*, foregut; *gc*, ganglionic cells; *nf*, nerve fibers; *nsc*, neurosecretory cells. (After Cazal and Rehm, from Kühn, 1955.)

In the overwhelming majority of insects, however, no external cause of any moult can be found, and the moults follow each other at intervals which appear to be determined entirely by internal processes in the animal. In many insects the body weight increases in a fixed proportion between two moults, often by a factor of two, and it would appear that a certain amount of synthesis has to be performed after each moult before the stimulus for a new moult is generated in the organism. However, even in cases where an external factor triggers the mechanism of moulting, it can be shown that the factor in question does not affect all parts of the body directly, but that it is mediated by the brain of the insect. If a larva of *Rhodnius* is decapitated within a day or two after feeding it does not moult, although it may remain alive for over a year. If, however, it is decapitated five or more days after a meal, moulting takes place; by that time a stimulus generated by the brain reaches beyond the level of decapitation and is able to spread throughout the body and cause the moult to proceed (Wigglesworth, 1954). A corresponding experiment in the moth *Platysamia* consists in activating one pupa by exposure to cold and then transplanting parts of the body of the activated pupa into an untreated pupa. The transplantation of the brain but not of other organs will cause the second pupa to moult and the adult moth to emerge, thus showing that the cold directly affects only the brain but that the rest of the body is stimulated to moulting through the mediation of the latter (Williams, 1946).

The question naturally arises as to how the brain affects the rest of the body. It is now established that, as with amphibian metamorphosis, the moulting and metamorphosis in insects is controlled by hormones, and that at least three organs of internal secretion are involved: the brain **(proto-cerebrum)**, the **corpora allata** and the **prothoracic gland.** In the brain a hormone is produced by two pairs of groups of **neurosecretory cells,** a median pair and a lateral pair (see Fig. 265). Behind the protocerebrum, alongside the dorsal aorta, there are to be found in most insects two pairs of bodies connected by nerve strands to the protocerebrum: first the **corpora cardiaca,** which are of the nature of nerve ganglia, and, more posteriorly, the **corpora allata,** consisting of secretory cells. The corpora allata may be fused into one body in some insects. The third endocrine gland, the **prothoracic gland,** is an irregular branching mass of glandular cells located in the thorax, in close association with the tracheal tubes (Fig. 266). The glandular cells of all three centers show regular secretory cycles preceding each moult, and the three types of secretion are necessary for the normal course of larval moults. The moult is initiated by the neurosecretory cells of the protocerebrum, but all that the hormone of the protocerebrum does is to activate the prothoracic gland. The prothoracic gland then produces a hormone which sets in motion the mechanism of moulting in the epidermis: the growth and proliferation of the epidermal cells, the shedding of the old cuticle and the production of the new one. The hormone produced by the prothoracic gland is called therefore the **moulting hormone** (or the growth and moulting hormone).

We have described above some of the evidence proving that the initial stimulus for moulting is given off by the brain (or the neurosecretory cells of the protocerebrum). One of the experiments consisted in transplanting the brain of an activated pupa of *Platysamia* into an untreated pupa, where-

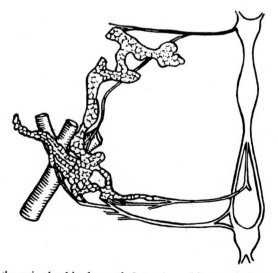

Fig. 266.   Prothoracic gland in the moth *Saturnia*, and its associations with the ventral nerve cord and the tracheal system. (After Lee, from Wigglesworth, 1954.)

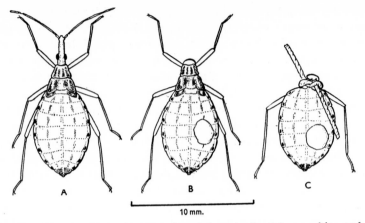

10 mm.

Fig. 267. Experiment showing necessity of thoracic gland for moulting and meta-morphosis. *A,* Normal larva of *Rhodnius. B,* Decapitated larva with an implant in the abdomen. *C,* Larva ligated through metathorax, with implant in isolated abdomen. (From Wigglesworth, 1954.)

upon the latter moulted and produced the moth. A variation of the same experiment has been used to prove that the secretion of the brain cannot act directly but only through the activation of the prothoracic gland. Instead of transplanting the activated brain into a whole pupa, it was implanted into the posterior half of a pupa cut in two (the cut surface was sealed with paraffin wax). Under these conditions the graft was powerless; no meta-morphosis took place. The reason for this is that the prothoracic gland is absent in the posterior half of the pupa. If in addition to the brain the pro-thoracic gland was also grafted, metamorphosis occurred (Williams, 1947, Wigglesworth, 1954). An analogous experiment has been performed on the bug *Rhodnius* (see Fig. 267). After the neurosecretory cells of the brain had been activated by the bug's having a meal of blood, the brain was trans-planted into the abdomen of a decapitated specimen. A decapitated larva is still in possession of the prothoracic gland, which could react to the im-planted brain and cause the moult to proceed. If, however, the activated brain was implanted into an isolated abdomen (Fig. 267, *c*), no moulting oc-curred. On the other hand moulting could be induced in the isolated ab-domen by the implantation of a thoracic gland (Wigglesworth, 1954).

The role of the brain and the thoracic gland as causative agents of moult-ing can also be demonstrated in insects in which the time of the moulting is not dependent on any particular external factor. If the brain is removed from caterpillars sufficiently early before the next expected moult, the caterpillars may remain alive for over two months but do not moult and do not pupate. The implantation of a brain from another caterpillar restores the ability of a brainless caterpillar to complete its development (Kühn and Piepho, 1936). Once the thoracic gland has become activated, the brain is no longer neces-sary for initiating the moult; only those parts moult (or pupate), however, to which the hormone of the thoracic gland has been able to gain access. If

a caterpillar in the last larval stage is constricted behind the thorax, the anterior part of the body will pupate, but the posterior part to which the moulting
hormone could not reach remains in the larval state (Fig. 268). A little
later, when the hormone has already spread throughout the body, a transverse constriction does not prevent pupation of the posterior end of the
caterpillar.

The hormones emitted by the protocerebral neurosecretory cells and the
thoracic gland induce an insect to moult but they do not determine whether
it will be a larval moult producing the next larval stage, the pupal moult,
converting the larva to pupa, or the imaginal moult, leading to the eclosion
of the imago. The third endocrine gland, the corpora allata, controls the
nature of the change that is going to take place at the time of the moulting.
Curiously enough the first two glands, the protocerebral neurosecretory
cells and the thoracic gland, when acting alone cause immediate metamorphosis—the development of the imago in hemimetabolous insects or of the
pupa in holometabolous insects.

It is possible to remove the corpora allata from caterpillars of moths.
Independently of the stage in which the operation is performed the caterpillars proceed to pupate at the next moult. In due course the moth emerges

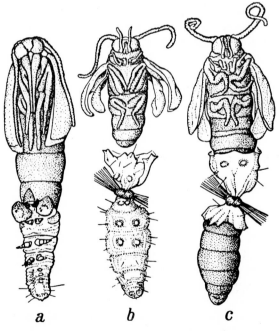

a   *b*   *c*

Fig. 268. Experiment showing necessity of thoracic gland for pupation. *a,* Normal
pupa of the meal moth *Ephestia kühniella. b,* Result of ligating the caterpillar before
the hormone of the thoracic gland has been released: the part posterior to the ligature
remains in the larval state. *c,* Result of ligating the caterpillar after the moulting hormone has been released: both parts pupate. (After Kühn and Piepho, from Kühn,
1955.)

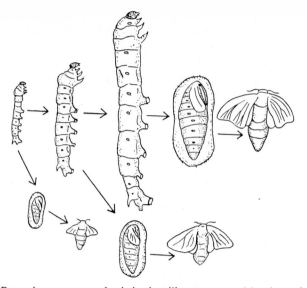

Fig. 269. Precocious metamorphosis in the silkworm, caused by the extirpation of the corpora allata in the third or fourth stage, respectively. The larva pupates normally after the fifth stage. (After Bounhiol, from Abeloos, 1956.)

from the pupa, although it may have reached only a fraction of the normal size (Fig. 269) (Bounhiol, 1937). Apparently the presence of the corpora allata is necessary to **prevent** metamorphosis, to keep the insect in the larval state. Accordingly, the secretion of the corpora allata has been called the **juvenile hormone.** Now at every larval moult the cells of the corpora allata show signs of secretory activity (swelling of cells, appearance and discharge of vacuoles, etc.), but no such activity is present during the pupal or imaginal moult. Accordingly it would seem that at the time metamorphosis takes place the corpora allata do not produce their secretion or at least are less active. That it is actually the absence of the juvenile hormone that is the

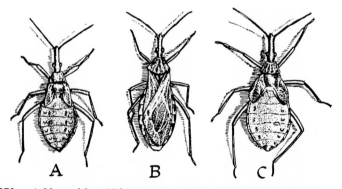

Fig. 270. *A*, Normal last (5th) stage nymph. *B*, Normal adult *Rhodnius*. *C*, Giant 6th stage nymph produced by implanting the corpus allatum from a 4th stage nymph into the abdomen of a 5th stage nymph. (From Wigglesworth, 1939.)

necessary condition for metamorphosis can be proved by implanting corpora allata from a young larva into the last stage larva which should be metamorphosing with the next moult. The larva may moult in due course, but under the influence of the juvenile hormone secreted by the graft, it is not transformed into an imago (in the case of a hemimetabolous insect) but instead produces an abnormally large larva (Fig. 270).

In the case of holometabolous insects the conditions are more complicated inasmuch as there are two moults accompanied by profound morphological changes—the pupal moult and the imaginal moult. The removal of the corpora allata from caterpillars causes the caterpillar to be transformed into a pupa. Some experiments, the details of which cannot be related here, indicate that the subsequent transformation of the pupa into the imago is probably connected with a further decrease of the juvenile hormone in the blood of the insect. After the ablation of the gland small amounts of the juvenile hormone could have still been present in circulation, but would have been used up by the time of the second moult (see Wigglesworth, 1954).

18–7. NATURE OF THE FACTORS CONTROLLING MOULTING AND METAMORPHOSIS IN INSECTS

In the previous paragraph it has been assumed that the agents produced by the thoracic gland and the corpora allata are hormones, that is, chemical substances emitted by the cells and circulating in the body fluids. This could be deduced from the fact that the effect of the gland does not depend on whether it is in its normal position, with all its connections to neighboring organs and to the nervous system intact, or on whether it had been transplanted to an abnormal site. Further evidence in favor of a diffusible substance being the means of action of the gland will now be presented.

An equivalent to the glands' independence of their position is the independence of the position of the reacting organs in the reaction. When moulting or metamorphosis occurs, not only do all parts of the body of the intact animal react together, but transplanted parts also do the same. Imaginal discs and other parts of the body may be transplanted between animals in different developmental stages, and they always moult and metamorphose together with the organs of the host and independently of their own age. A very elegant experiment of this kind carried out on the developing moth, *Ephestia kühniella,* consists in transplanting pieces of skin into the body cavity of another individual (Kühn, 1939; Piepho and Meyer, 1951; Kühn 1955). The edges of the implanted piece of skin curl so as to form a cyst, with the original distal surface of the skin turned inward. The proximal surface of the epidermis is bathed by the body fluids of the host and by the host's hormones if any are present in the body fluids. The necessary conditions are thus provided for the graft epidermis to react to any hormones circulating in the body of the host. It was found that with every moult of the host the cyst epidermis moulted also, discarding the old cuticle into the

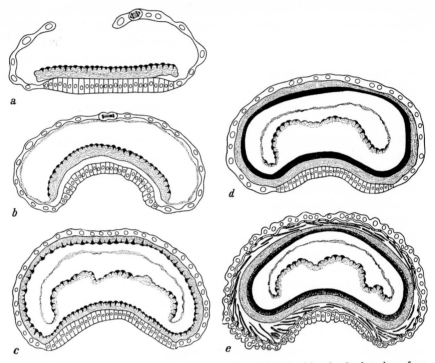

Fig. 271. Hormones in the development of a moth. Moulting in the interior of an implanted epidermal cyst under the influence of hormones circulating in the host's body. *a,* Cyst in the process of formation. *b, c,* Shedding of a larval cuticle. *d,* Production of a pupal cuticle. *e,* Pupal cuticle shed, an imaginal cuticle with scales produced underneath. (From Kühn, 1955.)

cavity of the cyst. Not only was the moulting of the graft simultaneous with that of the host, but the nature of the new cuticle was always the same as that of the host. When a larval moult occurred, the cyst epidermis produced a thin cuticle like the one that covered the body of the caterpillar. When the host pupated, the cyst produced a thick pupal cuticle. When the host metamorphosed into the adult moth, the epidermis of the cyst developed an imaginal cuticle with scales! All the successive cuticles could be seen later one inside the other, on sectioning the cyst (Fig. 271).

Even after reaching the stage of producing the cuticle of the adult moth, the epidermis does not lose its capacity for moulting, provided that the moulting hormone or both moulting and juvenile hormones are present in the surrounding fluid. A cyst which had gone through the pupal and imaginal moults may be excised from the first host and transplanted into a second one. If the second host is a caterpillar, the cyst will undergo a new moult simultaneously with the pupation of the host, will shed the imaginal cuticle with scales and will again produce a thick pupal cuticle. This may be followed, at the time of metamorphosis of the host, by a second imaginal cuticle with a new set of scales (see Fig. 272).

Apparently metamorphosis is fully reversible, at least in regard to the skin epidermis, and the nature of differentiations produced by the latter is solely dependent on the balance of hormones present in the blood. A reversal of metamorphosis, even a partial one, can occur, however, only under experimental conditions. In the normal life of an insect metamorphosis marks the end of morphogenesis and of growth (except for the growth of the gonads, which may continue in the adult). The reason for the cessation of further development is that the thoracic gland degenerates and breaks up after causing the last (imaginal) moult. With the thoracic gland gone, no other factors can reawaken the morphogenetic activity of the epidermis, and no further moulting can occur. This is also an explanation for the existence of a growth limit in insects which was referred to at the end of section 15–3.

In vertebrates the chemical nature of the agents emitted by the endocrine glands has been proved by preparing active extracts from the glands, containing chemically definable substances. The extraction of hormones from insects is much more difficult owing to the small mass of the endocrine glands. However, some success can be noted in this field also. The moulting hormone (the hormone of the thoracic gland) has been prepared in a chemically pure form by Butenandt and Karlson (1954). Five hundred kilograms of silkworm pupae were used to isolate 25 mg. of the pure substance, and 0.0075 mg. of this material was sufficient to cause the pupation of a fly larva. The chemical analysis of the substance gave the empirical formula $C_{18}H_{30}O_4$. The structural formula is not yet known.

The mode of action of the moulting hormone and the juvenile hormone has not yet been elucidated. Some indication of the physiological action of the former may be inferred from studying the metabolism of moth pupae

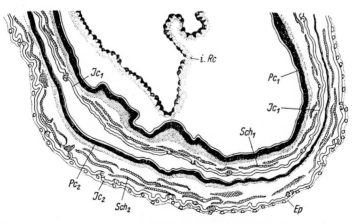

Fig. 272.   Control of metamorphosis by hormones in a moth (*Galleria melonella*). Reversal of metamorphosis in an epidermal cyst transplanted into a second, younger host. The consecutive cuticles are: larval (*i. Rc*); pupal (*Pc₁*); imaginal (*Ic₁*) with scales (*Sch₁*); then again pupal (*Pc₂*); and, lastly, imaginal for a second time (*Ic₂*), with a second generation of scales (*Sch₂*). *Ep*, Epidermis. (After Piepho and Meyer, from Kühn, 1955.)

in diapause and comparing them with the metabolism of activated pupae (the activation, as will be remembered, consists in the last instance in the release of the moulting hormone). At the beginning of diapause, respiration of the pupae greatly decreases and remains very low. After reactivation of the pupa oxygen consumption rises again. A similar curve is also given in the pupal stage of other insects, which do not require an external stimulus to proceed with their development. It is of some interest that not only does the amount of consumed oxygen change, but the mechanism of respiration seems to change also. Normally respiration of insects is dependent on cytochrome-c-oxidase and therefore it is very sensitive to metabolic poisons such as cyanide or carbon monoxide. The residual respiration in the middle of the pupal stage or in diapause, however, is only slightly sensitive to cyanide or carbon monoxide. Presumably this oxidation is dependent on the flavo-protein enzymes. It may be concluded that one of the aspects of the action of the moulting hormone is the production or activation of cytochrome-c-oxidase, which enables processes of metabolism to proceed at a higher rate with utilization of atmospheric oxygen. This conclusion is of interest as it falls in line with our previous statement that differentiation operates through the production of specific enzymatic patterns.

As to the juvenile hormone, it is known to increase metabolic rate. In this connection it might be mentioned that the corpora allata resume their activity in the imago after metamorphosis, and are found to be necessary for the growth of the oocytes in the ovaries.

18–8   FINAL REMARKS ON METAMORPHOSIS

The metamorphosis of amphibians and insects is an excellent example of the control of morphogenetic processes by hormones—a subject touched upon in section 16-4. The dependence of differentiation on diffusible chemical substances, as revealed in studies on metamorphosis, should be compared with the results of experiments on the influence of diffusible substances on the differentiation of cells in tissue cultures (section 14–4).

When comparing the interrelation of different hormones in insect and amphibian metamorphosis, one is astonished to find that these two groups of animals have developed causative mechanisms having a distinct general similarity. In both cases the transformation is initiated by a secreting organ closely associated with the brain: the hypophysis in amphibians, the neuro-secretory cells of the protocerebrum in insects. In both cases the secretion of this primary center does not act on the tissues directly, but stimulates the activity of a second endocrine gland: the thyroid gland in amphibians, the thoracic gland in insects. The juvenile hormone of insects has, however, no counterpart in amphibians: there is no hormone checking or preventing precocious development, or at least we are not aware of the existence of such an agent.

There is very little information as to the causative agents of metamorphosis in animals other than amphibians and insects, and we do not know whether their transformations are controlled by hormones. One would expect that at least in the cyclostomes the transformation of the larva (Ammocoetes) into the adult should be caused by hormones, but this has not so far been proved.

In the ascidian tadpole the absorption of the tail is in some way dependent on the anterior end of the body, since cutting off the anterior tip with the adhesive papillae will prevent the necrotization of the tail (Oka, 1943). Treating ascidian tadpoles with thyroid hormone accelerated the metamorphosis (Weiss, 1928), but this action is hardly specific, as the tadpole does not have a thyroid gland of its own, and treatment of the tadpoles with narcotics or even with distilled water has the same effect (Oka, 1958).

As to the metamorphosis of the larvae in the numerous groups of invertebrates it is not possible even to make any conjectures. The field is as yet completely unexplored.

CHAPTER **19**

Regeneration

The second mode of reawakening of the morphogenetic processes at an advanced stage of the ontogenetic cycle is by means of partial destruction of the system which has evolved as a result of the previous development. The animal organism possesses the ability to repair more or less extensive damage incurred by the body either accidentally in natural conditions or wilfully imposed by the experimenter. The damage repaired may be a wound which severs or partially destroys the tissues of the animal's body, or the damage may involve the loss of an organ or larger part of the body. These can sometimes be renewed, and in this case the process of repair is known as **regeneration.**

19–1   TYPICAL CASE OF REGENERATION: THE RENEWAL OF A LIMB IN A SALAMANDER

The limbs of newts and salamanders, both in the adult stage and in larvae, are capable of regeneration to a very high degree and have often been used for various experiments on the subject. Limbs, especially those of larvae, are sometimes bitten off by other members of the same species, or the limbs may be cut off at different levels for research purposes. Whatever the cause of the loss of a limb, the first stage of repair of the damage is that the epidermis from the edges of the wound starts spreading over the wound and very soon covers the open surface. The closing of the wound is relatively a very rapid process, and it is accomplished in the course of one or two days, depending on the size of the animal. During the next few days the epidermis covering the wound begins to bulge outward, becoming more or less conical in shape (Fig. 273). A mass of cells accumulates under the epidermis. These cells are in a state of active proliferation, and together with the epidermal covering they form what is known as the **regeneration blastema** or **regeneration bud.** The blastema grows rapidly, at first retaining its conical shape, but later it begins to be flattened dorsoventrally at the end. The flattened part is the rudiment of the carpus or tarsus, called the hand- or footplate, depending whether a forelimb or a hindlimb is regenerating. Soon the rudiments of the digits appear, separated by slight indentations at the edge of the plate (Fig. 273). In the meantime the mass of cells in the interior of the limb-bud becomes segregated to form the rudiments of the internal parts of the limb—the various bones of the limb skeleton and of the arm (or leg) muscles. These then undergo histological differentiation. The rudiments of the digits elongate, and the whole regenerating limb continues to grow until it attains the size of a normal limb. As the tissues of the regenerating limb differentiate, the limb resumes normal function (movement). Eventually the regenerated limb becomes completely indistinguishable from a normal limb. The time necessary for the completion of the process depends on the size and stage of development of the animal; regeneration is most rapid in small larvae, and here the regenerated limb may be complete after three weeks (in the case of an axolotl larva). In older and larger larvae the process may take a longer time, and in the adult salamander the completion of regeneration takes several months.

It is seen that the new limb is produced from a rudiment, the regeneration blastema, which, when formed, acquires the potentiality for development otherwise found only in the organ rudiments of the early embryo. There is the obvious difference, however, that the cells of the regeneration blastema develop into parts of one organ only (the limb), whereas the cells of the early embryo produce the whole animal. Reversion to a state somewhat similar to that of early embryonic cells obviously takes place in the formation of a regeneration blastema. How far this "rejuvenation" goes will be discussed later.

What is developed by regeneration supplements the residue of the animal's body (in our case by far the major part). The regenerated organ is thus an addition to the other parts. A type of regeneration exemplified by the renewal of limbs in salamanders is therefore called **epimorphosis** or **epimorphic regeneration** (Morgan, 1901). It is the common type found in higher animals.

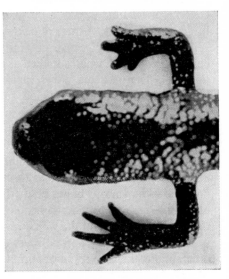

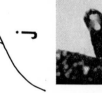

Fig. 273. Regeneration in the limbs of a newt, *Triturus cristatus. a, b, c, d, e, f,* Consecutive stages of regeneration of a forelimb amputated above the elbow. *g, h, i, j, k, l, m,* Stages of regeneration of a hindlimb amputated above the knee. (From Schwidefsky, 1934.)

19–2   REGENERATIVE ABILITY IN VARIOUS ANIMALS

Although present throughout the whole of the animal kingdom, the ability to regenerate lost parts differs both in scope and in its course in the various groups of animals.

In the coelenterates the regenerative ability is exceedingly high. It was the regeneration of the fresh water polyp, *Hydra,* first discovered by Trembley in 1740, that attracted the scientists' attention to this phenomenon. A hydra may be cut in two or more parts and each part will reconstitute itself into a new and complete individual of diminished size. The posterior end of a cut hydra regenerates the mouth and tentacles, the anterior part of the body regenerates the posterior end with the foot and adhesive disc. Even small sections of the body, comprising as little as 1/200 part of the original individual, can regenerate a complete whole. In the latter cases it is especially clear that the new individual is produced not by addition of parts to the remaining piece (epimorphosis) but by remodeling the whole available mass of cells into a new whole. This thus differs from the epimorphic regeneration of an amphibian limb. A type of repair involving a reorganization of the remaining part of the body of an animal is known as **morphallaxis,** or **morphallactic regeneration.** If only smaller parts are removed from the hydra, the remodeling of the remaining body is not so extensive, and with very small defects the regeneration approaches the epimorphic type, even though a typical regeneration bud cannot be clearly distinguished.

Other coelenterates regenerate to varying extents, the regenerative power being highest in the polypoid forms and much reduced in the medusoid forms.

The next group remarkable for their high regenerative ability are the planarians. Planarians may be cut across or lengthwise, and each part of the body will regenerate the missing half. Any part of the body may be replaced in this way: the head, the tail or the middle part with the pharynx. When the cut is made a regeneration blastema is formed at the cut surface, and the missing part is developed from the blastema. The remaining part is, however, reorganized on a diminished scale, so that the individual resulting from regeneration is smaller than the original one. The regeneration is thus carried out in a way that combines epimorphosis and morphallaxis. Other platyhelminths do not regenerate to any great extent.

The nemerteans have a high regenerative ability, and a complete worm may be formed even starting from very small fragments.

Many experiments have been carried out on regeneration in the annelids. Both polychaetes and oligochaetes regenerate anterior and posterior ends after an amputation. If an earthworm or other oligochaete is cut in two halves, the posterior half regenerates the anterior end with the mouth, and the anterior end regenerates a new posterior end. Two new individuals may thus be produced from the original one. In the majority of the annelids regeneration is, however, somewhat restricted; at the anterior cut surface only a limited number of segments are formed, the number being typical for every species (see Berrill, 1952). In the earthworm, *Allolobophora foetida,* this number is four or five. If five segments or less are cut off from the anterior end of the worm the regeneration is complete; if more than five segments are removed, only four or five segments will be regenerated, and the worm will thus not attain the same over-all number of segments as it had before the operation. If the cut is carried out behind the genital segments (tenth to fourteenth) only four or five anterior segments are regenerated; the genital organs are thus never renewed. On the other hand there is no restriction in the regeneration of posterior segments, and about as many are formed as have been removed. The process of regeneration is an epimorphosis, a regeneration bud being formed and the new parts developing at the expense of this bud.

In molluscs the regeneration is relatively poor. In gastropods eye stalks with eyes may be regenerated as well as parts of the head or of the foot. The whole head does not regenerate, and if the cerebral ganglia are removed together with a part of the head these will not regenerate. The arms of cephalopods may regenerate but not other parts of their body.

In nematodes the regenerative ability is very low; this may be connected with the high degree of differentiation of the cells of their bodies and the fixed limit to the total number of cells (*cf.* section 15–2). Only the closure of superficial wounds is still possible.

In arthropods regeneration is limited to the renewal of lost appendages, but this form of regeneration is fairly widespread. In most crustaceans the limbs may regenerate at any stage of development including the adult. In insects limb regeneration occurs only in the larval stages, and the regenerated limb often does not reach the size of a normal limb. The legs of crabs and some spiders are readily shed if seized by an enemy (or the experimenter). The legs break off at a preformed breaking point, across the second leg joint. There is here a constriction which is a modified joint. The leg is broken off by the violent contraction of a muscle (the extensor muscle of the leg). This self-mutilation is known as **autotomy.** Autotomy is probably a special adaptation helping the animal to escape being caught by a predator; if the predator gets hold of one limb, he succeeds in capturing only the limb but not the animal, the latter escaping at the expense of the loss of a limb.

After the amputation or loss of an appendage in an arthropod, the wound is covered by a chitinous plug. Underneath this a regeneration bud is formed which later reproduces the limb by way of epimorphic regeneration. The new limb does not, however, become apparent until the next moult. The regenerated limb is small at first, and attains normal proportion as a result of accelerated growth in the course of several moults.

Among the echinoderms, the starfishes, brittle stars and sea lilies can regenerate arms and parts of the disc. The arms appear to be lost rather often in natural conditions, as individuals regenerating one or more arms are found quite frequently. The Holothuroidea are capable of ejecting through the anus parts of their internal organs—the respiratory tree and the alimentary canal. These can be regenerated later.

In vertebrates the regenerative power is most spectacular in the urodele amphibians. In newts and salamanders, and especially in their larvae, not only limbs can regenerate (as described above) but also tails and external gills, and furthermore the upper and lower jaws. Parts of the eye can likewise regenerate, such as the lens and the retina. If most of the eye is removed it can be regenerated so long as a small part is left, and the new eye develops at the expense of the remaining fragment. In the anuran amphibians the faculty for regeneration is restricted to the larval stage of their development. Frog and toad tadpoles are able to regenerate their limbs and their tails. The legs of adult frogs and toads do not normally regenerate at all.

In fishes regeneration is very restricted. The fins can regenerate if cut off or damaged, but the tail (apart from the tailfin) does not regenerate.

The lizards are known to regenerate their tails. This regeneration follows autotomy. The tails are broken off at a preformed level near the base of the tail. To release the mechanism of autotomy the distal part of the tail must be injured or grabbed with such force as to cause the animal discomfort. After the tail is shed, a regeneration bud is formed on the wound surface, and this gives rise to a new tail. The latter, however, differs from the original tail; the vertebral column is of a simplified structure, and the scales covering the regenerated tail differ from the normal ones. The legs and even the digits in the same animals cannot regenerate completely, though very rudimentary structures are sometimes formed on the site of an amputation. Parts of the beak can be regenerated in birds, but otherwise their regenerative ability is rather poor.

In the mammals the regenerative ability is reduced to tissue regeneration, that is, the restoration of defects and lesions in various tissues, but not the restoration of lost organs. Tissue regeneration is often equivalent to wound healing. Thus skin wounds may be covered by newly formed epidermis and connective tissue. In the case of larger skin wounds, however, the newly formed connective tissue differs from the normal dermis and can be distinguished as scar tissue. The skeletal tissue has a high regenerative ability. Large defects in the bones, especially those of the limbs, can be made good. If parts of individual muscles are removed, the defects can be repaired by a proliferation taking place at the expense of the remaining part of the muscle. Lesions in the tendons can be replaced by connective tissue in the first place, the connective tissue later acquiring the structure and mechanical properties of the tendon. The tissues of some internal organs are able to proliferate to a very great extent and thus compensate for the loss of large parts of the organ. This is found to a striking degree in the case of the liver. The greater part of this organ may be removed; the remaining part then proliferates and restores the normal mass of liver parenchyme. The newly proliferated tissue does not, however, assume exactly the same shape as the liver before the operation.

From this survey it will be seen that on the whole representatives of the lower forms of animal life regenerate better—may restore their normal structure from smaller parts of the original individual than is possible for more highly organized animals. The rule is, however, not without numerous exceptions. Some exceptions may be explained by the high degree of histological differentiation in animals otherwise standing low in the scale of the animal kingdom (viz., the nematodes), a high degree of histological differentiation being antagonistic to the ability of cells to proliferate. In other cases this explanation does not hold; it does not tell us, for instance, why the fishes regenerate worse than the amphibians, although the latter are supposed to be derived from the former in the course of evolution.

With one and the same species of animal, the ability to regenerate may be greater in the earlier stages of the ontogenetic cycle. Legs may be regenerated in tadpoles but not in adult frogs. Adult insects cannot regenerate legs although the larvae or nymphs are capable of regeneration. It may be recalled in this connection that parts of early embryos are sometimes able to develop into whole animals whereas corresponding parts at a later stage are not even viable, as when the cleavage stage of an amphibian egg is compared with an adult amphibian, or even with an embryo after the end of gastrulation (see Chapter 6). It may be concluded that a renewal of the morphogenetic processes in regeneration occurs more easily if normal morphogenesis is still under way. If the morphogenetic processes of normal ontogenesis have come to a complete standstill, it is more difficult or even impossible to renew them again.

19–3   STIMULATION AND SUPPRESSION OF REGENERATION

Although the general level of regenerative ability of an animal is determined by its constitution, by the degree of differentiation of its tissues or the stages of ontogenetic development, and other factors, this ability may be increased or diminished by the environment or by the special treatment to which the animal is exposed.

The rate of regeneration is naturally dependent on temperature, as most biological processes are. Increase of temperature, up to a certain point, accelerates regeneration.

In *Planaria torva* regeneration is scarcely possible at a temperature of 3° C. Of six individuals kept at this temperature only one regenerated a head, and this was defective; the eyes and brain were not fully differentiated after six months. Regeneration was most rapid at 29.7° C.; at this temperature new heads developed in 4.6 days. A temperature of 31.5° C. was too high, and the heads regenerated after 8.5 days. A temperature of 32° C. proved to be lethal for the animals (after Lillie and Knowlton, from Morgan, 1901).

Food, on the other hand, does not affect regeneration very much. Even a fasting animal will regenerate at the expense of its own internal resources. In such diverse cases as rats regenerating parts of the liver, salamanders regenerating limbs or hydras and planarians regenerating parts of their

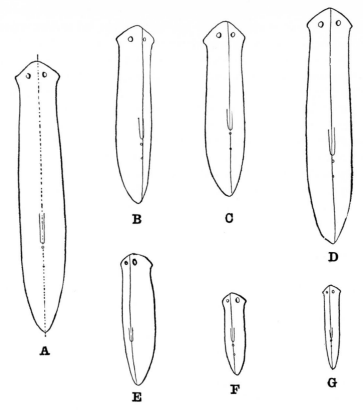

Fig. 274. Regeneration of a planarian which was cut in two, lengthwise (*A*). The left half was fed (*B, C, D*); the right half was kept without food (*E, F, G*). (From Morgan, 1901.)

body, depriving the animals of food does not prevent regeneration and may even accelerate it to a certain extent. If planarians are deprived of food for a long time, they can live by metabolizing constituents of their own body. The animal, of course, diminishes in size in consequence (degrowth, Needham, J., 1942). In this state a planarian can still regenerate. Although the over-all size decreases, the missing parts are gradually rebuilt so that a complete, even if very small, worm is eventually developed. Figure 274 shows the result of an experiment (Morgan, 1901) in which *Planaria lugubris* was cut lengthwise, and while the left half was fed, the right half was left entirely without food. Both regenerated. It appears, therefore, that regeneration is given top priority in the utilization of resources available to the organism. Although restriction of feeding seems to be favorable for regeneration, if anything, extreme degrees of emaciation by starving prevent regeneration except in organisms such as the planarian which is able to utilize its own body as a source of energy without deleterious results.

From both theoretical and practical viewpoints it would be very important to know whether it is possible to excite regeneration in organisms or their parts which do not normally regenerate. The legs of tailless amphibia are a very suitable object for experiments in this direction, since they regenerate in tadpoles but do not do so in adult frogs. The ability to regenerate disappears in the legs of tadpoles some time before metamorphosis sets in, at a stage when all the digits have been formed and the skeleton of the limb is in a state of chondrification (Polezhayev, 1946). At a later stage, when the cartilaginous skeleton of the limb is fully differentiated and the limb becomes bent in the knee joint, an amputated leg does not regenerate any more, but the wound is covered with skin and no regeneration bud develops. A regeneration bud may, however, be caused to be formed if after amputation the stump is traumatized by sticking a needle into it several times. This additional stimulus is sufficient to get the process of regeneration under way, and the regeneration bud subsequently develops into a limb in the usual manner (Polezhayev, 1946). In the adult frog traumatization with a needle does not appear to be sufficient to incite regeneration, but treatment of the wound after amputation of the leg with a hypertonic salt solution may cause regeneration (Fig. 275) (Rose, 1942). In all cases the essence of the treatment used for inciting regeneration is to increase the destruction of tissue beyond that caused by the amputation. The resulting dedifferentiation of tissues favors the formation of a regeneration bud (*cf.* sections 19–4 and 19–5).

The experiments on regeneration of legs in older tadpoles and adult frogs are very promising and give some hope of the possibility of inciting the regeneration of limbs in other animals (such as mammals) in which they do not normally regenerate. It must be remembered, however, that the legs of frogs present a special case, inasmuch as here we have an organ that normally regenerates during an earlier ontogenetic period.

The method applied to the legs of older tadpoles, i.e., traumatization by pricking with a needle, has been used to stimulate regeneration in the hydra (Tokin and Gorbunowa, 1934). Although a hydra is a paragon of high regenerative ability, the aboral tip of its body, that is, the foot with the adhe-

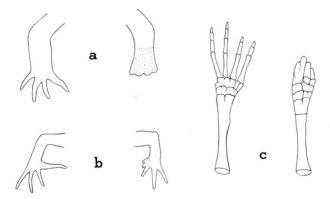

Fig. 275. Regeneration of limbs in the frog after chemical stimulation. *a*, In a one-year-old frog. *b*, In a frog one month after metamorphosis. Regenerated parts stippled. *c*, Skeletons of a normal and a regenerated forelimb. (From Rose, 1942.)

sive disc, if severed from the rest of the body does not normally regenerate a new hydra, but remains unchanged until it dies. Traumatization of an isolated foot in the manner described succeeded in causing it to reorganize itself into a complete new individual.

If the attempts to provoke regeneration where it does not naturally occur have been successful only in a few special cases, we possess some universally effective methods of preventing or inhibiting regeneration. One of the methods is by irradiating tissues with X-rays. If a leg or an arm of an adult newt is amputated, it readily regenerates. If, however, a newly formed regeneration bud of a limb is irradiated with X-rays, the development of the limb is either retarded or stopped altogether, depending on the amount of irradiation administered. There is some individual variation, but a dose of 5000 to 7000 r (r = roentgen, the unit of dosage of X-rays) suppresses regeneration in every case. An irradiated regeneration bud does not grow, and instead of developing the skeleton and muscle of the limb becomes filled with connective tissue. The bud may actually decrease in size, and become partially or wholly resorbed (Brunst and Scheremetjewa, 1933; Butler, 1933). The mechanism of the action of X-rays on the regeneration bud is obviously connected with the action of the rays on mitosis. X-rays are known to inhibit or suppress the mitotic activity of cells. If the cells of the regeneration bud cannot divide by mitosis, the growth of the rudiment becomes impossible; the number of cells available remains too small for the development and differentiation of the organ.

It is still more remarkable that the irradiation of a limb with X-rays may precede the amputation of the limb by months, and nevertheless make the regeneration impossible. A dose of 7000 r applied to a normal limb of an adult newt does not usually cause any visible effect. Neither in appearance nor in its function does an irradiated limb differ from a non-irradiated one. The treated animals may be kept for months without showing any deleterious effect of the irradiation. If, however, an irradiated limb is amputated, regeneration is completely lacking and the wound is instead covered by skin, leaving a permanent stump (Brunst, 1950).

It has been found that X-rays have the same effect on regeneration in all animals in which this effect has been tested. The effect is thus on some very fundamental property of living cells.

19–4 HISTOLOGICAL PROCESSES CONCERNED IN REGENERATION

Regeneration involves a complicated sequence of histological transformations in the stump of the amputated organ and later in the regeneration blastema.

The immediate effect of an amputation is that tissues and cells lying normally in the interior of the body emerge on the surface. Some of the cells are squashed, torn or otherwise destroyed, others become damaged by exposure to an unfavorable environment. The surface of the wound is thus covered with the debris of dead cells. In animals with a developed blood vessel system the blood from the damaged vessels flows onto the surface of the wound and there coagulates, thus stopping further loss of blood.

The next stage is the covering of the wound surface with epithelium. Skin epithelium spreads over the wound surface, penetrating underneath the blood clot, between it and the intact living connective tissue. The spread of the epithelium is due to ameboid movement of the cells, and does not involve growth at the edges of the wound. No mitoses are found in the epithelium at this time. The time needed for the epithelium to cover the wound surface depends on the size of the regenerating animal and the size of the wound, besides such external factors as temperature and others. In the salamander larva after the amputation of a leg the wound becomes closed by the epithelium in about one or two days. In invertebrates the closure of the wound may be assisted by the contraction of the subepidermal muscle layer, so that the surface which has to be covered by the epidermis is diminished.

After the closure of the wound a very important step in the process of regeneration sets in: dedifferentiation of the tissues adjoining the cut surface. This dedifferentiation has been best studied in vertebrates and is here perhaps most pronounced. Dedifferentiation proceeds in conformity with what has been said on this subject in section 14–4. The intercellular matrix of bone and cartilage becomes dissolved, and the cells come to lie freely under the epithelium which has covered the wound. The connective tissue fibers likewise disintegrate, and the connective tissue cells become indistinguishable morphologically from cells derived from the disintegration of the skeletal tissues. The muscles also undergo dedifferentiation, the myofibrils disappear, and the nucleo-cytoplasmic ratio greatly increases. The cells of all tissues become similar in appearance to embryonic cells. Whether this similarity is a superficial one or whether the cells acquire the properties of embryonic cells in every respect is a different matter, and will be dealt with in section 19–8.

The formation of the blastema or regeneration bud is the next step. Undifferentiated cells accumulate under the epidermis covering the wound, and together with it they form the regeneration bud, mentioned in section 19–1. There has been much argument as to the origin of the cells of the blastema. Two main theories on this subject have been proposed. According to one theory, the cells of the blastema are all of local origin, that is, derived from the tissues immediately adjoining the wound surface. Except for the epithelium covering the wound, the rest of the cells, according to this theory, are set free from the dedifferentiating connective tissue, skeleton, and so on. According to a second theory, the differentiated cells at the wound level have nothing to do with the formation of the blastema (except for the epithelium, which is obviously derived from the adjoining intact parts of the skin). The blastema is supposed to be derived from cells migrating to the regeneration site from more or less distant parts of the body by amoeboid movement, or brought there with the blood stream. These cells are supposed to be special "reserve" cells, which still possess a capacity for development that has been

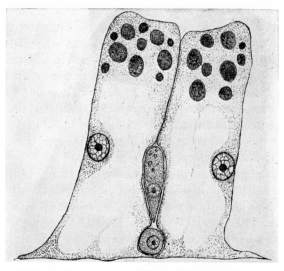

Fig. 276.   Interstitial cells between two differentiated epitheliomuscular endodermal cells in *Hydra*. (From McConnell, 1936.)

lost by the differentiated tissue cells. It is now known that neither of these two theories has universal application, but that both sources of cells of the blastema may occur in different animals. The problem thus resolves itself into finding what the local tissues contribute to the formation of the regeneration blastema and what is contributed by migratory cells in any given animal.

In the hydra the regeneration is not dependent on the tissues adjoining the cut surface, but it is performed by a special type of cell, scattered normally over most of the body and accumulating in the region of the wound. These cells are the "interstitial cells," so called because they are found lying in the intercellular spaces at the base of the ectodermal epithelium, and also in smaller numbers between the cells of the gut epithelium (Fig. 276). The cells are small with relatively very large nuclei (as in most undifferentiated cells) and basophilic cytoplasm. The part played by the interstitial cells in the life of the hydra is not restricted to processes of regeneration. The cnidoblasts, which serve for the immobilization and capture of the small animals on which the hydra feeds, can be used but once; that is, these cells degenerate after once having thrown out their thread. They must therefore be continuously replaced by new ones. The new cnidoblasts develop from interstitial cells. As the cnidoblasts are especially numerous on the tentacles of the hydra and are being used there most often, a regular stream of migrating interstitial cells serves to supply the tentacles. The interstitial cells are also the source of material for asexual reproduction of the hydra by budding (section 20–2), and lastly, the gametes (ova and spermatozoa) are derived from these same cells.

When a wound is inflicted on a hydra, it is first of all closed by the edges of the wound approximating to each other. Both ectoderm and endoderm participate in the initial closure of the wound. Next the interstitial cells start migrating and infiltrate the damaged region. Mitoses are observed in the interstitial cells even in remote parts of the body, and thus the supply of interstitial cells is increased. Eventually the masses of interstitial cells which have accumulated at the site of the wound build up whatever parts are missing.

In the planarians the processes of regeneration take a somewhat similar course. The wound is initially covered by the epidermis of the skin. The cells move tangentially over the wound surface without proliferating. This occurs in the first 24 hours after the infliction of the wound. The replacement of the lost parts occurs later at the expense of undifferentiated cells derived from the parenchyme (mesenchyme) of the worm. Whether these cells are "reserve" cells, as in the hydra, or are constituent cells of the normal connective tissue, which undergo dedifferentiation and become capable of participating in new morphogenetic processes, is not known for certain. They form the blastema under the sheet of epithelium, proliferate and later differentiate into the various organs (Fig. 277). The blastema, and so in the last instance the migrating parenchyme cells, give rise to most of the regenerating organs: the connective tissue, the pharynx, the nervous system, as well as the muscle of the body. The skin epithelium is derived from the skin at the edge of the wound, and also, it is claimed, the intestine in the regenerated part is derived from the cut edge of the old intestine (Bandier, 1937).

That the blastema is formed by cells migrating from remote parts of the body is definitely proved by the following experiments. If a planarian is irradiated with X-rays prior to wounding, no regeneration occurs, as we

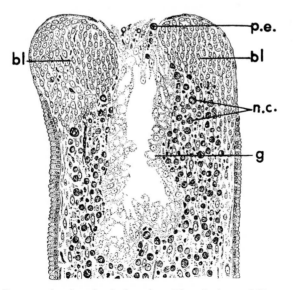

Fig. 277. Regeneration in a land planarian, *Rhynchodemus bilineatus. bl,* Regeneration blastema; *g,* gut; *n.c.,* necrotic cells; *p.e.,* proliferating endoderm cells. (From Bandier, 1937.)

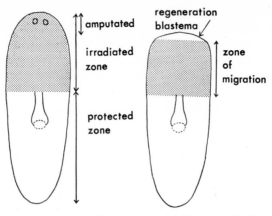

Fig. 278. Regeneration in a planarian after partial X-ray irradiation. (From Dubois and Wolff, 1947.)

know irradiation with X-rays makes the cells incapable of participating in the process of regeneration (see section 19–3). If, however, a small part of the worm is shielded from X-rays and then a part of its body is amputated, regeneration will proceed even if the wound is far away from the region which had not been irradiated. This can only mean that cells which had not been exposed to X-rays are capable of migrating long distances and then forming a healthy blastema on the cut surface (Fig. 278) (Dubois and Wolff, 1947).

In the annelids there is a special type of cell called neoblasts which serve for the formation of regenerating parts. The neoblasts are large cells found normally in the peritoneum of the posterior surface of the intersegmental septa. When the body of the worm is cut the neoblasts become activated and migrate toward the cut surface, following the course of the ventral nerve chain. The neoblasts accumulate under the epidermis which closes the wound and thus form the regeneration blastema. Not all the organs of the regenerated part are developed from the neoblasts, however, but only the mesodermal organs. The nervous system of the regenerated part is developed from the epidermis (from a thickening on its inner surface). Also the gut regenerates from the cut surface of the old gut. The competence of the neoblasts is thus more restricted than that of the mesenchyme cells in a planarian.

After treatment with X-rays the neoblasts are the first to suffer damage and degenerate, and they may be destroyed with weaker doses of the rays, which leave the other tissues intact. Irradiated worms will not regenerate (Zhinkin, 1934). Epidermis and endoderm of the gut, which normally contribute to the formation of the regenerate, are not visibly damaged by X-rays, but they do not produce new parts in the absence of neoblasts. This may be either a result of the failure of some stimulating action normally exercised by the neoblasts or a result of direct damage by the X-rays, making these tissues incapable of proliferation even if this does not interfere with their normal activities.

The origin of the cells forming the regeneration bud in vertebrates is naturally of special interest to us. This has been most exhaustively investigated in the case of regenerating legs of newts and salamanders. A study of microscopical sections of regenerating stumps of salamander legs does not allow of an unequivocal solution to the problem. Whereas a dedifferentiation of tissues adjoining the wound can easily be observed, it is very difficult to prove that the cells from dedifferentiating tissues actually form the regenerating blastema. This leaves the way open to those who would like to claim that the regeneration blastema has a different origin, that the cells of the blastema are brought to the site of regeneration with the blood stream. The question has been settled, however, by the application of local irradiation with X-rays (Butler, 1935; Brunst, 1950). We have seen earlier (section 19–3) that irradiation with sufficiently high dosage of X-rays suppresses the ability of salamanders to regenerate. A modification of this basic experiment could be used for the solution of the problem under discussion.

Two types of experiment have been applied. In the first experiment the whole animal was shielded from the X-rays (with a sheet of lead) and only one leg was exposed to the rays. The leg was then amputated inside the irradiated area. There was no regeneration (Brunst and Chérémétieva, 1936). In another experiment the whole animal was irradiated and an untreated leg was transplanted onto the irradiated body. Subsequently the leg was amputated (inside the untreated area) and it regenerated normally (Butler, 1935). It is thus obvious that the irradiation does not interfere with the powers for the regeneration of the animal as a whole, but only affects the cells which are directly exposed to the rays. If the regeneration cells had been brought to the site of regeneration from other parts of the body, the local irradiation of only one leg should not have prevented its regeneration. By careful application of the lead screen, it could be proved that even a very thin layer of normal (unirradiated) tissue is sufficient to ensure a normal regeneration of the leg (Fig. 279) (Scheremetjewa and Brunst, 1938).

The irradiation of hydrae or annelids in the first instance causes the degeneration of the interstitial cells or the neoblasts, and the ability to regenerate disappears as a consequence. It may be questioned in the case of newts and salamanders whether there is a special type of cell which is destroyed when the animals are made incapable of regeneration by the action of X-rays. The answer to this question should probably be a negative one; it is impossible to find by histological methods any cells in the tissues of the newt or salamander which could be identified as "reserve cells" or "undifferentiated cells."

At the time when the regeneration blastema is being formed, be this by way of migration of "reserve" cells from remote parts of the body or by dedifferentiation of local tissues, the mitotic activity of the cells is very low. While the cells dedifferentiate or migrate they do not divide. Once the blastema has been formed, however, numerous mitoses are found in it. In the

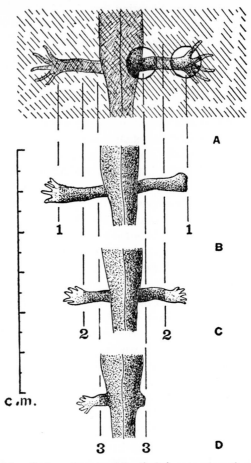

Fig. 279. Local irradiation of parts of the limb in a newt, and results of amputation at different levels in the irradiated limb (right) and in control limb (left). Drawing above (*A*) shows shielding plate covering the whole animal except for two circular openings through which the rays could pass. The limb fails to regenerate when amputated inside irradiated areas. (From Brunst, 1950.)

newt *Triturus pyrrhogaster* the formation of the blastema takes about two weeks, during which no mitoses are found. Then proliferation sets in rather suddenly (Litwiller, 1939). For a time the divisions of the cells outstrip growth, and this leads to a slight diminution of the size of the cells in the blastema. Later growth catches up with mitosis and both go hand in hand so that a very distinct period of growth follows. The growth rate is at its highest immediately after the blastema has been formed, and subsequently the rate of growth diminishes inversely proportionally to the time that has passed since the beginning of growth (Syngajewskaja, 1936). As the growth diminishes differentiation sets in; thus the inverse relationship between growth and differentiation holds good for regeneration as well as for normal development.

The reservation must be made that a spurt of mitotic activity and growth is not observed where regeneration is by way of morphallaxis, as in hydra. Although the interstitial cells are capable of mitotic division, the mitoses are spread diffusely throughout the whole body of the animal and not concentrated at the site of the wound (Kanajew, 1926).

The processes of histological differentiation taking place in the regenerating parts are as a rule similar to those occurring in normal ontogenetic development and need not be specially described here. In special cases, however, the processes may differ in two respects.

1. Certain tissues in regeneration may be developed from a source different from that in normal ontogenesis. An example of this has been mentioned previously; the regeneration cells of a planarian, which are derived from the parenchyme (mesenchyme) and are thus of a mesodermal nature, give rise in regeneration not only to the parenchyme and the muscles, but also to the nervous system and the pharynx. In normal development from the egg the nervous system and the pharynx are produced by the ectoderm. Similar cases are known to occur in some other animals.

2. Sometimes the regenerated structure is formed in a way different from the normal one. The notochord is the predecessor of the vertebral column during the normal development of a salamander larva. In an adult salamander the notochord is reduced and replaced by the bony vertebrae and it does not reappear in a regenerating tail. During regeneration the cartilages representing the rudiments of the vertebrae are formed around the spinal cord instead. The normal segmentation of the vertebral column is not restored in the regenerated tail of a lizard.

19–5   PHYSIOLOGICAL PROCESSES INVOLVED IN REGENERATION

The physiological state of regenerating animals allows us to distinguish two periods with a different type of metabolism: predominantly destructive metabolism in the first period and predominantly constructive metabolism in the second period. The period of destructive metabolism is naturally of greater significance in those animals in which there is considerable dedifferentiation at the beginning, as in amphibian regeneration. Most prominent is the abrupt increase in the activity of proteolytic enzymes. If a normal healthy tadpole tail is subjected to autolysis, a part of its own protein becomes decomposed by the action of the proteolytic enzymes contained in its tissues. This part can be measured by the amount of non-protein nitrogen produced in this way. The actual quantity observed in an experiment of this kind was 58 mg. per 100 gm. in 12 hours. A stump of an amputated tail under the same conditions produced 233 mg. per 100 gm. of non-protein nitrogen. In the stump of the amputated tail, which undergoes dedifferentiation, the activity (or quantity) of proteolytic enzymes was thus four times as great as in the normal tissue (Bromley and Orechowitsch, 1934).

The proteolytic enzyme that is mainly responsible for this increase is kathepsin. A direct measurement of the kathepsin activity, with gelatin as substrate, showed a rapid increase of the activity in the very first days after amputation. A maximum, which was more than twice the normal level, was reached three days after the amputation (Orechowitsch, Bromley and Kozmina, 1935). After this the activity of the kathepsin slowly decreased, reaching the normal level by the twelfth day. The activity of the dipeptidases was also found to have increased sharply during the first few days after the amputation, but the increase is of a longer duration, and the maximum of the increase is reached only by the eleventh day.

Through the activity of proteolytic enzymes the cells damaged by the operation are destroyed in the first instance, but the main result is the dedifferentiation of part of the healthy tissues adjacent to the wound surface: destruction of the intercellular substances and the highly differentiated parts of cells (such as the muscle fibrils). As result of these catabolic processes the amount of free amino acids increases to about double the amount in normal tissues—from 16.8 per cent to 35.1 per cent of the total nitrogen (Orechowitsch and Bromley, 1934). A qualitative change in the structure of the proteins and their products also takes place. This is shown by an increase in the amount of reduced sulfhydryl compounds, that is, compounds having a reduced —SH group, presumably owing to a reduction of the cystine in the tripeptide glutathione. This is important because reduced glutathione is known to increase the activity of kathepsin, and therefore this change may also be a contributing factor to the processes of dedifferentiation setting in after the infliction of a wound.

It is remarkable that catabolic processes after wounding are not necessarily restricted to the site of the injury but may occur throughout the whole body, as has been found in the case of mammals in which large amounts of proteins are broken down after the infliction of major wounds (see A. E. Needham, 1952).

Another phenomenon of a physiological nature following amputation is a change in the character of the oxidation in the tissues of the stump. The respiratory quotient, R.Q. (ratio of CO_2 produced to the amount of O_2 consumed), falls abruptly, reaching the level of 0.57 (Ryvkina, 1945). This may be considered a sign of incomplete oxidation in the tissues. Actually lactic acid was found to increase in regenerating tissue as compared with normal tissue. The amount of lactic acid found in the regeneration blastema of the limb of a newt was 39 mg. per 100 gm., as compared with 18 mg. per 100 gm. in normal limb tissue (Okuneff, 1933). As is well known, lactic acid is formed when anaerobic glycolysis replaces oxidation. The enzymes which in normal tissues carry on the oxidation, such as cytochrome oxidase, show a low level of activity in regenerating tissues. The changes concerning the R.Q. and the lactic acid content reach their maximum toward the time when the regeneration blastema is fully formed, and during the later stages—the stages of differentiation—both the lactic acid content and the R.Q. gradually return to normal. The absolute amount of oxygen consumed may fall in the early stages of regeneration, but in the differentiation phase it is above normal.

The changes in the metabolic processes taking place in the earlier phases of regeneration—accumulation of free amino acids, of lactic acid, of free —SH groups—all lead to a lowering of the pH in the tissues of the stump and the regeneration blastema. This could be measured directly in the case of limb regeneration of salamanders. In normal limb tissues the pH is 7.2. It becomes lower immediately after wounding, and reaches a minimum of 6.6 at the time when the blastema is being formed. After this the pH slowly rises, reaching the normal level when differentiation sets in (Okuneff, 1928).

What has been said suffices to show the main features of the predominantly destructive phase of regeneration: the period when the damaged cells are destroyed and dedifferentiation takes place. Some indications will have been noted of the metabolism of the constructive phase of regeneration. This is characterized by an increase of oxidation, by a return of the pH to normal, by a higher R.Q., by a decrease in the amount of free —SH groups and of lactic acid.

We have previously noted the connection that exists between growth and cytoplasmic ribonucleic acid. This connection has also been found to be valid in the case of regenerative growth. The cells of the regeneration blastema become very rich in cytoplasmic ribonucleic acid, even if the cells from which the blastema is derived do not originally contain large amounts of this substance. When the stage of differentiation of the regenerating part is reached, the amount of ribonucleic acid in the cells diminishes again (Roskin, cit. from Brachet, 1950b).

19–6   RELEASE OF REGENERATION

According to definition, regeneration is the replacement of lost parts. One could have expected therefore that the loss of some part of the body would be the adequate stimulus to set in motion the mechanism which restores the part, and thus restores the normal structure of the animal. This is by no means always the case. If a deep incision is made on the side of a salamander's limb, or on the side of the body in an earthworm or a planarian, a regeneration blastema may be formed on the cut surface. The blastema then proceeds to grow and develop into a new part, as in ordinary regeneration. In the case of a limb the new part thus developed will be the distal part of the limb, from the wound level outward. The development of the regenerating part proceeds just as if the entire distal part of the limb were cut off.

In the case of a planarian a lateral incision may cause the development from the wound surface of either a new head or a new tail, or both. If both a head and a tail are regenerated, the head forms from that part of the wound surface which faces anteriorly and the tail develops from the wound surface facing posteriorly. This results, of course, in the regenerated head lying more anterior than the regenerating tail. A somewhat similar reaction is produced by lateral incisions in the earthworm, with the restriction that lateral incisions near the head end of the worm give rise to additional heads, incisions in the middle part of the animal cause the development of both heads and tails, while incisions in the posterior part of the animal's body cause the formation of tails only. Another peculiarity in the case of the earthworm is that the incision must be deep enough to sever the ventral nerve chain if any regeneration at all is to take place (compare what will be said further, in section 19–7).

In each of the above mentioned cases the original parts of the animal (heads, tails, limbs) had not been removed, so that the regenerated parts are additional and therefore superfluous to the animal. The experiments allow us to draw the conclusion that not the absence of an organ but the presence of a wound is the stimulus for regeneration.

The development of a superfluous number of organs or parts of the body, as a result of regeneration, is called **super-regeneration.**

The clue given by super-regeneration has been followed up to analyze the stimulus leading to regeneration still further. It has been found that regeneration can be started even without inflicting an open wound. This has been performed by ligaturing a limb in a salamander (axolotl) (Nassonov, 1930). A tight ligature causes a considerable destruction of the tissues immediately affected by the pressure. The muscles and portions of the skeleton of the limb disintegrate, so that the part of the limb distal to the ligature becomes bent at an angle and is dragged about by the animal without being capable of movement of its own. The skin, however, turns out to be more resistant, and preserves its integrity. After some weeks the region of the limb just proximal to the ligature begins to swell, and it soon becomes evident that a regeneration blastema has been formed. The regeneration blastema then develops into the distal part of a new limb, although the old distal part of the limb is still present. With the processes of differentiation setting in, the skeleton of the old distal part of the limb may be joined again to the proximal skeleton.

The experiment with ligaturing the limb teaches us that the presence of an open wound is not essential for regeneration. What is really necessary is damage to the tissues of an organ that is capable of regeneration. Usually tissues are damaged by wounding, but if extensive damage to the tissues can be caused without an open wound, this suffices to start the sequence of processes leading eventually to regeneration.

Having reached so far we may suggest that damaging the tissues is necessary so that some substance or substances be released from the damaged and disintegrating tissues, which are the immediate cause of the processes that follow. Various types of experiments have been adduced in support of this concept.

It has been found that if a regeneration blastema of an axolotl limb is dried at low temperatures, so that the cells of the blastema are all killed, and this is then transplanted under the skin of an axolotl limb, it causes an outgrowth on the surface of the host limb, covered by skin and with a cartilaginous axis in the middle, comparable to a very rudimentary limb or at least to a digit. A similar outgrowth may be caused by an implanted piece of cartilage and also by introducing under the skin the products of alkaline hydrolysis of cartilage (Nassonov, 1936). These experiments immediately remind us of the experiments on the primary organizer in early amphibian development; there as here it was found that the stimulus for causing certain morphogenetic processes was not necessarily dependent on the integrity and vital activity of the cells of the inducing part, and that non-living substances could exert a similar action.

Another line of research consists in the treatment of the amputation surface with a solution of beryllium salt (beryllium nitrate). Beryllium nitrate applied to the amputation surface of a tadpole tail or of the limb of an *Ambystoma* larva completely suppresses regeneration (A. E. Needham, 1941, 1952). It has been suggested that beryllium in some way binds the substances released from the damaged cells at the wound surface that would have normally initiated the whole sequence of the processes of regeneration—the dedifferentiation in the first place and subsequently the formation of the regeneration blastema. Such an interpretation is supported by the following details of the experiments with beryllium.

First, the beryllium treatment must be carried out immediately after the amputation; an hour later the treatment is without any effect. This might mean that the substances released from the damaged cells have already started off the next step of the reaction (the processes of dedifferentiation) or at least created the conditions in which this next step inevitably follows.

Second, if after treatment of the wound surface the stump is again amputated 0.5 mm. proximal to the original cut, regeneration proceeds normally. Thus the action of the beryllium is only very local. If new cells are damaged they may in their turn release the same substances and "trigger off" regeneration. The treatment of normal tissues prior to amputation does not have any effect either; it is only after the cells are damaged that beryllium can in some way affect the results which could have been produced by the damaged cells. In this respect there is a profound difference between the inhibition of regeneration by X-rays (see section 19–3) and the inhibition of regeneration by beryllium nitrate. X-rays make the cells incapable of growth and reproduction, and this makes regeneration impossible. The demolition processes are not checked by the rays. On the other hand, beryllium does not impair the ability of the cells to grow and differentiate, but it checks all the stages of regeneration by preventing the initial steps of regeneration from occurring. It is thus very probable that it interferes with the specific factor releasing regeneration, and that this factor is a substance or substances given off by the damaged cells at the wound surface.

Other factors besides the substances from disintegrating cells have been shown to be of importance for starting regeneration. In colonies of hydroids the oxygen supply appears to be of major importance (Barth, 1940). If the concentration of oxygen in the surrounding sea water is below 1 cc. per liter, regeneration does not take place, even though the animals remain alive and will regenerate if they are placed again in normal sea water. The rate of regeneration, measured by the length of the hydranth blastema, or by the time necessary for the hydranth blastema to be formed, increases with increased concentration of oxygen, as can be seen from Table 20.

Table 20.   Effect of Varying Oxygen Tension on the Rate of Regeneration of the Hydroid *Tubularia* (after Barth, 1940)

Oxygen cc./liter	Length of Hydranth Primordium	Time for Primordium Formation (in Hours)
2.4	1072	36.1
3.2	1284	28.1
4.1	1370	26.8
4.8	1365	26.3
8.2	1640	24.5
11.3	1809	24.6
14.3	1840	24.1
16.5	1846	23.7
Control in open dish	1370	26.2

The data so far show that oxygen is a necessary condition for regeneration, and that it can affect the rate of the process. Further experiments show, however, that an increased oxygen supply can start the regeneration process where it would otherwise have not occurred. One of such experiments consists in cutting away the perisarc from the middle of the stem of a *Tubularia* hydroid. The perisarc normally decreases the amount of oxygen reaching the cells of the stem from the surrounding water; therefore cutting away of the perisarc increases the oxygen supply to the stem, with the result that it starts to proliferate and forms a hydranth at each end of the exposed region.

Another experiment consists in cutting out a piece of stem of *Tubularia* together with the perisarc and tying the perisarc with a thread at both ends. The section of stem enclosed thus inside the perisarc will not regenerate, because the oxygen supply to it is cut off. If, however, a bubble of oxygen is now injected into the gastric cavity of the stem, regeneration starts in the vicinity of the gas bubble.

Even though oxygen may serve as a stimulus for starting regeneration in hydroids, this is to be regarded as a special case, whereas the release of substances by wounding or destroying the cells probably plays a part in the regeneration of the most diverse animals and possibly even in the case of hydroids, when regeneration follows amputation of a hydroid or a cut through the stem of the colony.

19–7   REGENERATION AND THE NERVOUS SYSTEM

Among the factors concerned with regeneration the nervous system deserves particular attention, since it appears to have a special influence on regeneration. In the amphibians the early stages of regeneration cannot proceed normally in the absence of an adequate nerve supply to the region of the wound. If the nerves supplying the leg or the arm of a newt are destroyed simultaneously with the amputation of the limb or during the early stages of regeneration, the development of the regenerating limb is arrested, and the blastema ceases to grow or may even be resorbed (Fig. 280) (Schotté, 1923, 1926). If the nerves are cut before amputating the limb, processes of dedifferentiation set in, but instead of leading later to the formation of a limb blastema, they continue unchecked until most of the limb is destroyed. The constructive part of the regeneration does not start at all (Schotté and Butler, 1941, Schotté and Harland, 1943). The nervous system exerts its influence only on the earlier stages of regeneration. Once the regenerating limb has reached the stage when the differentiation commences, it can proceed with its development even in the absence of nerve supply (if the nerve is cut at that stage). The nerves are thus one of the necessary conditions for the growth of a regenerating part and even for the formation of the regeneration blastema.

The action of nerves, however, can go even further than that. In a newt it is possible to deviate a nerve from its normal position. In the case of limb nerves the experiment is done so that the nerve is transected at a distal level, then separated along part of its course, from the distal end to the shoulder or pelvic level. A cut is then made through the skin starting from the pelvis or the shoulder, and the nerve is placed in the cut, so that when the wound heals the free end of the nerve is in a position under the skin and away from the limb it normally supplies. It was found that if the end of the nerve was not too far from the base of the limb, a limb rudiment, similar to a regeneration blastema of the limb, was formed over the end of the nerve. In successful cases this grew out into a complete new limb (Locatelli, 1924). Thus the presence of a limb nerve may cause the formation of a limb at a spot where otherwise no limb development could have been expected.

That a stimulus causing growth and development can be exercised by a free nerve ending has also been shown in a somewhat different experiment. It has been mentioned previously that the limbs of an adult frog cannot normally regenerate, but that some degree of regeneration in adult limbs may be released by traumatization of the stump after the amputation of a limb (section 19–3). The sciatic nerve of a young metamorphosed frog was dissected as described above and deviated into the forelimb, which was then amputated. The presence of the sciatic nerve exerted the same action as traumatization of the limb stump; a blastema was formed and this developed

Fig. 280.   Arrest of regeneration in limbs with transected nerves (left limb in each pair). Limbs with intact nerves regenerate normally. (After Schotté, 1926.)

into an incomplete hand "with a tendency to hand or finger formation" (M. Singer, 1950).

The influence of the nervous system on regeneration is by no means restricted to the regeneration of limbs in amphibians. It is very obvious in annelids. After section of the body, when the neoblasts migrate to the surface of the wound to form a regeneration blastema, they follow the ventral nerve cord in their migration. If the nerve cord is excised some distance from the level at which an earthworm has been transected, no regeneration will occur at that level. A regeneration blastema may be formed, however, at the spot where the nerve cord ends, and a new anterior end of the worm will be formed some distance from the anterior cut surface (Morgan, 1901). The anterior end of the nerve cord can also be deflected similarly to the deviation of a limb nerve in a newt. The free end of the deflected nerve cord then causes the formation of a regeneration blastema and a new head (*cf.* Berrill, 1952). Even a simple section of the ventral nerve cord, without the removal of either end of the body, is sufficient to release regeneration in a polychaete (Okada, cit. after Berrill, 1952).

Much the same holds true for planarians; regeneration blastemas are formed in conjunction with cut nerve cords. The cerebral ganglion of a planarian excised with a piece of surrounding tissue and transplanted to a different body level causes the formation of a complete new head at the site of transplantation (Santos, 1931).

The experiments described so far prove beyond doubt that the influence of the nervous system is necessary for regeneration. They do not show, however, in what way the nervous system acts, whether it acts in the same way as when transmitting stimuli that release movement or secretion, or in some other way. An indication in this respect has been obtained in an experiment confronting a limb with nerve fibers of various origins (Weiss, 1950). A piece of nervous tissue taken from the brain or the spinal cord may be cultivated in the loose parenchyme filling the dorsal fin of urodele amphibians. After an initial partial degeneration the tissue starts growing, and nerve fibers are produced by the surviving nerve cells. If a young limb is simultaneously transplanted into the dorsal fin, connections may be established between the piece of nervous tissue and the limb; the outgrowing nerve fibers supply the skin and the muscles of the limb. The fibers in question have been found to be neither the normal motor nor the sensory fibers but are rather equivalent to the "association neurons" which connect different parts of the central nervous system. If the piece of nervous tissue supplying the nerves was derived from the spinal cord or the medulla, the limb muscles could perform contractions, either spontaneously or in answer to an irritation of the limb. The contractions were, however, entirely uncoordinated, in the nature of epileptic seizures rather than normal movements. No contractions occurred, however, if the nervous tissue was taken from the forebrain, diencephalon or midbrain. The transplanted limbs could then be amputated, and they regenerated no matter what part of brain or spinal cord supplied the nerves to the limbs. Nerves which cannot release muscle contractions are thus adequate for supporting regeneration. The conclusion is thus reached that probably any kind of nerve supply can be the source of influence necessary for regeneration, and that the nature of the influence is different from the ordinary transmission of stimuli by nerves.

It may be further suggested that it is hardly possible for the nerve fibers to be in contact with all the cells of a regeneration blastema while these are in motion taking up their positions under the wound surface. In these conditions the normal mechanism of transmission of impulses could hardly operate, and it is thus more likely that the nerve endings act through the release of some substance. What such a substance may be remains a problem for future research.

19–8   RELATION OF THE REGENERATING PARTS TO THE REMAINDER OF THE ORGAN AND TO THE ORGANISM AS A WHOLE

Considering the factors on which regeneration is dependent one can subdivide them in two groups:

1. The factors which are responsible for the regeneration taking place, and
2. The factors which determine that the right sort of organ regenerates.

So far we have been dealing mainly with the first group of factors. It has been shown, however, that sometimes a regenerating organ does not fit into the organization of the whole animal, as when a new limb regenerates without the original limb having been removed (section 19–6). Still, as a general rule, what regenerates corresponds to what has been lost. This can only mean that the position of the wound in some way determines the nature of the regenerating part. If the cut is through a limb, a limb will regenerate (if any regeneration occurs at all); if the cut is through a tail, a tail will regenerate. If the limb is cut at the lower arm level, parts of the lower arm, wrist and digits will develop; if the cut is at the upper arm level, the upper arm will regenerate as well. A part of the shoulder girdle may also be removed, and the whole of the arm together with the shoulder girdle will be renewed. If, however, the whole of the shoulder girdle together with the muscles of the shoulder is completely removed, the arm cannot regenerate any more. This shows that a remnant of the original organ must remain, so as to enable the organ to regenerate. The arm and the shoulder girdle with its muscles form one unit in respect of regeneration. Such a unit has been termed a "regeneration territory" (Guyenot and Ponse, 1930) or "regeneration field" (Weiss, 1926a). A similar regeneration territory is responsible for the regeneration of the tail in adult newts; so long as a small piece of the tail remains, regeneration is possible. If, however, the tail is amputated at the level of the last sacral vertebra, the whole "tail territory" is removed, and no regeneration takes place. The wound heals without any regeneration blastema being formed.

It is not inevitable that the regeneration should be completely repressed

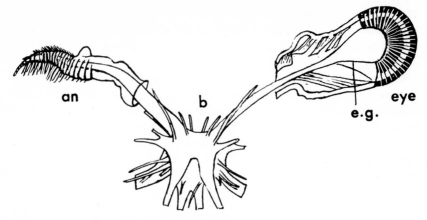

an   b   eye

e.g.

Fig. 281. Heteromorphic regeneration of an antenna in place of an amputated eye in *Palinurus. an,* Antenna, regenerated; *b,* brain; *e.g.,* eye ganglion. (After Herbst, from Hartmann: Allg. Biol., 1947.)

if the whole of a regeneration territory is destroyed. Regeneration may occur, but in some different way. In the shrimp, *Palinurus,* the eye may regenerate after being removed. The cut, however, must be made through the eye stalk, just proximal to the eye and distal to the nerve ganglion lying at the base of the eye (inside the stalk). If the cut is made at the base of the stalk, so that the ganglion is removed together with the eye, the eye will not regenerate. A regeneration blastema is formed, however, but instead of the lost eye, it develops into an antenna-like organ (Fig. 281). This phenomenon—a different organ developing from the one that has been removed—has been called **heteromorphosis.** The most plausible explanation of heteromorphosis in this case is that the ganglion and the eye constitute together one regeneration territory. If a part of the territory (the ganglion) remains intact, the complete system may be restored. If the whole territory is lost, it cannot be renewed at the expense of other parts of the body.

A different result, yet illustrating the same principle, has been observed in the regeneration from anterior cut surfaces in the earthworm. As shown previously (section 19–2), a new "head" is regenerated at the anterior end of the earthworm if it is cut off. This is possible, however, only if the cut is not too far from the anterior end of the animal. If the cut is beyond the middle of the worm, the posterior part cannot regenerate a new "head" any more. Regeneration takes place, but what is regenerated is a second tail. As this tail is at the anterior end of the sectioned worm, it is a case of heteromorphosis (Morgan, 1901). Again we see that a new anterior end of the animal can only be restored if not too much of the anterior body part has been removed. With more than half of the animal removed, the regeneration territory or regeneration field of the head is completely gone, and what regenerates can only be a tail end. There is an interesting peculiarity in this case which should be noted. By making deep lateral incisions in the body of the earthworm, it has been observed that the ability to regenerate a head diminishes gradually in an anteroposterior direction. The heads which are formed at the site of the wound become smaller and smaller, as the wound is made further away from the anterior end, and beyond the middle of the animal's body no head is regenerated although a tail may be formed.

The same is the case in the regeneration of planarians. If the body is cut transversely at different levels, the tail piece will regenerate a head only if the cut is not too far posterior. Conversely, a very short piece cut from the anterior end of the body of a planarian does not regenerate a tail at its posterior surface, but a second head. At the anterior end there is thus no competence for tail development present, or else the tendency to develop a head is so strong that the morphogenetic processes cannot proceed in any other way.

If the cut surface determines the nature of the organ or part of the body to be produced in regeneration, it may be questioned whether there are any special cells or tissues at the cut surface which are responsible for the course of regeneration. The problem may be approached experimentally by removing parts of the organ at the level of amputation one by one and observing the result. A classical experiment of this kind was performed by Weiss (1925). After making a slit through the skin and the muscle of the arm in the newt, Weiss removed the humerus. After such an operation the humerus is not restored. The wound healed, and then the limb was amputated through the upper arm. Owing to the first operation there was no bone or cartilage at the level of amputation. Nevertheless the regeneration proceeded in the normal way. A regeneration bud was formed and this developed all the parts of the limb distal to the level of amputation. The skeleton of the regenerated part was complete, including the distal part of the humerus. The proximal part of the humerus was lacking, as before (Fig. 282). The first conclusion that can be drawn from this experiment is that the tissues of the regenerated limb are not derived from corresponding tissues at the level of amputation; rather the cells of the regeneration blastema are to be considered as undifferentiated in so far as the various parts of the limb are concerned, and capable of fitting into any part of the limb.

This has also been confirmed in a very different way. The distal part of a newt's limb may be split lengthwise by a longitudinal cut and then each half amputated and allowed to regenerate. Two regeneration buds are then formed, one at the end of each portion of the limb. Each develops subsequently into a complete distal part, with a complete carpus or tarsus as the

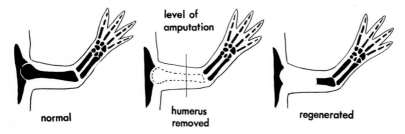

level of
amputation

normal

humerus
removed

regenerated

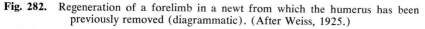

Fig. 282.   Regeneration of a forelimb in a newt from which the humerus has been previously removed (diagrammatic). (After Weiss, 1925.)

case may be, and not into half of one. Thus the cells derived from one half of a transverse section of the limb are capable of producing a complete organ (Weiss, 1926b). It is interesting to compare this result with the splitting of the early limb-bud of an embryo (see section 11–3). With regard to the animal as a whole the development of two limbs, where only one was amputated, is excessive; it is a case of super-regeneration.

Returning to the experiment with regeneration of a complete distal part of a limb in the absence of skeletal parts at the amputation level, we may note a second conclusion; the skeleton at the amputation level does not appear to be necessary for determining the nature of the regenerating organ. In another experiment the skin (corium) was shown similarly not to be necessary as a carrier of forces determining the specific nature of the regenerate. A strip of skin around the circumference of a newt's limb was removed and replaced by a section of the trachea turned inside out, so as to bring the tracheal epithelium into the position occupied normally by the epithelium of the limb and to put the connective tissue layer of the trachea in the place of the corium of the limb. After amputation through the "collar" made of tracheal wall, the regeneration proceeded normally, and the regenerated limb possessed a corium indistinguishable from that in normal limbs (Weiss, 1927a).

With the exclusion of the skeleton and the corium as carriers of the factors determining the course of regeneration, our attention turns naturally to the muscles and connective tissue of the limb. Excluding these tissues from the level of amputation cannot easily be done, but the action of muscles has been tested in a different way. Muscles from one regeneration territory (e.g., the tail) were transplanted in place of muscles of a different regeneration territory (e.g., the limb) and vice versa. If the organ was then amputated, the regeneration was no longer normal; the transplanted muscles exerted a specific influence on the nature of the regenerating organ. Thus if the arm muscles were removed as completely as possible and pieces of tail muscle were stuffed into the space between the arm skeleton and the skin, and then the limb was amputated, the distal part of the regenerating organ was no longer an arm but a structure resembling a tail (Liosner and Woronzowa, 1936).

If the cells of the regeneration blastema are capable of producing any part (any tissue) of the regenerating organ, the question further arises whether different kinds of organs may be produced from a regeneration blastema, other than the organ from the stump of which the blastema had been derived. Two opposing views have been held on this subject by different investigators. According to one view the cells of a regeneration blastema are completely undifferentiated and capable of developing into any part of the animal's body, except perhaps that mesodermal cells do not give rise to skin epithelium, though the transformation of epithelial cells into connective tissue cells has been considered as possible by a number of zoologists. In support of this view it is said that a very young blastema, corresponding in stage of development to those depicted in Fig. 273, *a*, may be transplanted to the amputation stump of a different organ, and then may develop in correspondence to its new position. Transplantations have been carried out to exchange the regeneration blastema of the fore- and the hindlimbs in the newt. An early forelimb blastema transplanted to the stump of an amputated hindlimb is said to be capable of developing into a hindlimb, and vice versa. Furthermore, the regeneration blastema of the tail has been transplanted to the shoulder region, and it was reported that it developed into a forelimb (Weiss, 1927b). As regeneration does not proceed in the absence of nerve supply, a limb nerve had to be diverted to the site of transplantation to make the transplanted blastema grow and differentiate. The latter procedure, however, introduces a serious source of error in the experiment; we have seen that the deviation of a nerve into the area around the basis of a limb alone, without the transplantation of a regeneration blastema, may induce the development of a supernumerary limb (experiments by Locatelli, section 19–7).

It may well have happened that the grafted regeneration blastema of the tail was gradually destroyed and replaced by local cells, which were activated for limb formation by the diverted nerve. The results of transplantation of regeneration buds between the fore- and hindlimbs have also been proved to be inconclusive; the mobilization of regeneration cells in the early stages to which the above experiments refer is not yet completed, and therefore the grafted cells could be replaced by local cells, with the result that the nature of the regenerated organ conforms to the position in which it develops. Careful experiments have proved that no change can be produced in the specific type of development of a regeneration blastema, no matter in what way it is transplanted, at least in the case of amphibian limb regeneration.

In planarians, on the other hand, the tail regeneration blastema has been grafted onto the anterior amputation surface, with the result that it developed into a new head (Gebhardt, 1926). This result has not yet been challenged, and it may be possible that the regeneration cells in lower animals are more plastic, retaining broader possibilities for differentiation. This is perhaps to be expected, as the regeneration territories in these animals are not strictly delimited; the competence to produce a tail or a head fades away gradually starting from the anterior or posterior end of the animals.

It is thus fairly safe to conclude that the capacities of regeneration cells have very definite limits, and that these limits are more narrow in the highly organized animals than in the lower forms of life. There is no question of the regeneration cells acquiring the same abilities for development as the egg, or the early cleavage cells. A further difference between the early embryonic cells and the regeneration cells is that the first may develop in complete isolation; the second can develop only in conjunction with the remainder of the organism, which provides the regeneration blastema with nourishment, supplies it with nerves, and exerts a certain degree of influence on the processes of differentiation of the regeneration blastema.

19–9 RECONSTITUTION FROM ISOLATED CELLS

Related to regeneration is the reaggregation of isolated cells into a new whole animal. This remarkable phenomenon has been discovered in sponges (H. V. Wilson, 1907). A sponge may be rubbed through bolting silk, so that the entire organization of the sponge is broken up, and the tissues of the animal are reduced to a pulp consisting of isolated cells and cell debris. If the pulp is allowed to stand, the isolated cells begin crawling about and aggregating themselves into larger masses. These masses then become organized into new sponges. Among the isolated cells one can distinguish the archeocytes, the collar cells and the dermal cells of the adult sponge. When the cells reaggregate, each type of cells sorts itself out and takes up in the aggregate a position, which belongs to it because of its specific properties; the dermal cells cover the whole aggregate from without, the collar cells join together and rebuild the collar cell chambers in the inside. The archeocytes take up their normal position; they also play an active part in the formation of the complex, owing to their greater ability for ameboid movement; their rate of movement is 0.6 to 3.5 μ per minute. Moving about, the archeocytes help the other cells to aggregate. About 2000 cells are necessary to produce a new individual (Galtsoff, 1925). The whole process, up to the opening of new oscula, takes about three weeks.

The phenomenon has been called **reconstitution.** Although reconstitution starts from individual cells, it has nothing in common with embryonic development, as the individual cells into which the sponge is broken up each retain their specific histologic character, and the whole process rests mainly on a rearrangement of the cells in space rather than on a progressive differentiation, although minor readjustments may possibly occur. The whole process can thus best be compared to regeneration by morphallaxis (see section 19–2).

Reconstitution by reaggregation of isolated cells as observed in adult sponges bears an obvious relationship to the reaggregation of embryonic cells which has been described in sections 5–8 and 14–4, and shows that the mechanisms of differential cellular affinity are present and active throughout the life of a metazoon, even if we do not always observe them. It would probably not be too far fetched to say that these mechanisms, after having participated in producing a multicellular organism, serve to maintain it and preserve its integrity.

CHAPTER **20**

Asexual Reproduction

Asexual reproduction is the development of a new individual without the participation of any stages of the sexual cycle, that is, without maturation and copulation of sex cells with concomitant reduction of the number of chromosomes in meiosis. As applied to multicellular animals, asexual reproduction means development of new individuals at the expense of **somatic** cells—cells of the **soma** or body, as opposed to generative or sex cells (gametes).

In this chapter we shall not be concerned with the evolution and adaptive significance of asexual reproduction; neither shall we deal with the factors regulating the alternation of sexual and asexual reproductive cycles. What will interest us is how a part of an already differentiated organism may again embark on an active process of morphogenesis. The part of the parental organism giving rise to a new individual in asexual reproduction may be called a **blastema,** and it always consists of a group of cells, whereas in sexual reproduction the new individual develops from one cell—the fertilized or parthenogenetically activated ovum. The development that starts from a blastema may be termed **blastogenesis** as opposed to **embryogenesis,** or development from the ovum. The individuals resulting from asexual reproduction are often referred to as **blastozooids;** individuals developing from an egg are then termed **oozooids.**

The blastozooids may have the same general organization as oozoids or they may even be indistinguishable from the latter. How this is achieved, in spite of the profoundly different initial stages, is the second major problem in relation to asexual reproduction.

20–1   OCCURRENCE AND FORMS OF ASEXUAL REPRODUCTION

Asexual reproduction takes on a variety of forms, largely because of the amount of tissue set aside for the production of a new individual and, correlated with this, the degree of organization of this tissue. Taking the size of the fragment and the degree of its organization as a guiding principle, we may subdivide the infinite multiplicity of modifications occurring in different Metazoa into three main types:

1. Fission—the new individual is formed from a relatively large portion of the body of the maternal organism, and differentiated organs and tissues or their parts are passed on to the offspring.
2. Budding—the new individual develops from a small outgrowth on the surface of the parent. No organs of the parent are passed on as such to the offspring, but the bud is supported by the parental organism at least during the initial stages of its development.
3. Gemmule formation—the new individual develops from groups of cells which become completely cut off from the maternal individual and disseminated, so that the development is, from the start, quite independent of the maternal body.

Fission.   The simplest form of fission is the separation of an adult individual into two parts of approximately equal size—similar to the binary fission in protozoans. In Metazoa it occurs in the most typical form in coelenterates and worms. In some corals (Anthozoa) fission occurs in the rather rare form of longitudinal fission, with the division plane in the long axis of the body. Fission in a plane passing along the main body axis also occurs in some brittle stars, as for instance in *Ophiactis savignyi* in which parental individuals break across the disc, and each half then proceeds to regenerate the missing half of the disc and three legs. In the worms (see Berrill, 1952) the planes of division are transverse to the body axis, so that the worm is divided into anterior and posterior halves. This type of reproduction is found in rhabdocoele turbellarians and in annelids, both polychaetes and oligochaetes. Both halves inherit from the parent the skin, a section of the alimentary canal, sections of the nerve cord and, in annelids, a number of mesodermal segments, with muscles and nephridia, and correspondingly

sections of parenchymatous mesoderm in rhabdocoeles. The posterior individual is at first devoid of a head and therefore lacks initially the supra-
esophageal ganglion, sense organs (eyes), tentacles and mouth parts, where
these occur. The head is produced either subsequent to division or, more
often, in preparation for the division, so that the posterior individual is fully
developed when the two blastozooids separate.

In rhabdocoeles and some oligochaetes, new fissions may be started by
one or both filial individuals even before they are separated from one another (Fig. 283). The result is a chain of blastozooids, some of them in

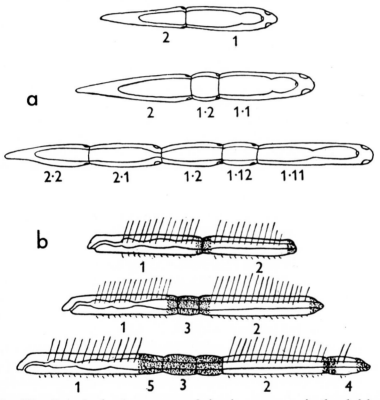

Fig. 283. Reproduction by transverse fission in worms: *a,* in the rhabdocoele
turbellarian *Stenostomum* (after Child); *b,* in the oligochaete *Nais* (after Stolte). The
numbers show the succession in the formation of individual blastozooids. (From
Berrill, 1952.)

different stages of reconstitution (degree of head development). The division may also be very unequal, so that of the two or several zooids one may
be clearly distinguishable as the parent and the other, or others, as offspring.

Cases in which subsequent transverse divisions occur prior to the earlier
divisions being completed lead to a special form of transverse fission, known
as **strobilation,** in which numerous transverse divisions occur more or less
simultaneously or in close succession, giving rise to a number of filial blastozooids. This is the classical method of propagation by which in Scyphozoa

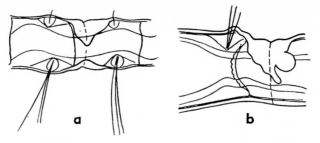

Fig. 284.   The initial stages of transverse division in annelids. *a, Pristina; b, Chaeto-gaster.* The epidermal thickenings are clearly indicated. (From Berrill, 1952.)

the asexual polypoid generation gives rise to the sexual medusoid generation. Similar multiple transverse division occurs in some polychaetes and also in ascidians (see Berrill, 1951). In the latter the part undergoing transverse division is the abdominal section of the body, which is devoid of the branchial chamber but contains the intestinal loop.

Both in worms and in ascidians the initial separation of the zooids is performed through the activity of the ectodermal epidermis. The epidermis forms a circular constriction which cuts inward, severs the internal organs and eventually cuts the body of the animal in two (Fig. 284) (Berrill, 1951, 1952).

Budding.   The most typical examples of budding are found in coelenterates and tunicates. Superficially the early bud appears as a small swelling or nodule on the lateral surface of the body of the parental animal. The nodule grows, takes shape and develops a mouth and a whorl of perioral tentacles in the case of the coelenterates, or in the tunicates develops a branchial chamber and other associated structures including the atrial cavity, and the oral and atrial siphons. The bud may become completely separated from the maternal body or remain permanently in connection with the latter, thus leading to the formation of a colony. In the fresh-water hydra, which normally exists in the form of single polyps, the daughter individuals may remain connected to the maternal body and occasionally even start forming secondary buds, but eventually this temporary colony splits into single individuals.

Not always are the buds directly formed on the body of the parental zooids. Often the buds develop on special outgrowths of the maternal animals; the outgrowths are then called **stolons.** These are typically present in many tunicates, but the branches of a hydrozoan colony on which the polyps develop have the same significance.

Gemmule Formation.   This form of asexual reproduction is found in fresh-water sponges and in bryozoans, and in both cases bodies are produced called **gemmules** in sponges and **statoblasts** in bryozoans, that can survive after the maternal individual (or rather maternal colony) dies off during an unfavorable season (winter in temperate countries, periods of drought in warmer climates). The gemmules and statoblasts are formed in the interior of the parental body from a number of undifferentiated cells, which become enclosed in a shell with special spicules in sponges (Fig. 285) or in a chitinous envelope in bryozoans. On destruction of the parental body the gemmules or statoblasts are set free, and after conditions have become favorable, the cells contained inside burst out and develop a new individual. For the gemmules of sponges such favorable conditions are created by the temperature rising above 16° C. (Brien, 1932).

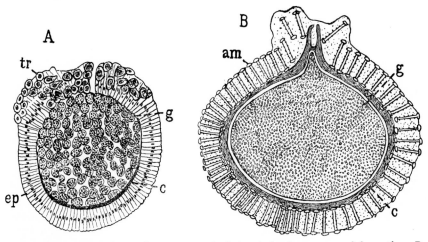

Fig. 285. Gemmule of the sponge *Ephydatia: A,* in the process of formation; *B,* in the mature stage. *am,* Amphidiscs; *c,* cuticular membrane; *ep,* epithelium; *g,* archeocytes; *tr,* trophocytes. (After Evans, from Bounoure, 1940.)

20–2   SOURCES OF CELLULAR MATERIAL IN ASEXUAL REPRODUCTION

The major problem with which we are faced in asexual reproduction is: what resources does a differentiated organism have in order to start building new parts in a new burst of morphogenetic activity?

The least difficulty in interpreting the development of new parts is encountered in asexual reproduction by fission.

It will be evident from what has been said before that reproduction by fission is closely related to regeneration, or, to put it another way: regeneration is an essential part of asexual reproduction by fission. In both cases a comparatively large part of the body is isolated and reconstructed into a new complete whole by the regeneration of the missing parts. There is every reason to believe that the actual mechanism of development of the missing parts is the same in the two cases. We have seen that the organs which are produced anew during regeneration in annelids and in turbellarians owe their origin in large measure to undifferentiated "reserve" cells: neoblasts in annelids, parenchyme cells in turbellarians (see pp. 485, 486). The same cells are available for the development of missing parts in daughter zooids. Neoblasts are actually reported to concentrate in the fission zone of oligochaetes. As in regeneration, the epidermis gives rise to the nervous system and sense organs in the new head of the posterior zooid.

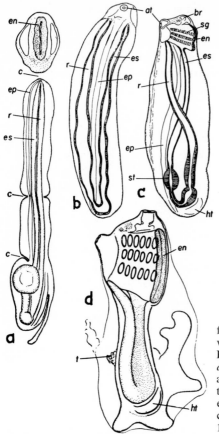

Fig. 286. Reproduction by transverse fission in the tunicate *Eudistoma*. *a,* Individual in the process of fission. *b, c,* Blastozooid regenerating the anterior end. *d,* Newly metamorphosed oozooid. *at,* atrial siphon; *br,* oral siphon; *c,* constrictions; *en,* endostyle; *ep,* epicardium; *es,* esophagus; *ht,* heart; *r,* rectum; *sg,* gill clefts; *t,* remnant of tail. (From Berrill, 1947.)

Where transverse fission occurs as a form of asexual reproduction in ascidians such as *Eudistoma* (Berrill, 1951), the daughter zooids consist of a section of the parent's body including epidermis, mesenchyme and a section of the intestinal tube, but do not include the branchial chamber and associated structures such as the atrial cavity, the nerve ganglia and the siphons. The missing parts are replaced by regeneration. In the process of regeneration a most important role is played by the **epicardium**—a mesodermal tube lying ventral to the intestine (see Fig. 286). The mesodermal lining of the epicardium has the properties of an undifferentiated reserve tissue, at the expense of which most of the internal organs in the anterior part of the new zooid develop. Remarkably enough, the branchial chamber does not regenerate at the expense of the remaining part of the endodermal alimentary canal but is also produced by the epicardium and secondarily becomes fused onto the endodermal esophagus.

In budding only very little of the organization of the parent is passed on to the daughter zooid. In a scyphozooid polyp the first sign of impending budding is the formation of a pocketlike evagination of the endodermal epithelium which pushes through the mesoglea and reaches the ectodermal layer.

The ectodermal layer now becomes activated and both ectoderm and endoderm form a conical protrusion on the surface of the body. This protrusion
gradually increases and elongates and becomes a cylindrical body with an
internal cavity which is in open communication with the gastric cavity of the
parent (Gilchrist, 1937). At a later stage a mouth is formed at the tip of the
bud, and a whorl of tentacles develops around it; the bud has become a new
polyp.

The budding in hydrozoid polyps is essentially similar. From this one
would tend to draw the conclusion that if not the pattern of organization then
at least the two main layers of the body, the ectoderm and the endoderm of
the daughter zooid, are directly taken over from the parental organism. The
majority of students of coelenterate development and reproduction believe,
however, that the process is not nearly so simple (see Hadzi, 1910; Schulze,
1918). We have seen that regeneration in the hydra is due to the activity of
so-called I-cells, small undifferentiated cells located between the larger functioning cells both of the ectoderm and endoderm. These cells serve for the
replacement of differentiated cells, especially of the cnidoblast cells, and in
cases of injury the I-cells form the regeneration blastema. The same I-cells
are supposed to be the main source for the formation of the bud in asexual
reproduction. The I-cells either accumulate locally and form the first thickening which is to become the bud, or possibly they infiltrate the bud in the
process of its formation and continue flowing into the bud and increasing its
size even after it has been formed. Eventually the whole or almost the whole
of the daughter zooid is built from the I-cells, the differentiated ectodermal
and endodermal cells taking very little, if any, part. In Hydromedusae
(*Lizzia claparedi, Rathkea octopunctata*—see Bounoure, 1940) it was observed that endoderm does not contribute to the formation of the bud at all.
The bud is at first represented by a nodule of cells in the ectodermal layer.
Some of the cells lying in the interior of the nodule become arranged in the
form of a vesicle, and it is from this vesicle that the endoderm of the daughter
medusa becomes developed (Fig. 287). This observation can be better

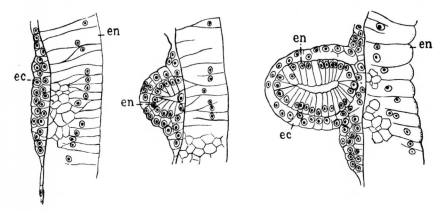

Fig. 287.   Development of a bud in the medusa *Rathkea*. *ec,* Ectoderm; *en,* endoderm.
(After Chun, from Bounoure, 1940.)

understood if it is accepted that in the hydrozoans a new zooid is derived neither from the endoderm nor from the ectoderm of the parent, but from the undifferentiated I-cells.

The budding in tunicates presents an almost infinite variety of forms but in the great majority of these animals, in particular in the ascidians, there is a very distinct common pattern which presents some remarkable features. The bud is essentially an outgrowth of the ectodermal epidermis, supported by another tissue which forms a hollow vesicle inside the outgrowth. In some cases parts of the gonads are included as a third element. The epidermis of tunicates is a very specialized tissue which produces the mantle, the cellulose layer covering the surface of the skin in these animals. In budding, the epidermis produces only more epidermis and provides the external covering for the blastozooid. The internal organs are developed from the above-mentioned vesicle formed inside the bud. Now it is most remarkable that the inner vesicle may be derived from a number of different tissues, namely:

(1) an outgrowth of the atrial cavity (*Botryllus*),
(2) an outgrowth from the pharyngeal wall at the posterior end of the endostyle (*Salpa, Pyrosoma*),
(3) an outgrowth of the epicardium, which was already mentioned before (*Distaplia*),
(4) a group of blood cells (*Botryllus*) (Oka and Watanabe, 1957).

All these types of development are found in cases where the bud is developed on the body of the parent or on a short stolon connected to the body.

(5) In some ascidians (*Clavellina, Perophora*) the buds develop on long rhizoidlike stolons. These consist only of epidermis and a longitudinal mesodermal septum, splitting the cavity of the stolon into two canals which allow for the backward and forward circulation of blood. Buds developing on such a stolon consist of the epidermal covering epithelium and the inner vesicle, which in this case is produced from the cells of the mesodermal septum (Fig. 288) (see Berrill, 1951).

Regardless of the origin of the internal vesicle, it produces a variety of organs and tissues: the alimentary canal with the branchial chamber, the atrial cavities, the heart, the nervous system (except in *Salpa*), and sometimes also the gonads. It is thus evident that several parts in the fully differentiated body of an ascidian retain very wide potentialities for development. These potentialities embrace practically the whole organization of the ascidian except the epidermis, and even of this limitation we cannot be sure, as the epidermis is always provided and so there would appear to be no incentive for the inner vesicle to produce the epidermis as well. It may well be that the potentiality for forming epidermis is not lacking, and then the cells of the inner vesicle would be truly **totipotent**—capable of producing every differentiation found in the animal species in question.

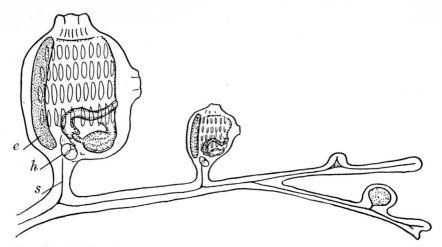

Fig. 288.   Budding from the stolon in the tunicate *Perophora*. *e*, Endostyle; *h*, heart; *s*, stolonial septum. (From Berrill, 1935.)

The existence of really totipotent cells is indisputable in the case of gemmule formation in the fresh-water sponges. The first rudiment of a gemmule in the interior of the sponge appears in the form of an accumulation of archeocytes—undifferentiated ameboid cells which are dispersed in between the differentiated elements of the body of a sponge. Around the central core consisting of archeocytes, other migrating cells are lodged and become arranged in the form of a columnar epithelium. The role of the epithelium is to produce the shell on the surface of the gemmule. This is strengthened by the action of special skeleton building cells—**scleroblasts**—which deposit in the shell peculiar spicules, the amphidiscs (see Fig. 285). Both the columnar epithelial cells and the scleroblasts are later withdrawn or they degenerate. A small opening, the micropyle, remains temporarily open in the shell. It has been claimed that feeder cells, **trophocytes,** enter the interior of the developing gemmule and pass on food reserves to the archeocytes and in their turn disappear. The archeocytes, on the other hand, accumulate in their cytoplasm food reserves consisting of glycoproteins in the form of discoidal platelets (see Fig. 291). When the development of the gemmule is completed, the archeocytes are the only living cells left in it, and all the tissues and cell types in the new sponge which develop eventually from the gemmule, including its sex cells, are derived from the archeocytes.

Summing up our review of the sources of materials for asexual reproduction, we see that the renewal of morphogenetic processes depends on the persistence, in the body of the adult animals, of cells which do not become as highly differentiated as the rest and are capable of developing into a variety of cell types. The role of these undifferentiated or "reserve" cells is the greater, the smaller the part of the parental body that is used in asexual propagation. In extreme cases the undifferentiated cells are "totipotent" and are in this respect equivalent to the egg cell, a similarity which may be further enhanced by the presence of deutoplasmic (food reserve) inclusions, as in the case of the gemmules of sponges.

Although asexual reproduction is a special form of morphogenesis which has been acquired, probably independently, only in some groups of the animal kingdom, it is evident from the above that it falls back on a very general property of all developing organisms: the potential equivalence of the daughter cells resulting from the cleavage of the fertilized egg. We have seen (section 4–5) that the cleavage cells retain the full complement of hereditary factors (genome) independently of the prospective significance of the individual cells, and that the nuclei are unrestricted in their potencies even after determination of the various areas of the embryo has set in (p. 93). The cytoplasm of cleavage cells may show differences in early stages, and later the elaboration of special cytoplasmic mechanisms (histological differentiation) may go so far that specific types of cells lose the ability to change the direction of their development. The loss of plasticity, however, is not inevitable, neither does it proceed at an equal tempo in all parts and in all cells of a differentiated organism. Cells which do not achieve a high degree of differentiation retain their plasticity, and, being in possession of the complete genome, can be made use of in asexual reproduction (see Bounoure, 1940).

20–3   COMPARISON OF BLASTOGENESIS AND EMBRYOGENESIS

In the general survey of ontogenetic development (section 1–2) we have found it useful to consider what tasks have to be performed by the embryo before the final condition (the development of the new adult individual) is achieved. If from this same viewpoint we compare embryogenesis (development of the egg) with blastogenesis (development from a blastema in asexual reproduction), we see at once that the task is very much simpler in the latter case. The process of producing a new individual is simplest in reproduction by fission, when the blastozooid is derived from half the parental organism and in this way is provided with a large proportion of the organs and parts which are necessary for making the new individual self-sufficient. What has to be done is the regeneration of missing parts. The whole mechanism of regeneration, as considered in Chapter 19, is brought into play, including the factors determining the regenerating parts (pp. 497–501). The remnant of the old individual determines the nature, position and orientation of the newly differentiated organs. The polarity and bilaterality of the parent organism prevail in the blastozooids.

The task of development is more complicated in the case of budding, since all organs and differentiated parts of the blastozooid have to be produced anew. Nevertheless the initial system, the bud, always has a higher degree of complexity than a fertilized egg or even than a blastula as it occurs in embryogenesis. A typical bud, as we have explained above, always consists of two layers of epithelial cells. The young zooid is thus already in possession of the concentric stratification of body layers, a condition which in embryogenesis is achieved only after gastrulation. It is very noteworthy, however, that the layers formed in the blastozooid do not necessarily correspond to the germinal layers developing in embryogenesis.

In the case of budding in coelenterates there is a closer correspondence between the outer and inner layer of the bud and the ectoderm and endoderm of the gastrula. The fate of the two layers is the same, but we have seen that the inner layer may be derived not from the endodermal epithelium of the parent, but from a thickening in the ectodermal epithelium (in all probability it is actually produced by the reserve I-cells).

When we turn to the buds of tunicates, we find that the inner vesicle corresponds to the endoderm neither in its origin nor its fate. It has been shown (p. 510) that although in some tunicates the inner vesicle may be derived from the endoderm of the parent (in the form of an outgrowth from the pharyngeal epithelium), it can be also derived from mesoderm (epicardium, mesodermal septum of the stolon, blood cells) or even ectoderm (lining of the atrial cavity). What is even more important, the inner vesicle gives rise to parts derived in embryogenesis from any of the three germinal layers. In spite of the diversity of origin of the inner vesicle, its later differentiation shows a considerable degree of uniformity in different tunicates. After the bud has grown to a certain degree (a minimal size, varying of course in different species, is essential) (Berrill, 1941), it becomes constricted at least partially from the parent zooid. Then folds start subdividing the inner vesicle into sections. (The following description refers to the development of the blastozooid in *Botryllus*.)

First of all two folds cut in from what will be the distal part of the new zooid, subdividing the inner vesicle into a median part, which will become the branchial chamber, and two lateral parts which give rise to the atrial cavity (Fig. 289) (in embryogenesis the atrial cavity develops as a pair of invaginations of the ectoderm which partially fuse and later have a common opening to the exterior). In addition to these three main subdivisions, further smaller pocketlike evaginations of the vesicle appear. One evagination near the anterior end of the central cavity gives rise to the nerve center (in embryogenesis the nerve ganglion is a remnant of the neural tube formed by infolding of the ectodermal neural plate, as in *Amphioxus* and in vertebrates). Two pockets at the posterior end give rise to the intestinal loop (including the esophagus and the stomach) and to the pericardium respectively. In embryogenesis the intestine is, of course, of endodermal origin and the pericardium develops from mesodermal mesenchyme. The initial state of the system in blastogenesis being not the same as that in embryogenesis, the course of development is different. The morphogenetic processes in budding appear to be simpler and more straightforward than in embryogenesis; the actively developing part, which is the inner vesicle (the epidermis of the bud is a differentiated tissue all the time, specialized in secreting the cellu-

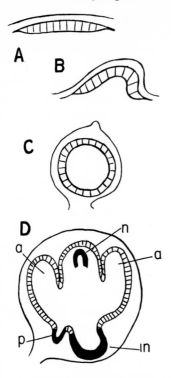

Fig. 289. Development of the bud in *Botryllus*. *a*, Rudiment of atrial cavity; *in*, rudiment of intestine; *n*, rudiment of ganglion; *p*, rudiment of pericardium. (From Berrill, 1951.)

lose mantle), proceeds directly to the formation of organ rudiments, omitting the stage of germ layer development.

In one further respect the morphogenetic processes in budding are simpler than in embryogenesis: the new individual inherits its polarity direct from the parent zooid. The point of attachment of the bud to the maternal body or the stolon always becomes the proximal end of the blastozooid.

In the development of gemmules the task of producing a new individual becomes most complicated and approaches that of the development of the egg. The special difficulties encountered are:

(1) The germination of the gemmule occurs after the death and decomposition of the parent animal, consequently the polarity of the new individual has to be worked out by itself and the parent body is no longer there to influence the polarity of the offspring.

(2) The complete homogeneity of the contents of the gemmule (in the case of the gemmules of sponges) deprives the new individual of any remnant of morphological organization. The structure of the new sponge has to be established by the interaction of practically independent cells.

As has been shown above (p. 511) the interior of the gemmule consists of only one type of cell—the archeocytes, which are rather large cells containing platelets of glycoprotein. Even before the germination of the gemmule some of the archeocytes become activated; they start dividing and in so doing give rise to smaller and smaller cells, very much like the blastomeres which diminish in size as cleavage progresses. The glycoprotein platelets gradually disappear; the nuclei become richer in chromatin, as is usual in actively growing and metabolizing cells. These small and active cells have been referred to as **histioblasts** (cells producing tissues) (Brien, 1932). When the gemmule germinates, the contents of the gemmule crawl out through the micropyle and form an irregular mass, surrounding the empty shell of the gemmule (Fig. 290, *a*). Both the histioblasts and the remaining glycoprotein-containing archeocytes leave the shell. Outside the shell, the divisions of archeocytes and their conversion into histioblasts continue (Fig. 291). The histioblasts now become arranged into an irregular meshwork (Fig. 290, *b*), cavities appear, and some of the histioblasts surrounding these cavities differentiate into choanocytes. Other histioblasts become epidermal cells, scleroblasts, pore cells or mesenchyme cells arranged in a typical way, so that the mass of cells soon becomes a small sponge, with a system of internal canals with ciliated chambers, ostia, etc.

It may be significant that the sponges are capable of reconstituting their structure after complete disaggregation, as has been described above, in

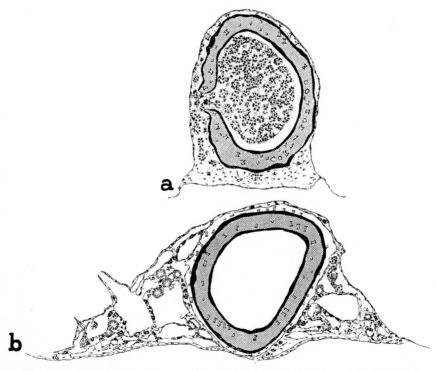

Fig. 290. Germination of gemmules. *a,* Contents of the gemmule of *Spongilla,* leaving the shell. *b,* Transformation of the contents of the gemmule of *Ephydatia* into a new sponge. (From Brien, 1932.)

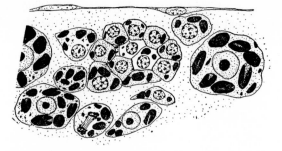

Fig. 291. Transformation of archeocytes into histioblasts during development of a new sponge from a gemmule. Archeocytes still with deutoplasmic inclusions; one is in mitosis. (From Brien, 1932.)

section 19–9. The development of the gemmule proceeds along very much the same lines; the development is direct in the extreme, the individual cells differentiating and taking up their positions in the whole.

A review of different forms of asexual reproduction reveals a principle of general significance. It shows that the complexity of morphogenetic processes is largely determined by the degree of difference that exists between the initial stage of development and the final condition. The initial stage in embryogenesis is a single cell; therefore a period of cleavage is necessary, which brings the system into a multicellular condition. In asexual reproduction the initial system is already multicellular and cleavage falls away. (Something resembling cleavage occurs, as indicated above, in the gemmules of sponges, where the archeocytes accumulate food reserves, similarly to oocytes in the ovary.) In asexual reproduction the initial system may have the cells arranged in more than one layer and this makes gastrulation dispensable. What remains of the main periods of development outlined in section 1–2 are: organogenesis, differentiation and growth.

A second very suggestive fact which emerges from a study of asexual reproduction is that given normal environment, the organization of the animal's body is entirely determined, in the last instance, by the hereditary constitution of the species-specific cells. The structure of the egg cell with its polarity and heterogeneous arrangement of cytoplasmic substances is a mechanism which provides for the orderly course of differentiation of the cells in a developing embryo. This mechanism is, however, dispensable; the same end product can be attained starting from a different initial constellation provided that the cells have the same hereditary constitution. In principle, the orderly organization of an animal's body should be attainable by the interaction of an assortment of different types of cells produced on the basis of species-specific hereditary potentialities (compare the reaggregation of cell suspensions, section 14–4); in practice this is possible only in relatively very simple biological systems.

References[*]

ABELOOS, M., 1956. Les métamorphoses. Armand Collin, Paris.

ABRAMS, R., 1951. Synthesis of nucleic acid purines in the sea urchin embryo. Exp. Cell Res. *2*, 235–242.

ADELMANN, H. B., 1932. The development of the prechordal plate and mesoderm of *Amblystoma punctatum*. J. Morph. *54*, 1–67.

————, 1936. The problem of cyclopia. I. II. Quart. Rev. Biol. *11*, 161–182, 284–304.

————, 1937. Experimental studies on the development of the eye. IV. The effect of the partial and complete excision of the prechordal substrate on the development of the eyes of *Amblystoma punctatum*. J. exp. Zool. *75*, 199–237.

ADLER, L., 1914. Metamorphosestudien an Batrachierlarven. Roux Arch. *39*, 21–45.

AFZELIUS, B. A., 1956. The ultrastructure of the cortical granules and their products in the sea urchin egg as studied with the electron microscope. Exp. Cell Res. *10*, 257–285.

————, 1957. Electron microscopy on the basophilic structures of the sea urchin egg. Zeit. Zellforsch. *45*, 660–675.

ALBAUM, H. G., and NESTLER, H. A., 1937. Xenoplastic ear induction between *Rana pipiens* and *Amblystoma punctatum*. J. exp. Zool. *75*, 1–9.

ALDERMAN, A. L., 1935. The determination of the eye in the anuran, *Hyla regilla*. J. exp. Zool. *70*, 205–232.

————, 1938. A factor influencing the bilaterality of the eye rudiment in *Hyla regilla*. Anat. Rec. *72*, 297–302.

* For brevity the title of the journal "Wilhelm Roux' Archiv für Entwicklungsmechanik der Organismen," previously "Archiv für Entwicklungsmechanik der Organismen," is given as "Roux Arch." Other abbreviations are according to Smith, W. A., Kent, F. L. and Stratton, G. B., 1952, World list of scientific periodicals. Butterworths, London.

ALFERT, M., and SWIFT, H., 1953. Nuclear DNA constancy: a critical evaluation of some exceptions reported by Lison and Pasteels. Exp. Cell Res. *5*, 455–460.

ALLEN, B. M., 1918. The results of thyroid removal in larvae of *Rana pipiens*. J. exp. Zool. *24*, 499–519.

———, 1929. The influence of the thyroid and hypophysis upon growth and development of amphibian larvae. Quart. Rev. Biol. *4*, 325–352.

———, 1938. The endocrine control of amphibian metamorphosis. Biol. Rev. *13*, 1–19.

ALLEN, E., 1923. Ovogenesis during sexual maturity. Amer. J. Anat. *31*, 439.

ANCEL, P., and VINTEMBERGER, P., 1948. Recherches sur le déterminisme de la symétrie bilaterale dans l'oeuf des amphibiens. Bull. biol. Suppl. *31*, 1–182.

ANDRES, G., 1953. Experiments on the fate of dissociated embryonic cells (chick) disseminated by the vascular route. Part II. Teratomas. J. exp. Zool. *122*, 507–540.

ANTHONY, R., 1913. Etude expérimentale des facteurs déterminant la morphologie cranienne des mammifères dépourvus de dents. C.R. Acad. Sci. Paris *157*, 649–650.

AREY, L. B., 1947. Developmental anatomy. 5th ed. Saunders, Philadelphia.

———, 1954. Developmental anatomy. 6th ed. Saunders, Philadelphia.

BAER, K. E. VON, 1828. Ueber Entwicklungsgeschichte der Tiere, Beobachtung und Reflexion. Königsberg.

BALINSKY, B. I., 1925. Transplantation des Ohrbläschens bei *Triton*. Roux Arch. *105*, 718–731.

———, 1931. Zur Dynamik der Extremitätenknospenbildung. Roux Arch. *123*, 565–648.

———, 1933. Das Extremitätenseitenfeld, seine Ausdehnung und Beschaffenheit. Roux Arch. *130*, 704–747.

———, 1939. Experiments on total extirpation of the whole entoderm in *Triton* embryos. C.R. Acad. Sci. URSS *23*, 196–198.

———, 1947. Kinematik des entodermalen Materials bei der Gestaltung der wichtigsten Teile des Darmkanals bei den Amphibien. Roux Arch. *143*, 126–166.

———, 1948. Korrelationen in der Entwicklung der Mund- und Kiemenregion und des Darmkanals bei Amphibien. Roux Arch. *143*, 365–395.

———, 1950. On the developmental processes in mammary glands and other epidermal structures. Trans. roy. Soc. Edinb. *62*, 1–31.

———, 1951. On the eye cup–lens correlation in some South African amphibians. Experientia *7*, 180.

———, 1958. On the factors controlling the size of the brain and eyes in anuran embryos. J. exp. Zool. *139*, 403–442.

BALKASCHINA, E. L., 1929. Ein Fall der Erbhomoösis (die Genovariation Aristopedia) bei *Drosophila melanogaster*. Roux Arch. *115*, 448–463.

BALTZER, F., 1940. Ueber erbliche letale Entwicklung und Austauschbarkeit artverschiedener Kerne bei Bastarden. Naturwissenschaften *28*, 177–187, 196–206.

———, 1941. Ueber die Pigmentierung merogonisch-haploider Bastarde zwischen der schwarzen und weissen Axolotlrasse. Verh. schweiz. naturf. Ges. 121 Jahresvers. Basel, 169–179.

BANDIER, J., 1937. Histologische Untersuchungen über die Regeneration von Landplanarien. Roux Arch. *135*, 316–348.

BANKI, O., 1929. Die Entstehung der äusseren Zeichen der bilateralen Symmetrie am Axolotlei nach Versuchen mit örtlicher Vitalfärbung. Verh. X. int. Zool. Kongr. Budapest.

BARTH, L. G., 1940. The process of regeneration in hydroids. Biol. Rev. *15*, 405–420.

———, and BARTH, L. J., 1954. The energetics of development. Columbia University Press,   New York.

BATAILLON, E., 1910. L'embryogenèse complète provoquée chez les Amphibiens par picûre de l'oeuf vierge, larves parthénogénétiques de *Rana fusca*. C. R. Acad. Sci. Paris. *150*, 996.

BAUTZMANN, H., 1926. Experimentelle Untersuchungen zur Abgrenzung des Organisationszentrums bei *Triton taeniatus*, mit einem Anhang: Ueber Induktion durch Blastulamaterial. Roux Arch. *108*, 283–321.

———, HOLTFRETER, J., SPEMANN, H., and MANGOLD, O., 1932. Versuche zur Analyse der Induktionsmittel in der Embryonalentwicklung. Naturwissenschaften 971–974.

BEATTY, R. A., 1957. Parthenogenesis and Polyploidy in Mammalian Development. University Press, Cambridge.

BEER, G. R. DE, 1947. The differentiation of neural crest cells into visceral cartilages and odontoblasts in *Amblystoma,* and a re-examination of the germ layer theory. Proc. roy. Soc. B *134,* 377–398.

BELLAIRS, R., 1953. Studies on the development of the foregut in the chick blastoderm. 2. The morphogenetic movements. J. Embr. exp. Morph. *1,* 369–385.

BERRILL, N. J., 1935. Studies on tunicate development. III. Differential retardation and acceleration. Phil. Trans. B *225,* 255–379.

————, 1941. Size and morphogenesis in the bud of *Botryllus.* Biol. Bull., Woods Hole *80,* 185–193.

————, 1947. The structure, development and budding of the ascidian, *Eudistoma.* J. Morph. *81,* 269–281.

————, 1951. Regeneration and budding in tunicates. Biol. Rev. *26,* 456–475.

————, 1952. Regeneration and budding in worms. Biol. Rev. *27,* 401–438.

BERTALANFFY, L. VON, 1948. Das organische Wachstum und seine Gesetzmässigkeiten. Experientia *4,* 255.

————, 1949. Problems of organic growth. Nature 156–158.

BIJTEL, J. H., 1931. Ueber die Entwicklung des Schwanzes bei Amphibien. Roux Arch. *125,* 448–486.

————, 1936. Die Mesodermbildungspotenzen der hinteren Medullarplattenbezirke bei *Amblystoma mexicanum* in bezug auf die Schwanzbildung. Roux Arch. *134,* 262–282.

BIRBECK, M. S. C., and MERCER, E. H., 1957. Electron microscopic, X-ray and birefringence studies on the proteins of the hair follicle. In SJÖSTRAND, F. S., and RHODIN, J.: Electron microscopy. Almqvist and Wiksell, Stockholm, 158–160.

BISHOP, M. W. H., and AUSTIN, C. R., 1957. Mammalian spermatozoa. Endeavour *16,* 137–150.

BOELL, E. J., 1945. Functional differentiation in embryonic development. II. Respiration and cytochrome oxidase activity in *Amblystoma punctatum.* J. exp. Zool. *100,* 331–352.

————, 1948. Biochemical differentiation during amphibian development. Ann. N.Y. Acad. Sci. *49,* 773–800.

————, 1955. Energy exchange and enzyme development during embryogenesis. In: WILLIER, B. H., WEISS, P. A., and HAMBURGER, V., Analysis of development. Saunders, Philadelphia, 520–555.

BONNET, C., 1745. Traité d'Insectologie, Paris. Quoted after Needham, J., 1959.

BORGHESE, E., 1950. The development *in vitro* of the submandibular and sublingual glands of *Mus musculus.* J. Anat. Lond. *84,* 287–302.

BOUNHIOL, J., 1937. La métamorphose des insectes serait inhibée dans leur jeune age par les corpora allata? C.R. Soc. Biol. Paris *126,* 1189–1191.

BOUNOURE, L., 1939. L'origine des cellules reproductrices et le problème de la lignée germinale. Gauthier-Villars, Paris.

————, 1940. Continuité germinale et reproduction agame. Gauthier-Villars, Paris.

BOYCOTT, A. E., DIVER, C., GARSTANG, S. L., and TURNER, F. M., 1930. The inheritance of sinistrality in *Limnaea peregra* (Mollusca, Pulmonata). Phil. Trans. B, *219,* 51–131.

BRACHET, A., 1935. Traité d'embryologie des vertébrés. 2nd ed., revised by DALCQ, A. and GÉRARD, P., Masson, Paris.

BRACHET, J., 1941. La localisation des acides pentosenucléiques dans les tissus animaux et les oeufs d'Amphibiens en voie de developpement. Arch. Biol., Paris *53,* 207–257.

————, 1947. The metabolism of nucleic acids during embryonic development. Cold Spr. Harb. Symp. quant. Biol. *12,* 18–27.

————, 1950a. Les caractéristiques biochémique de la compétence et de l'induction. Rev. Suisse Zool. *57,* Fascicule suppl. 1, 57–75.

————, 1950b. Chemical Embryology. Interscience Publishers, New York.

BRAMBELL, W. F. R., HEMMINGS, W. A. and HENDERSON, M., 1951. Antibodies and embryos. Athlone Press, London.

BRASH, J. C., 1929. The aetiology of irregularity and malocclusion of teeth. Publ. Dental Board, U. K.

BRAUER, A., 1902. Beiträge zur Kenntniss der Entwicklung und Anatomie der Gymnophionen. III, Die Entwicklung der Excretionsorgane. Zool. Jb. Abt. Anat. u. Ont., *16*, 1–176.

BRETSCHNEIDER, L. H., and RAVEN, C. P., 1951. Structural and topochemical changes in the egg cells of *Limnaea stagnalis* L. during oogenesis. Arch. néerl. Zool. *10*, 1–31.

BRIEN, P., 1932. Contribution a l' étude de la régénération naturelle chez les Spongillidae. Arch. Zool. exp. gén. *74*, 461–506.

BRIGGS, R., and KING, T. J., 1952. Transplantation of living nuclei from blastula cells into enucleated frogs' eggs. Proc. Nat. Acad. Sci. *38*, 455–463.

——, ——, 1953. Factors affecting the transplantability of nuclei of frog embryonic cells. J. exp. Zool. *122*, 485–506.

——, ——, 1957. Changes in the nuclei of differentiating endoderm cells as revealed by nuclear transplantation. J. Morph. *100*, 269–312.

BROMLEY, N. W., and ORECHOWITSCH, W. N., 1934. Ueber die Proteolyse in den regenerierenden Geweben. II. Die Aktivität der Gewebeprotease in verschiedenen Gebieten des Regenerats. Biochem. Z. *272*, 324–331.

BRUNST, V. V., 1950. Influence of X-rays on limb regeneration in urodele amphibians. Quart. Rev. Biol. *25*, 1–29.

——, and CHÉRÉMÉTIEVA, E. A., 1936. Sur la perte locale du pouvoir régénérateur chez le triton et l'axolotl causée par l'irradiation avec les rayons X. Arch. Zool. exp. gén. *78*, 57–67.

——, and SCHEREMETJEWA, E. A., 1933. Untersuchung des Einflusses von Röntgenstrahlen auf die Regeneration der Extremitäten beim Triton. Roux Arch. *128*, 181.

BRUYN, P. P. H. DE, 1945. The motion of migrating cells in tissue cultures of lymph nodes. Anat. Rec. *93*, 295–315.

BUEKER, E. D., 1947. Limb ablation experiments on the embryonic chick and its effect as observed on the mature nervous system. Anat. Rec. *97*, 157–174.

BURKE, V., SULLIVAN, N. P., PETERSEN, H., and WEED, R., 1944. Ontogenetic change in antigenic specificity of the organs of the chick. J. infect. Dis. *74*, 225–233.

BURR, H. S., 1916. The effect of removal of the nasal pits in *Amblystoma* embryos. J. exp. Zool. *20*, 27–57.

——, 1930. Hyperplasia in the brain of *Amblystoma*. J. exp. Zool. *55*, 171–191.

BUTENANDT, A., and KARLSON, P., 1954. Ueber die Isolierung eines Metamorphose-Hormons der Insecten in kristallisierter Form. Z. Naturf. *9b*, 389–391.

BUTLER, E. G., 1933. The effects of X-radiation on the regeneration of the fore limb of *Amblystoma* larvae. J. exp. Zool. *65*, 271–315.

——, 1935. Studies on limb regeneration in X-rayed *Amblystoma* larvae. Anat. Rec. *62*, 295–307.

BYTINSKI-SALZ, H., 1936. Lo sviluppo della coda negli anfibi. II. Alterazioni della correlazioni fra i territori costituenti l'abbozzo codale e comportamento dell' ectoderma della pinna caudale. R. C. Accad. Lincei. *24*, 82–88.

——, 1937. Trapianti di "organizzatore" nelle uova di Lampreda. Arch. ital. Anat. Embriol. *39*, 177–228.

CAMPENHOUT, E. VAN, 1935. Experimental researches on the origin of the acoustic ganglion in amphibian embryos. J. exp. Zool. *72*, 175–193.

CARTER, T. C., 1954. The genetics of luxate mice. IV. Embryology. J. Genet. *52*, 1–35.

CASTLE, W. E., and GREGORY, P. W., 1929. The embryological basis of size inheritance in the rabbit. J. Morph. *48*, 81.

CHANG, M. C., 1954. Development of parthenogenetic rabbit blastocysts induced by low temperature storage of unfertilized ova. J. exp. Zool. *125*, 127–149.

CHILD, C. M., 1936. Differential reduction of vital dyes in the early development of echinoderms. Roux Arch. *135*, 426–456.

——, 1941. Patterns and problems of development. University of Chicago Press, Chicago.

——, 1948. Exogastrulation by sodium azide and other inhibiting conditions in *Strongylocentrotus purpuratus*. J. exp. Zool. *107*, 1–38.

References

CLAYTON, R. M., 1951. Antigens in the developing newt embryo. Nature (Lond.) *168*, 120–121.

COLWIN, A. L., and COLWIN, L. H., 1957. Morphology of fertilization: Acrosome filament formation and sperm entry. In: TYLER, A., BORSTEL, R. C. VON, and METZ, C. B. The beginnings of embryonic development. American Association for the Advancement of Science, Washington.

———, ———, and PHILPOTT, D. E. 1957. Electron microscope studies of early stages of sperm penetration in *Hydroides hexagonus* (Annelida) and *Saccoglossus kowalewskii* (Enteropneusta) J. Biophys. Biochem. Cytol. *3*, 489–502.

CONKLIN, E. G., 1905. The orientation and cell-lineage of the ascidian egg. J. Acad. nat. Sci. Philad. Ser. 2, *13*.

———, 1931. The development of centrifuged eggs of ascidians. J. exp. Zool. *60*, 1–119.

———, 1932. The embryology of *Amphioxus*. J. Morph. *54*, 69–118.

COOPER, R. S., 1948. Antigens in development. J. exp. Zool. *107*, 397–433.

COPENHAVER, W. M., 1926. Experiments on the development of the heart of *Amblystoma punctatum*. J. exp. Zool. *43*, 321–371.

———, 1930. Results of heteroplastic transplantation of anterior and posterior parts of the heart rudiment in *Amblystoma* embryos. J. exp. Zool. *55*, 293–318.

———, 1933. Transplantation of heart and limb rudiments between *Amblystoma* and *Triton* embryos. J. exp. Zool. *65*, 131–157.

COSTELLO, D. P., 1945. Experimental studies of germinal localization in Nereis I. The development of isolated blastomeres. J. exp. Zool. *100*, 19–66.

———, 1948. Ooplasmic segregation in relation to differentiation. Ann. N.Y. Acad. Sci. *49*, 663–683.

DALCQ, A. M., 1954. Nouvelles données structurales et cytochimiques sur l'oeuf des mammifères. Rev. gén. Sci. pur. appl. *61*, 19–41.

———, and PASTEELS, J., 1937. Une conception nouvelle des bases physiologique de la morphogénèse. Arch. Biol. Liége, *48*, 121–147.

DAN, J. C., and WADA, S. K., 1955. Studies on the acrosome. IV. The acrosome reaction in some bivalve spermatozoa. Biol. Bull. Woods Hole *109*, 40–55.

DANFORTH, C. H., 1930. Developmental anomalies in a special strain of mice. Amer. J. Anat. *45*, 275–288.

DANIELLI, J. F., 1953. Cytochemistry—a critical approach. Chapman and Hall, London.

DANTSCHAKOFF, V., 1941. Der Aufbau des Geschlechts beim höheren Wirbeltier. Gustav Fischer, Jena.

DARLINGTON, C. D., 1944. Heredity, development, and infection. Nature (Lond.) *154*, 164–169.

DAVENPORT, C. B., 1895. Studies in morphogenesis. IV. A preliminary catalogue of the processes concerned in ontogeny. Bull. Mus. comp. Zool. Harv. *27*, 173–199.

DAVIDSON, J. N., 1947. Some factors influencing the nucleic acid content of cells and tissues. Cold Spr. Harb. Symp. quant. Biol. *12*, 50–59.

——— and LESLIE, I., 1950. A new approach in the biochemistry of growth and development. Nature (Lond.) *165*, 49–53.

DELAGE, Y., 1884. Evolution de la Sacculine. Arch. Z. expér. *2*, 417–736.

DETWILER, S. R., 1918. Experiments on the development of the shoulder girdle and the anterior limb of *Amblystoma punctatum*. J. exp. Zool. *25*, 499–538.

———, 1920. Experiments on the transplantation of limbs in *Amblystoma*. The formation of nerve plexuses and the function of the limbs. J. exp. Zool. *31*, 117–169.

———, 1926a. The effect of reduction of skin and muscle on the development of spinal ganglia. J. exp. Zool. *45*, 399–414.

———, 1926b. Experimental studies on morphogenesis in the nervous system. Quart. Rev. Biol. *1*, 61–86.

———, 1930. Observations upon the growth, function, and nerve supply of limbs when grafted to the head of salamander embryos. J. exp. Zool. *55*, 319–379.

———, 1934. An experimental study of spinal nerve segmentation in *Amblystoma* with reference to the pluri-segmental contribution to the brachial plexus. J. exp. Zool. *67*, 395–441.

———, 1936. Neuroembryology: An experimental study. Macmillan Co., New York.

DETWILER, S. R., 1949. The swimming capacity of *Amblystoma* larvae following reversal of the embryonic hindbrain. J. exp. Zool. *111*, 79–94

———, and DYKE, R. H. VAN, 1934. Further observations upon abnormal growth responses of spinal nerves in *Amblystoma* embryos. J. exp. Zool. *69*, 137–164.

DOLJANSKI, L., and ROULET, F., 1934. Zur Frage der Entstehung der bindegewebigen Strukturen. Roux Arch. *131*, 512–531.

DOLLANDER, A., 1953. Observations relatives a certaines propriétés du cortex de l'oeuf d'amphibien. Arch. Anat. micr. Morph. exp. *42*, 185–193.

DORRIS, F., 1935. The development of structure and function in the digestive tract of *Amblystoma punctatum*. J. exp. Zool. *70*, 491–527.

DRAGOMIROW, N., 1929. Ueber die Faktoren der embryonalen Entwicklung der Linse bei Amphibien. Roux Arch. *116*, 633–668.

———, 1933. Ueber Koordination der Teilprocesse in der embryonalen Morphogenese des Augenbechers. Roux Arch. *129*, 522–560.

———, 1936. Ueber Induktion sekundärer Retina im transplantierten Augenbecher bei *Triton* und *Pelobates*. Roux Arch. *134*, 716–737.

DREW, A. H., 1923. Growth and differentiation in tissue cultures. Brit. J. Exp. Path. *4*, 46–52.

DRIESCH, H., 1891. Entwicklungsmechanische Studien. I-II. Z. wiss. Zool. *53*, 160–182.

DUBOIS, F., and WOLFF, E., 1947. Sur une méthode d'irradiation localisée permettant de mettre en évidence la migration des cellules de régénération chez les Planaires. C. R. Soc. Biol. Paris *141*, 903–906.

DÜRKEN, B., 1911. Ueber frühzeitige Extirpation von Extremitätenanlagen beim Frosch. Z. wiss. Zool. *99*, 189–355.

DU SHANE, G. P., 1935. An experimental study of the origin of pigment cells in amphibia. J. exp. Zool. *72*, 1–31.

DUSPIVA, F., 1942. Die Verteilung der Peptidase auf Kern und Plasma bei Froschoozyten im Verlauf der zweiten Wachstumsperiode. Biol. Zbl. *62*, 403–431.

EBERT, J. D., 1950. An analysis of the effects of anti-organ sera on the development, in vitro, of the early chick blastoderm. J. exp. Zool. *115*, 351–377.

EKMAN, G., 1925. Experimentelle Beiträge zur Herzentwicklung der Amphibien. Roux Arch. *106*, 320–352.

ENDRES, H., 1895. Ueber Anstich- und Schnürversuche an Eiern von *Triton taeniatus*. Jber. Schles. Ges. vaterländ. Kultur. *73*.

EPHRUSSI, B., 1942. Chemistry of the "eye color hormones" of *Drosophila*. Quart. Rev. Biol. *17*, 327–338.

EVANS, H. M., and SWEZY, O., 1931. Ovogenesis and the normal follicular cycle in adult mammalia. Mem. Univ. Calif., *9*, 119.

———, SIMPSON, M. E., MEYER, R. K., and REICHERT, 1933. The growth and gonad-stimulating hormones of the anterior hypophysis. Mem. Univ. Calif. *11*, 1–446.

FALES, D. E., 1935. Experiments on the development of the pronephros of *Amblystoma punctatum*. J. exp. Zool. *72*, 147–173.

FAURÉ-FREMIET, E., 1925. La cinétique du developpement. Presses Universit. de France, Paris.

FELL, H. B., 1928. Experiments in vitro on the differentiation of cartilage and bone. Part I. Arch. Zellforsch. *7*, 390–412.

FICQ, A., 1954. Analyse de l'induction neurale chez les Amphibiens au moyen d'organisateurs marqués. J. Embr. exp. Morph. *2*, 194–203.

FILATOFF, D., 1916. The removal and transplantation of the auditory vesicle of the embryo of *Bufo* (the correlations at the formation of the cartilaginous skeleton). Russk. zool. Zh. *1*, 48–54.

FISCHEL, A., 1919. Ueber den Einfluss des Auges auf die Entwicklung und Erhaltung der Hornhaut. Klin. Mbl. Augenheilk. *62*, 1–5.

FISCHER, F. G., WEHMEIER, E., LEHMANN, H., JÜHLING, L., and HULTZSCH, K., 1935. Zur Kenntnis der Induktionsmittel in der Embryonal-Entwicklung. Ber. dtsch. chem. Ges. *68*, 1196–1199.

FLYNN, T. T., and HILL, J. P., 1939. The development of the Monotremata. IV. Growth of the ovarian ovum, maturation, fertilization, and early cleavage. Trans. zool. Soc. Lond. *24*, 445–582.

FRANK, G. M., 1925. Ueber Gesetzmässigkeiten in der Mitosenverteilung in den Gehirn-blasen im Zusammenhang mit Formbildungsprozessen. Arch. mikr. Anat. u. Entw. mech. *104*, 262–272.

FRASER, E. A., 1950. The development of the vertebrate excretory system. Biol. Rev. *25*, 159–187.

FRASER, R. C., 1954. Studies on the hypoblast of the young chick embryo. J. exp. Zool. *126*, 349–399.

GABRIEL, M. L., and FOGEL, S., 1955. Great experiments in biology. Prentice-Hall, Englewood Cliffs, N. J.

GAILLARD, P. J., 1942. Hormones regulating growth and differentiation in embryonic explants. Hermann and Co., Paris.

GALTSOFF, P. S., 1925. Regeneration after dissociation (an experimental study on sponges). I. Behaviour of dissociated cells of *Microciona prolifera* under normal and altered conditions. J. exp. Zool. *42*, 183–255.

GEBHARDT, H., 1926. Untersuchungen über die Determination bei Planarienregeneraten. Roux Arch. *107*, 684–726.

GILCHRIST, F. G., 1937. Budding and locomotion in the Scyphostomas of *Aurelia*. Biol. Bull. Woods Hole *72*, 99–124.

GINSBURG, A., 1953. The origin of bilateral symmetry in the eggs of acipenserid fishes. C.R. Acad. Sci. URSS *90*, 477–480.

———, and DETTLAFF, T., 1944. Experiments on transplantation and removal of organ rudiments in embryos of *Acipenser stellatus* in early developmental stages. C.R. Acad. Sci. URSS *44*, 209–212.

GLATTHAAR, E., and TÖNDURY, G., 1950. Untersuchungen an abortierten Früchten nach Rubeolaerkrankung der Mutter in der Frühschwangerschaft. Gynaecologia *129*, 315–320.

GLUECKSOHN-SCHOENHEIMER, S., 1943. The morphological manifestation of a domi-nant mutation in mice affecting tail and urogenital system. Genetics *28*, 341–348.

———, 1945. The embryonic development of mutants of the Sd-strain in mice. Gen-etics *30*, 29–38.

———, 1949. The effects of a lethal mutation responsible for duplications and twinning in mouse embryos. J. exp. Zool. *110*, 47–76.

GOERTTLER, K., 1928. Die Bedeutung der ventrolateralen Mesodermbezirke für die Herzanlage der Amphibienkeime. Anat. Anz. Erg. Heft *66*, 132–139.

GOETZ, R. H., 1938. On the early development of the Tenrecoidea (*Hemicentetes semi-spinosus*). Biomorphosis *1*, 67–79.

GOLDSCHMIDT, R., 1938. Physiological genetics. McGraw-Hill, New York.

———, 1945. The structure of podoptera, a homoeotic mutant of *Drosophila melano-gaster*. J. Morph. *77*, 71–103.

GOODRICH, E. S., 1930. Studies on the structure and development of vertebrates. Mac-millan, London.

GORBUNOVA, G. P., 1939. On the inducing properties of the medulla oblongata in amphibian embryos. C. R. Acad. Sci. URSS *23*, 298–301.

GORDON, C., 1936. The frequency of heterozygosis in free living populations of *Droso-phila melanogaster* and *Drosophila subobscura*. J. Genet. *33*, 25–60.

GRANT, P., 1953. Phosphate metabolism during oogenesis in *Rana temporaria*. J. exp. Zool. *124*, 513–543.

GREGORY, P. W., and CASTLE, W. E., 1931. Further studies on the embryological basis of size inheritance in the rabbit. J. exp. Zool. *59*, 199–211.

GROBSTEIN, C., 1953a. Analysis in vitro of the early organization of the rudiment of the mouse sub-mandibular gland. J. Morph. *93*, 19–44.

———, 1953b. Epithelio-mesenchymal specificity in the morphogenesis of mouse sub-mandibular rudiments in vitro. J. exp. Zool. *124*, 383–413.

———, 1953c. Morphogenetic interaction between embryonic mouse tissues separated by a membrane filter. Nature (Lond.) *172*, 869.

———, 1955. Inductive interaction in the development of the mouse metanephros. J. exp. Zool. *130*, 319–339.

———, 1957. Some transmission chracteristics of the tubule-inducing influence on mouse metanephrogenic mesenchyme. Exp. Cell Res. *13*, 575–587.

GROBSTEIN, C., and DALTON, A. J., 1957. Kidney tubule induction in mouse meta-nephrogenic mesenchyme without cytoplasmic contact. J. exp. Zool. *135*, 57–73.

GROSSER, O., 1945. Grundriss der Entwicklungsgeschichte des Menschen. Springer, Berlin.

GRUNEBERG, H., 1936. Gray-lethal, a new mutation in the house mouse. J. Hered. *27*, 105–109.

———, 1938. An analysis of the "pleiotropic" effects of a new lethal mutation in the rat (*Mus norvegicus*). Proc. roy. Soc. B *125*, 123–144.

———, 1952. The genetics of the mouse. Martinus Nijhoff, Hague.

GUARESCHI, C., 1935. Studi sulla determinazione dell' orechio interno degli anfibi anuri. Arch. ital. Anat. Embriol. *35*, 97–129.

GUDERNATSCH, F., 1912. Feeding experiments on tadpoles. Roux Arch. *35*, 457–483.

GUSTAFSON, T., 1950. Survey of the morphological action of the lithium ion and the chemical basis of its action. Rev. Suisse Zool. *57*, Suppl. 1, 77–92.

———, and HASSELBERG, I., 1951. Studies on enzymes in the developing sea urchin egg. Exp. Cell Res. *2*, 642–672.

———, and HJELTE, M., 1951. The amino-acid metabolism of the developing sea urchin egg. Exp. Cell Res. *2*, 474–490.

GUYÉNOT, E., and PONSE, K., 1930. Territoires de régénération et transplantations. Bull. biol. *64*, 251–287.

GUYER, M. F., and SMITH, E. A., 1918, 1920. Studies on cytolysins. I. II. J. exp. Zool. *26*, 65–82, *31*, 171–223.

HADORN, E., 1932. Ueber Organentwicklung und histologische Differenzierung in transplantierten merogonischen Bastardgeweben (*Triton palmatus* (♀) x *Triton cristatus* ♂). Roux Arch. *125*, 495–565.

HADZI, J., 1910. Die Entstehung der Knospe bei *Hydra*. Arb. zool. Inst. Univ. Wien, *18*.

HAEKEL, E., 1868. Natürliche Schöpfungsgeschichte. Berlin.

HALL, T. S., 1951. A source book in animal biology. McGraw-Hill, New York.

HAMBURGER, V., 1929. Experimentelle Beiträge zur Entwicklungsphysiologie der Nervenbahnen in der Froschextremität. Roux Arch. *119*, 47–99.

———, 1934. The effect of wing bud extirpation on the development of the central nervous system in chick embryos. J. exp. Zool. *68*, 449–494.

———, 1947. A manual of experimental embryology. (2nd imp.) University Press, Chicago.

———, 1956. Developmental correlations in neurogenesis. In: RUDNICK, D. (Editor), Cellular mechanisms in differentiation and growth. Princeton University Press, 191–212.

———, and HAMILTON, H. L., 1951. A series of normal stages in the development of the chick embryo. J. Morph. *88*, 49–92.

———, and LEVI-MONTALCINI, R., 1949. Proliferation, differentiation and degeneration in the spinal ganglia of the chick embryo under normal and experimental conditions. J. exp. Zool. *111*, 457–501.

HAMILTON, W. J., BOYD, J. D., and MOSSMAN, H. W., 1947. Human embryology. Heffer and Sons, Cambridge.

HARDING, C. V., HARDING, D., and PERELMAN, P., 1954. Antigens in sea urchin embryos. Exp. Cell Res. *6*, 202–210.

HARDY, M. H., 1952. The histochemistry of hair follicles in the mouse. J. Anat. Lond. *90*, 285–337.

———, 1953. Vaginal cornification of the mouse produced by oestrogens in vitro. Nature (Lond.) *172*, 1196.

HARRIS, H. A., 1933. Bone growth in health and disease. Oxford University Press, London.

HARRISON, R. G., 1904. Experimentelle Untersuchungen über die Entwicklung der Sinnesorgane der Seitenlinie bei den Amphibien. Arch. mikr. Anat. *63*, 35–149.

———, 1908. Embryonic transplantation and development of the nervous system. Anat. Rec. *2*, 385.

———, 1918. Experiments on the development of the fore limb of *Amblystoma*, a self-differentiating equipotential system. J. exp. Zool. *25*, 413–461.

———, 1921a. Experiments on the development of the gills in the amphibian embryo. Biol. Bull. Woods Hole *41*, 156–168.

———, 1921b. On relations of symmetry in transplanted limbs. J. exp. Zool. *32*, 1–136.

———, 1925a. The development of the balancer in *Amblystoma*, studied by the method of transplantation and in relation to the connective tissue problem. J. exp. Zool. *41*, 349–427.

———, 1925b. The effect of reversing the medio-lateral or transverse axis of the fore-limb bud in the salamander embryo (*Amblystoma punctatum*). Roux Arch. *106*, 469–502.

———, 1929. Correlation in the development and growth of the eye studied by means of heteroplastic transplantation. Roux Arch. *120*, 1–55.

———, 1935. Factors concerned in the development of the ear in *Amblystoma punctatum*. Anat. Rec. *64*, 38–39.

HARVEY, E. B., 1946. Structure and development of the clear quarter of the *Arbacia punctulata* egg. J. exp. Zool. *102*, 253–275.

———, 1956. The American *Arbacia* and other sea urchins. Princeton University Press, Princeton.

HAUSCHKA, T. S., 1951. Differentiation of skeletal structures from mouse embryo mince in the peritoneum of adult mice. Nature (Lond.) *168*, 1130.

HAYASHI, Y., 1955. Inductive effect of some fractions of tissue extracts after removal of pentose nucleic acid, tested on the isolated ectoderm of *Triturus*-gastrula. Embryologia 2, 145–162.

———, 1956. Morphogenetic effects of pentose nucleoprotein from the liver upon the isolated ectoderm. Embryologia *3*, 57–67.

———, 1958. The effects of pepsin and trypsin on the inductive ability of pentose nucleoprotein from guinea pig liver. Embryologia *4*, 33–53.

HELFF, O. M., 1928. Studies on amphibian metamorphosis. III. Physiol. Zool. *1*, 463–495.

HERBST, C., 1893. Experimentelle Untersuchungen über den Einfluss der veränderten chemischen Zusammensetzung des umgebenden Mediums auf die Entwicklung der Tiere. II. Mitt. zool. Sta. Neapel. *11*, 136–220.

HERRMANN, H., and NICHOLAS, J. S., 1948. Quantitative changes in muscle protein fractions during development. J. exp. Zool. *107*, 165–176.

———, ———, 1949. Nucleic acid content of whole homogenates and of fractions of developing rat muscle. J. exp. Zool. *112*, 341–360.

HERTIG, A. T., and ROCK, J., 1941. Two human ova of the pre-villous stage, having an ovulation age of about eleven and twelve days respectively. Contr. Embryol. Carneg. Instn. *29*, 127–156.

———, ———, 1945. Two human ova in the pre-villous stage, having a developmental age of about seven and nine days respectively. Contr. Embryol. Carneg. Instn. *31*, 67–84.

HERTWIG, O., 1906. Handbuch der vergleichenden und experimentellen Entwicklungslehre der Wirbeltiere. G. Fischer, Jena.

HEUSER, C. H., and CORNER, G. W., 1957. Developmental horizons in human embryos. Description of age group X, 4 to 12 somites. Contr. Embryol. Carneg. Instn. *36*, 31–39.

———, and STREETER, G. L., 1941. Development of the macaque embryo. Contr. Embryol. Carneg. Instn. *29*, 17–56.

HILL, J. P., 1918. Some observations on the early development of *Didelphis aurita*. Quart. J. micr. Sci. *63*, 91–140.

HÖRSTADIUS, S., 1928. Ueber die Determination des Keimes der Echinodermen. Acta zool. Stockh. *9*, 1–192.

———, 1935. Ueber die Determination im Verlaufe der Eiachse bei Seeigeln. Pubbl. Staz. zool. Napoli. *14*, 251–479.

———, 1944. Ueber die Folgen von Chordaexstirpation an spaeten Gastrulae und Neurulae von *Amblystoma punctatum*. Acta zool. Stockh. *25*, 75–87.

———, 1950. The neural crest. Oxford University Press, London.

———, 1952. Induction and inhibition of reduction gradients by the micromeres in the sea urchin egg. J. exp. Zool. *120*, 421–436.

HÖRSTADIUS, S., 1953a. Influence of implanted micromeres on reduction gradients and mitochondrial distribution in developing sea urchin eggs. J. Embr. exp. Morph. *1*, 257–259

————, 1953b. Vegetalization of the sea-urchin egg by dinitrophenol and animalization by trypsin and ficin. J. Embr. exp. Morph. *1*, 327–348.

————, 1955. Reduction gradients in animalized and vegetalized sea urchin eggs. J. exp. Zool. *129*, 249–256.

————, and SELLMAN, S., 1946. Experimentelle Untersuchungen über die Determination des knorpeligen Kopfskelettes bei Urodelen. Nova Acta So. Sci. Upsal. *13*.

————, and WOLSKY, A., 1936. Studien über die Determination der Bilateralsymmetrie des jungen Seeigelkeimes. Roux Arch. *135*, 69–113.

HOLTFRETER, J., 1934a. Der Einfluss thermischer, mechanischer und chemischer Eingriffe auf die Induzierfähigkeit von *Triton*-Keimteilen. Roux Arch. *132*, 225–306.

————, 1934b. Ueber die Verbreitung induzierender Substanzen und ihre Leistungen im *Triton*-Keim. Roux Arch. *132*, 307–383.

————, 1938a. Differenzierungspotenzen isolierter Teile der Urodelengastrula. Roux Arch. *138*, 522–656.

————, 1938b. Differenzierungspotenzen isolierter Teile der Anurengastrula. Roux Arch. *138*, 657–738.

————, 1939a. Gewebeaffinität, ein Mittel der embryonalen Formbildung. Arch. exp. Zellforsch. *23*, 169–209.

————, 1939b. Studien zur Ermittlung der Gestaltungsfaktoren in der Organentwicklung der Amphibien. I and II. Roux Arch. *139*, 110–190, 227–273.

————, 1943a. Properties and functions of the surface coat in amphibian embryos. J. exp. Zool. *93*, 251–323.

————, 1943b. A study of the mechanics of gastrulation. I. J. exp. Zool. *94*, 261–318.

————, 1943c. Experimental studies on the development of the pronephros. Rev. canad. Biol. *3*, 220–250.

————, 1946. Experiments on the formed inclusions of the amphibian egg. I. The effect of pH and electrolytes on yolk and lipochondria. J. exp. Zool. *101*, 355–405.

————, 1947a. Observations on the migration, aggregation and phagocytosis of embryonic cells. J. Morph. *80*, 25–56.

————, 1947b. Changes of structure and the kinetics of differentiating embryonic cells. J. Morph. *80*, 57–62.

————, 1947c. Neural induction in explants which have passed through a sublethal cytolysis. J. exp. Zool. *106*, 197–222.

HOLTZER, H., and DETWILER, S. R., 1953. An experimental analysis of the development of the spinal column. III. Induction of skeletogenous cells. J. exp. Zool. *123*, 335–368.

HORST, C. J. VAN DER, 1942. Early stages in the embryonic development of *Elephantulus*. S. Afr. J. Med. Sci. *7*, Biol. Suppl. 55–65.

HUNT, T. E., 1937. The origin of entodermal cells from the primitive streak of the chick embryo. Anat. Rec. *68*, 449–460.

HUXLEY, J. S., 1932. Problems of relative growth. Methuen and Co., London.

————, and BEER, G. R., DE, 1934. The elements of experimental embryology. University Press, Cambridge.

HYMAN, L. H., 1951. The invertebrates, Vol. 3. McGraw-Hill, New York.

JACOBSON, A. G., 1955. The roles of the optic vesicle and other head tissues in lens induction. Proc. Nat. Acad. Sci. *41*, 522–525.

JANSEN, M., 1920. On bone formation, its relation to tension and pressure. Manchester University Press, Manchester.

JOHANNSEN, O. A., and BUTT, F. H., 1941. Embryology of insects and myriapods. McGraw-Hill, New York.

JONES, H. O., and BREWER, J. I., 1941. A human embryo in the primitive-streak stage. Contr. Embryol. Carneg. Instn. *29*, 157–165.

JONES-SEATON, A., 1950. Étude de l'organisation cytoplasmique de l'oeuf des rongeurs, principalement quant à la basophilie ribonucléique. Arch. Biol. Paris, *61*, 291–444.

KAMER, J. C. VAN DE, 1949. Over de ontwikkeling de determinatie en de betekenis van de epiphyse en de paraphyse van de amphibiën. Dissertation. Van der Weil, Arnhem.

KANAJEW, J., 1926. Ueber die histologischen Vorgänge bei der Regeneration von *Pelmatohydra oligactis* Pall. Zool. Anz. *65*, 217–226.

KARASAKI, S., 1957. On the mechanism of the dorsalization in the ectoderm of *Triturus* gastrulae caused by precytolytic treatments. 1. Embryologia *3*, 317–334.

KEMP, N. E., 1953. Synthesis of yolk in oocytes of *Rana pipiens* after induced ovulation. J. Morph. *92*, 487–511.

——, 1956. Electron microscopy of growing oocytes of *Rana pipiens*. J. Biophys. Biochem. Cytol. *2*, 281–292.

KERR, J. G., 1919. Textbook of embryology. Vol. II. Vertebrata with the exception of mammalia. Macmillan, London.

KING, T. J., and BRIGGS, R., 1954. Transplantation of living nuclei of late gastrulae into enucleated eggs of *Rana pipiens*. J. Embr. exp. Morph. *2*, 73–80.

——, ——, 1956. Serial transplantation of embryonic nuclei. Cold Spr. Harb. Symp. quant. Biol. *21*, 271–290.

KITCHIN, I. C., 1949. The effect of notochordectomy in *Amblystoma mexicanum*. J. exp. Zool. *112*, 393–416.

KOGAN, R. E., 1939. Inducing action of the medulla oblongata on the trunk epithelium in amphibia. C.R. Acad. Sci. URSS *23*, 307–310.

KORSCHELT, E., 1936. Vergleichende Entwicklungsgeschichte der Tiere. G. Fischer, Jena.

KOWALEVSKY, A., 1866. Entwicklungsgeschichte der einfachen Ascidien. Mem. Acad. Sci. St. Petersb. *10*.

KÜHN, A., 1936. Versuche über die Wirkungsweise der Erbanlagen. Naturwissenschaften *24*, 1–10.

——, 1939. Zur Entwicklungsphysiologie der Schmetterlingsmetamorphose. Verh. 7 internat. Kongr. Ent. Berlin 780-796.

——, 1941. Ueber die Gen-Wirkkette der Pigmentbildung bei Insecten. Nachr. Akad. Wiss. Göttingen *6*, 231–261.

——, 1955. Vorlesungen über Entwicklungsphysiologie. Springer, Berlin.

——, and PIEPHO, H., 1936. Ueber hormonale Wirkungen bei der Verpuppung der Schmetterlinge. Nachr. Ges. Wiss., Göttingen *2*, 141–154.

KUHL, W., 1941. Untersuchungen über die Cytodynamik der Furchung und Frühentwicklung des Eies der weissen Maus. Abh. senckenb. naturf. Ges. *456*, 1–17.

LALLIER, R., 1956. Les ions de métaux lourds et le problème de la détermination embryonnaire chez les Echinodermes. J. Embr. exp. Morph. *4*, 265–278.

——, 1957. Recherches sur l'animalisation de l'oeuf de l'oursin *Paracentrotus lividus* par les dérivés polysulfoniques. Pubbl. Staz. zool. Napoli *30/2*, 185–209.

LEHMANN, F. E., 1937. Mesodermisierung des präsumptiven Chordamaterials durch Einwirkung von Lithiumchlorid auf die Gastrula von *Triton alpestris*. Roux Arch. *136*, 112–146.

——, 1945. Einführung in die physiologische Embryologie. Birkhäuser, Basel.

LENICQUE, P., HÖRSTADIUS, S., and GUSTAFSON, T., 1953. Change of distribution of mitochondria in animal halves of sea urchin eggs by the action of micromeres. Exp. Cell Res. *5*, 400–403.

LEVITT, M. M., 1932. On the post-embryonic growth of larvae of some Lepidoptera. Trav. Inst. Biol. Kiev. *5*, 451–468.

LEWIS, W. H., 1904. Experimental studies on the development of the eye in Amphibia. I. On the origin of the lens in *Rana palustris*. Amer. J. Anat. *3*, 505–536.

——, 1907. On the origin and differentiation of the otic vesicle in amphibian embryos. Anat. Rec. *1*, 141–145.

——, and HARTMANN, C. G., 1933. Early cleavage stages of the egg of the monkey (*Macacus rhesus*). Contr. Embryol. Carneg. Instn. *24*, 189–202.

LIEDKE, K. B., 1951. Lens competence in *Amblystoma punctatum*. J. exp. Zool. *117*, 573–591.

——, 1955. Studies on lens induction in *Amblystoma punctatum*. J. exp. Zool. *117*, 353–379.

LILLIE, F. R., 1919a. Problems of fertilization. University of Chicago Press, Chicago.

——, 1919b. The development of the chick. 2nd ed. Henry Holt, New York.

LINDAHL, P. E., 1933. Ueber "animalisierte" und "vegetativisierte" Seeigellarven. Roux Arch. *128*, 661–664.

LINDAHL, P. E., 1936. Zur Kenntnis der physiologischen Grundlagen der Determination im Seeigelkeim, Acta Zool. (Stockh.) *17*, 179–395.

LINDEGREN, C. C., and LINDEGREN, G., 1946. The cytogen theory. Cold Spr. Harb. Symp. quant. Biol. *11*, 115–129.

LIOSNER, L. D., and WORONZOWA, M. A., 1936. Regeneration des Organs mit transplantierten ortsfremden Muskelm. 2. Mitteilung. Zool. Anz. *115*, 55–58.

LITTLE, C. C., and BAGG, H. J., 1924. The occurrence of four inheritable morphological variations in mice and their possible relation to treatment with X-rays. J. exp. Zool. *41*, 45–92.

LITWILLER, R., 1939. Mitotic index and size in regenerating amphibian limbs. J. exp. Zool. *82*, 273–286.

LOCATELLI, P., 1924. Sulla formazione di arti sopranumerari. Boll. Soc. med-chir. Pavia *36*.

LOEB, J., 1913. Artificial parthenogenesis and fertilization. University of Chicago Press, Chicago.

LOGACHEV, E. D., 1956. On the mutual relations between the nucleus and the cytoplasm in growing egg-cells of Platyhelminths. C.R. Acad. Sci. URSS *111*, 507–509.

LOPASHOV, G. V., 1950. Experimental investigations of the sources of cellular material and condition of formation of the pectoral fins in teleost fishes. C.R. Acad. Sci. URSS *70*, 137–140.

———, 1956. Mechanisms of formation and origin of the choroid coat in the amphibian eye. C.R. Acad. Sci. URSS *109*, 653–656.

LØVTRUP, S., and WERDINIUS, B., 1957. Metabolic phases during amphibian embryogenesis. J. exp. Zool. *135*, 203–220.

LUTHER, W., 1935. Entwicklungsphysiologische Untersuchungen am Forellenkeim: die Rolle des Organizationszentrums bei der Entstehung der Embryonalanlage. Biol. Zbl. *55*, 114–137.

MACHEMER, H., 1929. Differenzierungsfähigkeit der Urnierenanlage von *Triton alpestris*. Roux Arch. *118*, 200–251.

———, 1932. Experimentelle Untersuchung über die Induktionsleistungen der oberen Urmundlippe in älteren Urodelenkeimen. Roux Arch. *126*, 391–456.

MAKAROV, P., 1957. Ueber ungelöste Probleme der gegenwärtigen Zytologie. Wiss. Zeitschr. der Martin-Luther Univ. Halle, Math.-Nat. Reihe, *6*, 549–568.

MANCHOT, E., 1929. Abgrenzung des Augenmaterials und anderer Teilbezirke der Medullarplatte; die Teilbewegungen wärend der Auffaltung (Farbmarkierungsversuche an Keimen von Urodelen). Roux Arch. *116*, 689–708.

MANGOLD, O., 1923. Transplantationsversuche zur Frage der Spezifität und der Bildung der Keimblätter bei *Triton*. Arch. micr. Anat. u. Entw. mech. *100*, 198–301.

———, 1928. Das Determinationsproblem. I. Das Nervensystem und die Sinnesorgane der Seitenlinie unter spezieller Berücksichtigung der Amphibien. Ergebn. Biol. *3*, 152–227.

———, 1929. Das Determinationsproblem. II. Die paarigen Extremitäten der Wirbeltiere in der Entwicklung. Ergebn. Biol., *5*, 290–404.

———, 1931a. Versuche zur Analyse der Entwicklung des Haftfadens bei Urodelen: ein beispiel für die Induktion artfremder Organe. Naturwissenschaften *19*, 905–911.

———, 1931b. Das Determinationsproblem. III. Das Wirbeltierauge in der Entwicklung und Regeneration. Ergebn. Biol. *7*, 193–403.

———, 1936. Experimente zur Analyse der Zusammenarbeit der Keimblätter. Naturwissenschaften *24*, 753–760.

———, 1949. Totale Keimblattchimären bei *Triton*. Naturwissenschaften *36*, 112–120.

———, and SPEMANN, H., 1927. Ueber Induktion von Medullarplatte durch Medullarplatte im jüngeren Keim, ein Beispiel homeogenetischer oder assimilatorischer Induktion. Roux Arch. *111*, 341–422.

MARTINI, E., 1912. Studien über die Konstanz histologischer Elemente. 3. *Hydatina senta*. Z. wiss. Zool. *102*, 425–645.

MARX, L., 1935. Bedingungen für die Metamorphose des Axolotls. Ergebn. Biol. *11*, 244–334.

McCONNELL, C. H., 1936. Mitosis in *Hydra*. Mitosis in the indifferent interstitial cells of *Hydra*. Roux Arch., *135*, 202–210.

McILWAIN, H., 1946. The magnitude of microbial reactions involving vitamin-like compounds. Nature (Lond.) *158*, 898–902.

McKEEHAN, M. S., 1951. Cytological aspects of embryonic lens induction in the chick. J. exp. Zool. *117*, 31–64.

——, 1956. The relative ribonucleic acid content of lens and retina during lens induction in the chick. Amer. J. Anat. *99*, 131–156.

——, 1958. Induction of portions of the chick lens without contact with the optic cup. Anat. Rec. *132*, 297–306.

McMASTER, R. D., 1955. Desoxy-ribose nucleic acid in cleavage and larval stages of the sea urchin. J. exp. Zool. *130*, 1–27.

MESTSCHERSKAIA, K. A., 1935. (cited after NEEDHAM, J., 1942.)

METZ, C. B., 1957. Specific egg and sperm substances and activation of the egg. In: TYLER, A., BORSTEL, R. C. VON, and METZ, C. B.: The beginnings of embryonic development. Amer. Ass. Adv. Sci. Washington.

MILAIRE, J., 1956. Contribution à l'étude morphologique et cytochimique des bourgeons de membres chez le rat. Arch. Biol. *67*, 297–391.

MILLARD, N., 1945. The development of the arterial system of *Xenopus laevis,* including experiments on the destruction of the larval aortic arches. Trans. roy. Soc. S. Afr. *30*, 217–234.

MINOT, C. S., 1891. Senescence and rejuvenation. J. Physiol. *12*, 97–153.

——, 1908. The problem of age, growth and death. London.

MIRSKY, A. E., and RIS, H., 1949. Variable and constant components of chromosomes. Nature (Lond.) *163*, 666–667.

MONROY, A., 1939. Sulla localizzazione delle celluli genitali primordiali in fasi precoci di sviluppo. Richerche sperimentali in Anfibi Anuri. Arch. ital. Anat. Embriol. *41*, 368–389.

MOOG, F., 1946. The physiological significance of the phosphomono-esterases. Biol. Rev. *21*, 41–59.

MOORE, A. B. C., 1950. The development of reciprocal androgenetic frog hybrids. Biol. Bull. Woods Hole *99*, 88–111.

MOORE, A. R., 1933. Is cleavage rate a function of the cytoplasm or of the nucleus? J. exp. Biol. *10*, 230–236.

MOORE, J. A., 1946. Studies in the development of frog hybrids. I. Embryonic development in the cross *Rana pipens* ♀ x *Rana sylvatica* ♂. J. exp. Zool. *101*, 173–213.

MORGAN, T. H., 1901. Regeneration. Macmillan Co., New York.

——, 1927. Experimental embryology. Columbia University Press, New York.

——, 1933. The formation of the antipolar lobe in *Ilyanassa.* J. exp. Zool. *64*, 433–467.

MOSCONA, A., 1952. Cell suspensions from organ rudiments of chick embryos. Exp. Cell Res. *3*, 535.

——, 1956. Development of heterotypic combinations of dissociated embryonic chick cells. Proc. Soc. Exp. Biol. Med. *92*, 410–416.

MOSSMAN, H. W., 1937. Comparative morphogenesis of the foetal membranes and accessory uterine structures. Contr. Embryol. Carneg. Instn. *26*, 133–246.

MUCHMORE, W. B., 1951. Differentiation of the trunk mesoderm in *Amblystoma maculatum.* J. exp. Zool. *118*, 137–185.

MÜLLER, F., 1864. Für Darwin. Leipzig.

MUNRO, A. F., 1939. Nitrogen excretion and arginase activity during amphibian development. Biochem. J. *33*, 1957–1965.

MURRAY, P. D. F., 1926. An experimental study of the development of the limbs of the chick. Proc. Linn. Soc. N.S.W. *51*, 187–263.

NAKAMURA, O., and TAHARA, Y., 1953. Formation of the stomach in Anura. Mem. Osaka Univ. Lib. Arts & Ed. *2*, 1–8.

NASSONOV, N. V., 1930. Die Regeneration der Axolotl-extremiteiten nach Ligaturanlegung. Roux Arch. *121*, 639–657.

——, 1936. Influence of various factors on morphogenesis following homotopical subcutaneous insertions of cartilage in the axolotl. C.R. Acad. Sci. URSS *4*, 97–100.

NAWAR, G., 1956. Experimental analysis of the origin of the autonomic ganglia in the chick embryo. Amer. J. Anat. *99*, 473–506.

NEEDHAM, A. E., 1941. Some experimental biological uses of the element beryllium (glucinum). Proc. Zool. Soc. A. *111*, 59–85.

———, 1952. Regeneration and wound-healing. Methuen, London.

NEEDHAM, J., 1931. Chemical embryology. University Press, Cambridge.

———, 1942. Biochemistry and morphogenesis. University Press, Cambridge.

———, 1959. A history of embryology, 2nd Ed. University Press, Cambridge.

NELSEN, O. E., 1953. Comparative embryology of the vertebrates. Blakiston, New York.

NICHOLAS, J. S., 1950. Development of contractility. Proc. Amer. phil. Soc. *94*, 175–183.

———, and RUDNICK, D., 1933. The development of embryonic rat tissues upon the chick chorio-allantois. J. exp. Zool. *66*, 193–261.

NIEUWKOOP, P. D., 1946. Experimental investigations on the origin and determination of the germ cells, and on the development of the lateral plates and germ ridges in urodeles. Arch. néerl. Zool. *8*, 1–205.

———, 1950. Causal analysis of the early development of the primordial germ cells and the germ ridges in urodeles. Arch. Anat. micr. Morph. exp. *39*, 257–268.

———, and FABER, J., 1956. Normal table of *Xenopus laevis* (Daudin). North Holland Publishing Co., Amsterdam.

———, ET AL. 1952. Activation and organization of the central nervous system in amphibians. J. exp. Zool. *120*, 1–108.

NIU, M. C., 1947. The axial organization of the neural crest, studied with particular reference to the pigmentary component. J. exp. Zool. *105*, 79–114.

———, 1956. New approaches to the problem of embryonic induction. In: RUDNICK, D. (Editor), Cellular mechanisms in differentiation and growth. University Press, Princeton, 155–171.

———, and TWITTY, V. C., 1953. The differentiation of gastrula ectoderm in medium conditioned by axial mesoderm. Proc. Nat. Acad. Sci. *39*, 985–989.

NORDENSKIÖLD, E., 1929. The history of biology. Kegan Paul, Trench and Trubner, London.

O'CONNOR, R. J., 1939. Experiments on the development of the amphibian mesonephros. J. Anat. Lond. *74*, 34–44.

OGI, K. I., 1957. Influence of sodium iodide and sodium thiocyanate upon the development of frog's embryos. Embryologia *3*, 221–236.

OKA, H., 1943. Metamorphosis of *Polycitor mutabilis* (Ascidiae compositae). Annot. zool. jap. *22*, 54–58.

———, 1958. Eksperimentaj studoj pri metamorfozo de ascidioj. Sciencaj Studoj Kopenhago, 217–220.

———, and WATANABE, H., 1957. Vascular budding. A new type of budding in *Botryllus*. Biol. Bull. Woods Hole *112*, 225–240.

OKADA, E. W., 1955. Isolationsversuche zur Analyse der Knorpelbildung aus Neuralleistenzellen bei Urodelenkeim. Mem. Coll. Sci., Univ. Kyoto, *22*, 23–28.

OKADA, T. S., 1955a. Experimental studies on the differentiation of the endodermal organs in amphibia. III. The relation between the differentiation of pharynx and head-mesenchyme. Mem. Coll. Sci., Univ. Kyoto, *22*, 17–22.

———, 1955b. Experimental studies on the differentiation of the endodermal organs in amphibia. IV. The differentiation of the intestine from the fore gut. Annot. zool. jap. *28*, 210–214.

OKUNEFF, N., 1928. Ueber einige physico-chemische Erscheinungen während der Regeneration. I. Messung der Wasserstoffionenkonzentration in regenerierenden Extremitäten des Axolotl. Biochem. Z. *195*, 421–427.

———, 1933. Ueber einige physiko-chemische Erscheinungen während der Regeneration. V. Ueber den Milchsäuregehalt regenerierender Axolotlextremitäten. Biochem. Z. *257*, 242–244.

OLSEN, M. W., and MARSDEN, S. J., 1954. Development of unfertilized turkey eggs. J. exp. Zool. *126*, 337–347.

OPPENHEIMER, J. M., 1936. Transplantation experiments on developing teleosts (*Fundulus* and *Perca*). J. exp. Zool. *72*, 409–437.

———, 1947. Organization of the teleost blastoderm. Quart. Rev. Biol. *22*, 105–118.

ORECHOWITSCH, W. N., and BROMLEY, N. W., 1934. Die histolysierenden Eigenschaften des Regenerationsblastems. Biol. Zbl. *54*, 524–535.

——, ——, and KUSMINA, N. A., 1935. Ueber die Proteolyse in den regenerierenden Geweben. III. Die Veränderung der Aktivität der Gewebeprotease während des Regenerationsprozesses der Organe von Amphibien. Biochem. Z. *277*, 186.

PALADE, G. E., and SIEKEWITZ, P., 1956. Liver microsomes. An integrated morphological and biochemical study. J. Biophys. Biochem. Cytol. *2*, 171–200.

PARKES, A. S., FIELDING, U., and BRAMBELL, F. W. R., 1927. Ovarian regeneration in the mouse after complete double ovariotomy. Proc. roy. Soc. B *101*, 328–354.

PASTEELS, J., 1937. Etudes sur la gastrulation des vertébrés méroblastiques. II. Reptiles. Arch. Biol. Paris *48*, 105–184.

——, 1940. Un aperçu comparatif de la gastrulation chez les chordés. Biol. Rev. *15*, 59–106.

——, 1945. On the formation of the primary entoderm of the duck (*Anas domestica*) and on the significance of the bilaminar embryo in birds. Anat. Rec. *93*, 5–21.

——, 1957. La formation de l'endophylle et de l'endoblaste vitelline chez les reptiles, chéloniens et lacertiliens. Acta anat. *30*, 601–612.

PATTEN, B. M., 1944. The embryology of the pig. 2nd ed. Blakiston, Philadelphia.

——, 1957. Early embrology of the chick. 4th ed. McGraw-Hill, New York.

——, 1958. Foundations of embryology. McGraw-Hill, New York.

PATTERSON, J. T., 1910. Studies on the early development of the hen's egg. I. History of the early cleavage and the accessory cleavage. J. Morph. *21*, 101–134.

PELTRERA, A., 1940. La capacita regolative dell' uovo di *Aplysia limacina* L. studiate con la centrifugazione e con le reazioni vitali. Pubbl. Staz. zool. Napoli *18*, 20.

PENNERS, A., 1924. Experimentelle Untersuchungen zum Determinationsproblem am Keim von *Tubifex rivulorum* Lam. I. Die Duplicitas cruciata und organbildende Keimbezirke. Arch. mikr. Anat. u. Entw. mech. *102*, 53–100.

——, 1925. Experimentelle Untersuchungen zum Determinationsproblem am Keim von *Tubifex rivulorum* Lam. II. Die Entwicklung teilweise abgetöteter Keime. Z. wiss. Zool. *127*, 1–140.

PERLMAN, P., and GUSTAFSON, T., 1948. Antigens in the egg and early development stages of the sea urchin. Experientia *4*, 481–483.

PIATT, J., 1948. Form and causality in neurogenesis. Biol. Rev. *23*, 1–45.

——, 1951. Transplantation experiments between pigmentless and pigmented eggs of *Ambystoma punctatum*. J. exp. Zool. *118*, 101–135.

PIEPHO, H., and MEYER, H., 1951. Reactionen der Schmetterlingshaut auf Häutungshormone. Biol. Zbl. *70*, 252–260.

PINCUS, G., 1936. The eggs of mammals. Macmillan, New York.

——, 1939. The comparative behaviour of mammalian eggs in vivo and in vitro. IV. The development of fertilized and artificially activated rabbit eggs. J. exp. Zool. *82*, 85–129.

——, and SHAPIRO, H., 1940. The comparative behaviour of mammalian eggs in vivo and in vitro. VII. Further studies on the activation of rabbit eggs. Proc. Amer. Phil. Soc., *83*, 631–647.

POLEZHAYEV, L. W., 1946. The loss and restoration of regenerative capacity in the limbs of tailless amphibia. Biol. Rev. *21*, 141–147.

POLLISTER, A. W., and MOORE, J. A., 1937. Tables for the normal development of *Rana sylvatica*. Anat. Rec., *68*, 486–496.

PRENTISS, C. W., and AREY, L. B., 1917. A laboratory manual and text-book of embryology. 2nd ed. Saunders, Philadelphia.

PRESCOTT, D. M., 1957. Relations between cell growth and cell division. In: RUDNICK, D. (Editor), Rhythmic and synthetic processes in growth. University Press, Princeton, 59–74.

PULLINGER, B. D., 1949. Squamous differentiation in mouse mammae: spontaneous and induced. Brit. J. Cancer *3*, 494–501.

RANZI, S., 1957. Early determination in development under normal and experimental conditions. In: TYLER, A., BORSTEL, R. C. VON, and METZ, C. B. The Beginnings of Embryonic Development. Amer. Ass. Adv. Sc., Washington, 291–318.

RANZI, S., and TAMINI, E., 1939. Die Wirkung von NaSCN auf die Entwicklung von Froschembryonen. Naturwissenschaften *27*, 566–567.

RAVEN, C. P., 1931. Zur Entwicklung der Ganglienleiste. I. Die Kinematik der Ganglienleistenentwicklung bei den Urodelen. Roux Arch. *125*, 210–292.

——, 1936. Zur Entwicklung der Ganglienleiste. V. Ueber die Differenzierung des Rumpfganglienleistenmaterials. Roux Arch. *134*, 122–146.

——, 1958. Morphogenesis: the analysis of molluscan development. Pergamon, London.

RAWLES, M. E., 1947. Origin of pigment cells from the neural crest in the mouse embryo. Physiol. Zool. *20*, 248–265.

——, 1948. Origin of melanophores and their role in development of the color patterns in vertebrates. Physiol. Rev. *28*, 383–408.

ROBERTSON, T. B., 1908. On the normal rate of growth of an individual and its biochemical significance. Roux Arch. *25*, 571–614.

ROCHE, V. DE, 1937. Differenzierungen von Geweben und ganzen Organen in Transplantaten der bastardmerogonischen Kombination *Triton alpestris* (♀) x *Triton palmatus* ♂. Roux Arch. *135*, 620–663.

ROSE, S. M., 1942. A method of inducing limb regeneration in adult Anura. Proc. Soc. exp. Biol. N.Y. *49*, 408–410.

ROTH, P. C. J., 1955. Les métamorphoses des batraciens. Dunod, Paris.

ROTHSCHILD, L., 1951. Sea urchin spermatozoa. Biol. Rev. *26*, 1–27.

——, 1956. Fertilization. Methuen, London.

ROTMANN, E., 1935. Der Anteil von Induktor und reagierenden Gewebe an der Entwicklung des Haftfadens. Roux Arch. *133*, 193–224.

ROUNDS, D. E., and FLICKINGER, R. A., 1958. Distribution of ribonucleoprotein during neural induction in the frog embryo. J. exp. Zool. *137*, 479–500.

ROUX, W., 1885. Ueber die Bestimmung der Hauptrichtungen des Froschembryo im Ei und über die erste Theilung des Froscheies. Breslauer ärztl. Zeitschr. 1–54.

——, 1888. Beiträge zur Entwicklungsmechanick des Embryo. 5. Ueber die künstliche Hervorbringung halber Embryonen durch Zerstörung einer der beiden ersten Furchungskugeln, sowie über die Nachentwicklung (Postgeneration) der fehlenden Körperhälfte. Virchows Arch. *64*, 113–154, 246–291.

——, 1905. Die Entwicklungsmechanick, ein neuer Zweig der biologischen Wissenschaft. Leipzig.

RUDNICK, D., 1948. Prospective areas and differentiation potencies in the chick blastoderm. Ann. N.Y. Acad. Sci. *49*, 761–772.

——, 1952. Development of the digestive tube and its derivatives. Ann. N.Y. Acad. Sci. *55*, 109–116.

——, and RAWLES, M. E., 1937. Differentiation of the gut in chorio-allantoic grafts from chick blastoderms. Physiol. Zool. *10*, 381–395.

RUGH, R., 1948. Experimental embryology. Burgess, Minneapolis.

RUNNSTRÖM, J., 1928. Plasmabau und Determination bei dem Ei von *Paracentrotus lividus* Lk. Roux Arch. *113*, 556–581.

——, 1952. The cell surface in relation to fertilization. Symp. Soc. exp. Biol. *6*, 39-88.

RUUD, G., 1929. Heteronom-orthotopische Transplantationen von Extremitätenanlagen bei Axolotlembryonen. Roux Arch. *118*, 308–351.

RYVKINA, D. E., 1945. Respiratory quotient in regenerating tissues. C.R. Acad. Sci. URSS *49*, 457–459.

SADOV, I. A., 1956. Micropyle formation in oocytes of Acipenseridae. C.R. Acad. Sci. URSS *111*, 1400–1402.

SANTOS, F., 1931. Studies on transplantation in Planaria. Physiol. Zool. *4*, 111–164.

SAUNDERS, J. W., 1948. The proximo-distal sequence of origin of the parts of the chick wing and the role of the ectoderm. J. exp. Zool. *108*, 363–403.

SCHEREMETJEWA, E. A., and BRUNST, V. V., 1938. Preservation of the regeneration capacity in the middle part of the limb of newt and its simultaneous loss in the distal and proximal parts of the same limb. Bull. Biol Méd. exp. URSS *6*, 723–724.

SCHMALHAUSEN, I. I., 1927. Beiträge zur quantitativen Analyse der Formbildung. I. Ueber Gesetzmässigkeiten des embryonalen Wachstums. Roux Arch. *109*, 455–512.

————, 1930a. Ueber Wachstumsformeln und Wachstumstheorien. Biol. Zbl. *50*, 292–307.

————, 1930b. Das Wachstumsgesetz als Gesetz der progressiven Differenzierung. Roux Arch. *123*, 153–178.

————, 1949. Factors of evolution. Blakiston, Philadelphia.

————, and BORDZILOWSKAJA, N., 1930. Das Wachstum niederer Organismen. I. Das individuelle Wachstum der Bakterien und Hefe. Roux Arch. *121*, 726.

————, and SYNGAJEWSKAJA, E., 1925. Studien über Wachstum und Differenzierung. I. Die individuelle Wachstumskurve von Paramaecium caudatum. Roux Arch. *105*, 711–717.

SCHMALHAUSEN, O. I., 1939. The role of the olfactory sac in the development of the cartilaginous capsule of the olfactory organ in urodeles. C.R. Acad. Sci. URSS *23*, 395–397.

————, 1950. A comparative experimental investigation of the early stages of development of the olfactory rudiments in amphibians. C.R. Acad. Sci. URSS *74*, 863–865.

SCHMIDT, G. A., 1933. Schnürungs- und Durchschneidungsversuche am Amphibienkeim. Roux Arch. *129*, 1–44.

SCHOTTÉ, O. E., 1923. Influence de la section tardive des nerfs sur les pattes de Tritons en régénération. C.R. Soc. Phys. Hist. nat. Genève *40*, 86–88.

————, 1926. Système nerveux et régénération chez le *Triton*. Rev. Suisse Zool. *33*, 1–211.

————, 1930. Der Determinationszustand der Anurengastrula im Transplantationsexperiment. Roux Arch. *122*, 663–664.

————, and BUTLER, E. G., 1941. Morphological effects of denervation and amputation of limbs in urodele larvae. J. exp. Zool. *87*, 279–322.

————, and HARLAND, M., 1943. Effects of denervation and amputation of hind limbs in Anuran tadpoles. J. exp. Zool. *93*, 453–493.

SCHULZE, P., 1918. Die Bedeutung der interstitiellen Zellen, etc. Sitz. Ber. Ges. Nat. Freunde, Berlin.

SCHWIDEFSKY, G., 1934. Entwicklung und determination der Extremitätenregenerate bei den Molchen. Roux Arch. *132*, 57–114.

SCHWIND, J., 1933. Tissue specificity at the time of metamorphosis in frog larvae. J. exp. Zool. *66*, 1–14.

SEIDEL, F., 1932. Die Potenzen der Furchungskerne im Libellenei und ihre Rolle bei der Aktivierund des Bildungszentrums. Roux Arch. *126*, 213–276.

SEVERINGHAUS, A. E., 1930. Gill development in *Amblystoma punctatum*. J. exp. Zool. *56*, 1–30.

SEWERTZOFF, A. N., 1931. Morphologische Gesetzmässigkeiten der Evolution. G. Fischer, Jena.

SHAVER, J. R., 1953. Studies on the initiation of cleavage in the frog egg. J. exp. Zool. *122*, 169–192.

SHEN, S. C., 1939. A quantitative study of amphibian neural tube induction with a water-soluble hydrocarbon. J. exp. Biol. *16*, 143–149.

SHUMWAY, W., 1940. Stages in the normal development of *Rana pipiens*. I. External form. Anat. Rec. *78*, 139–148.

SINGER, C., 1931. A short history of biology. Clarendon Press, Oxford.

SINGER, M., 1950. Induction of regeneration of the limb of the adult frog by augmentation of the nerve supply. Anat. Rec. *108*, 518–519.

SIRLIN, J. L., BRAHMA, S. K., and WADDINGTON, C. H., 1956. Studies on embryonic induction using radio-active tracers. J. Emb. exp. Morph. *4*, 248–253.

SMITH, P. E., and MacDOWELL, E. C., 1930. An hereditary anterior-pituitary deficiency in the mouse. Anat. Rec. *46*, 249–257.

SNELL, J. D. (Editor), 1941. Biology of the laboratory mouse. Dover, New York.

SOTELO, J. R., and PORTER, K. R., 1959. An electron microscope study of the rat ovum. J. Biophys. Biochem. Cytol. *5*, 327–342.

SPEMANN, H., 1901. Ueber Korrelationen in der Entwicklung des Auges. Verh. anat. Ges. Jena Verslg. Bonn *15*, 61–79.

————, 1901/1903. Entwicklungsphysiologische Studien am Tritonei. I, II, III. Roux Arch. *12*, 224–264, *15*, 448–534, *16*, 551–631.

SPEMANN, H., 1912a. Zur Entwicklung des Wirbeltierauges. Zool. Jahrb. Jena, Abt. allg. Zool. *32*, 1–98.

——, 1912b. Ueber die Entwicklung umgedrehter Hirnteile bei Amphibienembryonen. Zool. Jahrb. Jena Suppl. *15*, 1–48.

——, 1919. Experimentelle Forschungen zum Determinations– und Individualitätsproblem. Naturwissenschaften *32*, 1–33.

——, 1921. Die Erzeugung tierischer Chimären durch heteroplastische embryonale Transplantation zwischen *Triton cristatus* und *taeniatus*. Roux Arch. *48*, 533–570.

——, 1928. Die Entwicklung seitlicher und dorso-ventraler Keimhälften bei verzögerter Kernversorgung. Z. wiss. Zool. *132*, 105–134.

——, 1931. Ueber den Anteil von Implantat und Wirtskeim an der Orientierung und Beschaffenheit der induzierten Embryonalanlage. Roux Arch. *123*, 389–517.

——, 1936. Experimentelle Beiträge zu einer Theorie der Entwicklung. Springer, Berlin.

——, 1938. Embryonic development and induction. Yale University Press, New Haven.

——, and MANGOLD, H., 1924. Ueber Induktion von Embryonalanlagen durch Implantation artfremder Organisatoren. Arch. mikr. Anat. u. Entw. mech. *100*, 599–638.

——, and SCHOTTÉ, O., 1932. Ueber xenoplastische Transplantation als Mittel zur Analyse der embryonalen Induktion. Naturwissenschaften *20*, 463–467.

SPIEGELMAN, S., 1948. Differentiation as the controlled production of unique enzymatic patterns. Symp. Soc. exp. Biol. *2*, 286–325.

SPRATT, N. T., JR., 1946. Formation of the primitive streak in the explanted chick blastoderm marked with carbon particles. J. exp. Zool. *103*, 259–304.

——, 1947. Regression and shortening of the primitive streak in the explanted chick blastoderm. J. exp. Zool. *104*, 69–100.

——, 1948. Development of the early chick blastoderm on synthetic media. J. exp. Zool. *107*, 39–64.

SPRATT, R. E., 1954. Physiological mechanisms in development. Physiol. Rev. *34*, 1–24.

STÉPHAN, F., 1949. Les suppléances obtenues expérimentalement dans le système des arcs aortiques de l'embryon d'oiseau. C.R. Ass. Anat. *36*, 647.

STEVENS, L. C., 1954. The origin and development of chromatophores of *Xenopus laevis* and other anurans. J. exp. Zool. *125*, 221–246.

STÖHR. PH., JR., 1924. Experimentelle Studien an embryonalen Amphibienherzen. I. Ueber Explantation embryonaler Amphibienherzen. Arch Mikr. Anat. u. Entw. mech. *102*, 426–451.

STONE, L., 1926. Further experiments on the extirpation and transplantation of mesectoderm in *Amblystoma punctatum*. J. exp. Zool. *44*, 95–131.

STREETER, G. L., 1906. On the development of the membranous labyrinth and the acoustic and facial nerves in the human embryos. Amer. J. Anat., *6*, 139–165.

——, 1942, 1945, 1948, 1949, 1951. Developmental horizons in human embryos. Contr. Embryol. Carneg. Instn. *30*, 211-245. *31*, 29–64. *32*, 133–203. *33*, 149–167. *34*, 165–196.

STRÖER, W. F. H., 1933. Experimentelle Untersuchungen über die Mundentwicklung bei den Urodelen. Roux Arch. *130*, 131–186.

SWETT, F. H., 1926. On the production of double limbs in amphibians. J. exp. Zool. *44*, 419–473.

——, 1927. Differentiation of the amphibian limb. J. exp. Zool. *47*, 385–432.

——, 1937. Determination of limb-axes. Quart. Rev. Biol. *12*, 322–339.

SWIFT, H. H., 1950. The desoxyribose nucleic acid content of animal nuclei. Physiol. Zool. *23*, 169–198.

SYNGAJEWSKAJA, E., 1935. The individual growth of Protozoa: *Blepharisma lateritia* and *Actinophrys* sp. Trav. de l'Inst. Zool. Biol. Acad. Sci. Ukr. *8*, 151–157.

SYNGAJEWSKAJA, K., 1936. Die Wachstumsgeschwindigkeit bei der Regeneration der Extremitäten bei *Siredon pisciformis*. Zool. Jahrb. Abt. allg. Zool. *56*, 487–500.

——, 1937. Die reaktiven Eigenschaften des axialen Mesoderms. Trav. de l'Inst. Zool. Biol. Acad. Sci. Ukr., *17*, 41–59.

SZE, L. C., 1953. Changes in the amount of desoxyribonucleic acid in the development of *Rana pipiens*. J. exp. Zool. *122*, 577–601.

TAKAYA, H., 1955. Thermal influence upon the inducing specificity of the organizer. Proc. imp. Acad. Japan. *31*, 366–371.

TEN CATE, G., 1953. The intrinsic development of amphibian embryos. Dissertation. North Holland Publishing Co., Amsterdam.

——, and DOORENMAALEN, W. J. VAN, 1950. Analysis of the development of the eye-lens in chicken and frog embryos by means of the precipitin reaction. Proc. Konink. Nederl. Akad. van Wetensch. Amsterdam, *53*, 894–909.

TENNENT, D. H., 1914. The early influence of the spermatozoon upon the characters of Echinoid larvae. Pap. Tortugas Lab. *182*.

TERENTIEV, I. B., 1941. On the role played by the neural crest in the development of the dorsal fin in Urodela. C.R. Acad. Sci. URSS *31*, 91–94.

TERNI, T., 1943. Studio sperimentale della capacita pinnoformativa dei cercini midollari. Arch. ital. Anat. Embriol. *33*, 667–692.

THORELL, B., 1947. The relation of nucleic acids to the formation and differentiation of cellular proteins. Cold Spr. Harb. Symp. quant. Biol. *12*, 247–255.

TING, H. P., 1951. Diploid, androgenetic and gynogenetic haploid development in anuran hybridization. J. exp. Zool. *116*, 21–57.

TOIVONEN, S., 1940. Ueber die Leistungspezifität der abnormen Induktoren im Implantatversuch bei *Triton*. Ann. Acad. Sci. Fenn. A. *55*, 1–150.

——, 1945. Zur Frage der Induktion selbständiger Linsen durch abnorme Induktoren im Implantationsversuch bei *Triton*. Ann. Soc. zool-bot. Fenn. Vanamo. *11*, 1–28.

——, 1949. Zur Frage der Leistungsspezifität abnormer Induktoren. Experientia *5*, 323.

——, 1950. Stoffliche Induktoren. Rev. Suisse Zool. *57*, suppl. 1, 41–56.

——, 1953. Bone-marrow of the guinea-pig as a mesodermal inductor in implantation experiments with embryos of *Triturus*. J. Embr. exp. Morph. *1*, 97–104.

——, 1958. The dependence of the cellular transformation of the competent ectoderm on temporal relationships in the induction process. J. Embr. exp. Morph. *6*, 479–485.

——, and SAXÉN, L., 1955. Ueber die Induktion des Neuralrohrs bei Trituruskeimen als simultane Leistung des Leber- und Knochenmarkgewebes vom Meerschweinchen. Ann. Acad. Sci. Fenn. A. *30*, 1–29.

TOKIN, B., and GORBUNOWA, G., 1934. Untersuchungen über die Ontogenie der Zellen. Biol. Zh. *3*, 294–306.

TOWNES, P. L., and HOLTFRETER, J., 1955. Directed movements and selective adhesion of embryonic amphibian cells. J. exp. Zool. *128*, 53–120.

TWITTY, V. C., 1949. Developmental analysis of amphibian pigmentation. Symp. Soc. Study Develop. Growth. *9*, 133–161.

——, and NIU, M. C., 1948. Causal analysis of chromatophore migration. J. exp. Zool. *108*, 405–437.

TYLER, A., 1948. Fertilization and immunity. Physiol. Rev. *28*, 180–219.

——, BORSTEL, R. C. VON, and METZ, C. B. (Editors), 1957. The beginnings of embryonic development. Amer. Ass. Adv. Sc., Washington.

UMANSKI, E., 1935. Ueber die gegenseitige Vertretbarkeit präsumptiver Anlagen der Rückenmark- und Gehirnteile bei den Amphibien. Zool. Anz. *110*, 25–30.

VENDRELY, R., and VENDRELY, C., 1949. La teneur du noyau cellulaire en acide désoxyribonucléique à travers les organes, les individus et les espèces animales. Experientia *5*, 327–329.

VINCENTIIS, M. DE, 1954. Ulteriori indagini sull' organogenesi del cristallino. Riv. Biol. *36*.

VOGT, W., 1925. Gestaltungsanalyse am Amphibienkeim mit örtlicher Vitalfärbung. Vorwort über Wege und Ziele. I. Methodik und Wirkungsweise der ortlichen Vitalfärbung mit Agar als Farbträger. Roux Arch. *106*, 542–610.

——, 1929. Gestaltungsanalyse am Amphibienkeim mit örtlicher Vitalfärbung. II. Gastrulation und Mesodermbildung bei Urodelen und Anuren. Roux Arch. *120*, 385–706.

WADDINGTON, C. H., 1933. Induction by the endoderm in birds. Roux Arch. *128*, 502–521.

——, 1934. Experiments on embryonic induction. III. A note on inductions by chick primitive streak transplanted to the rabbit embryo. J. exp. Biol. *11*, 224–226.

——, 1936. Organizers in mammalian development. Nature (Lond.) *138*, 125.

——, 1938. The morphogenetic function of a vestigial organ in the chick. J. exp. Biol. *15*, 371–376.

——, 1939a. Order of magnitude of morphogenetic forces. Nature (Lond.) *144*, 637.

——, 1939b. An introduction to modern genetics. Allen and Unwin, London.

——, 1952. The epigenetics of birds. University Press, Cambridge.

——, and MULHERKAR, L., 1957. The diffusion of substances during embryonic induction in the chick. Proc. Zool. Soc., Calcutta, Mookerjee Memor. Vol., 141–147.

——, NEEDHAM, J., and BRACHET, J., 1936. Studies on the nature of the amphibian organization centre. III. The activation of the evocator. Proc. roy. Soc., B *120*, 173–198.

——, NEEDHAM, J., NOWINSKY, W. W., NEEDHAM, D. M., and LEMBERG, R., 1934. Active principle of the amphibian organization centre. Nature (Lond.) *134*, 103.

——, and SCHMIDT, G. A., 1933. Induction by heteroplastic grafts of the primitive streak in birds. Roux Arch. *128*, 522–563.

WAGNER, R. P., and MITCHELL, H. K., 1955. Genetics and metabolism. John Wiley and Sons, New York.

WARBURG, O., 1908. Beobachtungen über die Oxydationsprozesse im Seeigelei. Z. physiol. Chem. *57*, 1–16.

WATERMAN, A. J., 1936. Developmental capacities of transplanted hepatic, pancreatic and lung tissues of the rabbit embryo. Amer. J. Anat. *58*, 2–57.

WEBER, R., and BOELL, E. J., 1955. Ueber die Cytochromoxydaseaktivität der Mitochondrien von frühen Entwicklungsstadien des Krallenfrosches (*Xenopus laevis* Daud.). Rev. Suisse Zool. *62*, 260–268.

WEHMEIER, E., 1934. Versuche zur Analyse der Induktionsmittel bei der Medullarplatteninduktion von Urodelen. Roux Arch. *132*, 384–423.

WEISS, P., 1925. Unabhängigkeit der Extremitätenregeneration vom Skelett (bei *Triton cristatus*). Roux Arch. *104*, 359–394.

——, 1926a. Morphodynamik. Bornträger, Berlin.

——, 1926b. Ganzregenerate aus halbem Extremitätenquerschnitt. Roux Arch. *107*, 1–53.

——, 1927a. Die Herkunft der Haut im Extremitätenregenrat. Roux Arch. *109*, 584–610.

——, 1927b. Potenzprüfung am Regenerationsblastem. I. Extremitätenbildung aus Schwanzblastem im Extremitätenfeld bei *Triton*. Roux Arch. *111*, 317.

——, 1928. Experimentelle Untersuchungen zur Metamorphose der Ascidien; Beschleunigung des Metamorphoseeintrittes durch Thyreoideabehandlung der Larve. Biol. Zbl. *48*, 69–79.

——, 1929. Erzwingung elementarer Strukturverschiedenheiten am in vitro wachsenden Gewebe (Die Wirkung mechanischer Spannung auf Richtung und Intensität des Gewebewachstums und ihre Analyse). Roux Arch. *116*, 438–554.

——, 1934. In vitro experiments on the factors determining the course of the outgrowing nerve fiber. J. exp. Zool. *68*, 393–448.

——, 1939. Principles of development. Henry Holt, New York.

——, 1947. The problem of specificity in growth and development. Yale J. Biol. Med. *19*, 235–278.

——, 1950. The deplantation of fragments of nervous system in amphibians. I. Central reorganization and the formation of nerves. J. exp. Zool. *113*, 397–461.

——, 1953. Some introductory remarks on the cellular basis of differentiation. J. Embr. exp. Morph. *1*, 181–211.

——, 1955. Nervous system. In: WILLIER, B. H., WEISS, P. A., and HAMBURGER, V. Analysis of development. Saunders, Philadelphia, 346–401.

——, and ANDRES, G., 1952. Experiments on the fate of embryonic cells (chick) disseminated by the vascular route. J. exp. Zool. *121*, 449–488.

——, and JAMES, R., 1955. Skin metaplasia in vitro induced by brief exposure to vitamin A. Exp. Cell Res. Suppl. *3*, 381–394.

WEISSMANN, A., 1904. Vorträge über Descendenztheorie. Jena. 2 vols.

WETZEL, R., 1929. Untersuchungen am Hühnchen. Die Entwicklung des Keims während der ersten beiden Bruttage. Roux Arch. *119*, 118–321.

WHITE, E. L., 1948. An experimental study of the relationship between the size of the eye and the size of the optic tectum in the brain of the developing teleost, *Fundulus heteroclitus*. J. exp. Zool. *108*, 439–469.

WIEMAN, H. L., 1949. An introduction to vertebrate embryology. McGraw-Hill, New York.

WIGGLESWORTH, V. B., 1939. The principles of insect physiology. Methuen, London.

———, 1954. The physiology of insect metamorphosis. University Press, Cambridge.

WILLIAMS, C. M., 1946. Physiology of insect diapause: the role of the brain in the production and termination of pupal dormancy in the giant silkworm *Platysamia cercopia*. Biol. Bull. Woods Hole *90*, 234–243.

———, 1947. Physiology of insect diapause. II. Interaction between the pupal brain and prothoracic glands in the metamorphosis of the giant silkworm, *Platysamia cercopia*. Biol. Bull. Woods Hole *93*, 89–98.

WILLIER, B. H., 1952. Cells, feathers and colors. Bios *23*, 109–125.

———, and RAWLES, M. E., 1931. Developmental relations of heart and liver in chorio-allantoic grafts of whole chick blastoderms. Anat. Rec., *48*, 277–301.

———, WEISS, P. A., and HAMBURGER, V., 1955. Analysis of development. Saunders, Philadelphia.

WILSON, E. B., 1904. Experimental studies on germinal localization. I. The germ regions in the egg of *Dentalium*. II. Experiments on the cleavage-mosaic in *Patella* and *Dentalium*. J. exp. Zool. *1*, 1–72.

———, 1925. The cell in development and heredity. 3rd ed. Macmillan, New York.

WILSON, H. V., 1907. On some phenomena of coalescence and regeneration in sponges. J. exp. Zool. *5*, 245–258.

WITSCHI, E., 1948. Migration of the germ cells of human embryos from the yolk sac to the primitive gonadal folds. Contr. Embryol. Carneg. Instn. *32*, 67–80.

———, 1956. Development of vertebrates. Saunders, Philadelphia.

WITTEK, M., 1952. La vitellogénèse chez les Amphibiens. Arch. Biol. Paris. *63*, 133–198.

WOERDEMAN, M. W., 1933. Ueber den Glykogenstoffwechsel des Organisationszentrums in der Amphibiengastrulation. Proc. Koninkl. Akad. Wetensch., Amsterdam, *36*, 189–193.

WOLFF, C. F., 1759. Theoria generationis. (cited after Needham, J., 1931).

WRIGHT, S., 1941. The physiology of the gene. Physiol. Rev. *21*, 487–527.

———, 1945. Genes as physiological agents. Amer. Nat. *79*, 289–303.

———, and WAGNER, K., 1934. Types of subnormal development of the head from inbred strains of guinea pigs and their bearing on the classification of vertebrate monsters. Amer. J. Anat. *54*, 383–447.

YAMADA, T., 1937. Der Determinationszustand des Rumpfmesoderms im Molchkeim nach der Gastrulation. Roux Arch. *137*, 151–270.

———, 1938. Induktion der sekundären Embryonalanlage im Neunaugenkeim. Okajimas Folia anat. jap. *17*, 369–388.

———, 1950. Dorsalization of the ventral marginal zone of the *Triturus* gastrula. I. Ammonia-treatment of the medio-ventral marginal zone. Biol. Bull. Woods Hole *98*, 98–121.

———, 1958. Induction of specific differentiation by samples of proteins and nucleoproteins in the isolated ectoderm of *Triturus*-gastrulae. Experientia *14*, 81–87.

YNTEMA, C. L., 1950. An analysis of induction of the ear from foreign ectoderm in the salamander embryo. J. exp. Zool. *113*, 211–244.

ZHINKIN, L., 1934. Ueber die Wirkung der Röntgenstrahlen auf die Regeneration bei *Lumbriculus variegatus* Gr. Trav. Lab. Zool. exp. Morph. Anim. *3*, 71–97.

ZWILLING, E., 1940. An experimental analysis of the development of the anuran olfactory organ. J. exp. Zool. *84*, 291–323.

———, 1955. Ectoderm-mesoderm relationship in the development of the chick embryo limb bud. J. exp. Zool. *128*, 423–441.

———, 1956. Interaction between limb bud ectoderm and mesoderm in the chick embryo. IV. Experiments with a wingless mutant. J. exp. Zool. *132*, 241–253.

Index